Properties of Water and Steam in SI-Units

Zustandsgrößen von Wasser und Wasserdampf in SI-Einheiten

0-800 °C   0-1000 bar

Prepared by Ernst Schmidt †

Dr.-Ing. habil., Dr. rer. nat. h. c., LL. D. h. c.
VDI, M. I. Mech. E., em. o. Professor, Technische Hochschule München

Fourth, Enlarged Printing

Edited by Ulrich Grigull

Dr.-Ing., Dr.-Ing. E. h., em. o. Professor
Technische Universität München

1989

Springer-Verlag Berlin Heidelberg New York
R. Oldenbourg München

Including a Mollier h,s-Diagram and a T,s-Diagram
Mit einem Mollier h,s-Diagramm und einem T,s-Diagramm

The Mollier h,s-Diagram is also available separately:
Das Mollier h,s-Diagramm ist auch einzeln erhältlich:
ISBN 3-540-04659-3.
ISBN 0-387-04659-3

ISBN 3-540-09601-9 Springer-Verlag Berlin Heidelberg New York
ISBN 0-387-09601-9 Springer-Verlag New York Berlin Heidelberg
ISBN 3-540-04676-3 1. Auflage Springer-Verlag Berlin Heidelberg New York
ISBN 0-387-04676-3 1st edition Springer-Verlag New York Berlin Heidelberg

CIP-Titelaufnahme der Deutschen Bibliothek
Schmidt, Ernst:
Properties of water and steam in SI-units: 0–800°C, 0–1000 bar = Zustandsgrössen von Wasser und Wasserdampf in
SI-Einheiten/prepared by Ernst Schmidt. [These tab. were ed. by Ernst Schmidt in cooperation with American Soc.
of Mechan. Engineers, New York ... and with the concurrence of Associazione Termotecn. Italiana, Milano ...].–
4., enl. print./ed. by Ulrich Grigull.
Berlin; Heidelberg; New York: Springer; München: Oldenbourg, 1989
ISBN 3-540-09601-9 (Springer, Berlin...)
ISBN 0-387-09601-9 (Springer, New York...)

Printing: Color-Druck, Berlin. Bookbinding: Lüderitz & Bauer-GmbH, Berlin
2160/3020–54321 – Printed on acid-free paper

These tables were edited by   Diese Tafeln wurden herausgegeben von

Ernst Schmidt †

Professor, Dr.-Ing., Dr. rer. nat. h. c., LL. D. h. c.

in cooperation with   in Zusammenarbeit mit

American Society of Mechanical Engineers, New York

Japan Society of Mechanical Engineers, Tokyo

Verein Deutscher Ingenieure, Düsseldorf

and with the concurrence of   und mit Beteiligung von

Associazione Termotecnica Italiana, Milano

Centro Argentino de Ingenieros, Buenos Aires

Energiagazdàlkodasi Tumományos Egyesület, Budapest

Dansk Ingeniørforening, København

Den Norske Ingeniørforening, Oslo

The Institution of Engineers, Sydney

The South African Institution of Mechanical Engineers, Johannesburg

Koninklijke Instituut van Ingenieurs, 's-Gravenhage

Makina Mühendisleri Odasi, Ankara

Svenska Mekanisters Riksförening, Stockholm

Société Française des Thermiciens, Paris

Österreichischer Ingenieurs- und Architekten-Verein, Wien

Schweizerischer Ingenieur- und Architekten-Verein, Zürich

The Fourth, Enlarged   Der vierte, erweiterte
Printing has been edited by   Neudruck wurde herausgegeben von

Ulrich Grigull

Dr.-Ing., Dr.-Ing. E. h., em. o. Professor

Foreword to the Fourth, Enlarged Printing of 1989

During the last decade the international research into the properties of steam which has been encouraged and coordinated by the "International Association for the Properties of Steam" (IAPS) has produced new experimental data for the thermodynamic properties particulary in the critical area and at higher pressures. This expansion of the international data store, the "international input", has made it necessary to revise the Skeleton Tables 1963; these revisions were passed by the 10th International Conference on the Properties of Steam 1984 as the "IAPS Skeleton Tables 1985". These give the critically evaluated experimental data for the specific volume and the specific enthalpy with their tolerances from 0 °C to 800 °C and from 1 bar to 10 000 bar. Also the relevant data for the transport properties viscosity and thermal conductivity were revised and adjusted to our current state of knowledge.

A new equation is given for viscosity for industrial use. From this and the existent equations new tables can be calculated for viscosity, thermal conductivity and the Prandtl number. For the state of saturation a set of simple equations is given to calculate steam pressure, density, specific enthalpy and specific entropy; these equations are suitable for industrial use. The values for density and specific enthalpy calculated from these equations are identical with the equivalent values of the Skeleton Tables 1985.

All the above improvements and additions are contained in this reprint. The core of this volume, the thermodynamic properties calculated from "The 1967 IFC Formulation for Industrial Use" have been retained without modification, since this formulation has proved very succesful for practical calculations. This edition provides once more a consistent set of equations and tables for industrial use.

The editor wishes to thank Dipl.-Ing. P. Schiebener for the new calculation of the values of the transport properties at saturation and the Prandtl number.

Munich, October 1988   Ulrich Grigull

Vorwort zum Neudruck 1989

Die Erforschung der Eigenschaften des Wasserdampfes, die als internationale Gemeinschaftsaufgabe durch die „International Association for the Properties of Steam" (IAPS) angeregt und koordiniert wird, hat auch im letzten Jahrzehnt neue experimentelle Daten für die thermodynamischen Zustandsgrößen beigetragen, insbesondere im kritischen Bereich und bei höheren Drücken. Diese Erweiterung des internationalen Datenvorrats, des „international input", machte eine Revision der Rahmentafeln 1963 erforderlich, die als „IAPS Skeleton Tables 1985" von der 10th International Conference on the Properties of Steam 1984 verabschiedet wurden. Hierin sind kritisch ausgewertete experimentelle Daten für das spezifische Volumen und die spezifische Enthalpie mit den zugehörigen Toleranzen von 0 °C bis 800 °C und von 1 bar bis 10 000 bar angegeben. Auch die entsprechenden Daten für die Transportgrößen Viskosität und Wärmeleitfähigkeit wurden revidiert und dem Stand des Wissens angepaßt.

Für die Viskosität wurde eine neue Gleichung für den industriellen Gebrauch angegeben. Mit dieser und den schon bestehenden Gleichungen lassen sich neue Tafeln für Viskosität, Wärmeleitfähigkeit und Prandt-Zahl berechnen. Für den Sättigungsbereich wurde ein Satz einfacher Gleichungen zur Berechnung des Dampfdrucks, der Dichte, der spezifischen Enthalpie und der spezifischen Entropie mitgeteilt, der besonders für industrielle Berechnungen geeignet ist. Die mit diesen Gleichungen berechneten Werte für die Dichte und die spezifische Enthalpie sind identisch mit den entsprechenden Werten der Rahmentafeln 1985.

Alle erwähnten Verbesserungen und Erweiterungen wurden in diese Neuauflage aufgenommen. Der Kern dieses Tafelwerks, die nach „The 1967 IFC Formulation for Industrial Use" berechneten thermodynamischen Zustandsgrößen, wurde unverändert übernommen, da diese Formulation sich für Berechnungen in der Praxis voll bewährt hat. So steht mit dieser Ausgabe wiederum ein konsistenter Satz von Gleichungen und Tafeln für den industriellen Gebrauch zur Verfügung.

Der Herausgeber dankt Herrn Dipl.-Ing. P. Schiebener für die Neuberechnung der Sättigungswerte der Transportgrößen und der Prandtl-Zahl.

München, im Oktober 1988   Ulrich Grigull

Foreword to the Third, Enlarged Printing of 1982

In recent years, the "International Association for the Properties of Steam", founded in 1968, has dealt with transport properties of water substance, in addition to the traditional thermodynamic equilibrium properties.

Thus, the 1979 edition of the "Properties of Water and Steam" included tables and plots of thermal conductivity, viscosity and Prandtl-Number of water substance according to the present state of knowledge.

A special working group of the "Association" has been concerned in more recent times with those properties which govern the electrochemical behavior of water substance. These properties are important for corrosion problems, in the chemistry of steam generators and cycles, and in applications to geohydrothermal energy and to oceanography; they can serve also to improve the understanding of the behavior of aqueous electrolytes, and the structure of water substance.

Equations and tables of the
    static dielectric constant
    (static permittivity) and
    ion product

have been published in releases of the "Association", following their approval by the "Association". These equations are included in the present 1982 Edition. The equations, tables and diagrams for the remaining state properties are included unchanged from the 1979 edition or the 1981 corrected reprint.

Munich, April 1982

Ulrich Grigull

Vorwort zum Neudruck 1982

Die „International Association for the Properties of Steam", gegründet 1968, hat in den letzten Jahren neben den traditionellen Zustandsgrößen des thermodynamischen Gleichgewichts auch die Transportgrößen von Wasser und Wasserdampf behandelt. So wurden in den „Neudruck 1979" dieser Tafeln auch Zahlentafeln und Diagramme der Wärmeleitfähigkeit, der Viskosität und der Prandtl-Zahl von Wasser und Wasserdampf nach dem aktuellen Stand des Wissens aufgenommen.

Eine besondere Arbeitsgruppe der „Association" hat sich in letzter Zeit mit jenen Zustandsgrößen beschäftigt, die für das elektrochemische Verhalten des Wassers maßgebend sind. Diese Zustandsgrößen sind wichtig für Probleme der Korrosion, der Chemie der Dampfkessel und Kreisläufe, der Verwendung geohydrothermaler Energie und der Ozeanographie; sie dienen dem Verständnis des Verhaltens wässeriger Elektrolyte und der Strukturaufklärung.

Für die Zustandsgrößen
    statische Dielektrizitätskonstante
    (statische Permittivität) und
    Ionenprodukt

sind Gleichungen und Zahlentafeln in Verlautbarungen der „Association" veröffentlicht, nachdem sie internationale Anerkennung gefunden hatten. Diese Gleichungen und Zahlentafeln sind in den vorliegenden „Neudruck 1982" aufgenommen. Die Gleichungen, Zahlentafeln und Diagramme für die übrigen Zustandsgrößen sind unverändert aus dem „Neudruck 1979" bzw. aus dem berichtigten „Neudruck 1981" übernommen worden.

München, im April 1982

Ulrich Grigull

Foreword to the Second, Revised and Updated, Printing of 1979

The great demand for the "Properties of Water and Steam in SI-Units", prepared by Prof. Dr.-Ing. ERNST SCHMIDT in 1969 and meanwhile out of print, made a second, revised and updated, printing of these tables necessary. This second printing had to be reviewed mainly with regard to changes of the 1969-edition caused by the latest decisions of the International Conferences on the Properties of Steam and its working groups.

The main part of the edition of 1969 containing the equilibrium properties of water and steam was based on the sets of equations of the "1976 IFC Formulation for Industrial Use", which were adopted to this reprint without changes. These tabulated values still satisfy the needs of the industrial user of today. Significant changes were made in the chapters on thermal conductivity, dynamic viscosity, Prandtl number, and surface tension. These properties were revised in the period from 1972 to 1978. Improved Skeleton Tables and equations fitted to the actual state of knowledge have been published and internationally standardized by the vote of the members of the "International Association for the Properties of Steam". These values were taken into this new edition replacing the old ones.

Special thanks for their assistance in preparing the reprint of the "Properties" are due to Prof. Dr.-Ing. J. STRAUB and Dipl.-Ing. K. SCHEFFLER, Munich.

Munich, May 1979

Ulrich Grigull

Vorwort zum Neudruck 1979

Die anhaltende Nachfrage nach den „Properties of Water and Steam in SI-Units", die 1969 von ERNST SCHMIDT herausgegeben wurden und inzwischen vergriffen sind, machten einen Neudruck dieses Werkes erforderlich. Dieser Neudruck war vor allem daraufhin durchzusehen, ob durch inzwischen zustande gekommene Beschlüsse der Internationalen Konferenzen über die Eigenschaften des Wasserdampfes und deren Arbeitsgruppen wesentliche Änderungen gegenüber dem Stand von 1969 eingetreten waren.

Die eigentlichen Zustandsgrößen von Wasser und Wasserdampf, die den Hauptteil der „Properties" von 1969 ausmachten, beruhten auf den Gleichungssätzen „The 1967 IFC Formulation for Industrial Use", die unverändert in diesen Neudruck aufgenommen werden konnten, da sie den Bedürfnissen der industriellen Praxis auch heute noch voll entsprechen.

Wesentlich verändert werden mußten die Angaben über die Transportgrößen Wärmeleitfähigkeit, dynamische Viskosität und Prandtl-Zahl sowie über die Oberflächenspannung. Diese Größen wurden im Zeitraumvon 1972 bis 1978 neu bearbeitet. Verbesserte, dem Stand des Wissens angepaßte Rahmentafeln und Gleichungen wurden veröffentlicht und durch Zustimmung der Mitglieder der „International Association for the Properties of Steam" zum internationalen Standard erhoben. Diese Werte sind in diese Neuausgabe anstelle der früheren aufgenommen.

Für die Mitwirkung an dieser Neuausgabe der „Properties" wird den Herren Professor Dr.-Ing. J. STRAUB und Dipl.-Ing. K. SCHEFFLER, beide in München, herzlich gedankt.

München, im Mai 1979

Ulrich Grigull

Foreword

These tables of the properties of water and steam were calculated in their entirety by using a set of equations accepted by the members of the Sixth International Conference on the Properties of Steam under the title "The 1967 IFC Formulation for Industrial Use". Deviations from the figures computed became necessary for smoothing only in small areas of isobaric specific heat capacity.

For the most part, the tables on transport properties were also based on internationally accepted sets of equations. The quantity symbols and unit symbols used are those found in international standards.

The 1967 IFC Formulation for Industrial Use was prepared by the International Formulation Committee (IFC) formed in 1963 by the Sixth International Conference on the Properties of Steam held in New York. The IFC was composed of representatives from the following countries:

Czechoslovak Socialist Republic
Federal Republic of Germany
Japan
Union of Soviet Socialist Republics
United Kingdom of Great Britain and Northern Ireland
United States of America

These Tables are mainly intended for use in industry. They are edited in concurrence with the major engineering institutions in a large number of countries and represent a valuable contribution towards international cooperation. These tables take the place of a new edition of the former version B (published in 1963) of the "VDI-Wasserdampftafeln".

Thanks are due to the IFC members and all those who by their theoretical and experimental investigations have contributed to deepening our knowledge of the properties of steam.

Special thanks for their assistance in preparing the tables are due to Dipl.-Ing. J. BACH, Munich, Prof. Dr.-Ing. U. GRIGULL, Munich, Dr.-Ing. F. MAYINGER, Nuremberg, Dipl.-Ing. R. MEYER-PITTROFF, Munich, Dipl.-Ing. M. REIMANN, Munich, Dr. K. R. SCHMIDT, Erlangen, Dr.-Ing. D. SCHWARZ, Francfort/Main, and Dr.-Ing. H. TRATZ, Erlangen.

Grateful acknowledgement is also made of the computer facilities made available by Siemens AG, Erlangen.

Munich, January 1969

Ernst Schmidt

Vorwort

Die hier vorgelegten Tafeln der Zustandsgrößen von Wasser und Wasserdampf wurden vollständig mit Hilfe eines Satzes von Gleichungen berechnet, der unter dem Namen „The 1967 IFC Formulation for Industrial Use" von den Mitgliedern der Sechsten Internationalen Konferenz über die Eigenschaften des Wasserdampfes angenommen wurde. Abweichungen von den berechneten Zahlenwerten im Sinne einer Glättung waren nur in kleinen Bereichen der isobaren spezifischen Wärmekapazität erforderlich.

Auch den Tabellen der Transportgrößen lagen größtenteils international angenommene Gleichungssätze zugrunde. Die Bezeichnung der Größen und Einheiten folgt dem derzeitigen Stand der internationalen Normung.

Die 1967 IFC Formulation for Industrial Use wurde vom International Formulation Committee (IFC) aufgestellt, das von der Sechsten Internationalen Konferenz über die Eigenschaften des Wasserdampfes 1963 in New York eingesetzt worden war. Das IFC bestand aus Vertretern aus folgenden Ländern:

Bundesrepublik Deutschland
Japan
Tschechoslowakische Sozialistische Republik
Union der Sozialistischen Sowjetrepubliken
Vereinigtes Königreich von Großbritannien und Nordirland
Vereinigte Staaten von Amerika

Die hier vorgelegten Tafeln sind vorzugsweise für den Gebrauch in der Industrie bestimmt. Sie sind gemeinsam mit den maßgebenden Ingenieurvereinen aus zahlreichen Ländern herausgegeben, die damit einen dankenswerten Beitrag zur internationalen Zusammenarbeit geleistet haben. Diese Tafeln treten an die Stelle einer Neuauflage der früheren Ausgabe B (1963) der "VDI-Wasserdampftafeln".

Den Mitgliedern des IFC und allen denen, die durch experimentelle und theoretische Untersuchungen zur Vertiefung unserer Kenntnis der Eigenschaften des Wasserdampfes beigetragen haben, gebührt Dank.

Für die Mitwirkung bei der Vorbereitung der Tafeln wird insbesondere gedankt den Herren Dipl.-Ing. J. BACH, München, Prof. Dr.-Ing. U. GRIGULL, München, Dr.-Ing. F. MAYINGER, Nürnberg, Dipl.-Ing. R. MEYER-PITTROFF, München, Dipl.-Ing. M. REIMANN, München, Dr. K. R. SCHMIDT, Erlangen, Dr.-Ing. D. SCHWARZ, Frankfurt/Main, und Dr.-Ing. H. TRATZ, Erlangen.

Die Firma Siemens AG, Erlangen, stellte dankenswerter Weise ihre elektronischen Rechenanlagen zur Verfügung.

München, im Januar 1969

Ernst Schmidt

Contents

Inhaltsverzeichnis

Contents

A. Introduction

1. Generalities

Skeleton Tables for thermodynamic properties were established at the very first International Conference on the Properties of Steam (London 1929). It was a grid of points of state, defined by pressure and temperature, to which were added the values of specific volume and specific enthalpy, established by averaging and extrapolation. The tolerances of each value provide a measure of the reliability of our knowledge.

Later Conferences improved upon these Skeleton Tables and expanded their ranges of pressure and temperature according to the progress in experimental technique. The Skeleton Tables 1985 shown in the Tables B 4 were agreed at the Tenth International Conference on the Properties of Steam in Moscow 1984 [277].

The magnitudes of the tolerances have been so chosen so that all experimental results regarded as trustworthy lie within them. Thus the tolerances in the skeleton tables are at the same time an indication of the uncertainty of their values. The tolerances cannot be defined more precisely (for instance by the use of statistical theory of errors) because of the small number of experimental data points scattered over large regions.

The units used in these tables are those of the International System of Units (SI) with their decimal fractions and multiples. The unit for the thermodynamic temperature and temperature differences in the Kelvin (symbol K).

When Celsius temperatures were used, the unit degree Celsius (symbol °C) is understood as a special name for Kelvin.

To satisfy industrial demands the bar (symbol bar) was chosen as the derived unit for the pressure. The conversion into the SI-unit Pascal can easily be done by the relation

$$1 \text{ bar} = 0{,}1 \text{ MPa}.$$

With increased use of digital computers for complicated calculations, especially in the design and the optimisation of power stations, it has become necessary to develop a formulation of the properties of water and steam for industrial use.

For this purpose the Sixth International Conference on the Properties of Steam (1963 in New York) formed the International Formulation Committee (IFC). The result of its work is: "The 1967 IFC Formulation for Industrial Use" [261].

This internationally recognized formulation given in paragraph CI consists of a set of equations and corresponds to the state of our knowledge in early 1967. It comprises the whole area of the skeleton tables and the values derived from it are everywhere within the tolerances of the skeleton table and are thermodynamically consistent, i.e. they satisfy the known relationships among the several parameters of state.

The values of specific volume, specific enthalpy and specific entropy of water and steam in the following

A. Einführung

1. Allgemeines

Schon die Erste Internationale Konferenz über die Eigenschaften des Wasserdampfs (London 1929) stellte Rahmentafeln (Skeleton Tables) für thermodynamische Zustandsgrößen auf. Es handelte sich um ein Gitter von Zustandspunkten, beschrieben durch Druck und Temperatur, zu denen Zustandswerte des spezifischen Volumens und der spezifischen Enthalpie ermittelt wurden. Die zu jedem Wert gehörenden Toleranzen waren ein Maß für die Zuverlässigkeit unseres Wissens.

Spätere Konferenzen verbesserten diese Rahmentafeln und erweiterten die Bereiche von Druck und Temperatur nach dem Fortschritt der experimentellen Technik. Die in den Tafeln B 4 wiedergegebenen Rahmentafeln 1985 wurden auf der Zehnten Internationalen Konferenz über die Eigenschaften des Wasserdampfs (Moskau 1984) verabschiedet [277].

Die Größe der Toleranzen ist so bemessen, daß alle als vertrauenswürdig angesehenen Versuchsergebnisse in ihrem Bereich liegen. Die Toleranzen der Rahmentafeln sind also zugleich ein Maß für die Unsicherheit der Rahmentafelwerte. Bei der geringen Zahl der über große Bereiche verteilten Versuchswerte ist eine genauere Definition, etwa im Sinne der statistischen Fehlertheorie nicht möglich.

In den hier vorgelegten Tafeln werden die Einheiten des Internationalen Einheitensystems (SI) und dessen dezimale Teile und Vielfache benutzt. Für die Einheit der thermodynamischen Temperatur und der Temperaturdifferenzen wird der Name Kelvin (Einheitenzeichen K) verwendet.

Bei der Angabe von Celsius-Temperaturen wird die Einheit Grad Celsius (Einheitenzeichen °C) als besonderer Name für das Kelvin verwendet.

Als abgeleitete Einheit für den Druck wurde das im industriellen Bereich bewährte Bar (Einheitenzeichen bar) beibehalten. Zur SI-Einheit Pascal besteht die einfache Beziehung

$$1 \text{ bar} = 0{,}1 \text{ MPa}.$$

Mit dem zunehmenden Gebrauch von digitalen Rechenanlagen für komplizierte Berechnungen, besonders bei der Auslegung und Optimierung von Kraftanlagen, wurde es nötig, eine für den Gebrauch der Industrie geeignete Formulierung der thermodynamischen Eigenschaften des Wassers und Wasserdampfes zu entwickeln.

Zu diesem Zweck wurde von der Sechsten Internationalen Konferenz über die Eigenschaften des Wasserdampfes (1963 in New York) das International Formulation Committee (IFC) ins Leben gerufen, das die „The 1967 IFC Formulation for Industrial Use" ausarbeitete [261].

Diese international anerkannte, in Abschnitt CI wiedergegebene Formulation besteht aus einem Satz von Gleichungen; sie umfaßt den ganzen in der Rahmentafel angegebenen Bereich und liefert Werte, die überall innerhalb der Toleranzen der Rahmentafel liegen und thermodynamisch konsistent sind, d. h. den bekannten Beziehungen zwischen den verschiedenen Zustandsgrößen genügen.

Die in der folgenden Dampftafel bis zu Temperaturen von 800 °C und Drücken von 1000 bar angegebenen

tables up to temperatures of 800 °C and pressures of 1000 bar are calculated with the help of this formulation, they are given to a larger number of significant figures than the accuracy of the experimental data and the range of tolerances warrant in order to facilitate interpolation.

The spacing of temperature normally is 10 K. For a small region in the neighbourhood of the critical state this was reduced to 1 K in order to permit a better interpolation of the values, which here vary sharply with temperature (Table B 3a).

The specific isobaric heat capacity was derived from the formulation with the help of the thermodynamic relations. However, in a small supercritical region at high pressures slight graphical smoothing was necessary. The values of the specific heat capacity are given in Table B 5 and represented in a diagram as a function of temperature with pressure as parameter.

The Eighth International Conference on the Properties of Steam 1974 authorized a working group to revise the existing tables for the transport properties dynamic viscosity and. thermal conductivity because of proved discrepancies between the tables and newer measurements [186, 187]. New Skeleton Tables have been constructed [198, 203], new equations for the whole region from 0 to 800 °C and up to 1000 bar have been developed [104, 170, 175, 208, 254] and these results have been summarized in releases [264, 265] and therefore internationally recognized.

The Tables and equations in Chapter B 4 and C II represent a release of the 10th International Conference on the Properties of Steam [277]. The diagrams of the transport properties both versus temperature with pressure as variable and in the p, T-plane were plotted with the aid of the standardized equations for industrial use [278, 279].

The Table of the Prandtl number $Pr = \eta c_p / \lambda$ was calculated using the equations for industrial use for viscosity [278] and for thermal conductivity [279] and the "1967 IFC Formulation for Industrial Use" [261].

The p, T-diagram of the isentropic exponent k was iterated by the canonical equation of state of R. POLLAK [182], which was chosen for reasons of easier computation.

The new representation of the surface tension σ is included in chapter B in the form of the internationally recognized Skeleton Table and equation [263], completed by a diagram for σ and the Laplace coefficient.

The text is printed in German and English. A h,s- and a T,s-diagram are enclosed as supplements.

thermodynamischen Zustandsgrößen spezifisches Volumen, spezifische Enthalpie und spezifische Entropie von Wasser und Dampf sind mit dieser Formulation gerechnet. Sie sind in den Tabellen mit größerer Stellenzahl angegeben als es die Genauigkeit der Versuchswerte und der Spielraum der Toleranzen rechtfertigt, um das Interpolieren zu erleichtern.

Die Stufung der Temperatur beträgt normalerweise 10 K. Für einen kleinen Bereich in der Nähe des kritischen Zustandes wurde in einer besonderen Tabelle (Tafel B 3a) die Temperaturstufung auf 1 K verfeinert, um eine bessere Interpolation der hier mit der Temperatur stark veränderlichen Werte zu ermöglichen.

Die spezifische isobare Wärmekapazität wurde mit Hilfe thermodynamischer Beziehungen aus der Formulation berechnet. In einem kleinen überkritischen Bereich bei hohen Drücken war es nötig, graphisch zu glätten. Die Werte der spezifischen Wärmekapazität sind in Tafel B 5 enthalten und in einem Diagramm über der Temperatur mit dem Druck als Parameter dargestellt.

Die achte internationale Wasserdampfkonferenz 1974 setzte eine Arbeitsgruppe mit der Aufgabe ein, die durch neuere Arbeiten erwiesene Unstimmigkeit zwischen den Rahmentafeln der Transportgrößen dynamische Viskosität und Wärmeleitfähigkeit und weiteren neuen Messungen [186, 187] durch eine revidierte Darstellung dieser Größen zu beseitigen. Es wurden Rahmentafeln aufgestellt [198, 203], neue Gleichungen für das durch die Zustandsgleichungen abgedeckte Gebiet von 0 bis 800 °C und bis 1000 bar entwickelt [104, 170, 175, 208, 254] und diese Ergebnisse zusammengefaßt als Empfehlung (Release) [264, 265] als internationaler Standard vorgeschlagen. Die im Abschnitt B 4 und C II aufgeführten Tafeln und Gleichungen geben eine Empfehlung der 10th International Conference on the Properties of Steam wieder [277]; die Diagramme der Transportgrößen über der Temperatur mit dem Druck als Parameter und in der p, T-Ebene mit der Transportgröße als Parameter wurden mit den für den industriellen Gebrauch standardisierten Gleichungen der Transportgrößen [278, 279] gezeichnet.

Die Tafel der Prandtl-Zahl $Pr = \eta c_p / \lambda$ wurde mit Hilfe der Gleichungen für industriellen Gebrauch für die Viskosität [278] und für die Wärmeleitfähigkeit [279] sowie der „1967 IFC Formulation for Industrial Use" berechnet [261].

Das p,T-Diagramm des Isentropenexponenten k wurde unter Verwendung der kanonischen Zustandsgleichung von R. POLLAK [182] iterativ ermittelt und gezeichnet. Diese Gleichung wurde aus Gründen der vereinfachten Iteration gewählt.

Die Oberflächenspannung σ wurde ebenfalls neu dargestellt. Ein Rahmentafelentwurf sowie eine Gleichung, beide international anerkannt [263], sind im Abschnitt B aufgenommen, ergänzt durch ein Diagramm für σ sowie für den aus der Oberflächenspannung berechneten Laplace-Koeffizienten.

Der Text dieser Neuauflage ist in deutsch und englisch abgefaßt. Als Beilage sind ein h,s- und ein T,s-Diagramm angefügt.

2. Thermophysical Properties Used in the Tables with their Symbols and Units
Thermophysikalische Zustandsgrößen der Tafeln mit ihren Symbolen und Einheiten

Quantity	Größe	Symbol	Unit/Einheit
temperature	Temperatur	T	K
Celsius temperature	Celsius-Temperatur	t	°C

$$t = (T - T_0), \quad T_0 = 273{,}15 \text{ K}$$

Quantity	Größe	Symbol	Unit/Einheit
pressure	Druck	p	bar
specific volume	spezifisches Volumen	v	m³/kg; dm³/kg
density	Dichte	ϱ	kg/m³
specific enthalpy	spezifische Enthalpie	h	kJ/kg
specific heat of evaporation	spezifische Verdampfungswärme	r	kJ/kg
specific entropy	spezifische Entropie	s	kJ/kg K
specific isobaric heat capacity	spezifische isobare Wärmekapazität	c_p	kJ/kg K
dynamic viscosity	dynamische Viskosität	η	10^{-6} kg/s m = µPa s
thermal conductivity	Wärmeleitfähigkeit	λ	W/K m, mW/K m
Prandtl number	Prandtl-Zahl	Pr	– – –

$$Pr = \eta c_p / \lambda$$

Quantity	Größe	Symbol	Unit/Einheit
surface tension	Oberflächenspannung	σ	mN/m
Laplace coefficient	Laplace-Koeffizient	a	m

$$a^2 = \sigma/g(\varrho' - \varrho'')$$

Quantity	Größe	Symbol	Unit/Einheit
isentropic exponent	Isentropenexponent	k	– – –

$$k = -(v/p)(\partial p/\partial v)_s$$

Quantity	Größe	Symbol	Unit/Einheit
static dielectric constant	statische Dielektrizitätskonstante	ε	As/Vm = F/m
ion product	Ionenprodukt	K_w	(mol/kg)²

in state of saturation	im Sättigungszustand
for water index '	für Wasser Index '
for steam index ''	für Dampf Index ''

3. Thermodynamic Relations

Differential relations exist between the quantities of state T, p, v, h, and s, resulting from the laws of thermodynamics. Using the so-called thermodynamic potentials

specific free energy or Helmholtz function f,
specific free enthalpy or Gibbs function $g = f + p\,v$
hey read:

3. Thermodynamische Beziehungen

Zwischen den Zustandsgrößen T, p, v, h und s bestehen Differentialbeziehungen, die sich aus den Hauptsätzen der Thermodynamik ergeben. Bei Benutzung der sogenannten thermodynamischen Potentiale

spezifische freie Energie oder Helmholtz-Funktion f,
spezifische freie Enthalpie oder
Gibbs-Funktion $g = f + p\,v$ lauten sie:

$$s = -(\partial g/\partial T)_p = -(\partial f/\partial T)_v, \quad v = +(\partial g/\partial p)_T, \quad p = -(\partial f/\partial v)_T, \quad h = g + T\,s = f + p\,v + T\,s. \quad (1)$$

Equations of state which satisfy these differential relations are called thermodynamically consistent.

Zustandsgleichungen, welche diese Differentialbeziehungen erfüllen, bezeichnet man als thermodynamisch konsistent.

4. Definitions and Physical Constants

The temperature T is defined by the International Practical Temperature Scale of 1968 (IPTS 68) [272].

The internal energy and the entropy of liquid water at the triple point are zero by definition (resolution of the 5th International Conference on the Properties of Steam, London 1956).

Molecular data of H_2O

molar (universal) gas constant

molar mass

specific gas constant

The molar mass and the specific gas constant refer to water of normal isotopic composition (natural abundance) [275].

Critical Data of H_2O [276]

4. Definitionen und physikalische Konstanten

Die Temperatur T ist durch die Internationale Praktische Temperaturskala von 1968 (IPTS 68) definiert [272].

Die innere Energie und die Entropie von flüssigem Wasser beim Tripelpunkt sind nach Definition Null (Beschluß der 5th International Conference on the Properties of Steam, London 1956).

Molekulare Werte von H_2O

molare (universelle) Gaskonstante

$$R = (8{,}314\,510 \pm 0{,}000\,070)\ \text{J/mol K [273]},$$

molare Masse

$$M_i(H_2O) = 18{,}0152\ \text{kg/kmol [274]},$$

spezifische Gaskonstante

$$R_i(H_2O) = R/M_i(H_2O) = 0{,}461\,527\ \text{kJ/kg K}.$$

Die molare Masse und die spezifische Gaskonstante beziehen sich auf Wasser natürlicher Isotopenzusammensetzung [275].

Kritische Werte von H_2O [276]

$$T_c = (647{,}14 + \delta)\ \text{K}; \quad t_c = (373{,}99 + \delta)\ °\text{C},$$
$$\delta = 0{,}00 \pm 0{,}1,$$
$$p_c = (220{,}64 + 2{,}7\delta \pm 0{,}05)\ \text{bar},$$
$$\varrho_c = (322 \pm 3)\ \text{kg/m}^3,$$
$$v_c = (3{,}11 \pm 0{,}03)\ \text{dm}^3/\text{kg}.$$

5. Bibliography — Literaturverzeichnis

Steam Tables and Diagrams — Dampftafeln und Diagramme

[a] CALLENDAR, G. S., and A. C. EGERTON: The 1939 Callendar Steam Tables, London 1944.
[b] BAIN, R. W.: N.E.L.-Steam Tables, Edinburgh 1964.
[c] DZUNG, L. S., u. W. ROHRBACH: Enthalpie-Entropie-Diagramme für Wasserdampf und Wasser, Berlin/Göttingen/Heidelberg: Springer 1955.
[d] FAXÉN, O. H.: Thermodynamic Tables in the Metric System for Water and Steam, Stockholm: Nordisk Rotogravyr 1953.
[e] KEENAN, J. H., and F. G. KEYES: Thermodynamic Properties of Steam, New York/London 1936.
[f] KNOBLAUCH, O., E. RAISCH, H. HAUSEN u. W. KOCH: Tabellen und Diagramme für Wasserdampf, 2. Aufl., München und Berlin 1932.
[g] MOLLIER, R.: Neue Tabellen und Diagramme für Wasserdampf, 7. Aufl., Berlin 1932.
[h] Revised Steam Tables and Diagrams of the Japanes Soc. of Mech. Eng. Tokyo 1955.
[i] SCHMIDT, E.: VDI-Wasserdampftafeln, 6. Aufl., Berlin/Göttingen/Heidelberg, München: Springer und Oldenbourg 1963. Ausgabe A (kcal, at), Ausgabe B (Joule, bar).
[k] 1967 Steam Tables. The Electrical Research Association, London: Eduard Arnold (Publishers) Ltd. 1967.
[l] WUKALOWITSCH, M. P.: Thermodynamische Eigenschaften des Wassers und des Wasserdampfes, 6. Aufl. Moskau, Berlin: VEB Verlag Technik 1958.
[m] The 1967 ASME Steam Tables. The American Society of Mechanical Engineers, New York 1967.
[n] HAAR, L., J. S. GALLAGHER and G. S. KELL: NBS/NRC Steam Tables. Washington: Hemisphere 1984.
[o] GRIGULL, U., J. STRAUB and P. SCHIEBENER: Steam Tables in SI-Units (English-German). Berlin/Heidelberg/New York: Springer 1984.
[p] HAAR, L., J. S. GALLAGHER and G. S. KELL: NBS/NRC Wasserdampftafeln (in German). Berlin/Heidelberg/New York 1988.

Special Reports — Einzelarbeiten

[1] AMIRCHANOFF, CH. I., u. A. M. KERIMOFF: Investigation of heat capacity at continuous volume of water and steam using direct method along saturation line, including critical point. Teploenergetika 10 (1963) Nr. 8, S. 64/69. Ref.: BWK 15 (1963) Nr. 11, S. 544.
[2] AMIRCHANOFF, CH. I., u. A. M. KERIMOFF: An Experimental Study of the Heat Capacity C_v of Water and Steam in the Supercritical State. Teploenergetika 10 (1963) Nr. 9, S. 61/65. Ref.: BWK 16 (1964) Nr. 1, S. 40/41.
[3] ANGUS, S., and D. M. NEWITT: The measurement of the specific enthalpy of steam at pressures of 60 to 1000 bar and temperatures of 400 to 700 °C. Philos. Trans. Royal Soc., Ser. A 259 (1966) Nr. 1098.
[4] BACH, J., u. U. GRIGULL: Eine Interpolationsgleichung für die Wärmeleitfähigkeit von überhitztem Wasserdampf. Brennstoff – Wärme – Kraft (BWK) 18 (1966) Nr. 3, S. 125/127.
[5] BAEHR, H. D., u. H. THÜRMER: Eine Zustandsgleichung für Wasser im kritischen und überkritischen Gebiet. Brennstoff – Wärme – Kraft (BWK) 17 (1965) 20/25.
[6] BAIN, R. W.: The preparation of steam tables with the aid of a digital computer. J. mech. Engng. Sci. 8 (4) (1961) 289/294.
[7] BAIN, R. W., E. A. S. PATON and A. S. SCRIMGEOUR: An equation of state for steam. NEL Report No. 7. East Kilbride, Glasgow: National Engineering Laboratory 1961.
[8] BRUGES, E. A.: Transport Properties of Water and Steam — Viscosity. Techn. Report No. 9 (A) Sept. University of Glasgow 1963.
[9] BRUGES, E. A., B. LATTO and A. K. RAY: New correlations and tables of the coefficient of viscosity of water and steam up to 1000 bar and 1000 °C. Int. J. Heat and Mass Transfer 9 (1966) Nr. 5, S. 465/480.
[10] CALLENDAR, G. S., and A. EGERTON: An Experimental Study of the Enthalpy of Steam. Philos. Trans. Roy. Soc. [London], Series A 1 252 (1959/60) 133/164.
[11] CALLENDAR, G. S., and A. C. EGERTON: Tables of total heat of steam. ERA Report J/T 173. Leatherhead, Surrey: Electrical Research Association, 1958.
[12] CHALLONER, A. R., and R. W. POWELL: Thermal conductivity of liquids, new determinations for seven liquids and appraisal of existing values. Proc. Roy. Soc. A 238 (1956) 90/106.
[13] FRIEDMAN, A. G., and L. HAAR: High speed machine computation of ideal gas thermodynamic functions. 1. Isotopic water molecules. J. chem. Phys. 22 (12) (1954) 2051/2058.
[14] FRITZ, W., u. H. POLTZ: Absolutbestimmung der Wärmeleitfähigkeit von Flüssigkeiten. Int. J. Heat and Mass Transfer 5 (1962) 307/316.
[15] GRIGULL, U., u. J. BACH: Die Oberflächenspannung und verwandte Zustandsgrößen des Wassers. Brennstoff – Wärme – Kraft (BWK) 18 (1966) Nr. 2, S. 73/75.
[16] GRIGULL, U., F. MAYINGER u. J. BACH: Viskosität, Wärmeleitfähigkeit und Prandtl-Zahl von Wasser und Wasserdampf. Wärme- und Stoffübertragung. Bd. 1 (1968) 15/34.
[17] HAVLIČEK, J., and L. MIŠKOVSKI: Research at the Masaryk Academy in Prague on the physical properties of water and steam. Helv. phys. Acta 9 (1936), 161/207.
[18] HAYWOOD, R. W.: The Sixth International Conference on the Properties of Steam (Official Release), New York, Oktober 1963.

[19] HENNING, F.: Rahmentafeln für Wasser und Wasserdampf nebst Erläuterungen. Ergebnisse der 3. Internationalen Dampftafelkonferenz. VDI-Z. 79 (1935) Nr. 45, S. 1359/1362.

[20] HOLSER, W. T., and G. C. KENNEDY: Properties of water. Part IV. Pressure-volume-temperature relations of water in the range 100—400 °C and 100—1400 bars. Amer. J. Sci. 256 (10) (1958) 744/754.

[21] HOLSER, W. T., and G. C. KENNEDY: Properties of water. Part V. Pressure-volume-temperature relations of water in the range 400—1000 °C and 100—1400 bars. Amer. J. Sci. 257 (1) (1959) 71/77.

[22] JAKOB, M.: Rahmentafeln für Wasserdampf nebst Erläuterungen. Beschlüsse der Internationalen Dampftafel-Konferenz in London. Z. VDI 73 (1929) 1856/1858.

[23] JAKOB, M.: Rahmentafeln für Wasserdampf nebst Erläuterungen. Ergebnisse der 2. Internationalen Dampftafel-Konferenz in Berlin. Z. VDI 75 (1931) 187/188.

[24] JUZA, J.: Equation of state for saturated superheated steam. Prague: Mechanical Engineering Research Institute, Czechoslovak Academy of Sciences, 1962.

[25] JUZA, J.: Equation of State for Water and Steam in the Region of Critical Point — Part I and Part II, Critical Point and its Immediate Vicinity.

[26] JUZA, J.: Equation of State for Water and Steam in the Range from —20 to +900 °C, from 0 to 100000 Bar — Part II.

[27] JUZA, J., V. KMONIČEK u. K. SCHOVANEC: The Joule-Thomson Effect in Light and Heavy Water for the Range of 1.2 to 1.8 bar, 130 to 190 °C.

[28] JUZA, J.: Equations for Thermodynamic Properties of Water and Steam Suitable for Electronic Computers (2nd corrected edition).
(Die Arbeiten [24 bis [28] sind 1963 herausgegeben worden von der Tschechoslowakischen Akademie der Wissenschaften, Prag 6, Puschkinplatz 9.)

[29] KELL, G. S., u. E. WHALLEY: The PVT Properties of Water. I. Experimental Methods, II. Compressions in the Range 0° to 150 °C up to 1026 Bars Nat. Res. Council, Ottawa, N.R.C. No. 7607 u. 7608 (1963).

[30] KELL, G. S., G. E. McLAURIN u. E. WHALLEY: The PVT Properties of Water. III. The Thermal Expansion from 80 to 150 °C. Nat. Res. Council, Ottawa. N.R.C. No. 7609 (1963).

[31] KELL, G. S., u. E. WHALLEY: A Proposal for the International Skeleton Tables of the Properties of Liquid Water in the Range 0—150 °C, 0—1000 bar. Nat. Res. Council, Ottawa (1963).

[32] KENNEDY, G. C., W. L. KNIGHT and W. T. HOLSER: Properties of water, Part III. Specific volume of liquid water to 100 °C and 1400 bars. Amer. J. Sci. 256 (8) (1958) 590/595.

[33] KESTIN, J., B. H. SAGE, J. H. KEENAN and F. G. KEYES: Comparisons and comments on existing steam data with formulations. Paper submitted at the Moscow meeting of the International Co-ordinating Committee on the Properties of Steam, July 1958.

[34] KESTIN, J., and J. H. WHITELAW: Transport Properties of Water and Steam. Brown University, June 1963.

[35] KESTIN, J., J. H. WHITELAW and T. F. ZIEN: Thermal Conductivity of Superheated Steam. Brown University, Providence, July 1963.

[36] KEYES, F. G., L. B. SMITH and H. T. GERRY: The specific volume of steam in the saturated and superheated condition together with derived values of the enthalpy, heat capacity and Joule-Thomson coefficients. Part IV. Steam research program. Proc. Amer. Acad. Arts Sci. 70 (8) (1936) 319/364.

[37] KEYES, F. G.: Intercomparisons of Observational Data for Steam with Various Formulated Data, Boston, August 1963.

[38] KEYES, F. G., and R. G. VINES: Report of the U.S.A. Commission on Properties of Steam to the Sixth International Conference — Thermal Conductivity, New York 1963.

[39] KIRILLIN, V. A., and S. A. ULYBIN: An analysis of the accuracy and a summary table of experimental values of specific volume of water and steam obtained at the Moscow Power Institute (in Russian). Teploenergetika 6 (9) (1959) 3/7.

[40] KIRILLIN, V. A., L. I. RUMJANZEW and W. N. SUBAREW: Experimental Research of Specific Volumes of Water and Steam of High Parameters. Beitrag der russischen Dampftafelkommission zu det V. Internationalen Dampftafelkonferenz, London 1956, S. 37.

[41] LAWSON, A. W., R. LOWELL and A. L. JAIN: Thermal conductivity of water at high pressure. J. Chem. Phys. 30 (1959) 643.

[42] MAYINGER, F., E. SCHMIDT u. H. TRATZ: Neue Zustandsgleichungen für Wasserdampf unter Berücksichtigung ihrer Verwendung in elektronischen Rechenanlagen. Brennstoff – Wärme – Kraft (BWK) 14 (1962) 261/266 und 360.

[43] MAYINGER, F.: Messungen der Viskosität von Wasser und Wasserdampf bis zu 700 °C und 800 at. Int. J. Mass and Heat Transfer 5 (1962) 807/824.

[44] MAYINGER, F., u. U. GRIGULL: Viskosität und Wärmeleitfähigkeit des Wasserdampfs. Neue internationale Rahmentafeln und ihre Auswertung. Brennstoff – Wärme – Kraft (BWK) 17 (1965) 53/60.

[45] MOSZYNSKI, J. R.: Measurement of the viscosity of steam. Trans. ASME 83 C (1961) 111.

[46] NEWITT, D. M., and S. ANGUS: Notes and tables on the experimental determination of the specific enthalpy of steam; ERA Report I/T 181. British Electrical and Allied Industries Research Association, Leatherhead, Surrey 1961.

[47] OSBORNE, N. S., and C. H. MEYERS: A formula and tables for the pressure of saturated water vapour in the range 0 to 374 °C. J. Res. Nat. Bur. Stand. 13 (1934) 1/20.

[48] OSBORNE, N. S., H. F. STIMSON and D. C. GINNINGS: Thermal properties of saturated water and steam. J. Res. Nat. Bur. Stand. 23 (1939) 261/270.

[49] POWEL, R. W., and A. R. CHALLONER: The Thermal Conductivity of Water. An Investigation of a Reported Anomaly. Phil. Mag. 46 (1959) 1183/1186.

[50] RASSKASOW, D. S., u. A. E. SCHEINDLIN: Spezifische Wärme c_p von Wasser und Wasserdampf im überkritischen Gebiet. Teploenergetika 4 (1957) Nr. 11, S. 81/83. Bericht darüber in Brennstoff – Wärme – Kraft (BWK) 10 (1958) 345.

[51] RAY, A. K.: Measurement of the Cinematic Viscosity of Steam at High Pressures and Temperatures. University of Glasgow 1963.

[52] RIVKIN, S. L., u. T. S. ACHUNDOFF: An Experimental Study of the Specific Volumes of Water at Temperatures of 374.15 up to 500 °C and at pressures up to 600 kg/cm². Teploenergetika 10 (1963) Nr. 9, S. 66/71. Ref.: BWK 16 (1964) Nr. 2, S. 99.

[53] Schmidt, E., u. W. Sellschopp: Wärmeleitfähigkeit des Wassers bei Temperaturen bis zu 270 °C. Forsch. Ing.-Wes. 3 (1932) 277.

[54] Schmidt, E., u. W. Leidenfrost: Wärmeleitzahl-Messungen an Wasser, Äthenglykylol-Wasser-Mischungen und Kaliumchlorid-Lösungen im Temperaturbereich von 0 bis 100 °C. Forsch. Ing.-Wes. 21 (1955) 176.

[55] Schmidt, E.: Vierte Internationale Dampftafel-Konferenz. VDI-Z. 97 (1955) Nr. 23, 796/798.

[56] Schmidt, E.: Fünfte Internationale Dampftafel-Konferenz. Brennstoff – Wärme – Kraft (BWK) 9 (1957) Nr. 9, S. 432/435.

[57] Schmidt, E.: Internationale Dampftafel-Konferenz in Moskau. BWK 10 (1958) Nr. 12, S. 553/556.

[58] Schmidt, E., u. K. Traube: Gradients of density of fluids near their critical states in the field of gravity. In: Progress in International Research on Thermodynamic and Transport Properties, New York u. London 1962, S. 193/205.

[59] Schmidt, E.: Auf dem Weg zur internationalen Wasserdampftafel. BWK 15 (1963) Nr. 1, S. 15/19.

[60] Schmidt, E.: Verhandlungen und Ergebnisse der Sechsten Internationalen Konferenz über die Eigenschaften des Wasserdampfes. Die neuen Rahmentafeln 1963 für die thermodynamischen Eigenschaften des Wasserdampfes. BWK 16 (1964) Nr. 7, S. 322/330.

[61] Scheindlin, A. E., E. E. Spielrein u. W. W. Sytschoff: Spezifische Wärme c_p des Wasserdampfes. Teploenergetika 6 (1959) Nr. 12, S. 80/83. Ref.: BWK 13 (1961) 282/283.

[62] Scheindlin, A. E., W. W. Sytschoff, M. M. Chilal u. N. I. Gorbunova: Experimental Investigation of the Enthalpy of Water and Steam at Temperatures up to 390 °C and Pressures up to 500 kg/cm². Teploenergetika 10 (1963) Nr. 9, S. 76/81. Ref.: BWK 16 (1964) Nr. 2, S. 100.

[63] Schwarz, D.: Entwicklung einer Zustandsgleichung für überhitzten Wasserdampf. Fortschritts-Ber. VDI-Z. Reihe 6, 10 (1967) 1/100.

[64] Sigwart, K.: Messungen der Zähigkeit von Wasser und Wasserdampf bis ins kritische Gebiet. Forsch. Ing.-Wes. 7 (1936) 125.

[65] Sirota, A. M., and D. L. Timrot: Experimental investigation of the specific heat of steam in the sub-critical region (in Russian). Teploenergetika 3 (7) (1956) 16/23.

[66] Sirota, A. M.: Specific heat and enthalpy of steam at supercritical pressures (in Russian). Teploenergetika 5 (7) (1958) 10/13.

[67] Sirota, A. M., and B. K. Maltsew: An experimental investigation on the specific heat of water at temperatures from 10—500 °C and pressures up to 500 kg/cm² (in Russian). Teploenergetika 6 (9) (1959) 7/15.

[68] Sirota, A. M., and P. E. Beliakova: On the calorific properties of water under pressures up to 500 kg/cm² and temperatures up to 300 °C (in Russian). Teploenergetika 6 (10) (1959) 67/70.

[69] Sirota, A. M., and B. K. Maltsew: Experimental data for the specific heat of steam under pressures 300 to 500 atm and temperatures 500—600 °C (in Russian). Teploenergetika 7 (10) (1960) 67/68.

[70] Sirota, A. M., P. E. Beliakova, N. B. Wargaftik, A. A. Tarzimanow and O. N. Oleschuk: Tables of the Thermal Conductivity and Heat Capacity of Water and Steam (engl. Übers.). All-Union Wärmetechn. Inst. (V. T. I.) 1963.

[71] Sirota, A. M., B. K. Maltsew and A. Y. Grischkoff: An Experimental Study of the Heat Capacity of Water at High Pressures. Teploenergetika 10 (1963) Nr. 9, S. 57/60. Ref.: BWK 16 (1964) Nr. 1, S. 39/40.

[72] Smith, L. B., and F. G. Keyes: Steam Research Program, Part III: The Volumes of Unit Mass of Liquid Water and their Correlation as a Function of Pressure and Temperature. Proc. Amer. Acad. of Arts and Sciences 69 (1934) 285/312.

[73] Straub J.: Optische Bestimmung von Dichteschichtungen im kritischen Zustand. Chem. Ing. Techn. 39 (1967) 291/96.

[74] Suguwara, S., T. Sato, T. Minamiyama and J. Yata: On the Equation of State of Steam. J.C.P.S. Report No. 11.

[75] Sytschew, W. W.: Der Joule-Thomson-Koeffizient für Wasser und Wasserdampf auf der Sättigungslinie (russ.). Teploenergetika 8 (1962) Nr. 1, S. 66/69.

[76] Sytschew, W. W.: Analysis of existing heat capacity C data of water and steam at the saturation line. Teploenergetika 10 (1963) Nr. 7, S. 68.

[77] Tanaka, K., et al.: Comment on the Report „Transport Properties of Water and Steam" by J. Kestin and J. H. Whitelaw. 1. Okt. 1963.

[78] Tanaka, K., et al.: Viscosity of Steam at High Pressures and High Temperatures (The 2nd Report). J.C.P.S. Report No. 10.

[79] Tanishita, I., and K. Watanabe: Experimental Study on the Specific Volume of Steam at High Temperatures and Pressure. (Second Report: Temperature Range 600—900 °C under the Pressures up to 900 kg/cm²). J.C.P.S. Report No. 9.

(Die Arbeiten [76] bis [78] wurden herausgegeben vom Resources Research Institute, Kawaguchi.)

[80] Timrot, D. L.: Determination of the viscosity of steam and water at high temperatures and pressures. J. Phys. U.S.S.R. 2 (1940) 419.

[81] Timrot, D. L., u. N. B. Wargaftik: Wärmeleitfähigkeit des Wassers. J. techn. Phys. (U.S.S.R.) 10 (1940) 1063.

[82] Tratz, H.: Neue Zustandsgleichungen für flüssiges Wasser und eine Gleichung der Dampfdruckkurve. Brennstoff – Wärme – Kraft (BWK) 14 (1962) 379/383 u. 504.

[83] Traube, K.: Messungen von Dichteschichtungen in der Umgebung des kritischen Zustandes (gasförmig-flüssig) nach einem optischen Verfahren. VDI-Forschungsheft Nr. 487, Düsseldorf 1961.

[84] Vesper, H.: Näherungsgleichungen für die Zustandsgrößen des Wassers und des Dampfes an den Grenzkurven zur Verwendung in elektronischen Rechenmaschinen. Brennstoff – Wärme – Kraft (BWK) 15 (1963) Nr. 1, S. 20/23 u. 108.

[85] Wargaftik, N. B., and O. N. Oleschuk: Experimental investigation of the heat conduction of water. Teploenergetika 6 (1959) 70. Bericht darüber in Brennstoff – Wärme – Kraft (BWK) 13 (1961) 37/38.

[86] Wargaftik, N. B., u. A. A. Tarsimanow: Auswertung von Messungen der Wärmeleitfähigkeit des Wasserdampfes (russ.). Teploenergetika 8 (1961) 5/S. Übersetzung in Arch. Energiewirtsch. 15 (1961) 983/991. Bericht darüber in Brennstoff – Wärme – Kraft (BWK) 14 (1962) 244/245.

[87] Wargaftik, N. B.: Corrections to the Draft Skeleton Tables for Thermal Conductivity of Water and Steam (engl. Übers.). Moskauer Aeronaut. Inst. u. All-Union Wärmetechn. Inst. (V. T. I.) Sept. 1963.

[88] Whitelaw, J. H.: Viscosity of steam at supercritical pressures. J. Mech. Eng. Sci. 2 (1960) 288.

[89] Wukalowitsch, M. P., V. N. Zubarew and P. G. Prusakow: Experimental investigation of the enthalpy of steam (in Russian). Teploenergetika 5 (7) (1958) 22/26.

[90] Wukalovitsch, M. P., V. N. Zubarew, A. A. Alexandrow and Y. Y. Kalinin: Experimental determination of the specific volume of water under pressures up to 1200 kgf/cm² at 300 °C (in Russian). Teploenergetika 6 (10) (1959) 74/77.

[91] Wukalowitsch, M. P., B. W. Dsampow, D. S. Rasskasow u. S. A. Remisow: Neue Tafeln für die spezifischen Volume des Wassers bei 1200 ata und 300 °C. Teploenergetika 7 (1960) Nr. 7, S. 4/12. Bericht darüber in Brennstoff – Wärme – Kraft (BWK) 12 (1960) 492/498.

[92] Wukalowitsch, M. P., B. W. Dsampow, D. S. Rasskasow u. S. A. Remisow: Tabellen der Wärmekapazität c_p des Wassers und Wasserdampfes (Tafel nach neuesten Messungen). Teploenergetika 8 (1961) Nr. 12, S. 70/77. Bericht darüber in Brennstoff – Wärme – Kraft (BWK) 14 (1962) 242/244.

[93] Wukalowitsch, M. P., B. W. Dsampow u. D. S. Rasskasow: Eigenschaften des Wassers und des Wasserdampfes bis zu Drücken von 1000 ata im Temperaturbereich von 300 bis 1000 °C (Tafeln nach neuesten Messungen). Teploenergetika 8 (1961) Nr. 7, S. 48/49.

[94] Wukalowitsch, M. P., V. N. Zubarew and A. A. Alexandrow: Experimental determination of the specific volume of steam in the temperature range 400 °C to 650 °C and for pressures up to 1200 kgf/cm² (in Russian). Teploenergetika 8 (10) (1961) 79/86.

[95] Wukalowitsch, M. P., V. N. Zubarew and A. A. Alexandrow: Experimental determination of the specific volume of steam in the temperature range 700 °C to 900 °C and for pressures up to 1200 kgf/cm² (in Russian). Teploenergetika 9 (1) (1962) 49/51.

[96] Wukalowitsch, M. P., u. L. I. Tscherneeva: Experimental Investigations of the Heat Conductivity Coefficient of Steam up to 660 °C and 1500 kg/cm². Teploenergetika 10 (1963) Nr. 9, S. 71/76. Ref.: BWK 16 (1964) Nr. 2, S. 100/101.

[97] Yamada, Y., u. S. Nagai: Measurement of Enthalpy of Steam at High Temperature and Pressure. J.C.P.S. Report No. 7.
(Die Arbeiten [97] und [98] wurden herausgegeben von The Japan Society of Mechanical Engineers, Nihon Kikaku Kyokai Bldg., 89, Akasaka Hitotsugi-cho, Minato-ku, Tokyo.)

[98] Yamada, Y.: Thermodynamically consistent values of v, i and s of steam in the neighbourhood of the Critical Point (Second Report). J.C.P.S. Report No. 8.

[99] Zagoruchenko, V. A.: Comparison of the Skeleton Tables for Steam Project with Experimental Thermal and Calorimetric Data. Teploenergetika 10 (1963) Nr. 9, S. 54.

[100] Zens, R.: Ein Programmsystem für die elektronische Berechnung von Kreisprozessen bei Dampfturbinenanlagen. Siemens-Z. 37 (1936) Nr. 7 u. 8.

[101] Zens, R.: Näherungsgleichung für die Berechnung isentroper Wärmegefälle des überhitzten Wasserdampfes mit einer Datenverarbeitungsanlage. Siemens-Z. 38 (1964) Nr. 1, S. 32/36.

Appendix to the Bibliography — Anhang zum Literaturverzeichnis

[102] AGAJEV, N. A., and A. D. JUSIBOVA: The Viscosity of Ordinary and Heavy Water at High Pressures in the Temperature Range of 0 °C to 150 °C (in Russian). Doklady Akademii Nauk SSSR **180** (1968) 334—337.

[103] ALEXANDROV, A. A.: Draft of the Skeleton Table for the Thermal Conductivity of Water and Steam. Moscow Power Institute (1974).

[104] ALEXANDROV, A. A., A. I. IVANOV and A. B. MATVEEV: Draft of the Skeleton Tables of the Dynamic Viscosity of Water and Steam (II). Sov. Nat. Com. Prop. Steam, Moskau (1974).

[105] AMIRCHANOV, CH. I., and A. P. ADAMOV: Thermal Conductivity of Steam near and above the Critical Region (in Russian). Teploenergetica **9** (1963) 69—72.

[106] AMIRCHANOV, CH. I., and A. P. ADAMOV: Heat Conductivity of Water and Steam at Temperatures of 350 to 460 °C and Pressures of 200 to 1000 kgf/cm². Report C-10 to the 7th ICPS, Tokyo (1968).

[107] AMIRCHANOV, CH. I., A. P. ADAMOV and U. B. MAGOMEDOV: Experimentelle Untersuchung der Wärmeleitfähigkeit von Wasser bei Temperaturen von 25—350 °C und Drücken von 0,1—245,3 MPa (russ.). Akadem. Nauk SSSR, Dagestanische Abteilung (1974).

[108] BACH, J., u. U. GRIGULL: Instationäre Messung der Wärmeleitfähigkeit mit optischer Registrierung. Wärme- und Stoffübertragung **3** (1970) 44—57.

[109] BARNETT, S. C., T. W. JACKSON and R. H. WHITESIDE: An Investigation of the Viscosity of Steam at High Pressures. ASME — Paper 63-WA-240 (1963).

[110] BASU, R. S., and J. V. SENGERS: Thermal Conductivity of Steam in the Critical Region. Proc. 7th Symp. Thermoph. Prop., ASME (1977) 822—830.

[111] BINGHAM, E. C., u. G. F. WHITE: Die Viskosität von Wasser. Z. phys. Chemie **80** (1912) 670—686.

[112] BINGHAM, E. C.: Fluidity and Plasticity. McGraw-Hill Book Co. 1922, 26/27, 48/49, 339—341.

[113] BOHR, N.: Determination of the Surface Tension of Water by the Method of Jet-Vibration. Proc. Roy. Soc. London **82** (1909) 146.

[114] BONILLA, C. F., S. J. WANG and H. WEINER: The Viscosity of Steam, Heavy-Water Vapor, and Argon at Atmospheric Pressure up to High Temperatures. Trans. ASME **78** (1956) 1285—1289.

[115] BONILLA, C. F., R. D. BROOKS and H. WEINER: The Viscosity of Steam and of Nitrogen at Atmospheric Pressure and High Temperatures. ASME-Ing. Mech. Eng.-Heat Transfer Symp. (1951) 167.

[116] BRAIN, T. J. S.: New Thermal Conductivity Measurements for Argon Nitrogen and Steam. Int. J. Heat Mass Transfer **10** (1967) 737—744.

[117] BRAIN, T. J. S.: Thermal Conductivity of Steam at Atmospheric Pressure., J. Mech. Engin. Science **11** (1969) 392—401.

[118] BRAUNE, H., u. R. LINKE: Über die innere Reibung einiger Gase und Dämpfe. Z. phys. Chemie **A 149** (1930) 195—215.

[119] BRIDGMAN, P. W.: The Effect of Pressure on the Viscosity of Fourty-three Pure Liquids. Proc. Amer. Acad. Arts Sciences **61** (1926) 57—99.

[120] BURY, P., R. TUFEU and B. LE NEINDRE: Thermal Conductivity Coefficient of Steam up to 500 °C and 500 bar. Report Laborat. Interat. Molec. Hautes Press. CNRS Bellevue, France (1973).

[121] CASTELLI, V. J., and E. M. STANLEY: Thermal Conductivity of Distilled Water as Function of Pressure and Temperature. J. Chem. Engin. Data **19** (1974) 8—11.

[122] CHERNEEVA, L. I.: The Experimental Research of the Water Heat Conduction under Pressure from 100 to 1000 bar and Temperatures from 100 to 350 °C. Report C-9 to the 7th ICPS, Tokyo (1968).

[123] CINI, R., G. LOGLIO and A. FICALBI: Temperature Dependence of the Surface Tension of Water by the Equilibrium Ring Method. J. Colloid. Int. Sci. **41**, 2 (1972) 287.

[124] COCKETT, A. H., and A. FERGUSON: The Surface Tension of Water and Heavy Water. Phil. Mag. J. Sci., 70th Serie (1939) 685.

[125] COE, J. R., and T. B. GODFREY: Viscosity of Water. J. Appl. Physics **15** (1944) 625—626.

[126] DUDZIAK, K. H., u. E. U. FRANCK: Messungen der Viskosität des Wassers bis 560 °C und 3500 bar. Ber. Bunsengesellsch. **70** (1966) 1120—1128.

[127] EICHER, L. D., and B. J. ZWOLINSKI: High Precision Viscosity of Supercooled Water and Analysis of the Extended Range Temperature Coefficient. J. Phys. Chem. **75** (1971) 2016—2024.

[128] GEDDES, J. A.: The Fluidity of Dioxane-Water Mixtures. J. Amer. Chem. Society **55** (1933) 4832—4837.

[129] GEIER, H., u. K. SCHÄFER: Wärmeleitfähigkeit von reinen Gasen und Gasgemischen zwischen 0 und 1200 °C. Allg. Wärmetechnik **10** (1961) 70—75.

[130] GITTENS, G. J.: Variation of Surface Tension of Water with Temperature. J. Colloid. Int. Sci. **30**, 3 (1968) 406.

[131] GRIGULL, U., M. REIMANN and K. SCHEFFLER: Data Survey of the Viscosity of Pure Water. Invited Paper, Proc. 8th ICPS, France, 1 (1974) 113.

[132] HALLET, J.: The Temperature Dependence of the Viscosity of Supercooled Water. Proc. Phys. Soc. **82** (1963) 1046—1059.

[133] HARDY, R. C., and R. L. COTTINGTON: Viscosity of Deuterium Oxide and Water in the Range 5 °C to 125 °C. J. Research NBS **42** (1949) 573—578.

[134] HARKINS, W. D., and F. E. BROWN: The Determination of Surface Tension (Free Surface Energy), and the Weight of Falling Drops: The Surface Tension of Water and Benzene by the Capillary Height Method. J. Am. Chem. Soc. **41** (1919) 499.

[135] HAWKINS, G. A., H. L. SOLBERG and A. A. POTTER: The Viscosity of Water and Superheated Steam. Trans. ASME **57** (1935) 395—401.

[136] HAWKINS, G. A., H. L. SOLBERG and A. A. POTTER: The Viscosity of Superheated Steam Trans. ASME **62** (1940) 677—688.

[137] HEIKS, J. R. et al.: The Density, Surface Tension and Viscosity of Deuterium Oxide at Elevated Temperatures. J. Phys. Chem. **58** (1954) 488.

[138] HEYDWEILLER, A.: Die innere Reibung einiger Flüssigkeiten oberhalb ihres Siedepunktes. Wied. Ann. Phys. Chemie **59** (1896) 193—212.

[139] HÖPPLER, F.: Über neuere Messungen der Zähigkeit des Wassers. Z. angew. Phys. **4** (1952) 297—300.

[140] HORNE, R. A., and D. S. JOHNSON: The Viscosity of Water under Pressure. J. Phys. Chem. **70** (1966) 2182—2190.

[141] HOSKING, R.: The Viscosity of Water. Philos. Mag. **18** (1909) 260—263.

[142] HOUGH, E. W., B. B. WOOD Jr. and M. J. RZASA: Adsorption at Water-Helium-Methane and -Nitrogen. Interfaces at Pressures to 15,000 PSIA. J. Phys. Chem. **56** (1952) 996.

[143] JACKSON, T. W., B. LATTO, C. E. WILLBANKS and J. W. HODGSON: The Viscosity of Steam to 10 000 PSIA. Adv. Thermophys. Prop. Extr. Temp. Press. ASME (1965) 221.

[144] JÄGER, F. M.: Über die Temperaturabhängigkeit der molekularen freien Oberflächenenergie von Flüssigkeiten im Temperaturbereich von — 80 bis + 1650 °C. Z. Anorg. Allgem. Chem. **100** (1917) 53.

[145] JASPERS, J. J.: The Surface Tension of Pure Liquid Compounds. J. Phys. Chem. Ref. Data 1, 4 (1972) 841

[146] KALÄHNE, A.: Über die Benutzung stehender Capillarwellen auf Flüssigkeiten als Beugungsgitter und die Oberflächenspannung von Wasser und Quecksilber. Ann. Phys. 7 (1902) 440.

[147] KEENAN, J. H., F. G. KEYES, P. G. HILL and J. G. MOORE: Steam Tables. New York: J. Wiley 1969.

[148] KERIMOV, A. M., N. A. AGAJEV and A. A. ABASADE: The Experimental Investigation of Viscosity of Water Substance at Temperatures from 100 °C to 275 °C and at High Pressures (in Russian). Teploenergetica **18** (1969) 87—90.

[149] KESTIN, J., and P. D. RICHARDSON: The Viscosity of Superheated Steam up to 275 °C — A Refined Determination. J. Heat Transfer, Trans. ASME **85** (1963) 295—302.

[150] KESTIN, J., and H. E. WANG: The Viscosity of Superheated Steam up to 270 °C. Physica **26** (1960) 575—584.

[151] KEYES, F. G., and D. J. SANDELL: New Measurements of the Heat Conductivity of Steam and Nitrogen. Trans. ASME (1950) 767—778.

[152] KEYES, F. G., and R. G. VINES: The Thermal Conductivity of Steam. Int. J. Heat Mass. Transf. 7 (1964) 33—40.

[153] KJELLAND-FOSTERUD, E.: Determination of the Viscosity of Steam at Super-Critical Pressures. J. Mech. Eng. Science 1 (1959) 30—38.

[154] KOROSI, A., and B. M. FABUSS: Viscosity of Liquid Water from 25 °C to 150 °C — Measurements in Pressurized Glass Capillary Viscometer. Anal. Chem. **40** (1968) 157—163.

[155] KORSON, L., W. DROST-HANSEN and F. J. MILLERO: Viscosity of Water at Various Temperatures. J. Phys. Chem. **73** (1969) 34—39.

[156] KUNDT, A., u. E. WARBURG: Über Reibung und Wärmeleitung verdünnter Gase. Ann. Phys. Chem. **155** (1875) 337—365/525—550.

[157] Landolt-Börnstein: Zahlenwerte und Funktionen aus Physik, Chemie, Astronomie, Geophysik und Technik, 6. Aufl., Bd. 2/3. Tl., Berlin/Göttingen/Heidelberg: Springer 1956, 421.

[158] LATTO, B.: Viscosity of Steam at Atmospheric Pressure. Int. J. Heat Mass Transfer 8 (1965) 689—720.

[159] LE NEINDRE, B., P. BURY, R. TUFEU, P. JOHANNIN et B. VODAR: Résultats expérimentaux sur la Conductivité Thermique de l'Eau lourde en Phase liquide, jusqu'à une Température de 370 °C. Report C-1 to the 7th ICPS, Tokyo (1968).

[160] LE NEINDRE, B., P. BURY, R. TUFEU, P. JOHANNIN et B. VODAR: Mesure de la Conductivité Thermique de la Vapeur d'Eau et de la Vapeur d'Eau Lourde jusqu'à la Pression de Saturation et 330 °C. Report C-2 to the 7th ICPS, Tokyo (1968).

[161] LE NEINDRE, B., R. TUFEU, P. BURY and J. V. SENGERS: Thermal Conductivity of Carbon Dioxide and Steam in the Supercritical Region. Ber. Bunsen Ges. Phys. Chem. **77** (1973) 263—275.

[162] LEROUX, M. P.: Détermination du coefficient de viscosité de l'eau en valeur absolue. Ann. de Phys., ser. 10, **4** (1925) 163—247.

[163] LEROUX, M. P.: Détermination du coefficient de viscosité de l'eau en valeur absolue. Comptes rendus **180** (1925) 914—916.

[164] LIVINGTON, J., R. MORGAN and A. McD. McAFEE: The Weight of a Falling Drop and the Laws of Tate. J. Am. Chem. Soc. **33**, 8 (1911) 1275.

[165] MALJAROV, G. A.: Determination of the Viscosity of Water at a Temperature of 20 °C (in Russian). Trudi WNIIM 37 (1959) 125—139.

[166] MASHIROV, V. E.: Measurements cited in A. A. TARZIMANOV, V. S. LOZOWOI, Report C-8 to the 7th ICPS, Tokyo 1968.

[167] MASSOUDI, R., and A. D. KING Jr.: Effect of Pressure on the Surface Tension of Water. Adsorption of Low Molecular Weight Gases on Water at 25 °C. J. Phys. Chem. **78**, 22 (1974) 2262.

[168] MASTERTON, W. L., J. BIANCHI and E. J. SLOWINSKI Jr.: Surface Tension and Adsorption in Gas-Liquid Systems at Moderate Pressures. J. Phys. Chem. **67** (1963) 615.

[169] MINAMIYAMA, T., and J. YATA: Measurement of Thermal Conductivity of Liquid Water Proceedings 8th ICPS, Gies, France 1974. Part II submitted to Working Group II of IAPS (1975).

[170] MINAMIYAMA, T., and J. YATA: A Formulation of Thermal Conductivity. Kyoto Institute of Technology, Matsugasaki, Sakyo-ku, Kyoto, Japan (1976).

[171] MILVERTON, S. W.: An Experimental Investigation of Thermal Conduction through Vapours. Proc. Royal Society London A **150** (1935) 287—308.

[172] MOSER, H.: Der Absolutwert der Oberflächenspannung des reinen Wassers nach der Bügelmethode und seine Abhängigkeit von der Temperatur. Ann. Phys. **82**, 4 (1927) 993.

[173] MOSZYNSKI, J. R.: The Viscosity of Steam and Water at Moderate Temperatures and Pressures. J. Heat Transfer, Trans. ASME **83** (1961) 111—124.

[174] NAGASHIMA, A., and M. IKEDA: Available Input of the Viscosity of Water and Steam. Keio University, Yokohama (1973).

[175] NAGASHIMA, A., M. IKEDA and I. TANISHITA: Correlation of Viscosity for Water and Steam. Submitted to the 8th ICPS, Giens (1974).

[176] NAGASHIMA, A.: Viscosity of Water Substance — New International Formulation and its Background. Phys. Chem. Ref. Data 6 (1977) 4, 1133—1166.

[177] NOACK, K.: Über den Einfluß von Temperatur und Konzentration auf die Fluidität von Flüssigkeitsgemischen. Wied. Ann. Phys. Chem. 27 (1886) 289—300.

[178] NOVAK, E. S., and R. J. GROSH: An Investigation of Certain Thermodynamic and Transport Properties of Water and Water Vapor in the Critical Region. Argonne Nat. Laborat., Sub-Contr. 31-109-38-704 (1959).

[179] OLTERMANN, G.: Messungen der Viskosität von Wasser und Wasserdampf in der Nähe seines kritischen Zustandes. Lehrst./Institut Verfahrenstechnik, TU Hannover (1976), Dissertation TU Hannover (1977).

[180] OSBORNE, H. H.: An Experimental Investigation of the Viscosity of Steam. PhD-Thesis, Purdue University (1958).

[181] POISEUILLE, J. L. M.: cited in HOSKING (1909), Mém. Savants Étrangers 9 (1846) 423.

[182] POLLAK, R.: Zustandsgleichung für Wasser und Wasserdampf. BWK 5 (1975), Dissertation, Ruhr-Universität Bochum (1974).

[183] PULUJ, J.: Über die Reibung der Dämpfe. Sitz. Ber. d. math.-naturw. Klasse d. kais. Akad. d. Wiss. Wien 78 (1878) 279—313.

[184] RASTORGUJEW, JU. L., and W. W. PUGATSCH: Experimental Investigation of the Heat Conductivity of Water at High Pressure (in Russian). Teploenergetica 4 (1970) 77—79.

[185] RAY, A. K.: A New Determination of the Kinematic Viscosity of Steam at Supercritical Pressures and Temperatures. J. Mech. Engng. Science 6 (1964) 137—143.

[186] REIMANN, M., N. ROSNER, K. SCHEFFLER and R. MEYER-PITTROFF: Dynamic Viscosity of Water Substance. A Comparison of New Equations and Measurements. Institut A f. Thermodynamik, TU München (1973).

[187] REIMANN, M., N. ROSNER, K. SCHEFFLER and R. MEYER-PITTROFF: Thermal Conductivity of Water Substance. A Comparison of New Equations and Measurements. Institut A f. Thermodynamik, TU München (1973).

[188] REIMANN, M., K. SCHEFFLER and U. GRIGULL: Temperature and Saturation Pressure of Water Substance Measured up to the Critical Point. Proceedings 8th ICPS, Giens, France (1974).

[189] RICHARDS, T. W., and E. K. CARVER: A Critical Study of the Capillary Rise Method of Determining Surface Tension, with Data for Water, Benzene, Toluene, Chloroform, Carbon, Tetrachloride, Ether and Dimethyl Aniline. J. Am. Chem. Soc. 43 (1921) 827.

[190] RICHARDS, T. W., C. L. SPEYERS and E. K. CARVER: The Determination of Surface Tension with Very Small Volumes of Liquids, and the Surface Tensions of Octanes and Xylenes at several Temperatures. J. Am. Chem. Soc., 46 (1924) 1196.

[191] RIVKIN, S. L., A. S. LEVIN and L. B. ISRAILEVSKI: Experimental Determination of the Dynamic Viscosity of Superheated Steam up to 450 °C and 350 bar (in Russian). Report B-9 to the 7th ICPS, Tokyo (1968).

[192] RIVKIN, S. L., A. S. LEVIN and L. B. ISRAILEVSKI: Experimental Study of the Viscosity of Steam at Temperatures up to 450 °C and Pressures up to 350 bar (in Russian). Teploenergetica 15 (1968) 74—78.

[193] RIVKIN, S. L., A. S. LEVIN and L. B. ISRAILEVSKI: Experimental Investigation of Viscosity of Water Substance at Temperatures from 250 to 375 °C and Pressures up to 500 bar (in Russian). Teploenergetica 17 (1970—11) 79—81.

[194] RIVKIN, S. L., A. S. LEVIN and L. B. ISRAILEVSKI: The Dynamic Viscosity of Steam near the Saturation Line (in Russian). Teploenergetica 17 (1970—8) 88—91.

[195] RIVKIN, S. L., A. S. LEVIN and L. B. ISRAILEVSKI: An Investigation of the Coefficient of the Dynamic Viscosity of Water and Steam (in Russian). Teplo-i Massoperenos 7 (1972) 61—71.

[196] RIVKIN, S. L., A. S. LEVIN and L. B. ISRAILEVSKI: Experimental Investigation of Viscosity of Water Substance at Temperatures from 250 to 375 °C and Pressures up to 1000 bar (in Russian). Insvestija Akad. Nauk USSR Minsk 1 (1972) 33—37.

[197] RIVKIN, S. L., A. S. LEVIN and L. B. ISRAILEVSKI: Experimental Investigation of the Viscosity of Water in the Supercritical Region at Pressures up to 500 bar and Temperatures up to 500 °C (in Russian). Teploenergetica 8 (1973) 11/12.

[198] ROSNER, N., M. REIMANN, K. SCHEFFLER and U. GRIGULL: Draft of the Skeleton Table for Dynamic Viscosity of Water and Steam. Lehrstuhl A f. Thermodynamik, TU München (1975), presented to the IAPS meeting of Special Committee.

[199] ROSCOE, R., and W. BAINBRIDGE: Viscosity Dermination by the Oscillating Vessel Method, II: The Viscosity of Water at 20 °C. Proc. Phys. Soc. 72 (1958) 585—595.

[200] SATO, T., T. MINAMIYAMA, J. YATA and T. OKA: Measurements of the Viscosity of Steam at Various Temperatures and Pressures. Report B-4 to the 7th ICPS, Tokyo (1968).

[201] SCHEFFLER, K., N. ROSNER and M. REIMANN: Available Input of Dynamic Viscosity of Water Substance. Institut A f. Thermodynamik, TU München (1973).

[202] SCHEFFLER, K., N. ROSNER and M. REIMANN: Available Input of the Thermal Conductivity of Water Substance. Institut A f. Thermodynamik, TU München (1974).

[203] SCHEFFLER, K., N. ROSNER and U. GRIGULL: Draft of the Skeleton Table Proposal for Thermal Conductivity of Water Substance. Lehrstuhl A f. Thermodynamik, TU München (1975).

[204] SCHEFFLER, K., and J. STRAUB: The Dynamic Viscosity of Water Substance. A Comparison Between the IFC 1967 and the IFC 1968. Lehrstuhl A f. Thermodynamik, TU München (1976).

[205] SCHEFFLER, K., J. STRAUB and U. GRIGULL: The Dynamic Viscosity of Water Substance. Proceedings of the 7th Symposium on Thermophysical Properties. ASME (1977) 684—694.

[206] SCHEFFLER, K., N. ROSNER, J. STRAUB and U. GRIGULL: Der internationale Standard der dynamischen Viskosität von Wasser und Wasserdampf. BWK 30 (1978) 2, 73.

[207] SCHEFFLER, K., N. ROSNER, J. STRAUB and U. GRIGULL: Die Wärmeleitfähigkeit von Wasser und Wasserdampf eingereicht zur Veröffentlichung BWK (1979).

[208] SENGERS, J. V., and R. S. BASU: A New Equation for the Thermal Conductivity of Water and Steam. Nat. Bur. Standards, Washington (1976).

[209] SHIFRIN, A. S.: The Viscosity of Steam at Atmospheric Pressure (in Russian). Teploenergetica 6 (1959) 22—27

[210] SHUGAJEV, W.: The Viscosity of Steam at High Pressures (in Russian). Zhurn. Esp. Techn. Fiz. 3 (1933) 3—9

[211] SHUGAJEV, W., and S. SOROKIN: The Viscosity of Steam at High Pressures (in Russian). Zhurn. Fiz. 9 (1939) 939—941.

[212] SIROTA, A. M., W. I. LATUNIN u. G. M. BELJAEVA: Experimentelle Untersuchung der maximalen Wärmeleitfähigkeit von Wasser bei kritischem Zustand (russ.). Teploenergetica 8 (1973) 6—11.

[213] SIROTA, A. M., W. I. LATUNIN u. G. M. BELJAEVA: Experimentelle Untersuchung der maximalen Wärmeleitfähigkeit von Wasser bei kritischem Zustand (russ.). Proc. 8th ICPS, Giens, France (1974).

[214] SIROTA, A. M., W. I. LATUNIN u. G. M. BELJAEVA: Experimentelle Untersuchung der maximalen Wärmeleitfähigkeit von Wasser bei kritischem Zustand (russ.). Teploenergetica 10 (1974) 52—58.

[215] SIROTA, A. M., W. I. LATUNIN u. G. M. BELJAEVA: Experimentelle Untersuchung der maximalen Wärmeleitfähigkeit von Wasser bei kritischem Zustand (russ.). Teploenergetika 1 (1976) 61—67.

[216] Sirota, A. M., W. I. Latunin u. G. M. Beljaeva: Experimentelle Untersuchung der maximalen Wärmeleitfähigkeit von Wasser bei kritischem Zustand (russ.). Teploenergetica 5 (1976) 70—75.

[217] Sirota, A. M., W. I. Latunin u. G. M. Beljaeva: Experimentelle Untersuchung der maximalen Wärmeleitfähigkeit von Wasser bei kritischem Zustand (russ.). Teploenergetika 6 (1976) 84—88.

[218] Slotte, K. F.: Über die innere Reibung einiger Lösungen und die Reibungskonstante des Wassers bei verschiedenen Temperaturen. Ann. Phys. u. Chem. 20 (1883) 257—267.

[219] Slowinski Jr., E. J., E. E. Gates and C. E. Waring: The Effect of Pressure on the Surface Tension of Liquids. J. Phys. Chem. 61 (1956) 808.

[220] Smith, C. J.: An Experimental Study of Viscous Properties of Water Vapour. Proc. Roy. Soc. A 106 (1924) 83—96.

[221] Speyerer, H.: Die Bestimmung der Zähigkeit des Wasserdampfes. Z. VDI 69 (1925) 747—752.

[222] Sprung, A.: Experimentelle Untersuchung über die Flüssigkeitsreibung bei Salzlösungen. Pogg. Ann. Phys. Chem. 159 (1876) 1—35.

[223] Ssementschenko, W. K., u. E. A. Davidowskaja: Über das oberflächliche Aussalzen durch Elektrolyte. Kolloid-Z. 73/1 (1935) 24.

[224] Straub, J., N. Rosner and U. Grigull: Proposal for the Representation of the Surface Tension of Water. Submitted to IAPS, Schliersee, Germany (1975).

[225] Straub, J., N. Rosner and U. Grigull: Representation of the Surface Tension of Saturated Water. Submitted to IAPS, Ottawa, Canada (1975).

[226] Sudgen, S.: The Dermination of Surface Tension from the Rise in Capillary Tubes. J. Chem. Soc. London 119/2 (1921) 1483.

[227] Swindells, J. F., J. R. Coe and T. B. Godfrey: Absolute Viscosity of Water at 20 °C. J. Res. NBS 48 (1952) 1—31.

[228] Tanaka, K., S. Sasaki and H. Hattori: Study of Measurements of Viscosity of Steam at High Temperatures (in Japanese). Nippon Kikai Kalekai ISME 31 (1965) 1837—1846.

[229] Tanishita, I., A. Nagashima and S. Yamaguchi: Viscosity Measurements of Water and Steam at High Temperatures and High Pressures. Bull. ISME 12 (1969) 1467—1478.

[230] Tarzimanow, A. A.: Thermal Conductivity of Steam Near the Saturation Line (in Russian). Teploenergetica 7 (1962) 73—77.

[231] Tarzimanow, A. A., and W. S. Lozowoi: Experimental Investigation of the Heat Conductivity of Water at High Pressures. Report C-8 to the 7th ICPS, Tokyo (1968).

[232] Tarzimanow, A. A., and W. S. Lozowoi: Unpublished Measurements cited in [233], (1970).

[233] Tarzimanow, A. A., and M. Zajnullin: Results of Measurements of the Heat Conductivity of Water Vapour at Pressures up to 1000 bar (in Russian). Teploenergetica 8 (1973) 2—6.

[234] Thakur, D. K., and K. Hickman: Surface Tension of Water at 100 °C. J. Colloid. Int. Sci. 50/3 (1974) 525.

[235] Thorpe, T. E., and J. W. Rodger: cited in: Coe (1944), cited in: Bingham (1922), Roy. Soc. Phil. Trans., 185 A (1894) 397—710.

[236] Timrot, D. L., and N. B. Vargaftik: Determination of the Dependence of Thermal Conductivity of Steam on the Temperature (in Russian). Izvestiya VTI 9 (1935).

[237] Timrot, D. L., and N. B. Vargaftik: The Thermal Conductivity and Viscosity of Steam at High Temperatures and Pressures (in Russian). J. Phys. 2 (1939) 101—111.

[238] Timrot, D. L., u. N. B. Vargaftik: Wärmeleitfähigkeit des Wassers (russ.). J. Techn. Phys. USSR 10 (1940) 1063.

[239] Timrot, D. L., and A. W. Khlopkina: Experimental Determination of the Viscosity of Water and Steam at Elevated Parameters (in Russian). Teploenergetica 10 (1963) 64—67.

[240] University of Glasgow: The Viscosity of Steam in the Pressure Range 200 to 1000 at and in the Temperature Range from 377 to 540 °C (1958): cited in [178] (1959) 163.

[241] Vogel, H.: Über die Viskosität einiger Gase und ihre Temperaturabhängigkeit bei tiefen Temperaturen. Ann. Phys. 43 (1914) 1235—1251.

[242] Vargaftik, N. B., and O. M. Oleschuk: The Dependence of Thermal Conductivity of Gases on Temperature. Isvestiya VTI 9 (1946).

[243] Vargaftik, N. B., u. E. V. Smirnova: Über die Abhängigkeit der Wärmeleitfähigkeit von Wasserdampf von der Temperatur (russ.). J. Techn. Phys. 6 (1956) 1251—1261.

[244] Vargaftik, N. B., and A. A. Tarzimanow: Experimental Investigation of the Thermal Conductivity of Water Vapour under High Parameters (in Russian). Teploenergetika 9 (1959) 15—21.

[245] Vargaftik, N. B., and O. M. Oleschuk: Experimental Investigation of the Thermal Conductivity of Water (in Russian). Teploenergetika 10 (1959) 70—74.

[246] Vargaftik, N. B., and A. A. Tarzimanow: Experimental Investigation of the Thermal Conductivity of Water Vapour (in Russian). Teploenergetika 7 (1960) 12—16.

[247] Vargaftik, N. B., and N. H. Zimina: The Thermal Conductivity of Steam at High Temperatures. Teploenergetika 11 (1964) 84—86.

[248] Vargaftik, N. V., L. D. Voljak and B. N. Volkov: Temperature Dependence of Surface Tension of Water (in Russian). Results of the Confederation of the Physical-Technical Thermodynamic Conference, USSR (1969).

[249] Vargaftik, N. B., N. A. Vanicheva and L. B. Jakusch: Thermal Conductivity of D_2O in Gaseous Phys. Ing. Phys. J. 25 (1973) 336—340.

[250] Vargaftik, N. B., L. D. Voljak and B. N. Volkov: Investigating the Surface Tension of H_2O and D_2O at Near-Critical Temperatures. Teploenergetika 20 (1973) 80.

[251] Venart, J. E. S.: The Thermal Conductivity of Water and Steam. University of Glasgow, Techn. Report 14 (1965).

[252] Vines, R. G.: Measurement of the Thermal Conductivity of Gases at High Temperatures. Trans. ASME, J. Heat Transfer (1960) 48—52.

[253] Voljak, L. D.: Determination of Temperature Dependence of Surface Tension of Water (in Russian). Dokl. Akad. Nauk. USSR 47/2 (1950) 307.

[254] Watson, J. T. R.: (1) Zero Density Equation for Steam. (2) Proposed Representative Equation for the Thermal Conductivity of Water-Substance outside the Critical Region. Properties of Fluids Division, National Engineering Laboratory, East Kilbride, Glasgow, Scotland (1977).

[255] WHITELAW, J. H.: The Determination of the Kinematic Viscosity of Steam at Supercritical Pressures and Temperatures University of Glasgow, Techn. Rep. 1 (1960).
[256] WHITELAW, J. H.: New Viscosity Data on Steam and Nitrogen. University of Glasgow. Techn. Rep. 3 (1960).
[257] WHITE, TWINING: cited in COE (1944). Ann. Chem. Journ. 50 (1913) 380.
[258] WILKINSON, M. C.: Extended Use of and Comments on the Drop-Weight (Drop-Volume) Technique for the Determination of Surface Tension and Interfacial Tensions. J. Colloid Int. Sci. 40/1 (1972) 14.
[259] ZHDANOV, A. G.: Capillary Viscosimeter for Determination of the Viscosity of Steam in a Closed Circle (in Russian). Inz. Fiz. Zhurn. 15 (1968) 1120—1123.
[260] International Critical Tables 4 (1928).
[261] ,,The IFC — 1967 Formulation for Industrial Use". Issued by the International Formulation Committee of the 6th Conference on the Properties of Steam (1967), New York, ASME.
[262] "The IFC — 1968 Formulation for General and Scientific Use". New York, ASME (1968).
[236] Release on Surface Tension of Water Substances. Issued by the Intern. Assoc. Prop. Steam, Sept. 1975.
[264] Release on Dynamic Viscosity of Water Substance. Issued by the International Association for the Properties of Steam, Sept. 1975.
[265] Release on Thermal Conductivity of Water Substance. Issued by the International Association for the Properties of Steam. Secretary: Dr. H. J. WHITE, Nat. Bur. Stand., Washington D. C. (1977).
[266] Release on Static Dielectric Constant of Water Substance, September 1977. Issued by the International Association for the Properties of Steam. Available from Dr. HOWARD J. WHITE, Jr., Executive Secretary IAPS. Office of Standard Reference Data, Administration Building, Room A 523 National Bureau of Standards, Washington, D. C. 20234, USA.
[267] Codata Bulletin 11, December 1973.
[268] JUZA, J.: Rozpravy Československé Akademie Véd. 76 (1966) 1—143.
[269] UEMATSU, M., and E. U. FRANCK: Static Dielectric Constant of Water and Steam. J. Phys. Chem., Reference Data 9 (1980) 1291—1306.
HEGER, K., M. UEMATSU and E. U. FRANCK: The Static Dielectric Constant of Water at High Pressures and Temperatures to 500 MPa and 550 °C. Ber. Bunsenges. Phys. Chem. 84 (1980) 758—762.
[270] Release on the Ion Product of Water Substance, May 1980. Issued by the International Association for the Properties of Steam. Available as [266].
[271] MARSHALL, W. L., and E. U. FRANCK: Ion Product of Water Substance, 0—1000 °C, 1—10000 Bars. New International Formulation and Its Background. J. Phys. Chem. Reference Data 10 (1981) 295—304.
[272] Bekanntmachung über Temperaturskalen. PTB Mitteilungen 87 (1977) 497—510. — PRESTON-THOMAS, H.: The International Practical Temperature Scale of 1968, Amended Edition of 1975. Metrologia 12 (1976) 7—17.
[273] 1986 Recommended Values of the Fundamental Physical Constants. Codata Newsletter October 1986. See also PTB Mitteilungen 97 (1987) 498—507.
[274] Atomic Weights of the Elements 1983. Pure Applied Chem. 56 (1984) 653—674.
[275] KELL, G. S.: J. Phys. Chem. Ref. Data 6 (1977) 1109.
[276] Release on IAPS Statement 1983 of Values of Temperature, Pressure and Density of Ordinary and Heavy Water Substance at their Respective Critical Points. May 1983.
[277] Release on The IAPS Skeleton Tables 1985 for The Thermodynamic Properties of Ordinary Water Substance. November 1985.
[278] Release on The IAPS Formulation 1985 for the Viscosity of Ordinary Water Substance. November 1985.
[279] Release on The IAPS Formulation for the Thermal Conductivity of Ordinary Water Substance. November 1985.
[280] IAPS Supplementary Release on Saturation Properties of Ordinary Water Substance. September 1986.
[281] SAUL, A. and W. WAGNER: International Equations for the Saturation Properties of Ordinary Water Substance. J. Phys. Chem. Ref. Data 16 (1987) 893—899.

B. Tables of the Properties of Water and Steam

Tafeln der Eigenschaften von Wasser und Wasserdampf

Table 1. State of Saturation (Temperature Table)

Sättigungszustand (Temperaturtafel)

t	T	p	v'	v''	ϱ''	h'	h''	r	s'	s''
°C	K	bar	m³/kg	m³/kg	kg/m³	kJ/kg	kJ/kg	kJ/kg	kJ/kg K	kJ/kg K
0,00	273,15	0,006108	0,0010002	206,3	0,004847	−0,04	2501,6	2501,6	−0,0002	9,1577
0,01	273,16	0,006112	0,0010002	206,2	0,004851	0,00	2501,6	2501,6	0,0000	9,1575
1	274,15	0,006566	0,0010001	192,6	0,005192	4,17	2503,4	2499,2	0,0152	9,1311
2	275,15	0,007055	0,0010001	179,9	0,005558	8,39	2505,2	2496,8	0,0306	9,1047
3	276,15	0,007575	0,0010001	168,2	0,005946	12,60	2507,1	2494,5	0,0459	9,0785
4	277,15	0,008129	0,0010000	157,3	0,006358	16,80	2508,9	2492,1	0,0611	9,0526
5	278,15	0,008718	0,0010000	147,2	0,006795	21,01	2510,7	2489,7	0,0762	9,0269
6	279,15	0,009345	0,0010000	137,8	0,007258	25,21	2512,6	2487,4	0,0913	9,0015
7	280,15	0,010012	0,0010001	129,1	0,007748	29,41	2514,4	2485,0	0,1063	8,9762
8	281,15	0,010720	0,0010001	121,0	0,008267	33,60	2516,2	2482,6	0,1213	8,9513
9	282,15	0,011472	0,0010002	113,4	0,008816	37,80	2518,1	2480,3	0,1361	8,9265
10	283,15	0,012270	0,0010003	106,4	0,009396	41,99	2519,9	2477,9	0,1510	8,9020
11	284,15	0,013116	0,0010003	99,91	0,01001	46,19	2521,7	2475,5	0,1658	8,8776
12	285,15	0,014014	0,0010004	93,84	0,01066	50,38	2523,6	2473,2	0,1805	8,8536
13	286,15	0,014965	0,0010006	88,18	0,01134	54,57	2525,4	2470,8	0,1952	8,8297
14	287,15	0,015973	0,0010007	82,90	0,01206	58,75	2527,2	2468,5	0,2098	8,8060
15	288,15	0,017039	0,0010008	77,98	0,01282	62,94	2529,1	2466,1	0,2243	8,7826
16	289,15	0,018168	0,0010010	73,38	0,01363	67,13	2530,9	2463,8	0,2388	8,7593
17	290,15	0,019362	0,0010012	69,09	0,01447	71,31	2532,7	2461,4	0,2533	8,7363
18	291,15	0,02062	0,0010013	65,09	0,01536	75,50	2534,5	2459,0	0,2677	8,7135
19	292,15	0,02196	0,0010015	61,34	0,01630	79,68	2536,4	2456,7	0,2820	8,6908
20	293,15	0,02337	0,0010017	57,84	0,01729	83,86	2538,2	2454,3	0,2963	8,6684
21	294,15	0,02485	0,0010019	54,56	0,01833	88,04	2540,0	2452,0	0,3105	8,6462
22	295,15	0,02642	0,0010022	51,49	0,01942	92,23	2541,8	2449,6	0,3247	8,6241
23	296,15	0,02808	0,0010024	48,62	0,02057	96,41	2543,6	2447,2	0,3389	8,6023
24	297,15	0,02982	0,0010026	45,93	0,02177	100,59	2545,5	2444,9	0,3530	8,5806
25	298,15	0,03166	0,0010029	43,40	0,02304	104,77	2547,3	2442,5	0,3670	8,5592
26	299,15	0,03360	0,0010032	41,03	0,02437	108,95	2549,1	2440,2	0,3810	8,5379
27	300,15	0,03564	0,0010034	38,81	0,02576	113,13	2550,9	2437,8	0,3949	8,5168
28	301,15	0,03778	0,0010037	36,73	0,02723	117,31	2552,7	2435,4	0,4088	8,4959
29	302,15	0,04004	0,0010040	34,77	0,02876	121,48	2554,5	2433,1	0,4227	8,4751
30	303,15	0,04241	0,0010043	32,93	0,03037	125,66	2556,4	2430,7	0,4365	8,4546
31	304,15	0,04491	0,0010046	31,20	0,03205	129,84	2558,2	2428,3	0,4503	8,4342
32	305,15	0,04753	0,0010049	29,57	0,03382	134,02	2560,0	2425,9	0,4640	8,4140
33	306,15	0,05029	0,0010053	28,04	0,03566	138,20	2561,8	2423,6	0,4777	8,3939
34	307,15	0,05318	0,0010056	26,60	0,03759	142,38	2563,6	2421,2	0,4913	8,3740
35	308,15	0,05622	0,0010060	25,24	0,03961	146,56	2565,4	2418,8	0,5049	8,3543
36	309,15	0,05940	0,0010063	23,97	0,04172	150,74	2567,2	2416,4	0,5184	8,3348
37	310,15	0,06274	0,0010067	22,76	0,04393	154,91	2569,0	2414,1	0,5319	8,3154
38	311,15	0,06624	0,0010070	21,63	0,04624	159,09	2570,8	2411,7	0,5453	8,2962
39	312,15	0,06991	0,0010074	20,56	0,04865	163,27	2572,6	2409,3	0,5588	8,2772
40	313,15	0,07375	0,0010078	19,55	0,05116	167,45	2574,4	2406,9	0,5721	8,2583
41	314,15	0,07777	0,0010082	18,59	0,05379	171,63	2576,2	2404,5	0,5854	8,2395
42	315,15	0,08198	0,0010086	17,69	0,05652	175,81	2577,9	2402,1	0,5987	8,2209
43	316,15	0,08639	0,0010090	16,84	0,05938	179,99	2579,7	2399,7	0,6120	8,2025
44	317,15	0,09100	0,0010094	16,04	0,06236	184,17	2581,5	2397,3	0,6252	8,1842
45	318,15	0,09582	0,0010099	15,28	0,06546	188,35	2583,3	2394,9	0,6383	8,1661

t	T	p	v'	v''	ϱ''	h'	h''	r	s'	s''
45	318,15	0,09582	0,0010099	15,28	0,06546	188,35	2583,3	2394,9	0,6383	8,1661
46	319,15	0,10086	0,0010103	14,56	0,06869	192,53	2585,1	2392,5	0,6514	8,1481
47	320,15	0,10612	0,0010107	13,88	0,07206	196,71	2586,9	2390,1	0,6645	8,1302
48	321,15	0,11162	0,0010112	13,23	0,07557	200,89	2588,6	2387,7	0,6776	8,1125
49	322,15	0,11736	0,0010117	12,62	0,07922	205,07	2590,4	2385,3	0,6906	8,0950
50	323,15	0,12335	0,0010121	12,05	0,08302	209,26	2592,2	2382,9	0,7035	8,0776
51	324,15	0,12961	0,0010126	11,50	0,08697	213,44	2593,9	2380,5	0,7164	8,0603
52	325,15	0,13613	0,0010131	10,98	0,09108	217,62	2595,7	2378,1	0,7293	8,0432
53	326,15	0,14293	0,0010136	10,49	0,09535	221,80	2597,5	2375,7	0,7422	8,0262
54	327,15	0,15002	0,0010140	10,02	0,09979	225,98	2599,2	2373,2	0,7550	8,0093
55	328,15	0,15741	0,0010145	9,579	0,1044	230,17	2601,0	2370,8	0,7677	7,9926
56	329,15	0,16511	0,0010150	9,159	0,1092	234,35	2602,7	2368,4	0,7804	7,9759
57	330,15	0,17313	0,0010156	8,760	0,1142	238,53	2604,5	2365,9	0,7931	7,9595
58	331,15	0,18147	0,0010161	8,381	0,1193	242,72	2606,2	2363,5	0,8058	7,9431
59	332,15	0,19016	0,0010166	8,021	0,1247	246,91	2608,0	2361,1	0,8184	7,9269
60	333,15	0,19920	0,0010171	7,679	0,1302	251,09	2609,7	2358,6	0,8310	7,9108
61	334,15	0,2086	0,0010177	7,353	0,1360	255,28	2611,4	2356,2	0,8435	7,8948
62	335,15	0,2184	0,0010182	7,044	0,1420	259,46	2613,2	2353,7	0,8560	7,8790
63	336,15	0,2286	0,0010188	6,749	0,1482	263,65	2614,9	2351,3	0,8685	7,8633
64	337,15	0,2391	0,0010193	6,469	0,1546	267,84	2616,6	2348,8	0,8809	7,8477
65	338,15	0,2501	0,0010199	6,202	0,1612	272,02	2618,4	2346,3	0,8933	7,8322
66	339,15	0,2615	0,0010205	5,948	0,1681	276,21	2620,1	2343,9	0,9057	7,8168
67	340,15	0,2733	0,0010211	5,706	0,1752	280,40	2621,8	2341,4	0,9180	7,8015
68	341,15	0,2856	0,0010217	5,476	0,1826	284,59	2623,5	2338,9	0,9303	7,7864
69	342,15	0,2984	0,0010223	5,256	0,1903	288,78	2625,2	2336,4	0,9426	7,7714
70	343,15	0,3116	0,0010228	5,046	0,1982	292,97	2626,9	2334,0	0,9548	7,7565
71	344,15	0,3253	0,0010235	4,846	0,2063	297,16	2628,6	2331,5	0,9670	7,7417
72	345,15	0,3396	0,0010241	4,656	0,2148	301,35	2630,3	2329,0	0,9792	7,7270
73	346,15	0,3543	0,0010247	4,474	0,2235	305,55	2632,0	2326,5	0,9913	7,7124
74	347,15	0,3696	0,0010253	4,300	0,2326	309,74	2633,7	2324,0	1,0034	7,6979
75	348,15	0,3855	0,0010259	4,134	0,2419	313,94	2635,4	2321,5	1,0154	7,6835
76	349,15	0,4019	0,0010266	3,976	0,2515	318,13	2637,1	2318,9	1,0275	7,6693
77	350,15	0,4189	0,0010272	3,824	0,2615	322,33	2638,7	2316,4	1,0395	7,6551
78	351,15	0,4365	0,0010279	3,680	0,2718	326,52	2640,4	2313,9	1,0514	7,6410
79	352,15	0,4547	0,0010285	3,541	0,2824	330,72	2642,1	2311,4	1,0634	7,6271
80	353,15	0,4736	0,0010292	3,409	0,2933	334,92	2643,8	2308,8	1,0753	7,6132
81	354,15	0,4931	0,0010299	3,283	0,3046	339,11	2645,4	2306,3	1,0871	7,5995
82	355,15	0,5133	0,0010305	3,162	0,3163	343,31	2647,1	2303,8	1,0990	7,5858
83	356,15	0,5342	0,0010312	3,046	0,3283	347,51	2648,7	2301,2	1,1108	7,5722
84	357,15	0,5557	0,0010319	2,935	0,3407	351,71	2650,4	2298,7	1,1225	7,5588
85	358,15	0,5780	0,0010326	2,829	0,3535	355,92	2652,0	2296,5	1,1343	7,5454
86	359,15	0,6011	0,0010333	2,727	0,3667	360,12	2653,6	2293,1	1,1460	7,5321
87	360,15	0,6249	0,0010340	2,630	0,3803	364,32	2655,3	2290,9	1,1577	7,5189
88	361,15	0,6495	0,0010347	2,536	0,3942	368,53	2656,9	2288,4	1,1693	7,5058
89	362,15	0,6749	0,0010354	2,447	0,4087	372,73	2658,5	2285,8	1,1809	7,4928
90	363,15	0,7011	0,0010361	2,361	0,4235	376,94	2660,1	2283,2	1,1925	7,4799
91	364,15	0,7281	0,0010369	2,279	0,4388	381,15	2661,7	2280,6	1,2041	7,4670
92	365,15	0,7561	0,0010376	2,200	0,4545	385,36	2663,4	2278,0	1,2156	7,4543
93	366,15	0,7849	0,0010384	2,125	0,4707	389,56	2665,0	2275,4	1,2271	7,4416
94	367,15	0,8146	0,0010391	2,052	0,4873	393,78	2666,6	2272,8	1,2386	7,4291
95	368,15	0,8453	0,0010399	1,982	0,5045	397,99	2668,1	2270,2	1,2501	7,4166
96	369,15	0,8769	0,0010406	1,915	0,5221	402,20	2669,7	2267,5	1,2615	7,4042
97	370,15	0,9094	0,0010414	1,851	0,5402	406,42	2671,3	2264,9	1,2729	7,3919
98	371,15	0,9430	0,0010421	1,789	0,5589	410,63	2672,9	2262,2	1,2842	7,3796
99	372,15	0,9776	0,0010429	1,730	0,5780	414,85	2674,4	2259,6	1,2956	7,3675
100	373,15	1,0133	0,0010437	1,673	0,5977	419,06	2676,0	2256,9	1,3069	7,3554
101	374,15	1,0500	0,0010445	1,618	0,6180	423,28	2677,6	2254,3	1,3182	7,3434
102	375,15	1,0878	0,0010453	1,566	0,6388	427,50	2679,1	2251,6	1,3294	7,3315
103	376,15	1,1267	0,0010461	1,515	0,6601	431,73	2680,7	2248,9	1,3406	7,3196
104	377,15	1,1668	0,0010469	1,466	0,6821	435,95	2682,2	2246,3	1,3518	7,3078
105	378,15	1,2080	0,0010477	1,419	0,7046	440,17	2683,7	2243,6	1,3630	7,2962

Table 1. State of Saturation (Temperature Table) (Continuation)
Sättigungszustand (Temperaturtafel) (Fortsetzung)

t	T	p	v'	v''	ϱ''	h'	h''	r	s'	s''
105	378,15	1,2080	0,0010477	1,419	0,7046	440,17	2683,7	2243,6	1,3630	7,2962
106	379,15	1,2504	0,0010485	1,374	0,7277	444,40	2685,3	2240,9	1,3742	7,2845
107	380,15	1,2941	0,0010494	1,331	0,7515	448,63	2686,8	2238,2	1,3853	7,2730
108	381,15	1,3390	0,0010502	1,289	0,7758	452,85	2688,3	2235,4	1,3964	7,2615
109	382,15	1,3852	0,0010510	1,249	0,8008	457,08	2689,8	2232,7	1,4074	7,2501
110	383,15	1,4327	0,0010519	1,210	0,8265	461,32	2691,3	2230,0	1,4185	7,2388
111	384,15	1,4815	0,0010527	1,173	0,8528	465,55	2692,8	2227,3	1,4295	7,2275
112	385,15	1,5316	0,0010536	1,137	0,8798	469,78	2694,3	2224,5	1,4405	7,2164
113	386,15	1,5832	0,0010544	1,102	0,9075	474,02	2695,8	2221,8	1,4515	7,2052
114	387,15	1,6362	0,0010553	1,069	0,9359	478,26	2697,2	2219,0	1,4624	7,1942
115	388,15	1,6906	0,0010562	1,036	0,9650	482,50	2698,7	2216,2	1,4733	7,1832
116	389,15	1,7465	0,0010571	1,005	0,9948	486,74	2700,2	2213,4	1,4842	7,1723
117	390,15	1,8039	0,0010579	0,9753	1,025	490,98	2701,6	2210,7	1,4951	7,1614
118	391,15	1,8628	0,0010588	0,9463	1,057	495,23	2703,1	2207,9	1,5060	7,1507
119	392,15	1,9233	0,0010597	0,9184	1,089	499,47	2704,5	2205,1	1,5168	7,1399
120	393,15	1,9854	0,0010606	0,8915	1,122	503,72	2706,0	2202,2	1,5276	7,1293
121	394,15	2,0492	0,0010615	0,8655	1,155	507,97	2707,4	2199,4	1,5384	7,1187
122	395,15	2,1145	0,0010625	0,8405	1,190	512,22	2708,8	2196,6	1,5491	7,1082
123	396,15	2,1816	0,0010634	0,8162	1,225	516,47	2710,2	2193,7	1,5599	7,0977
124	397,15	2,2504	0,0010643	0,7928	1,261	520,73	2711,6	2190,9	1,5706	7,0873
125	398,15	2,3210	0,0010652	0,7702	1,298	524,99	2713,0	2188,0	1,5813	7,0769
126	399,15	2,3933	0,0010662	0,7484	1,336	529,25	2714,4	2185,2	1,5919	7,0666
127	400,15	2,4675	0,0010671	0,7273	1,375	533,51	2715,8	2182,3	1,6026	7,0564
128	401,15	2,5435	0,0010681	0,7069	1,415	537,77	2717,2	2179,4	1,6132	7,0462
129	402,15	2,6215	0,0010691	0,6872	1,455	542,04	2718,5	2176,5	1,6238	7,0361
130	403,15	2,7013	0,0010700	0,6681	1,497	546,31	2719,9	2173,6	1,6344	7,0261
131	404,15	2,7831	0,0010710	0,6497	1,539	550,58	2721,3	2170,7	1,6449	7,0161
132	405,15	2,8670	0,0010720	0,6319	1,583	554,85	2722,6	2167,8	1,6555	7,0061
133	406,15	2,9528	0,0010730	0,6146	1,627	559,12	2723,9	2164,8	1,6660	6,9962
134	407,15	3,041	0,0010740	0,5980	1,672	563,40	2725,3	2161,9	1,6765	6,9864
135	408,15	3,131	0,0010750	0,5818	1,719	567,68	2726,6	2158,9	1,6869	6,9766
136	409,15	3,223	0,0010760	0,5662	1,766	571,96	2727,9	2155,9	1,6974	6,9669
137	410,15	3,317	0,0010770	0,5511	1,815	576,24	2729,2	2153,0	1,7078	6,9572
138	411,15	3,414	0,0010780	0,5364	1,864	580,53	2730,5	2150,0	1,7182	6,9475
139	412,15	3,513	0,0010790	0,5222	1,915	584,81	2731,8	2147,0	1,7286	6,9380
140	413,15	3,614	0,0010801	0,5085	1,967	589,10	2733,1	2144,0	1,7390	6,9284
141	414,15	3,717	0,0010811	0,4952	2,019	593,40	2734,3	2140,9	1,7493	6,9190
142	415,15	3,823	0,0010821	0,4823	2,073	597,69	2735,6	2137,9	1,7597	6,9095
143	416,15	3,931	0,0010832	0,4698	2,129	601,99	2736,9	2134,9	1,7700	6,9001
144	417,15	4,042	0,0010843	0,4577	2,185	606,29	2738,1	2131,8	1,7803	6,8908
145	418,15	4,155	0,0010853	0,4460	2,242	610,60	2739,3	2128,7	1,7906	6,8815
146	419,15	4,271	0,0010864	0,4346	2,301	614,90	2740,6	2125,7	1,8008	6,8723
147	420,15	4,389	0 0010875	0 4236	2,361	619,21	2741,8	2122,6	1,8110	6,8631
148	421,15	4,510	0,0010886	0,4129	2,422	623,52	2743,0	2119,5	1,8213	6,8539
149	422,15	4,634	0,0010897	0,4025	2,484	627,83	2744,2	2116,3	1,8315	6,8448
150	423,15	4,760	0,0010908	0,3924	2,548	632,15	2745,4	2113,2	1,8416	6,8358
151	424,15	4,889	0,0010919	0,3827	2,613	636,47	2746,5	2110,1	1,8518	6,8268
152	425,15	5,021	0,0010930	0,3732	2,679	640,79	2747,7	2106,9	1,8620	6,8178
153	426,15	5,155	0,0010941	0,3640	2,747	645,12	2748,9	2103,8	1,8721	6,8089
154	427,15	5,293	0,0010953	0,3551	2,816	649,45	2750,0	2100,6	1,8822	6,8000
155	428,15	5,433	0,0010964	0,3464	2,886	653,78	2751,2	2097,4	1,8923	6,7911
156	429,15	5,577	0,0010976	0,3380	2,958	658,11	2752,3	2094,2	1,9023	6,7823
157	430,15	5,723	0,0010987	0,3299	3,032	662,45	2753,4	2091,0	1,9124	6,7735
158	431,15	5,872	0,0010999	0,3219	3,106	666,79	2754,5	2087,7	1,9224	6,7648
159	432,15	6,025	0,0011011	0,3142	3,182	671,13	2755,6	2084,5	1,9325	6,7561
160	433,15	6,181	0,0011022	0,306S	3,260	675,47	2756,7	2081,3	1,9425	6,7475

Table 1. State of Saturation (Temperature Table) (Continuation)
Sättigungszustand (Temperaturtafel) (Fortsetzung)

t	T	p	v'	v''	ϱ''	h'	h''	r	s'	s''
160	433,15	6,181	0,0011022	0,3068	3,260	675,47	2756,7	2081,3	1,9425	6,7475
161	434,15	6,339	0,0011034	0,2995	3,339	679,82	2757,8	2078,0	1,9525	6,7389
162	435,15	6,502	0,0011046	0,2924	3,420	684,18	2758,9	2074,7	1,9624	6,7303
163	436,15	6,667	0,0011058	0,2856	3,502	688,53	2759,9	2071,4	1,9724	6,7218
164	437,15	6,836	0,0011070	0,2789	3,586	692,89	2761,0	2068,1	1,9823	6,7133
165	438,15	7,008	0,0011082	0,2724	3,671	697,25	2762,0	2064,8	1,9923	6,7048
166	439,15	7,183	0,0011095	0,2661	3,758	701,62	2763,1	2061,4	2,0022	6,6964
167	440,15	7,362	0,0011107	0,2600	3,847	705,99	2764,1	2058,1	2,0121	6,6880
168	441,15	7,545	0,0011119	0,2540	3,937	710,36	2765,1	2054,7	2,0219	6,6796
169	442,15	7,731	0,0011132	0,2482	4,029	714,74	2766,1	2051,3	2,0318	6,6713
170	443,15	7,920	0,0011145	0,2426	4,123	719,12	2767,1	2047,9	2,0416	6,6630
171	444,15	8,114	0,0011157	0,2371	4,218	723,50	2768,0	2044,5	2,0515	6,6548
172	445,15	8,311	0,0011170	0,2317	4,316	727,89	2769,0	2041,1	2,0613	6,6465
173	446,15	8,511	0,0011183	0,2265	4,415	732,28	2769,9	2037,7	2,0711	6,6384
174	447,15	8,716	0,0011196	0,2215	4,515	736,67	2770,9	2034,2	2,0809	6,6302
175	448,15	8,924	0,0011209	0,2165	4,618	741,07	2771,8	2030,7	2,0906	6,6221
176	449,15	9,137	0,0011222	0,2117	4,723	745,47	2772,7	2027,3	2,1004	6,6140
177	450,15	9,353	0,0011235	0,2071	4,829	749,88	2773,6	2023,8	2,1102	6,6059
178	451,15	9,574	0,0011248	0,2025	4,937	754,28	2774,5	2020,2	2,1199	6,5979
179	452,15	9,798	0,0011262	0,1981	5,048	758,70	2775,4	2016,7	2,1296	6,5899
180	453,15	10,027	0,0011275	0,1938	5,160	763,12	2776,3	2013,1	2,1393	6,5819
181	454,15	10,259	0,0011289	0,1896	5,274	767,54	2777,1	2009,6	2,1490	6,5739
182	455,15	10,496	0,0011302	0,1855	5,391	771,96	2778,0	2006,0	2,1587	6,5660
183	456,15	10,738	0,0011316	0,1815	5,509	776,39	2778,8	2002,4	2,1683	6,5581
184	457,15	10,983	0,0011330	0,1776	5,629	780,82	2779,6	1998,8	2,1780	6,5503
185	458,15	11,233	0,0011344	0,1739	5,752	785,26	2780,4	1995,2	2,1876	6,5424
186	459,15	11,488	0,0011358	0,1702	5,877	789,70	2781,2	1991,5	2,1972	6,5346
187	460,15	11,747	0,0011372	0,1666	6,003	794,15	2782,0	1987,8	2,2068	6,5268
188	461,15	12,010	0,0011386	0,1631	6,132	798,60	2782,8	1984,2	2,2164	6,5191
189	462,15	12,278	0,0011401	0,1596	6,264	803,06	2783,5	1980,5	2,2260	6,5113
190	463,15	12,551	0,0011415	0,1563	6,397	807,52	2784,3	1976,7	2,2356	6,5036
191	464,15	12,829	0,0011430	0,1531	6,533	811,98	2785,0	1973,0	2,2451	6,4959
192	465,15	13,111	0,0011444	0,1499	6,671	816,45	2785,7	1969,3	2,2547	6,4883
193	466,15	13,398	0,0011459	0,1468	6,812	820,92	2786,4	1965,5	2,2642	6,4806
194	467,15	13,690	0,0011474	0,1438	6,955	825,40	2787,1	1961,7	2,2738	6,4730
195	468,15	13,987	0,0011489	0,1408	7,100	829,88	2787,8	1957,9	2,2833	6,4654
196	469,15	14,289	0,0011504	0,1380	7,248	834,37	2788,4	1954,1	2,2928	6,4578
197	470,15	14,596	0,0011519	0,1352	7,398	838,86	2789,1	1950,2	2,3023	6,4503
198	471,15	14,909	0,0011534	0,1324	7,551	843,36	2789,7	1946,4	2,3117	6,4428
199	472,15	15,226	0,0011549	0,1298	7,706	847,86	2790,3	1942,5	2,3212	6,4353
200	473,15	15,549	0,0011565	0,1272	7,864	852,37	2790,9	1938,6	2,3307	6,4278
201	474,15	15,877	0,0011581	0,1246	8,025	856,88	2791,5	1934,6	2,3401	6,4203
202	475,15	16,210	0,0011596	0,1221	8,188	861,40	2792,1	1930,7	2,3496	6,4128
203	476,15	16,549	0,0011612	0,1197	8,354	865,92	2792,7	1926,7	2,3590	6,4054
204	477,15	16,893	0,0011628	0,1173	8,522	870,45	2793,2	1922,8	2,3684	6,3980
205	478,15	17,243	0,0011644	0,1150	8,694	874,99	2793,8	1918,8	2,3778	6,3906
206	479,15	17,598	0,0011660	0,1128	8,868	879,53	2794,3	1914,7	2,3872	6,3832
207	480,15	17,959	0,0011676	0,1106	9,045	884,07	2794,8	1910,7	2,3966	6,3759
208	481,15	18,326	0,0011693	0,1084	9,225	888,62	2795,3	1906,6	2,4060	6,3686
209	482,15	18,699	0,0011709	0,1063	9,408	893,17	2795,7	1902,6	2,4153	6,3612
210	483,15	19,077	0,0011726	0,1042	9,593	897,74	2796,2	1898,5	2,4247	6,3539
211	484,15	19,462	0,0011743	0,1022	9,782	902,31	2796,6	1894,3	2,4340	6,3466
212	485,15	19,852	0,0011760	0,1003	9,974	906,87	2797,1	1890,2	2,4434	6,3394
213	486,15	20,249	0,0011777	0,09834	10,17	911,45	2797,5	1886,0	2,4527	6,3321
214	487,15	20,651	0,0011794	0,09646	10,37	916,03	2797,9	1881,8	2,4620	6,3249
215	488,15	21,060	0,0011811	0,09463	10,57	920,63	2798,3	1877,6	2,4713	6,3176

Table 1. State of Saturation (Temperature Table) (Continuation)
Sättigungszustand (Temperaturtafel) (Fortsetzung)

t	T	p	v'	v''	ϱ''	h'	h''	r	s'	s''
215	488,15	21,060	0,0011811	0,09463	10,57	920,63	2798,3	1877,6	2,4713	6,3176
216	489,15	21,475	0,0011829	0,09283	10,77	925,23	2798,6	1873,4	2,4806	6,3104
217	490,15	21,896	0,0011846	0,09107	10,98	929,83	2799,0	1869,1	2,4900	6,3032
218	491,15	22,324	0,0011864	0,08936	11,19	934,44	2799,3	1864,9	2,4992	6,2961
219	492,15	22,758	0,0011882	0,08768	11,41	939,06	2799,6	1860,6	2,5085	6,2889
220	493,15	23,198	0,0011900	0,08604	11,62	943,67	2799,9	1856,2	2,5178	6,2817
221	494,15	23,645	0,0011918	0,08443	11,84	948,30	2800,2	1851,9	2,5271	6,2746
222	495,15	24,099	0,0011936	0,08286	12,07	952,93	2800,5	1847,5	2,5363	6,2674
223	496,15	24,560	0,0011954	0,08132	12,30	957,58	2800,7	1843,1	2,5456	6,2603
224	497,15	25,027	0,0011973	0,07982	12,53	962,22	2800,9	1838,7	2,5548	6,2532
225	498,15	25,501	0,0011992	0,07835	12,76	966,89	2801,2	1834,3	2,5641	6,2461
226	499,15	25,982	0,0012010	0,07691	13,00	971,55	2801,4	1829,8	2,5733	6,2390
227	500,15	26,470	0,0012029	0,07550	13,24	976,21	2801,5	1825,3	2,5825	6,2319
228	501,15	26,965	0,0012048	0,07412	13,49	980,88	2801,7	1820,8	2,5918	6,2249
229	502,15	27,467	0,0012068	0,07277	13,74	985,57	2801,8	1816,3	2,6010	6,2178
230	503,15	27,976	0,0012087	0,07145	14,00	990,26	2802,0	1811,7	2,6102	6,2107
231	504,15	28,493	0,0012107	0,07016	14,25	994,96	2802,1	1807,1	2,6194	6,2037
232	505,15	29,016	0,0012127	0,06889	14,52	999,68	2802,2	1802,5	2,6286	6,1967
233	506,15	29,547	0,0012147	0,06765	14,78	1004,4	2802,2	1797,9	2,6378	6,1896
234	507,15	30,086	0,0012167	0,06643	15,05	1009,1	2802,3	1793,2	2,6469	6,1826
235	508,15	30,632	0,0012187	0,06525	15,33	1013,8	2802,3	1788,5	2,6562	6,1756
236	509,15	31,186	0,0012207	0,06408	15,61	1018,6	2802,3	1783,8	2,6653	6,1686
237	510,15	31,747	0,0012228	0,06294	15,89	1023,3	2802,3	1779,0	2,6745	6,1616
238	511,15	32,317	0,0012249	0,06182	16,18	1028,1	2802,3	1774,2	2,6837	6,1546
239	512,15	32,893	0,0012270	0,06073	16,47	1032,8	2802,3	1769,4	2,6928	6,1476
240	513,15	33,478	0,0012291	0,05965	16,76	1037,6	2802,2	1764,6	2,7020	6,1406
241	514,15	34,071	0,0012312	0,05860	17,06	1042,4	2802,1	1759,8	2,7112	6,1336
242	515,15	34,672	0,0012334	0,05757	17,37	1047,2	2802,0	1754,9	2,7203	6,1266
243	516,15	35,281	0,0012355	0,05656	17,68	1052,0	2801,9	1749,9	2,7295	6,1196
244	517,15	35,898	0,0012377	0,05558	17,99	1056,8	2801,8	1745,0	2,7386	6,1127
245	518,15	36,523	0,0012399	0,05461	18,31	1061,6	2801,6	1740,0	2,7478	6,1057
246	519,15	37,157	0,0012422	0,05366	18,64	1066,4	2801,4	1735,0	2,7569	6,0987
247	520,15	37,799	0,0012444	0,05272	18,97	1071,2	2801,2	1730,0	2,7661	6,0917
248	521,15	38,449	0,0012467	0,05181	19,30	1076,1	2801,0	1724,9	2,7752	6,0848
249	522,15	39,108	0,0012490	0,05092	19,64	1080,9	2800,7	1719,8	2,7843	6,0778
250	523,15	39,776	0,0012513	0,05004	19,99	1085,8	2800,4	1714,6	2,7935	6,0708
251	524,15	40,452	0,0012536	0,04918	20,33	1090,7	2800,1	1709,5	2,8026	6,0639
252	525,15	41,137	0,0012560	0,04833	20,69	1095,5	2799,8	1704,3	2,8118	6,0569
253	526,15	41,831	0,0012584	0,04750	21,05	1100,4	2799,5	1699,1	2,8209	6,0499
254	527,15	42,534	0,0012608	0,04669	21,42	1105,3	2799,1	1693,8	2,8300	6,0429
255	528,15	43,246	0,0012632	0,04590	21,79	1110,2	2798,7	1688,5	2,8392	6,0359
256	529,15	43,967	0,0012656	0,04511	22,17	1115,1	2798,3	1683,2	2,8483	6,0290
257	530,15	44,697	0,0012681	0,04435	22,55	1120,1	2797,9	1677,8	2,8574	6,0220
258	531,15	45,437	0,0012706	0,04360	22,94	1125,0	2797,4	1672,4	2,8666	6,0150
259	532,15	46,185	0,0012731	0,04286	23,33	1130,0	2796,9	1667,0	2,8757	6,0080
260	533,15	46,943	0,0012756	0,04213	23,73	1134,9	2796,4	1661,5	2,8848	6,0010
261	534,15	47,711	0,0012782	0,04142	24,14	1139,9	2795,9	1656,0	2,8940	5,9940
262	535,15	48,488	0,0012808	0,04073	24,55	1144,9	2795,3	1650,4	2,9031	5,9869
263	536,15	49,275	0,0012834	0,04004	24,97	1149,9	2794,7	1644,8	2,9123	5,9799
264	537,15	50,071	0,0012861	0,03937	25,40	1154,9	2794,1	1639,2	2,9214	5,9729
265	538,15	50,877	0,0012887	0,03871	25,83	1159,9	2793,5	1633,6	2,9306	5,9658
266	539,15	51,693	0,0012914	0,03806	26,27	1165,0	2792,8	1627,8	2,9397	5,9588
267	540,15	52,519	0,0012942	0,03743	26,72	1170,0	2792,1	1622,1	2,9489	5,9517
268	541,15	53,355	0,0012969	0,03680	27,17	1175,1	2791,4	1616,3	2,9580	5,9446
269	542,15	54,202	0,0012997	0,03619	27,63	1180,1	2790,6	1610,5	2,9672	5,9375
270	543,15	55,058	0,0013025	0,03559	28,10	1185,2	2789,9	1604,6	2,9763	5,9304

t	T	p	v'	v''	ϱ''	h'	h''	r	s'	s''
270	543,15	55,058	0,0013025	0,03559	28,10	1185,2	2789,9	1604,6	2,9763	5,9304
271	544,15	55,925	0,0013053	0,03500	28,57	1190,3	2789,1	1598,7	2,9855	5,9233
272	545,15	56,802	0,0013082	0,03442	29,06	1195,4	2788,2	1592,8	2,9947	5,9162
273	546,15	57,689	0,0013111	0,03385	29,55	1200,6	2787,4	1586,8	3,0039	5,9091
274	547,15	58,587	0,0013141	0,03329	30,04	1205,7	2786,5	1580,7	3,0131	5,9019
275	548,15	59,496	0,0013170	0,03274	30,55	1210,9	2785,5	1574,7	3,0223	5,8947
276	549,15	60,415	0,0013200	0,03220	31,06	1216,0	2784,6	1568,6	3,0315	5,8876
277	550,15	61,346	0,0013231	0,03166	31,58	1221,2	2783,6	1562,4	3,0406	5,8804
278	551,15	62,287	0,0013261	0,03114	32,11	1226,4	2782,6	1556,2	3,0499	5,8731
279	552,15	63,239	0,0013292	0,03063	32,65	1231,6	2781,5	1549,9	3,0591	5,8659
280	553,15	64,202	0,0013324	0,03013	33,19	1236,8	2780,4	1543,6	3,0683	5,8586
281	554,15	65,176	0,0013356	0,02963	33,75	1242,1	2779,3	1537,2	3,0775	5,8513
282	555,15	66,162	0,0013388	0,02914	34,31	1247,3	2778,1	1530,8	3,0868	5,8440
283	556,15	67,158	0,0013420	0,02867	34,88	1252,6	2777,0	1524,3	3,0960	5,8367
284	557,15	68,167	0,0013453	0,02820	35,47	1257,9	2775,7	1517,8	3,1053	5,8294
285	558,15	69,186	0,0013487	0,02773	36,06	1263,2	2774,5	1511,3	3,1146	5,8220
286	559,15	70,218	0,0013520	0,02728	36,66	1268,5	2773,2	1504,6	3,1239	5,8146
287	560,15	71,261	0,0013554	0,02683	37,27	1273,9	2771,8	1498,0	3,1331	5,8072
288	561,15	72,315	0,0013589	0,02639	37,89	1279,2	2770,5	1491,2	3,1424	5,7997
289	562,15	73,382	0,0013624	0,02596	38,52	1284,6	2769,1	1484,5	3,1518	5,7923
290	563,15	74,461	0,0013659	0,02554	39,16	1290,0	2767,6	1477,6	3,1611	5,7848
291	564,15	75,551	0,0013695	0,02512	39,81	1295,4	2766,2	1470,7	3,1704	5,7773
292	565,15	76,654	0,0013732	0,02471	40,48	1300,9	2764,6	1463,8	3,1798	5,7697
293	566,15	77,769	0,0013769	0,02430	41,15	1306,3	2763,1	1456,8	3,1891	5,7621
294	567,15	78,897	0,0013806	0,02390	41,83	1311,8	2761,5	1449,7	3,1985	5,7545
295	568,15	80,037	0,0013844	0,02351	42,53	1317,3	2759,8	1442,6	3,2079	5,7469
296	569,15	81,189	0,0013882	0,02313	43,24	1322,8	2758,2	1435,4	3,2173	5,7392
297	570,15	82,355	0,0013921	0,02275	43,96	1328,3	2756,4	1428,1	3,2268	5,7315
298	571,15	83,532	0,0013960	0,02238	44,69	1333,9	2754,7	1420,8	3,2362	5,7237
299	572,15	84,723	0,0014000	0,02201	45,43	1339,4	2752,9	1413,4	3,2457	5,7159
300	573,15	85,927	0,0014041	0,02165	46,19	1345,0	2751,0	1406,0	3,2552	5,7081
301	574,15	87,144	0,0014082	0,02129	46,96	1350,7	2749,1	1398,5	3,2647	5,7003
302	575,15	88,374	0,0014123	0,02094	47,75	1356,3	2747,2	1390,9	3,2742	5,6924
303	576,15	89,617	0,0014166	0,02060	48,54	1362,0	2745,2	1383,2	3,2837	5,6844
304	577,15	90,873	0,0014208	0,02026	49,36	1367,7	2743,2	1375,5	3,2933	5,6765
305	578,15	92,144	0,0014252	0,01993	50,18	1373,4	2741,1	1367,7	3,3029	5,6685
306	579,15	93,427	0,0014296	0,01960	51,02	1379,1	2739,0	1359,8	3,3125	5,6604
307	580,15	94,725	0,0014341	0,01928	51,88	1384,9	2736,8	1351,9	3,3221	5,6523
308	581,15	96,036	0,0014387	0,01896	52,75	1390,7	2734,6	1343,9	3,3318	5,6442
309	582,15	97,361	0,0014433	0,01864	53,64	1396,5	2732,3	1335,8	3,3415	5,6360
310	583,15	98,700	0,0014480	0,01833	54,54	1402,4	2730,0	1327,6	3,3512	5,6278
311	584,15	100,05	0,0014527	0,01803	55,46	1408,3	2727,6	1319,4	3,3609	5,6195
312	585,15	101,42	0,0014576	0,01773	56,40	1414,2	2725,2	1311,0	3,3707	5,6111
313	586,15	102,80	0,0014625	0,01743	57,36	1420,1	2722,7	1302,6	3,3805	5,6028
314	587,15	104,20	0,0014675	0,01714	58,33	1426,1	2720,2	1294,1	3,3903	5,5943
315	588,15	105,61	0,0014726	0,01686	59,33	1432,1	2717,6	1285,5	3,4002	5,5858
316	589,15	107,04	0,0014778	0,01657	60,34	1438,1	2714,9	1276,8	3,4101	5,5772
317	590,15	108,48	0,0014831	0,01629	61,37	1444,2	2712,2	1268,0	3,4200	5,5686
318	591,15	109,93	0,0014885	0,01602	62,43	1450,3	2709,4	1259,1	3,4300	5,5599
319	592,15	111,40	0,0014939	0,01575	63,50	1456,4	2706,6	1250,2	3,4400	5,5512
320	593,15	112,89	0,0014995	0,01548	64,60	1462,6	2703,7	1241,1	3,4500	5,5423
321	594,15	114,39	0,0015052	0,01522	65,72	1468,8	2700,7	1231,9	3,4601	5,5334
322	595,15	115,91	0,0015109	0,01496	66,86	1475,1	2697,6	1222,6	3,4702	5,5244
323	596,15	117,44	0,0015168	0,01470	68,03	1481,3	2694,5	1213,2	3,4804	5,5154
324	597,15	118,99	0,0015228	0,01445	69,23	1487,7	2691,3	1203,6	3,4906	5,5062
325	598,15	120,56	0,0015289	0,01419	70,45	149,0	2688,0	1194,0	3,5008	5,4969

Table 1. State of Saturation (Temperature Table) (Continuation)
Sättigungszustand (Temperaturtafel) (Fortsetzung)

t	T	p	v'	v''	ϱ''	h'	h''	r	s'	s''
325	598,15	120,56	0,0015289	0,01419	70,45	1494,0	2688,0	1194,0	3,5008	5,4969
326	599,15	122,14	0,0015352	0,01395	71,70	1500,4	2684,6	1184,2	3,5111	5,4876
327	600,15	123,73	0,0015416	0,01370	72,97	1506,9	2681,1	1174,2	3,5215	5,4781
328	601,15	125,35	0,0015481	0,01346	74,28	1513,4	2677,6	1164,2	3,5319	5,4685
329	602,15	126,98	0,0015547	0,01322	75,62	1519,9	2673,9	1154,0	3,5423	5,4588
330	603,15	128,63	0,0015615	0,01299	76,99	1526,5	2670,2	1143,6	3,5528	5,4490
331	604,15	130,29	0,0015684	0,01276	78,39	1533,2	2666,3	1133,1	3,5634	5,4391
332	605,15	131,97	0,0015755	0,01253	79,83	1539,9	2662,3	1122,5	3,5740	5,4290
333	606,15	133,67	0,0015827	0,01230	81,30	1546,6	2658,3	1111,7	3,5847	5,4188
334	607,15	135,38	0,0015902	0,01208	82,81	1553,4	2654,1	1100,7	3,5955	5,4084
335	608,15	137,12	0,0015978	0,01185	84,36	1560,3	2649,7	1089,5	3,6063	5,3979
336	609,15	138,87	0,0016055	0,01163	85,95	1567,2	2645,3	1078,1	3,6172	5,3872
337	610,15	140,64	0,0016135	0,01142	87,58	1574,2	2640,7	1066,6	3,6282	5,3764
338	611,15	142,42	0,0016217	0,01120	89,26	1581,2	2636,0	1054,8	3,6392	5,3653
339	612,15	144,23	0,0016301	0,01099	90,99	1588,3	2631,2	1042,9	3,6504	5,3541
340	613,15	146,05	0,0016387	0,01078	92,76	1595,5	2626,2	1030,7	3,6616	5,3427
341	614,15	147,89	0,0016476	0,01057	94,58	1602,7	2621,0	1018,3	3,6729	5,3312
342	615,15	149,76	0,0016567	0,01037	96,46	1610,0	2615,7	1005,7	3,6844	5,3194
343	616,15	151,64	0,0016661	0,01016	98,39	1617,5	2610,3	992,8	3,6959	5,3074
344	617,15	153,54	0,0016758	0,009962	100,4	1624,9	2604,7	979,7	3,7075	5,2952
345	618,15	155,45	0,0016858	0,009763	102,4	1632,5	2598,9	966,4	3,7193	5,2828
346	619,15	157,39	0,0016961	0,009566	104,5	1640,2	2593,0	952,8	3,7311	5,2702
347	620,15	159,35	0,0017067	0,009371	106,7	1648,0	2586,9	938,9	3,7431	5,2574
348	621,15	161,33	0,0017178	0,009178	109,0	1655,8	2580,7	924,8	3,7553	5,2444
349	622,15	163,33	0,0017292	0,008988	111,3	1663,8	2574,2	910,4	3,7676	5,2311
350	623,15	165,35	0,0017411	0,008799	113,6	1671,9	2567,7	895,7	3,7800	5,2177
351	624,15	167,39	0,0017532	0,008609	116,2	1680,4	2560,7	880,4	3,7933	5,2036
352	625,15	169,45	0,0017661	0,008420	118,8	1689,3	2553,5	864,2	3,8071	5,1893
353	626,15	171,54	0,0017796	0,008232	121,5	1698,4	2546,1	847,7	3,8209	5,1746
354	627,15	173,64	0,0017937	0,008045	124,3	1707,5	2538,4	830,9	3,8349	5,1596
355	628,15	175,77	0,0018085	0,007859	127,2	1716,6	2530,4	813,8	3,8489	5,1442
356	629,15	177,92	0,0018241	0,007674	130,3	1725,9	2522,1	796,2	3,8629	5,1283
357	630,15	180,09	0,0018406	0,007490	133,5	1735,2	2513,5	778,3	3,8772	5,1121
358	631,15	182,29	0,0018580	0,007306	136,9	1744,7	2504,6	759,9	3,8915	5,0953
359	632,15	184,51	0,0018764	0,007123	140,4	1754,3	2495,2	740,9	3,9061	5,0780
360	633,15	186,75	0,0018959	0,006940	144,1	1764,2	2485,4	721,3	3,9210	5,0600
361	634,15	189,02	0,0019167	0,006757	148,0	1774,2	2475,2	701,0	3,9362	5,0414
362	635,15	191,31	0,0019388	0,006573	152,1	1784,6	2464,4	679,8	3,9518	5,0220
363	636,15	193,62	0,0019626	0,006388	156,5	1795,3	2453,0	657,8	3,9679	5,0017
364	637,15	195,96	0,0019882	0,006201	161,3	1806,4	2440,9	634,6	3,9846	4,9804
365	638,15	198,33	0,0020160	0,006012	166,3	1818,0	2428,0	610,0	4,0021	4,9579
366	639,15	200,72	0,0020464	0,005819	171,9	1830,2	2414,1	583,9	4,0205	4,9339
367	640,15	203,13	0,0020802	0,005621	177,9	1843,2	2399,0	555,7	4,0401	4,9081
368	641,15	205,57	0,0021181	0,005416	184,6	1857,3	2382,4	525,1	4,0613	4,8801
369	642,15	208,04	0,0021618	0,005201	192,3	1872,8	2363,9	491,1	4,0846	4,8492
370	643,15	210,54	0,0022136	0,004973	201,1	1890,2	2342,8	452,6	4,1108	4,8144
371	644,15	213,06	0,0022778	0,004723	211,7	1910,5	2317,9	407,4	4,1414	4,7738
372	645,15	215,62	0,0023636	0,004439	225,3	1935,6	2286,9	351,4	4,1794	4,7240
373	646,15	218,20	0,0024963	0,004084	244,9	1970,5	2244,0	273,5	4,2325	4,6559
374	647,15	220,81	0,0028407	0,003458	289,2	2046,3	2155,0	108,6	4,3487	4,5166
374,15	647,30	221,20	0,00317	0,00317	315,5	2107,4		0,0	4,4429	

Critical Constants adopted for the Preparation of the 1967 IFC Formulation for Industrial Use [261]. See also Section C I 1.3

Kritische Werte, die der 1967 IFC Formulation for Industrial Use zugrunde liegen [261]. Siehe auch Abschnitt C I 1.3.

temperature — Temperatur . . 374,15 °C
pressure — Druck . . . . . . 221,20 bar
spec. volume — spez. Volumen 0,00317 m³/kg

spec. enthalpy — spez. Enthalpie 2107,4 kJ/kg
spec. entropy — spez. Entropie . . 4,4429 kJ/kg K

Table 2. State of Saturation (Pressure Table)
Sättigungszustand (Drucktafel)

p	t	v'	v''	ϱ''	h'	h''	r	s'	s''
bar	°C	m³/kg	m³/kg	kg/m³	kJ/kg	kJ/kg	kJ/kg	kJ/kg K	kJ/kg K
0,010	6,9828	0,0010001	129,20	0,007739	29,34	2514,4	2485,0	0,1060	8,9767
0,015	13,036	0,0010006	87,98	0,01137	54,71	2525,5	2470,7	0,1957	8,8288
0,020	17,513	0,0010012	67,01	0,01492	73,46	2533,6	2460,2	0,2607	8,7246
0,025	21,096	0,0010020	54,26	0,01843	88,45	2540,2	2451,7	0,3119	8,6440
0,030	24,100	0,0010027	45,67	0,02190	101,00	2545,6	2444,6	0,3544	8,5785
0,035	26,694	0,0010033	39,48	0,02533	111,85	2550,4	2438,5	0,3907	8,5232
0,040	28,983	0,0010040	34,80	0,02873	121,41	2554,5	2433,1	0,4225	8,4755
0,045	31,035	0,0010046	31,14	0,03211	129,99	2558,2	2428,2	0,4507	8,4335
0,050	32,898	0,0010052	28,19	0,03547	137,77	2561,6	2423,8	0,4763	8,3960
0,055	34,605	0,0010058	25,77	0,03880	144,91	2564,7	2419,8	0,4995	8,3621
0,060	36,183	0,0010064	23,74	0,04212	151,50	2567,5	2416,0	0,5209	8,3312
0,065	37,651	0,0010069	22,02	0,04542	157,64	2570,2	2412,5	0,5407	8,3029
0,070	39,025	0,0010074	20,53	0,04871	163,38	2572,6	2409,2	0,5591	8,2767
0,075	40,316	0,0010079	19,24	0,05198	168,77	2574,9	2406,2	0,5763	8,2523
0,080	41,534	0,0010084	18,10	0,05523	173,86	2577,1	2403,2	0,5925	8,2296
0,085	42,689	0,0010089	17,10	0,05848	178,69	2579,2	2400,5	0,6079	8,2082
0,090	43,787	0,0010094	16,20	0,06171	183,28	2581,1	2397,9	0,6224	8,1881
0,095	44,833	0,0010098	15,40	0,06493	187,65	2583,0	2395,3	0,6361	8,1691
0,10	45,833	0,0010102	14,67	0,06814	191,83	2584,8	2392,9	0,6493	8,1511
0,11	47,710	0,0010111	13,42	0,07454	199,68	2588,1	2388,4	0,6738	8,1177
0,12	49,446	0,0010119	12,36	0,08089	206,94	2591,2	2384,3	0,6963	8,0872
0,13	51,062	0,0010126	11,47	0,08722	213,70	2594,0	2380,3	0,7172	8,0592
0,14	52,574	0,0010133	10,69	0,09351	220,02	2596,7	2376,7	0,7367	8,0334
0,15	53,997	0,0010140	10,02	0,09977	225,97	2599,2	2373,2	0,7549	8,0093
0,16	55,341	0,0010147	9,433	0,1060	231,59	2601,6	2370,0	0,7721	7,9869
0,17	56,615	0,0010154	8,911	0,1122	236,93	2603,8	2366,9	0,7883	7,9658
0,18	57,826	0,0010160	8,445	0,1184	241,99	2605,9	2363,9	0,8036	7,9460
0,19	58,982	0,0010166	8,027	0,1246	246,83	2607,9	2361,1	0,8182	7,9272
0,20	60,086	0,0010172	7,650	0,1307	251,45	2609,9	2358,4	0,8321	7,9094
0,21	61,145	0,0010178	7,307	0,1368	255,88	2611,7	2355,8	0,8453	7,8925
0,22	62,162	0,0010183	6,995	0,1430	260,14	2613,5	2353,3	0,8581	7,8764
0,23	63,139	0,0010189	6,709	0,1490	264,23	2615,2	2350,9	0,8702	7,8611
0,24	64,082	0,0010194	6,447	0,1551	268,18	2616,8	2348,6	0,8820	7,8464
0,25	64,992	0,0010199	6,204	0,1612	271,99	2618,3	2346,4	0,8932	7,8323
0,26	65,871	0,0010204	5,980	0,1672	275,67	2619,9	2344,2	0,9041	7,8188
0,27	66,722	0,0010209	5,772	0,1732	279,24	2621,3	2342,1	0,9146	7,8058
0,28	67,547	0,0010214	5,579	0,1793	282,69	2622,7	2340,0	0,9248	7,7933
0,29	68,347	0,0010219	5,398	0,1852	286,05	2624,1	2338,1	0,9346	7,7812
0,30	69,124	0,0010223	5,229	0,1912	289,30	2625,4	2336,1	0,9441	7,7695
0,32	70,615	0,0010232	4,922	0,2032	295,55	2628,0	2332,4	0,9623	7,7474
0,34	72,029	0,0010241	4,650	0,2150	301,48	2630,4	2328,9	0,9795	7,7266
0,36	73,374	0,0010249	4,408	0,2269	307,12	2632,6	2325,5	0,9958	7,7070
0,38	74,658	0,0010257	4,190	0,2387	312,50	2634,8	2322,3	1,0113	7,6884
0,40	75,886	0,0010265	3,993	0,2504	317,65	2636,9	2319,2	1,0261	7,6709
0,45	78,743	0,0010284	3,576	0,2796	329,64	2641,7	2312,0	1,0603	7,6307
0,50	81,345	0,0010301	3,240	0,3086	340,56	2646,0	2305,4	1,0912	7,5947
0,55	83,737	0,0010317	2,964	0,3374	350,61	2649,9	2299,3	1,1194	7,5623
0,60	85,954	0,0010333	2,732	0,3661	359,93	2653,6	2293,6	1,1454	7,5327
0,65	88,021	0,0010347	2,535	0,3945	368,62	2656,9	2288,3	1,1696	7,5055
0,70	89,959	0,0010361	2,365	0,4229	376,77	2660,1	2283,3	1,1921	7,4804
0,75	91,785	0,0010375	2,217	0,4511	384,45	2663,0	2278,6	1,2131	7,4570
0,80	93,512	0,0010387	2,087	0,4792	391,72	2665,8	2274,0	1,2330	7,4352
0,85	95,152	0,0010400	1,972	0,5071	398,63	2668,4	2269,8	1,2518	7,4147
0,90	96,713	0,0010412	1,869	0,5350	405,21	2670,9	2265,6	1,2696	7,3954
0,95	98,204	0,0010423	1,777	0,5627	411,49	2673,2	2261,7	1,2865	7,3771

Table 2. State of Saturation (Pressure Table) (Continuation)   Sättigungszustand (Drucktafel) (Fortsetzung)

p	t	v'	v''	ϱ''	h'	h''	r	s'	s''
1,0	99,632	0,0010434	1,694	0,5904	417,51	2675 4	2257,9	1,3027	7,3598
1,1	102,32	0,0010455	1,549	0,6455	428,84	2679,6	2250,8	1,3330	7,3277
1,2	104,81	0,0010476	1,428	0,7002	439,36	2683,4	2244,1	1,3609	7,2984
1,3	107,13	0,0010495	1,325	0,7547	449,19	2687,0	2237,8	1,3868	7,2715
1,4	109,32	0,0010513	1,236	0,8088	458,42	2690,3	2231,9	1,4109	7,2465
1,5	111,37	0,0010530	1,159	0,8628	467,13	2693,4	2226,2	1,4336	7,2234
1,6	113,32	0,0010547	1,091	0,9165	475,38	2696,2	2220,9	1,4550	7,2017
1,7	115,17	0,0010563	1,031	0,9700	483,22	2699,0	2215,7	1,4752	7,1813
1,8	116,93	0,0010579	0,9772	1,023	490,70	2701,5	2210,8	1,4944	7,1622
1,9	118,62	0,0010594	0,9290	1,076	497,85	2704,0	2206,1	1,5127	7,1440
2,0	120,23	0,0010608	0,8854	1,129	504,70	2706,3	2201,6	1,5301	7,1268
2,1	121,78	0,0010623	0,8459	1,182	511,29	2708,5	2197,2	1,5468	7,1105
2,2	123,27	0,0010636	0,8098	1,235	517,62	2710,6	2193,0	1,5627	7,0949
2,3	124,71	0,0010650	0,7768	1,287	523,73	2712,6	2188,9	1,5781	7,0800
2,4	126,09	0,0010663	0,7465	1,340	529,64	2714,5	2184,9	1,5929·	7,0657
2,5	127,43	0,0010675	0,7184	1,392	535,34	2716,4	2181,0	1,6071	7,0520
2,6	128,73	0,0010688	0,6925	1,444	540,87	2718,2	2177,3	1,6209	7,0389
2,7	129,98	0,0010700	0,6684	1,496	546,24	2719,9	2173,6	1,6342	7,0262
2,8	131,20	0,0010712	0,6460	1,548	551,44	2721,5	2170,1	1,6471	7,0140
2,9	132,39	0,0010724	0,6251	1,600	556,51	2723,1	2166,6	1,6595	7,0023
3,0	133,54	0,0010735	0,6056	1,651	561,43	2724,7	2163,2	1,6716	6,9909
3,1	134,66	0,0010746	0,5872	1,703	566,23	2726,1	2159,9	1,6834	6,9799
3,2	135,75	0,0010757	0,5700	1,754	570,90	2727,6	2156,7	1,6948	6,9693
3,3	136,82	0,0010768	0,5538	1,806	575,46	2729,0	2153,5	1,7059	6,9589
3,4	137,86	0,0010779	0,5385	1,857	579,92	2730,3	2150,4	1,7168	6,9489
3,5	138,87	0,0010789	0,5240	1,908	584,27	2731,6	2147,4	1,7273	6,9392
3,6	139,86	0,0010799	0,5103	1,960	588,53	2732,9	2144,4	1,7376	6,9297
3,7	140,83	0,0010809	0,4974	2,011	592,69	2734,1	2141,4	1,7476	6,9205
3,8	141,78	0,0010819	0,4851	2,062	596,77	2735,3	2138,6	1,7574	6,9116
3,9	142,71	0,0010829	0,4734	2,113	600,76	2736,5	2135,7	1,7670	6,9028
4,0	143,62	0,0010839	0,4622	2,163	604,67	2737,6	2133,0	1,7764	6,8943
4,1	144,52	0,0010848	0,4516	2,214	608,51	2738,7	2130,2	1,7856	6,8860
4,2	145,39	0,0010858	0,4415	2,265	612,27	2739,8	2127,5	1,7945	6,8779
4,3	146,25	0,0010867	0,4318	2,316	615,97	2740,9	2124,9	1,8033	6,8700
4,4	147,09	0,0010876	0,4226	2,366	619,60	2741,9	2122,3	1,8120	6,8623
4,5	147,92	0,0010885	0,4138	2,417	623,16	2742,9	2119,7	1,8204	6,8547
4,6	148,73	0,0010894	0,4053	2,467	626,67	2743,9	2117,2	1,8287	6,8473
4,7	149,53	0,0010903	0,3972	2,518	630,11	2744,8	2114,7	1,8368	6,8401
4,8	150,31	0,0010911	0,3894	2,568	633,50	2745,7	2112,2	1,8448	6,8330
4,9	151,08	0,0010920	0,3819	2,619	636,83	2746,6	2109,8	1,8527	6,8260
5,0	151,84	0,0010928	0,3747	2,669	640,12	2747,5	2107,4	1,8604	6,8192
5,2	153,33	0,0010945	0,3611	2,769	646,53	2749,3	2102,7	1,8754	6,8059
5,4	154,76	0,0010961	0,3485	2,870	652,76	2750,9	2098,1	1,8899	6,7932
5,6	156,16	0,0010977	0,3367	2,970	658,81	2752,5	2093,7	1,9040	6,7809
5,8	157,52	0,0010993	0,3257	3,070	664,69	2754,0	2089,3	1,9176	6,7690
6,0	158,84	0,0011009	0,3155	3,170	670,42	2755,5	2085,0	1,9308	6,7575
6,2	160,12	0,0011024	0,3059	3,270	676,01	2756,9	2080,8	1,9437	6,7464
6,4	161,38	0,0011039	0,2968	3,369	681,46	2758,2	2076,8	1,9562	6,7357
6,6	162,60	0,0011053	0,2883	3,469	686,78	2759,5	2072,7	1,9684	6,7252
6,8	163,79	0,0011068	0,2803	3,568	691,98	2760,8	2068,8	1,9802	6,7150
7,0	164,96	0,0011082	0,2727	3,667	697,06	2762,0	2064,9	1,9918	6,7052
7,2	166,10	0,0011096	0,2655	3,766	702,03	2763,2	2061,1	2,0031	6,6956
7,4	167,21	0,0011110	0,2587	3,866	706,90	2764,3	2057,4	2,0141	6,6862
7,6	168,30	0,0011123	0,2522	3,964	711,68	2765,4	2053,7	2,0249	6,6771
7,8	169,37	0,0011137	0,2461	4,063	716,35	2766,4	2050,1	2,0354	6,6683
8,0	170,41	0,0011150	0,2403	4,162	720,94	2767,5	2046,5	2,0457	6,6596
8,2	171,44	0,0011163	0,2347	4,261	725,43	2768,5	2043,0	2,0558	6,6511
8,4	172,45	0,0011176	0,2294	4,360	729,85	2769,4	2039,6	2,0657	6,6429
8,6	173,44	0,0011188	0,2243	4,458	734,19	2770,4	2036,2	2,0753	6,6348
8,8	174,41	0,0011201	0,2195	4,557	738,45	2771,3	2032,8	2,0848	6,6269

Table 2. State of Saturation (Pressure Table) (Continuation)   Sättigungszustand (Drucktafel) (Fortsetzung)

p	t	v'	v''	ϱ''	h'	h''	r	s'	s''
9,0	175,36	0,0011213	0,2148	4,655	742,64	2772,1	2029,5	2,0941	6,6192
9,2	176,29	0,0011226	0,2104	4,754	746,76	2773,0	2026,2	2,1033	6,6116
9,4	177,21	0,0011238	0,2061	4,852	750,82	2773,8	2023,0	2,1122	6,6042
9,6	178,12	0,0011250	0,2020	4,950	754,81	2774,6	2019,8	2,1210	6,5969
9,8	179,01	0,0011262	0,1981	5,049	758,74	2775,4	2016,7	2,1297	6,5898
10,0	179,88	0,0011274	0,1943	5,147	762,61	2776,2	2013,6	2,1382	6,5828
10,5	182,02	0,0011303	0,1855	5,392	772,03	2778,0	2005,9	2,1588	6,5659
11,0	184,07	0,0011331	0,1774	5,637	781,13	2779,7	1998,5	2,1786	6,5497
11,5	186,05	0,0011359	0,1700	5,883	789,92	2781,3	1991,3	2,1977	6,5342
12,0	187,96	0,0011386	0,1632	6,127	798,43	2782,7	1984,3	2,2161	6,5194
12,5	189,81	0,0011412	0,1569	6,372	806,69	2784,1	1977,4	2,2338	6,5051
13,0	191,61	0,0011438	0,1511	6,617	814,70	2785,4	1970,7	2,2510	6,4913
13,5	193,35	0,0011464	0,1457	6,862	822,49	2786,6	1964,2	2,2676	6,4780
14,0	195,04	0,0011489	0,1407	7,106	830,08	2787,8	1957,7	2,2837	6,4651
14,5	196,69	0,0011514	0,1360	7,351	837,46	2788,9	1951,4	2,2993	6,4526
15,0	198,29	0,0011539	0,1317	7,596	844,67	2789,9	1945,2	2,3145	6,4406
15,5	199,85	0,0011563	0,1275	7,840	851,69	2790,8	1939,2	2,3292	6,4289
16,0	201,37	0,0011586	0,1237	8,085	858,56	2791,7	1933,2	2,3436	6,4175
16,5	202,86	0,0011610	0,1201	8,330	865,27	2792,6	1927,3	2,3576	6,4065
17,0	204,31	0,0011633	0,1166	8,575	871,84	2793,4	1921,5	2,3713	6,3957
17,5	205,72	0,0011656	0,1134	8,820	878,28	2794,1	1915,9	2,3846	6,3853
18,0	207,11	0,0011678	0,1103	9,065	884,58	2794,8	1910,3	2,3976	6,3751
18,5	208,47	0,0011701	0,1074	9,310	890,75	2795,5	1904,7	2,4103	6,3651
19,0	209,80	0,0011723	0,1047	9,555	896,81	2796,1	1899,3	2,4228	6,3554
19,5	211,10	0,0011744	0,1020	9,801	902,75	2796,7	1893,9	2,4349	6,3459
20,0	212,37	0,0011766	0,09954	10,05	908,59	2797,2	1888,6	2,4469	6,3367
20,5	213,63	0,0011787	0,09716	10,29	914,32	2797,7	1883,4	2,4585	6,3276
21,0	214,85	0,0011809	0,09489	10,54	919,96	2798,2	1878,2	2,4700	6,3187
21,5	216,06	0,0011830	0,09272	10,78	925,50	2798,6	1873,1	2,4812	6,3100
22,0	217,24	0,0011850	0,09065	11,03	930,95	2799,1	1868,1	2,4922	6,3015
22,5	218,41	0,0011871	0,08867	11,28	936,32	2799,4	1863,1	2,5030	6,2931
23,0	219,55	0,0011892	0,08677	11,52	941,60	2799,8	1858,2	2,5136	6,2849
23,5	220,68	0,0011912	0,08495	11,77	946,80	2800,1	1853,3	2,5241	6,2769
24,0	221,78	0,0011932	0,08320	12,02	951,93	2800,4	1848,5	2,5343	6,2690
24,5	222,87	0,0011952	0,08152	12,27	956,98	2800,7	1843,7	2,5444	6,2612
25,0	223,94	0,0011972	0,07991	12,51	961,96	2800,9	1839,0	2,5543	6,2536
25,5	225,00	0,0011991	0,07835	12,76	966,87	2801,2	1834,3	2,5640	6,2461
26,0	226,04	0,0012011	0,07686	13,01	971,72	2801,4	1829,6	2,5736	6,2387
26,5	227,06	0,0012031	0,07541	13,26	976,50	2801,6	1825,0	2,5831	6,2315
27,0	228,07	0,0012050	0,07402	13,51	981,22	2801,7	1820,5	2,5924	6,2244
27,5	229,07	0,0012069	0,07268	13,76	985,88	2801,9	1816,0	2,6016	6,2173
28,0	230,05	0,0012088	0,07139	14,01	990,48	2802,0	1811,5	2,6106	6,2104
28,5	231,01	0,0012107	0,07014	14,26	995,03	2802,1	1807,1	2,6195	6,2036
29,0	231,97	0,0012126	0,06893	14,51	999,53	2802,2	1802,6	2,6283	6,1969
29,5	232,91	0,0012145	0,06776	14,76	1004,0	2802,2	1798,3	2,6370	6,1903
30	233,84	0,0012163	0,06663	15,01	1008,4	2802,3	1793,9	2,6455	6,1837
31	235,67	0,0012200	0,06447	15,51	1017,0	2802,3	1785,4	2,6623	6,1709
32	237,45	0,0012237	0,06244	16,02	1025,4	2802,3	1776,9	2,6786	6,1585
33	239,18	0,0012274	0,06053	16,52	1033,7	2802,3	1768,6	2,6945	6,1463
34	240,88	0,0012310	0,05873	17,03	1041,8	2802,1	1760,3	2,7101	6,1344
35	242,54	0,0012345	0,05703	17,54	1049,8	2802,0	1752,2	2,7253	6,1228
36	244,16	0,0012381	0,05541	18,05	1057,6	2801,7	1744,2	2,7401	6,1115
37	245,75	0,0012416	0,05389	18,56	1065,2	2801,4	1736,2	2,7547	6,1004
38	247,31	0,0012451	0,05244	19,07	1072,7	2801,1	1728,4	2,7689	6,0896
39	248,84	0,0012486	0,05106	19,58	1080,1	2800,8	1720,6	2,7829	6,0789
40	250,33	0,0012521	0,04975	20,10	1087,4	2800,3	1712,9	2,7965	6,0685
41	251,80	0,0012555	0,04850	20,62	1094,6	2799,9	1705,3	2,8099	6,0583
42	253,24	0,0012589	0,04731	21,14	1101,6	2799,4	1697,8	2,8231	6,0482
43	254,66	0,0012623	0,04617	21,66	1108,5	2798,9	1690,3	2,8360	6,0383
44	256,05	0,0012657	0,04508	22,18	1115,4	2798,3	1682,9	2,8487	6,0286

Table 2. State of Saturation (Pressure Table) (Continuation)   Sättigungszustand (Drucktafel) (Fortsetzung)

p	t	v'	v''	ϱ''	h'	h''	r	s'	s''
45	257,41	0,0012691	0,04404	22,71	1122,1	2797,7	1675,6	2,8612	6,0191
46	258,75	0,0012725	0,04304	23,24	1128,8	2797,0	1668,3	2,8735	6,0097
47	260,07	0,0012758	0,04208	23,76	1135,3	2796,4	1661,1	2,8855	6,0004
48	261,37	0,0012792	0,04116	24,29	1141,8	2795,7	1653,9	2,8974	5,9913
49	262,65	0,0012825	0,04028	24,83	1148,2	2794,9	1646,8	2,9091	5,9824
50	263,91	0,0012858	0,03943	25,36	1154,5	2794,2	1639,7	2,9206	5,9735
51	265,15	0,0012891	0,03861	25,90	1160,7	2793,4	1632,7	2,9320	5,9648
52	266,37	0,0012924	0,03782	26,44	1166,8	2792,6	1625,7	2,9431	5,9561
53	267,58	0,0012957	0,03707	26,98	1172,9	2791,7	1618,8	2,9541	5,9476
54	268,76	0,0012990	0,03633	27,52	1178,9	2790,8	1611,9	2,9650	5,9392
55	269,93	0,0013023	0,03563	28,07	1184,9	2789,9	1605,0	2,9757	5,9309
56	271,09	0,0013056	0,03495	28,62	1190,8	2789,0	1598,2	2,9863	5,9227
57	272,22	0,0013089	0,03429	29,16	1196,6	2788,0	1591,4	2,9967	5,9146
58	273,35	0,0013121	0,03365	29,72	1202,3	2787,0	1584,7	3,0071	5,9066
59	274,46	0,0013154	0,03303	30,27	1208,0	2786,0	1578,0	3,0172	5,8986
60	275,55	0,0013187	0,03244	30,83	1213,7	2785,0	1571,3	3,0273	5,8908
61	276,63	0,0013219	0,03186	31,39	1219,3	2784,0	1564,7	3,0372	5,8830
62	277,70	0,0013252	0,03130	31,95	1224,8	2782,9	1558,0	3,0471	5,8753
63	278,75	0,0013285	0,03076	32,51	1230,3	2781,8	1551,5	3,0568	5,8677
64	279,79	0,0013317	0,03023	33,08	1235,7	2780,6	1544,9	3,0664	5,8601
65	280,82	0,0013350	0,02972	33,65	1241,1	2779,5	1538,4	3,0759	5,8527
66	281,84	0,0013383	0,02922	34,22	1246,5	2778,3	1531,9	3,0853	5,8452
67	282,84	0,0013415	0,02874	34,79	1251,8	2777,1	1525,4	3,0946	5,8379
68	283,84	0,0013448	0,02827	35,37	1257,0	2775,9	1518,9	3,1038	5,8306
69	284,82	0,0013481	0,02782	35,95	1262,2	2774,7	1512,5	3,1129	5,8233
70	285,79	0,0013513	0,02737	36,53	1267,4	2773,5	1506,0	3,1219	5,8162
71	286,75	0,0013546	0,02694	37,12	1272,5	2772,2	1499,6	3,1308	5,8090
72	287,70	0,0013579	0,02652	37,70	1277,6	2770,9	1493,3	3,1397	5,8020
73	288,64	0,0013611	0,02611	38,29	1282,7	2769,6	1486,9	3,1484	5,7949
74	289,57	0,0013644	0,02572	38,89	1287,7	2768,3	1480,5	3,1571	5,7880
75	290,50	0,0013677	0,02533	39,48	1292,7	2766,9	1474,2	3,1657	5,7811
76	291,41	0,0013710	0,02495	40,08	1297,6	2765,5	1467,9	3,1742	5,7742
77	292,31	0,0013743	0,02458	40,68	1302,6	2764,2	1461,6	3,1827	5,7673
78	293,21	0,0013776	0,02422	41,29	1307,4	2762,8	1455,5	3,1911	5,7606
79	294,09	0,0013809	0,02387	41,90	1312,3	2761,3	1449,1	3,1994	5,7538
80	294,97	0,0013842	0,02353	42,51	1317,1	2759,9	1442,8	3,2076	5,7471
81	295,84	0,0013876	0,02319	43,12	1321,9	2758,4	1436,6	3,2158	5,7404
82	296,70	0,0013909	0,02286	43,74	1326,6	2757,0	1430,3	3,2239	5,7338
83	297,55	0,0013942	0,02254	44,36	1331,4	2755,5	1424,1	3,2320	5,7272
84	298,39	0,0013976	0,02223	44,98	1336,1	2754,0	1417,9	3,2399	5,7207
85	299,23	0,0014009	0,02193	45,61	1340,7	2752,5	1411,7	3,2479	5,7141
86	300,06	0,0014043	0,02163	46,24	1345,4	2750,9	1405,5	3,2558	5,7076
87	300,88	0,0014077	0,02133	46,87	1350,0	2749,4	1399,3	3,2636	5,7012
88	301,70	0,0014111	0,02105	47,51	1354,6	2747,8	1393,2	3,2713	5,6948
89	302,51	0,0014145	0,02077	48,15	1359,2	2746,2	1387,0	3,2790	5,6884
90	303,31	0,0014179	0,02050	48,79	1363,7	2744,6	1380,9	3,2867	5,6820
91	304,10	0,0014213	0,02023	49,44	1368,3	2743,0	1374,7	3,2943	5,6757
92	304,89	0,0014247	0,01996	50,09	1372,8	2741,4	1368,6	3,3018	5,6694
93	305,67	0,0014281	0,01971	50,74	1377,2	2739,7	1362,5	3,3093	5,6631
94	306,44	0,0014316	0,01945	51,40	1381,7	2738,0	1356,3	3,3168	5,6568
95	307,21	0,0014351	0,01921	52,06	1386,1	2736,4	1350,2	3,3242	5,6506
96	307,97	0,0014385	0,01897	52,73	1390,6	2734,7	1344,1	3,3315	5,6444
97	308,73	0,0014420	0,01873	53,40	1395,0	2733,0	1338,0	3,3388	5,6382
98	309,48	0,0014455	0,01849	54,07	1399,3	2731,2	1331,9	3,3461	5,6321
99	310,22	0,0014490	0,01827	54,75	1403,7	2729,5	1325,8	3,3534	5,6259
100	310,96	0,0014526	0,01804	55,43	1408,0	2727,7	1319,7	3,3605	5,6198
102	312,42	0,0014597	0,01760	56,80	1416,7	2724,2	1307,5	3,3748	5,6076
104	313,86	0,0014668	0,01718	58,19	1425,2	2720,6	1295,3	3,3889	5,5955
106	315,27	0,0014741	0,01678	59,60	1433,7	2716,9	1283,1	3,4029	5,5835
108	316,67	0,0014814	0,01639	61,03	1442,2	2713,1	1270,9	3,4167	5,5715

35

Table 2. State of Saturation (Pressure Table) (Continuation)   Sättigungszustand (Drucktafel) (Fortsetzung)

p	t	v'	v''	ϱ''	h'	h''	r	s'	s''
110	318,05	0,0014887	0,01601	62,48	1450,6	2709,3	1258,7	3,4304	5,5595
112	319,40	0,0014962	0,01564	63,94	1458,9	2705,4	1246,5	3,4440	5,5476
114	320,74	0,0015037	0,01528	65,43	1467,2	2701,5	1234,3	3,4575	5,5358
116	322,06	0,0015113	0,01494	66,93	1475,4	2697,4	1222,0	3,4708	5,5239
118	323,36	0,0015190	0,01461	68,46	1483,6	2693,3	1209,7	3,4840	5,5121
120	324,65	0,0015268	0,01428	70,01	1491,8	2689,2	1197,4	3,4972	5,5002
122	325,91	0,0015346	0,01397	71,59	1499,9	2684,9	1185,0	3,5102	5,4884
124	327,17	0,0015426	0,01366	73,19	1508,0	2680,6	1172,6	3,5232	5,4765
126	328,40	0,0015507	0,01337	74,81	1516,0	2676,1	1160,1	3,5361	5,4646
128	329,62	0,0015589	0,01308	76,46	1524,0	2671,6	1147,6	3,5488	5,4527
130	330,83	0,0015672	0,01280	78,14	1532,0	2667,0	1135,0	3,5616	5,4408
132	332,02	0,0015756	0,01252	79,85	1540,0	2662,3	1122,3	3,5742	5,4288
134	333,19	0,0015842	0,01226	81,59	1547,9	2657,4	1109,5	3,5868	5,4168
136	334,36	0,0015928	0,01200	83,36	1555,8	2652,5	1096,7	3,5993	5,4047
138	335,51	0,0016017	0,01174	85,16	1563,7	2647,5	1083,8	3,6118	5,3925
140	336,64	0,0016106	0,01150	86,99	1571,6	2642,4	1070,7	3,6242	5,3803
142	337,76	0,0016197	0,01125	88,86	1579,5	2637,1	1057,6	3,6366	5,3679
144	338,87	0,0016290	0,01102	90,77	1587,4	2631,8	1044,4	3,6490	5,3555
146	339,97	0,0016385	0,01079	92,71	1595,3	2626,3	1031,0	3,6613	5,3431
148	341,06	0,0016481	0,01056	94,69	1603,1	2620,7	1017,6	3,6736	5,3305
150	342,13	0,0016579	0,01034	96,71	1611,0	2615,0	1004,0	3,6859	5,3178
152	343,19	0,0016679	0,01012	98,77	1618,9	2609,2	990,3	3,6981	5,3051
154	344,24	0,0016782	0,009914	100,9	1626,8	2603,3	976,5	3,7104	5,2922
156	345,28	0,0016886	0,009707	103,0	1634,7	2597,3	962,6	3,7226	5,2793
158	346,31	0,0016993	0,009505	105,2	1642,6	2591,1	948,5	3,7348	5,2663
160	347,33	0,0017103	0,009308	107,4	1650,5	2584,9	934,3	3,7471	5,2531
162	348,34	0,0017216	0,009114	109,7	1658,5	2578,5	920,0	3,7594	5,2399
164	349,33	0,0017331	0,008925	112,0	1666,5	2572,1	905,6	3,7717	5,2267
166	350,32	0,0017447	0,008738	114,4	1674,5	2565,5	891,0	3,7842	5,2132
168	351,30	0,0017570	0,008553	116,9	1683,0	2558,6	875,6	3,7974	5,1994
170	352,26	0,0017696	0,008371	119,5	1691,7	2551,6	859,9	3,8107	5,1855
172	353,22	0,0017826	0,008191	122,1	1700,4	2544,4	844,1	3,8240	5,1713
174	354,17	0,0017961	0,008014	124,8	1709,0	2537,1	828,1	3,8372	5,1570
176	355,11	0,0018101	0,007839	127,6	1717,6	2529,5	811,9	3,8504	5,1425
178	356,04	0,0018247	0,007667	130,4	1726,2	2521,8	795,6	3,8635	5,1278
180	356,96	0,0018399	0,007498	133,4	1734,8	2513,9	779,1	3,8765	5,1128
182	357,87	0,0018556	0,007330	136,4	1743,4	2505,8	762,3	3,8896	5,0975
184	358,77	0,0018721	0,007165	139,6	1752,1	2497,4	745,3	3,9028	5,0820
186	359,67	0,0018893	0,007001	142,8	1760,9	2488,8	727,9	3,9160	5,0661
188	360,55	0,0019072	0,006839	146,2	1769,7	2479,8	710,1	3,9294	5,0498
190	361,43	0,0019260	0,006678	149,8	1778,7	2470,6	692,0	3,9429	5,0332
192	362,30	0,0019458	0,006517	153,4	1787,8	2461,1	673,3	3,9566	5,0160
194	363,16	0,0019666	0,006358	157,3	1797,0	2451,1	654,1	3,9706	4,9983
196	364,02	0,0019886	0,006198	161,3	1806,5	2440,7	634,2	3,9849	4,9801
198	364,86	0,0020120	0,006038	165,6	1816,3	2429,8	613,5	3,9996	4,9610
200	365,70	0,0020370	0,005877	170,2	1826,5	2418,4	591,9	4,0149	4,9412
202	366,53	0,0020639	0,005714	175,0	1837,0	2406,2	569,2	4,0308	4,9204
204	367,36	0,0020931	0,005548	180,2	1848,1	2393,3	545,1	4,0474	4,8984
206	368,17	0,0021252	0,005379	185,9	1859,9	2379,3	519,5	4,0651	4,8750
208	368,98	0,0021610	0,005205	192,1	1872,5	2364,2	491,7	4,0841	4,8498
210	369,78	0,0022015	0,005023	199,1	1886,3	2347,6	461,3	4,1048	4,8223
212	370,58	0,0022488	0,004831	207,0	1901,5	2328,9	427,4	4,1279	4,7917
214	371,37	0,0023061	0,004624	216,3	1919,0	2307,4	388,4	4,1543	4,7569
216	372,15	0,0023793	0,004392	227,7	1940,0	2281,5	341,6	4,1861	4,7154
218	372,92	0,0024832	0,004115	243,0	1967,2	2248,0	280,8	4,2276	4,6622
220	373,69	0,0026714	0,003728	268,3	2011,1	2195,6	184,5	4,2947	4,5799
221,20	374,15	0,00317		315,5	2107,4		0,0	4,4429	

Table 3. Water and Superheated Steam

Wasser und überhitzter Dampf

t_s: temperature of saturation, Sättigungstemperatur,
v'', h'', s'': for saturated steam, für gesättigten Dampf

t	0,01 bar $t_s = 6,983$ °C			0,02 bar $t_s = 17,51$ °C			0,03 bar $t_s = 24,10$ °C			0,04 bar $t_s = 28,98$ °C		
	v'' 129,2	h'' 2514,4	s'' 8,9767	v'' 67,01	h'' 2533,6	s'' 8,7246	v'' 45,67	h'' 2545,6	s'' 8,5785	v'' 34,80	h'' 2554,5	s'' 8,4755
°C	v m³/kg	h kJ/kg	s kJ/kg K	v m³/kg	h kJ/kg	s kJ/kg K	v m³/kg	h kJ/kg	s kJ/kg K	v m³/kg	h kJ/kg	s kJ/kg K
0	0,0010002	— 0,0	— 0,0002	0,0010002	— 0,0	— 0,0002	0,0010002	— 0,0	— 0,0002	0,0010002	— 0,0	— 0,0002
10	130,6	2520,0	8,9966	0,0010003	42,0	0,1510	0,0010003	42,0	0,1510	0,0010003	42,0	0,1510
20	135,2	2538,6	9,0611	67,58	2538,3	8,7404	0,0010017	83,9	0,2963	0,0010017	83,9	0,2963
30	139,9	2557,2	9,1236	69,90	2556,9	8,8031	46,58	2556,7	8,6152	34,92	2556,4	8,4818
40	144,5	2575,9	9,1842	72,21	2575,6	8,8637	48,12	2575,4	8,6760	36,08	2575,2	8,5426
50	149,1	2594,6	9,2430	74,52	2594,4	8,9226	49,67	2594,2	8,7349	37,24	2593,9	8,6016
60	153,7	2613,3	9,3001	76,84	2613,1	8,9797	51,21	2612,9	8,7922	38,40	2612,7	8,6589
70	158,3	2632,1	9,3556	79,15	2631,9	9,0353	52,75	2631,7	8,8478	39,56	2631,6	8,7146
80	163,0	2650,9	9,4096	81,46	2650,7	9,0894	54,30	2650,6	8,9019	40,71	2650,4	8,7688
90	167,6	2669,7	9,4622	83,77	2669,6	9,1421	55,84	2669,5	8,9546	41,87	2669,3	8,8216
100	172,2	2688,6	9,5136	86,08	2688,5	9,1934	57,38	2688,4	9,0060	43,03	2688,3	8,8730
110	176,8	2707,6	9,5636	88,39	2707,4	9,2435	58,92	2707,3	9,0562	44,18	2707,2	8,9232
120	181,4	2726,5	9,6125	90,70	2726,4	9,2924	60,46	2726,3	9,1051	45,34	2726,2	8,9721
130	186,0	2745,6	9,6603	93,01	2745,5	9,3402	62,00	2745,4	9,1529	46,50	2745,3	9,0200
140	190,7	2764,6	9,7070	95,32	2764,5	9,3870	63,54	2764,5	9,1997	47,65	2764,4	9,0668
150	195,3	2783,7	9,7527	97,63	2783,7	9,4327	65,08	2783,6	9,2454	48,81	2783,5	9,1125
160	199,9	2802,9	9,7975	99,94	2802,8	9,4775	66,62	2802,8	9,2902	49,96	2802,7	9,1573
170	204,5	2822,1	9,8413	102,25	2822,1	9,5213	68,16	2822,0	9,3341	51,12	2821,9	9,2012
180	209,1	2841,4	9,8843	104,55	2841,3	9,5643	69,70	2841,3	9,3771	52,27	2841,2	9,2443
190	213,7	2860,7	9,9265	106,86	2860,6	9,6065	71,24	2860,6	9,4193	53,43	2860,5	9,2865
200	218,4	2880,1	9,9679	109,17	2880,0	9,6479	72,78	2880,0	9,4607	54,58	2879,9	9,3279
210	223,0	2899,5	10,0085	111,48	2899,4	9,6885	74,32	2899,4	9,5013	55,73	2899,4	9,3685
220	227,6	2919,0	10,0484	113,79	2918,9	9,7284	75,85	2918,9	9,5412	56,89	2918,8	9,4084
230	232,2	2938,5	10,0876	116,10	2938,5	9,7677	77,39	2938,4	9,5805	58,04	2938,4	9,4476
240	236,8	2958,1	10,1262	118,40	2958,0	9,8062	78,93	2958,0	9,6190	59,20	2958,0	9,4862
250	241,4	2977,7	10,1641	120,71	2977,7	9,8441	80,47	2977,7	9,6570	60,35	2977,6	9,5241
260	246,0	2997,4	10,2014	123,02	2997,4	9,8814	82,01	2997,4	9,6943	61,51	2997,3	9,5615
270	250,7	3017,2	10,2381	125,33	3017,2	9,9182	83,55	3017,1	9,7310	62,66	3017,1	9,5982
280	255,3	3037,0	10,2743	127,64	3037,0	9,9543	85,09	3037,0	9,7672	63,81	3036,9	9,6344
290	259,9	3056,9	10,3099	129,94	3056,9	9,9899	86,63	3056,8	9,8028	64,97	3056,8	9,6700
300	264,5	3076,8	10,3450	132,25	3076,8	10,0251	88,17	3076,8	9,8379	66,12	3076,8	9,7051
310	269,1	3096,8	10,3796	134,56	3096,8	10,0597	89,70	3096,8	9,8725	67,28	3096,8	9,7397
320	273,7	3116,9	10,4137	136,87	3116,9	10,0938	91,24	3116,9	9,9066	68,43	3116,8	9,7738
330	278,4	3137,0	10,4473	139,17	3137,0	10,1274	92,78	3137,0	9,9403	69,58	3137,0	9,8075
340	283,0	3157,2	10,4805	141,48	3157,2	10,1606	94,32	3157,2	9,9735	70,74	3157,2	9,8407
350	287,6	3177,5	10,5133	143,79	3177,5	10,1934	95,86	3177,4	10,0062	71,89	3177,4	9,8735
360	292,2	3197,8	10,5457	146,10	3197,8	10,2257	97,40	3197,8	10,0386	73,05	3197,7	9,9058
370	296,8	3218,2	10,5776	148,41	3218,2	10,2577	98,94	3218,1	10,0705	74,20	3218,1	9,9378
380	301,4	3238,6	10,6091	150,71	3238,6	10,2892	100,47	3238,6	10,1021	75,35	3238,6	9,9693
390	306,0	3259,1	10,6403	153,02	3259,1	10,3204	102,01	3259,1	10,1333	76,51	3259,1	10,0005
400	310,7	3279,7	10,6711	155,33	3279,7	10,3512	103,55	3279,7	10,1641	77,66	3279,7	10,0313
410	315,3	3300,4	10,7016	157,64	3300,4	10,3817	105,09	3300,3	10,1945	78,82	3300,3	10,0617
420	319,9	3321,1	10,7317	159,94	3321,1	10,4118	106,63	3321,1	10,2246	79,97	3321,0	10,0918
430	324,5	3341,9	10,7614	162,25	3341,8	10,4415	108,17	3341,8	10,2544	81,12	3341,8	10,1216
440	329,1	3362,7	10,7909	164,56	3362,7	10,4710	109,71	3362,7	10,2838	82,28	3362,7	10,1510
450	333,7	3383,6	10,8200	166,87	3383,6	10,5001	111,24	3383,6	10,3129	83,43	3383,6	10,1802
460	338,4	3404,6	10,8488	169,17	3404,6	10,5289	112,78	3404,6	10,3418	84,59	3404,6	10,2090
470	343,0	3425,7	10,8773	171,48	3425,6	10,5574	114,32	3425,6	10,3703	85,74	3425,6	10,2375
480	347,6	3446,8	10,9056	173,79	3446,8	10,5857	115,86	3446,8	10,3985	86,89	3446,7	10,2657
490	352,2	3468,0	10,9335	176,10	3468,0	10,6136	117,40	3467,9	10,4265	88,05	3467,9	10,2937
500	356,8	3489,2	10,9612	178,41	3489,2	10,6413	118,94	3489,2	10,4541	89,20	3489,2	10,3214

Table 3. Water and Superheated Steam (Continuation) Wasser und überhitzter Dampf (Fortsetzung)

t	0,01 bar $t_s = 6,983$ °C			0,02 bar $t_s = 17,51$ °C			0,03 bar $t_s = 24,10$ °C			0,04 bar $t_s = 28,98$ °C		
	v'' 129,2	h'' 2514,4	s'' 8,9767	v'' 67,01	h'' 2533,6	s'' 8,7246	v'' 45,67	h'' 2545,6	s'' 8,5785	v'' 34,80	h'' 2554,5	s'' 8,4755
°C	v	h	s	v	h	s	v	h	s	v	h	s
500	356,8	3489,2	10,9612	178,41	3489,2	10,6413	118,94	3489,2	10,4541	89,20	3489,2	10,3214
510	361,4	3510'6	10,9886	180,71	3510,5	10,6687	120,47	3510,5	10,4815	90,36	3510,5	10,3488
520	366,0	3531,9	11,0157	183,02	3531,9	10,6958	122,01	3531,9	10,5087	91,51	3531,9	10,3759
530	370,7	3553,4	11,0426	185,33	3553,4	10,7227	123,55	3553,4	10,5356	92,66	3553,4	10,4028
540	375,3	3574,9	11,0693	187,64	3574,9	10,7494	125,09	3574,9	10,5622	93,82	3574,9	10,4295
550	379,9	3596,5	11,0957	189,94	3596,5	10,7758	126,63	3596,5	10,5886	94,97	3596,5	10,4559
560	384,5	3618,2	11,1218	192,25	3618,2	10,8019	128,17	3618,2	10,6148	96,12	3618,2	10,4820
570	389,1	3640,0	11,1478	194,56	3640,0	10,8279	129,71	3639,9	10,6408	97,28	3639,9	10,5080
580	393,7	3661,8	11,1735	196,87	3661,8	10,8536	131,24	3661,8	10,6665	98,43	3661,7	10,5337
590	398,4	3683,7	11,1990	199,17	3683,6	10,8791	132,78	3683,6	10,6920	99,59	3683,6	10,5592
600	403,0	3705,6	11,2243	201,48	3705,6	10,9044	134,32	3705,6	10,7173	100,74	3705,6	10,5845
610	407,6	3727,6	11,2494	203,79	3727,6	10,9295	135,86	3727,6	10,7423	101,89	3727,6	10,6095
620	412,2	3749,7	11,2742	206,10	3749,7	10,9543	137,40	3749,7	10,7672	103,05	3749,7	10,6344
630	416,8	3771,9	11,2989	208,40	3771,9	10,9790	138,94	3771,8	10,7919	104,20	3771,8	10,6591
640	421,4	3794,1	11,3234	210,71	3794,1	11,0035	140,47	3794,1	10,8163	105,36	3794,1	10,6836
650	426,0	3816,4	11,3476	213,02	3816,4	11,0277	142,01	3816,4	10,8406	106,51	3816,4	10,7078
660	430,7	3838,7	11,3717	215,33	3838,7	11,0518	143,55	3838,7	10,8647	107,66	3838,7	10,7319
670	435,3	3861,2	11,3956	217,64	3861,2	11,0757	145,09	3861,2	10,8886	108,82	3861,1	10,7558
680	439,9	3883,7	11,4194	219,94	3883,7	11,0995	146,63	3883,6	10,9123	109,97	3883,6	10,7796
690	444,5	3906,2	11,4429	222,25	3906,2	11,1230	148,17	3906,2	10,9359	111,12	3906,2	10,8031
700	449,1	3928,9	11,4663	224,56	3928,8	11,1464	149,70	3928,8	10,9593	112,28	3928,8	10,8265
710	453,7	3951,6	11,4895	226,87	3951,5	11,1696	151,24	3951,5	10,9825	113,43	3951,5	10,8497
720	458,3	3974,3	11,5125	229,17	3974,3	11,1926	152,78	3974,3	11,0055	114,59	3974,3	10,8727
730	463,0	3997,1	11,5354	231,48	3997,1	11,2155	154,32	3997,1	11,0284	115,74	3997,1	10,8956
740	467,6	4020,0	11,5581	233,79	4020,0	11,2382	155,86	4020,0	11,0511	116,89	4020,0	10,9183
750	472,2	4043,0	11,5807	236,10	4043,0	11,2608	157,40	4043,0	11,0736	118,05	4043,0	10,9409
760	476,8	4066,0	11,6031	238,40	4066,0	11,2832	158,94	4066,0	11,0960	119,20	4066,0	10,9633
770	481,4	4089,1	11,6253	240,71	4089,1	11,3054	160,47	4089,1	11,1183	120,35	4089,1	10,9855
780	486,0	4112,3	11,6474	243,02	4112,2	11,3275	162,01	4112,2	11,1404	121,51	4112,2	11,0076
790	490,7	4135,5	11,6693	245,33	4135,5	11,3494	163,55	4135,5	11,1623	122,66	4135,5	11,0295
800	495,3	4158,7	11,6911	247,63	4158,7	11,3712	165,09	4158,7	11,1841	123,82	4158,7	11,0513

Table 3. Water and Superheated Steam (Continuation) Wasser und überhitzter Dampf (Fortsetzung)

t	0,05 bar $t_s = 32,90$ °C			0,06 bar $t_s = 36,18$ °C			0,07 bar $t_s = 39,03$ °C			0,08 bar $t_s = 41,53$ °C		
	v'' 28,19	h'' 2561,6	s'' 8,3960	v'' 23,74	h'' 2567,5	s'' 8,3312	v'' 20,53	h'' 2572,6	s'' 8,2767	v'' 18,10	h'' 2577,1	s'' 8,2296
°C	v	h	s	v	h	s	v	h	s	v	h	s
0	0,0010002	— 0,0	—0,0002	0,0010002	— 0,0	—0,0002	0,0010002	— 0,0	—0,0002	0,0010002	— 0,0	—0,0002
10	0,0010003	42,0	0,1510	0,0010003	42,0	0,1510	0,0010003	42,0	0,1510	0,0010003	42,0	0,1510
20	0,0010017	83,9	0,2963	0,0010017	83,9	0,2963	0,0010017	83,9	0,2963	0,0010017	83,9	0,2963
30	0,0010043	125,7	0,4365	0,0010043	125,7	0,4365	0,0010043	125,7	0,4365	0,0010043	125,7	0,4365
40	28,85	2574,9	8,4390	24,04	2574,7	8,3543	20,60	2574,5	8,2826	0,0010078	167,5	0,5721
50	29,78	2593,7	8,4981	24,81	2593,5	8,4135	21,26	2593,3	8,3418	18,60	2593,1	8,2797
60	30,71	2612,6	8,5555	25,59	2612,4	8,4709	21,92	2612,2	8,3993	19,18	2612,0	8,3372
70	31,64	2631,4	8,6113	26,36	2631,2	8,5267	22,59	2631,1	8,4552	19,76	2630,9	8,3932
80	32,56	2650,3	8,6655	27,13	2650,1	8,5810	23,25	2650,0	8,5095	20,34	2649,8	8,4476
90	33,49	2669,2	8,7183	27,90	2669,1	8,6339	23,91	2668,9	8,5624	20,92	2668,8	8,5005
100	34,42	2688,1	8,7698	28,68	2688,0	8,6854	24,58	2687,9	8,6140	21,50	2687,8	8,5521
110	35,34	2707,1	8,8200	29,45	2707,0	8,7356	25,24	2706,9	8,6642	22,08	2706,8	8,6024
120	36,27	2726,1	8,8690	30,22	2726,0	8,7846	25,90	2725,9	8,7133	22,66	2725,8	8,6515
130	37,19	2745,2	8,9168	30,99	2745,1	8,8325	26,56	2745,0	8,7612	23,24	2744,9	8,6994
140	38,12	2764,3	8,9636	31,76	2764,2	8,8793	27,22	2764,1	8,8080	23,82	2764,1	8,7463
150	39,04	2783,4	9,0094	32,53	2783,4	8,9251	27,88	2783,3	8,8539	24,39	2783,2	8,7921
160	39,97	2802,6	9,0542	33,30	2802,6	8,9700	28,54	2802,5	8,8987	24,97	2802,4	8,8370
170	40,89	2821,9	9,0982	34,07	2821,8	9,0139	29,20	2821,8	8,9427	25,55	2821,7	8,8809
180	41,81	2841,2	9,1412	34,84	2841,1	9,0569	29,86	2841,0	8,9857	26,13	2841,0	8,9240
190	42,74	2860,5	9,1834	35,61	2860,4	9,0992	30,52	2860,4	9,0279	26,71	2860,3	8,9662
200	43,66	2879,9	9,2248	36,38	2879,8	9,1406	31,18	2879,8	9,0694	27,28	2879,7	9,0077
210	44,59	2899,3	9,2654	37,15	2899,3	9,1812	31,84	2899,2	9,1100	27,86	2899,2	9,0483
220	45,51	2918,8	9,3054	37,92	2918,8	9,2212	32,50	2918,7	9,1500	28,44	2918,7	9,0883
230	46,43	2938,3	9,3446	38,69	2938,3	9,2604	33,16	2938,3	9,1892	29,02	2938,2	9,1275
240	47,36	2957,9	9,3832	39,46	2957,9	9,2990	33,82	2957,9	9,2278	29,59	2957,8	9,1661
250	48,28	2977,6	9,4211	40,23	2977,6	9,3369	34,48	2977,5	9,2657	30,17	2977,5	9,2041
260	49,20	2997,3	9,4584	41,00	2997,3	9,3742	35,14	2997,3	9,3031	30,75	2997,2	9,2414
270	50,13	3017,1	9,4952	41,77	3017,1	9,4110	35,80	3017,0	9,3398	31,33	3017,0	9,2781
280	51,05	3036,9	9,5313	42,54	3036,9	9,4472	36,46	3036,9	9,3760	31,90	3036,8	9,3143
290	51,97	3056,8	9,5670	43,31	3056,8	9,4828	37,12	3056,7	9,4116	32,48	3056,7	9,3500
300	52,90	3076,7	9,6021	44,08	3076,7	9,5179	37,78	3076,7	9,4467	33,06	3076,7	9,3851
310	53,82	3096,7	9,6367	44,85	3096,7	9,5525	38,44	3096,7	9,4813	33,64	3096,7	9,4197
320	54,74	3116,8	9,6708	45,62	3116,8	9,5866	39,10	3116,8	9,5155	34,21	3116,8	9,4538
330	55,67	3137,0	9,7045	46,39	3136,9	9,6203	39,76	3136,9	9,5491	34,79	3136,9	9,4875
340	56,59	3157,1	9,7377	47,16	3157,1	9,6535	40,42	3157,1	9,5823	35,37	3157,1	9,5207
350	57,51	3177,4	9,7704	47,93	3177,4	9,6863	41,08	3177,4	9,6151	35,94	3177,3	9,5535
360	58,44	3197,7	9,8028	48,70	3197,7	9,7186	41,74	3197,7	9,6475	36,52	3197,7	9,5858
370	59,36	3218,1	9,8347	49,47	3218,1	9,7506	42,40	3218,1	9,6794	37,10	3218,1	9,6178
380	60,28	3238,6	9,8663	50,23	3238,5	9,7821	43,06	3238,5	9,7110	37,67	3238,5	9,6493
390	61,21	3259,1	9,8975	51,00	3259,1	9,8133	43,72	3259,0	9,7422	38,25	3259,0	9,6805
400	62,13	3279,7	9,9283	51,77	3279,6	9,8441	44,38	3279,6	9,7730	38,83	3279,6	9,7113
410	63,05	3300,3	9,9587	52,54	3300,3	9,8746	45,04	3300,3	9,8034	39,41	3300,3	9,7418
420	63,98	3321,0	9,9888	53,31	3321,0	9,9047	45,70	3321,0	9,8335	39,98	3321,0	9,7719
430	64,90	3341,8	10,0186	54,08	3341,8	9,9344	46,35	3341,8	9,8633	40,56	3341,8	9,8016
440	65,82	3362,7	10,0480	54,85	3362,6	9,9639	47,01	3362,6	9,8927	41,14	3362,6	9,8311
450	66,74	3383,6	10,0772	55,62	3383,6	9,9930	47,67	3383,5	9,9219	41,71	3383,5	9,8602
460	67,67	3404,6	10,1060	56,39	3404,5	10,0218	48,33	3404,5	9,9507	42,29	3404,5	9,8890
470	68,59	3425,6	10,1345	57,16	3425,6	10,0504	48,99	3425,6	9,9792	42,87	3425,6	9,9176
480	69,51	3446,7	10,1627	57,93	3446,7	10,0786	49,65	3446,7	10,0074	43,45	3446,7	9,9458
490	70,44	3467,9	10,1907	58,70	3467,9	10,1065	50,31	3467,9	10,0354	44,02	3467,9	9,9737
500	71,36	3489,2	10,2184	59,47	3489,2	10,1342	50,97	3489,2	10,0631	44,60	3489,1	10,0014
510	72,28	3510,5	10,2458	60,24	3510,5	10,1616	51,63	3510,5	10,0905	45,18	3510,5	10,0288
520	73,21	3531,9	10,2729	61,01	3531,9	10,1888	52,29	3531,9	10,1176	45,75	3531,9	10,0560
530	74,13	3553,4	10,2998	61,77	3553,4	10,2157	52,95	3553,4	10,1445	46,33	3553,3	10,0829
540	75,05	3574,9	10,3265	62,54	3574,9	10,2423	53,61	3574,9	10,1712	46,91	3574,9	10,1095
550	75,98	3596,5	10,3529	63,31	3596,5	10,2687	54,27	3596,5	10,1976	47,48	3596,5	10,1359

Table 3. Water and Superheated Steam (Continuation) Wasser und überhitzter Dampf (Fortsetzung)

t	0,05 bar t_s = 32,90 °C			0,06 bar t_s = 36,18 °C			0,07 bar t_s = 39,03 °C			0,08 bar t_s = 41,53 °C		
	v'' 28,19	h'' 2561,6	s'' 8,3960	v'' 23,74	h'' 2567,5	s'' 8,3312	v'' 20,53	h'' 2572,6	s'' 8,2767	v'' 18,10	h'' 2577,1	s'' 8,2296
°C	v	h	s	v	h	s	v	h	s	v	h	s
550	75,98	3596,5	10,3529	63,31	3596,5	10,2687	54,27	3596,5	10,1976	47,48	3596,5	10,1359
560	76,90	3618,2	10,3790	64,08	3618,2	10,2949	54,93	3618,2	10,2237	48,06	3618,2	10,1621
570	77,82	3639,9	10,4050	64,85	3639,9	10,3208	55,59	3639,9	10,2497	48,64	3639,9	10,1881
580	78,75	3661,7	10,4307	65,62	3661,7	10,3466	56,25	3661,7	10,2754	49,22	3661,7	10,2138
590	79,67	3683,6	10,4562	66,39	3683,6	10,3721	56,91	3683,6	10,3009	49,79	3683,6	10,2393
600	80,59	3705,6	10,4815	67,16	3705,6	10,3973	57,56	3705,6	10,3262	50,37	3705,5	10,2646
610	81,51	3727,6	10,5066	67,93	3727,6	10,4224	58,22	3727,6	10,3513	50,95	3727,6	10,2896
620	82,44	3749,7	10,5314	68,70	3749,7	10,4473	58,88	3749,7	10,3761	51,52	3749,7	10,3145
630	83,36	3771,8	10,5561	69,47	3771,8	10,4719	59,54	3771,8	10,4008	52,10	3771,8	10,3392
640	84,28	3794,1	10,5806	70,24	3794,0	10,4964	60,20	3794,0	10,4253	52,68	3794,0	10,3636
650	85,21	3816,3	10,6049	71,01	3816,3	10,5207	60,86	3816,3	10,4496	53,25	3816,3	10,3879
660	86,13	3838,7	10,6289	71,77	3838,7	10,5448	61,52	3838,7	10,4736	53,83	3838,7	10,4120
670	87,05	3861,1	10,6529	72,54	3861,1	10,5687	62,18	3861,1	10,4976	54,41	3861,1	10,4359
680	87,98	3883,6	10,6766	73,31	3883,6	10,5924	62,84	3883,6	10,5213	54,98	3883,6	10,4597
690	88,90	3906,2	10,7001	74,08	3906,2	10,6160	63,50	3906,2	10,5448	55,56	3906,2	10,4832
700	89,82	3928,8	10,7235	74,85	3928,8	10,6394	64,16	3928,8	10,5682	56,14	3928,8	10,5066
710	90,75	3951,5	10,7467	75,62	3951,5	10,6626	64,82	3951,5	10,5914	56,72	3951,5	10,5298
720	91,67	3974,3	10,7697	76,39	3974,3	10,6856	65,48	3974,3	10,6144	57,29	3974,3	10,5528
730	92,59	3997,1	10,7926	77,16	3997,1	10,7085	66,14	3997,1	10,6373	57,87	3997,1	10,5757
740	93,51	4020,0	10,8153	77,93	4020,0	10,7312	66,80	4020,0	10,6600	58,45	4020,0	10,5984
750	94,44	4043,0	10,8379	78,70	4043,0	10,7537	67,46	4043,0	10,6826	59,02	4043,0	10,6210
760	95,36	4066,0	10,8603	79,47	4066,0	10,7761	68,11	4066,0	10,7050	59,60	4066,0	10,6433
770	96,28	4089,1	10,8825	80,24	4089,1	10,7984	68,77	4089,1	10,7272	60,18	4089,1	10,6656
780	97,21	4112,2	10,9046	81,01	4112,2	10,8204	69,43	4112,2	10,7493	60,75	4112,2	10,6877
790	98,13	4135,4	10,9265	81,77	4135,4	10,8424	70,09	4135,4	10,7712	61,33	4135,4	10,7096
800	99,05	4158,7	10,9483	82,54	4158,7	10,8642	70,75	4158,7	10,7930	61,91	4158,7	10,7314

Table 3. Water and Superheated Steam (Continuation) Wasser und überhitzter Dampf (Fortsetzung)

	0,09 bar $t_s = 43,79$ °C			0,10 bar $t_s = 45,83$ °C			0,12 bar $t_s = 49,45$ °C			0,14 bar $t_s = 52,57$ °C		
t	v'' 16,20	h'' 2581,1	s'' 8,1881	v'' 14,67	h'' 2584,8	s'' 8,1511	v'' 12,36	h'' 2591,2	s'' 8,0872	v'' 10,69	h'' 2596,7	s'' 8,0334
°C	v	h	s	v	h	s	v	h	s	v	h	s
0	0,0010002	− 0,0	−0,0002	0,0010002	− 0,0	−0,0002	0,0010002	− 0,0	−0,0002	0,0010002	− 0,0	−0,0002
10	0,0010002	42,0	0,1510	0,0010002	42,0	0,1510	0,0010002	42,0	0,1510	0,0010002	42,0	0,1510
20	0,0010017	83,9	0,2963	0,0010017	83,9	0,2963	0,0010017	83,9	0,2963	0,0010017	83,9	0,2963
30	0,0010043	125,7	0,4365	0,0010043	125,7	0,4365	0,0010043	125,7	0,4365	0,0010043	125,7	0,4365
40	0,0010078	167,5	0,5721	0,0010078	167,5	0,5721	0,0010078	167,5	0,5721	0,0010078	167,5	0,5721
50	16,53	2592,9	8,2248	14,87	2592,7	8,1757	12,38	2592,2	8,0905	0,0010121	209,3	0,7035
60	17,04	2611,8	8,2824	15,34	2611,6	8,2334	12,77	2611,2	8,1483	10,94	2610,8	8,0763
70	17,56	2630,7	8,3384	15,80	2630,6	8,2894	13,16	2630,2	8,2045	11,28	2629,9	8,1325
80	18,08	2649,7	8,3929	16,27	2649,5	8,3439	13,55	2649,2	8,2591	11,61	2648,9	8,1872
90	18,59	2668,6	8,4458	16,73	2668,5	8,3969	13,94	2668,2	8,3122	11,94	2668,0	8,2404
100	19,11	2687,6	8,4974	17,20	2687,5	8,4486	14,33	2687,3	8,3639	12,28	2687,0	8,2922
110	19,62	2706,7	8,5478	17,66	2706,6	8,4989	14,71	2706,3	8,4143	12,61	2706,1	8,3427
120	20,14	2725,7	8,5969	18,12	2725,6	8,5481	15,10	2725,4	8,4635	12,94	2725,2	8,3920
130	20,65	2744,8	8,6449	18,59	2744,7	8,5961	15,49	2744,6	8,5116	13,27	2744,4	8,4401
140	21,17	2764,0	8,6917	19,05	2763,9	8,6430	15,87	2763,7	8,5585	13,60	2763,6	8,4871
150	21,68	2783,2	8,7376	19,51	2783,1	8,6888	16,26	2782,9	8,6044	13,93	2782,8	8,5330
160	22,20	2802,4	8,7825	19,98	2802,3	8,7338	16,64	2802,2	8,6494	14,26	2802,0	8,5780
170	22,71	2821,6	8,8265	20,44	2821,6	8,7777	17,03	2821,4	8,6934	14,59	2821,3	8,6220
180	23,22	2840,9	8,8695	20,90	2840,9	8,8208	17,41	2840,8	8,7365	14,93	2840,7	8,6652
190	23,74	2860,3	8,9118	21,36	2860,2	8,8631	17,80	2860,1	8,7788	15,26	2860,0	8,7074
200	24,25	2879,7	8,9532	21,83	2879,6	8,9045	18,19	2879,5	8,8202	15,59	2879,5	8,7489
210	24,76	2899,1	8,9939	22,29	2899,1	8,9452	18,57	2899,0	8,8609	15,92	2898,9	8,7897
220	25,28	2918,6	9,0339	22,75	2918,6	8,9852	18,96	2918,5	8,9009	16,25	2918,4	8,8296
230	25,79	2938,2	9,0731	23,21	2938,2	9,0244	19,34	2938,1	8,9402	16,58	2938,0	8,8689
240	26,31	2957,8	9,1117	23,67	2957,8	9,0630	19,73	2957,7	8,9788	16,91	2957,6	8,9075
250	26,82	2977,5	9,1497	24,14	2977,4	9,1010	20,11	2977,4	9,0168	17,24	2977,3	8,9455
260	27,33	2997,2	9,1870	24,60	2997,2	9,1383	20,50	2997,1	9,0541	17,57	2997,0	8,9829
270	27,84	3017,0	9,2237	25,06	3016,9	9,1751	20,88	3016,9	9,0909	17,90	3016,8	9,0196
280	28,36	3036,8	9,2599	25,52	3036,8	9,2113	21,27	3036,7	9,1271	18,23	3036,7	9,0558
290	28,87	3056,7	9,2956	25,98	3056,7	9,2469	21,65	3056,6	9,1627	18,56	3056,6	9,0915
300	29,38	3076,6	9,3307	26,45	3076,6	9,2820	22,04	3076,6	9,1978	18,89	3076,5	9,1266
310	29,90	3096,7	9,3653	26,91	3096,6	9,3167	22,42	3096,6	9,2325	19,22	3096,5	9,1613
320	30,41	3116,7	9,3994	27,37	3116,7	9,3508	22,81	3116,7	9,2666	19,55	3116,6	9,1954
330	30,92	3136,9	9,4331	27,83	3136,8	9,3845	23,19	3136,8	9,3003	19,88	3136,8	9,2291
340	31,44	3157,1	9,4663	28,29	3157,0	9,4177	23,58	3157,0	9,3335	20,21	3157,0	9,2623
350	31,95	3177,3	9,4991	28,75	3177,3	9,4504	23,96	3177,3	9,3663	20,54	3177,2	9,2951
360	32,46	3197,7	9,5315	29,22	3197,6	9,4828	24,35	3197,6	9,3986	20,87	3197,6	9,3274
370	32,98	3218,0	9,5634	29,68	3218,0	9,5148	24,73	3218,0	9,4306	21,20	3218,0	9,3594
380	33,49	3238,5	9,5950	30,14	3238,5	9,5463	25,12	3238,4	9,4621	21,53	3238,4	9,3910
390	34,00	3259,0	9,6261	30,60	3259,0	9,5775	25,50	3259,0	9,4933	21,86	3258,9	9,4221
400	34,51	3279,6	9,6569	31,06	3279,6	9,6083	25,88	3279,6	9,5241	22,19	3279,5	9,4530
410	35,03	3300,3	9,6874	31,52	3300,2	9,6388	26,27	3300,2	9,5546	22,52	3300,2	9,4834
420	35,54	3321,0	9,7175	31,99	3321,0	9,6689	26,65	3320,9	9,5847	22,85	3320,9	9,5135
430	36,05	3341,8	9,7473	32,45	3341,7	9,6986	27,04	3341,7	9,6145	23,18	3341,7	9,5433
440	36,57	3362,6	9,7767	32,91	3362,6	9,7281	27,42	3362,6	9,6439	23,51	3362,5	9,5727
450	37,08	3383,5	9,8058	33,37	3383,5	9,7572	27,81	3383,5	9,6730	23,84	3383,5	9,6019
460	37,59	3404,5	9,8347	33,83	3404,5	9,7860	28,19	3404,5	9,7019	24,16	3404,4	9,6307
470	38,10	3425,6	9,8632	34,29	3425,6	9,8146	28,58	3425,5	9,7304	24,49	3425,5	9,6592
480	38,62	3446,7	9,8914	34,76	3446,7	9,8428	28,96	3446,7	9,7586	24,82	3446,6	9,6875
490	39,13	3467,9	9,9194	35,22	3467,9	9,8707	29,35	3467,8	9,7866	25,15	3467,8	9,7154
500	39,64	3489,1	9,9471	35,68	3489,1	9,8984	29,73	3489,1	9,8143	25,48	3489,1	9,7431
510	40,16	3510,5	9,9745	36,14	3510,5	9,9258	30,12	3510,4	9,8417	25,81	3510,4	9,7705
520	40,67	3531,9	10,0016	36,60	3531,9	9,9530	30,50	3531,8	9,8688	26,14	3531,8	9,7977
530	41,18	3553,3	10,0285	37,06	3553,3	9,9799	30,89	3553,3	9,8957	26,47	3553,3	9,8246
540	41,69	3574,9	10,0552	37,53	3574,9	10,0065	31,27	3574,8	9,9224	26,80	3574,8	9,8512
550	42,21	3596,5	10,0816	37,99	3596,5	10,0329	31,66	3596,4	9,9488	27,13	3596,4	9,8776

Table 3. Water and Superheated Steam (Continuation) Wasser und überhitzter Dampf (Fortsetzung)

t	0,09 bar $t_s = 43,79$ °C			0,10 bar $t_s = 45,83$ °C			0,12 bar $t_s = 49,45$ °C			0,14 bar $t_s = 52,57$ °C		
	v'' 16,20	h'' 2581,1	s'' 8,1881	v'' 14,67	h'' 2584,8	s'' 8,1511	v'' 12,36	h'' 2591,2	s'' 8,0872	v'' 10,69	h'' 2596,7	s'' 8,0334
°C	v	h	s	v	h	s	v	h	s	v	h	s
550	42,21	3596,5	10,0816	37,99	3596,5	10,0329	31,66	3596,4	9,9488	27,13	3596,4	9,8776
560	42,72	3618,1	10,1077	38,45	3618,1	10,0591	32,04	3618,1	9,9749	27,46	3618,1	9,9038
570	43,23	3639,9	10,1337	38,91	3639,9	10,0851	32,42	3639,9	10,0009	27,79	3639,8	9,9297
580	43,75	3661,7	10,1594	39,37	3661,7	10,1108	32,81	3661,7	10,0266	28,12	3661,7	9,9555
590	44,26	3683,6	10,1849	39,83	3683,6	10,1363	33,19	3683,6	10,0521	28,45	3683,5	9,9810
600	44,77	3705,5	10,2102	40,29	3705,5	10,1616	33,58	3705,5	10,0774	28,78	3705,5	10,0062
610	45,29	3727,6	10,2353	40,76	3727,5	10,1866	33,96	3727,5	10,1025	29,11	3727,5	10,0313
620	45,80	3749,6	10,2601	41,22	3749,6	10,2115	34,35	3749,6	10,1273	29,44	3749,6	10,0562
630	46,31	3771,8	10,2848	41,68	3771,8	10,2362	34,73	3771,8	10,1520	29,77	3771,8	10,0809
640	46,82	3794,0	10,3093	42,14	3794,0	10,2606	35,12	3794,0	10,1765	30,10	3794,0	10,1053
650	47,34	3816,3	10,3336	42,60	3816,3	10,2849	35,50	3816,3	10,2008	30,43	3816,3	10,1296
660	47,85	3838,7	10,3577	43,06	3838,7	10,3090	35,89	3838,7	10,2249	30,76	3838,6	10,1537
670	48,36	3861,1	10,3816	43,53	3861,1	10,3329	36,27	3861,1	10,2488	31,09	3861,1	10,1776
680	48,87	3883,6	10,4053	43,99	3883,6	10,3567	36,66	3883,6	10,2725	31,42	3883,6	10,2014
690	49,39	3906,2	10,4288	44,45	3906,2	10,3802	37,04	3906,2	10,2961	31,75	3906,1	10,2249
700	49,90	3928,8	10,4522	44,91	3928,8	10,4036	37,43	3928,8	10,3194	32,08	3928,8	10,2483
710	50,41	3951,5	10,4754	45,37	3951,5	10,4268	37,81	3951,5	10,3426	32,41	3951,5	10,2715
720	50,93	3974,3	10,4985	45,83	3974,3	10,4498	38,19	3974,2	10,3657	32,74	3974,2	10,2945
730	51,44	3997,1	10,5213	46,30	3997,1	10,4727	38,58	3997,1	10,3885	33,07	3997,1	10,3174
740	51,95	4020,0	10,5440	46,76	4020,0	10,4954	38,96	4020,0	10,4113	33,40	4020,0	10,3401
750	52,46	4043,0	10,5666	47,22	4042,9	10,5180	39,35	4042,9	10,4338	33,73	4042,9	10,3627
760	52,98	4066,0	10,5890	47,68	4066,0	10,5404	39,73	4066,0	10,4562	34,06	4065,9	10,3851
770	53,49	4089,1	10,6112	48,14	4089,1	10,5626	40,12	4089,0	10,4784	34,39	4089,0	10,4073
780	54,00	4112,2	10,6333	48,60	4112,2	10,5847	40,50	4112,2	10,5005	34,72	4112,2	10,4294
790	54,52	4135,4	10,6552	49,06	4135,4	10,6066	40,89	4135,4	10,5225	35,05	4135,4	10,4513
800	55,03	4158,7	10,6770	49,53	4158,7	10,6284	41,27	4158,7	10,5443	35,38	4158,7	10,4731

Table 3. Water and Superheated Steam (Continuation) Wasser und überhitzter Dampf (Fortsetzung)

t	0,16 bar $t_s = 55{,}34\ °C$			0,18 bar $t_s = 57{,}83\ °C$			0,20 bar $t_s = 60{,}09\ °C$			0,22 bar $t_s = 62{,}16\ °C$		
	v'' 9,433	h'' 2601,6	s'' 7,9869	v'' 8,445	h'' 2605,9	s'' 7,9460	v'' 7,650	h'' 2609,9	s'' 7,9094	v'' 6,995	h'' 2613,5	s'' 7,8764
°C	v	h	s	v	h	s	v	h	s	v	h	s
0	0,0010002	— 0,0	— 0,0002	0,0010002	— 0,0	— 0,0002	0,0010002	— 0,0	— 0,0002	0,0010002	— 0,0	— 0,0002
10	0,0010002	42,0	0,1510	0,0010002	42,0	0,1510	0,0010002	42,0	0,1510	0,0010002	42,0	0,1510
20	0,0010017	83,9	0,2963	0,0010017	83,9	0,2963	0,0010017	83,9	0,2963	0,0010017	83,9	0,2963
30	0,0010043	125,7	0,4365	0,0010043	125,7	0,4365	0,0010043	125,7	0,4365	0,0010043	125,7	0,4365
40	0,0010078	167,5	0,5721	0,0010078	167,5	0,5721	0,0010078	167,5	0,5721	0,0010078	167,5	0,5721
50	0,0010121	209,3	0,7035	0,0010121	209,3	0,7035	0,0010121	209,3	0,7035	0,0010121	209,3	0,7035
60	9,570	2610,5	8,0137	8,502	2610,1	7,9585	0,0010171	251,1	0,8310	0,0010171	251,1	0,8310
70	9,862	2629,5	8,0701	8,763	2629,2	8,0150	7,883	2628,8	7,9656	7,163	2628,5	7,9208
80	10,155	2648,6	8,1249	9,023	2648,3	8,0699	8,117	2648,0	8,0206	7,376	2647,7	7,9759
90	10,446	2667,7	8,1782	9,282	2667,4	8,1233	8,351	2667,1	8,0740	7,589	2666,9	8,0294
100	10,737	2686,8	8,2301	9,541	2686,5	8,1752	8,585	2686,3	8,1261	7,802	2686,0	8,0816
110	11,028	2705,9	8,2807	9,800	2705,7	8,2258	8,818	2705,5	8,1768	8,014	2705,2	8,1323
120	11,319	2725,0	8,3300	10,059	2724,8	8,2752	9,051	2724,6	8,2262	8,226	2724,4	8,1818
130	11,609	2744,2	8,3781	10,317	2744,0	8,3234	9,283	2743,8	8,2744	8,438	2743,7	8,2301
140	11,899	2763,4	8,4251	10,575	2763,2	8,3705	9,516	2763,1	8,3215	8,649	2762,9	8,2772
150	12,189	2782,6	8,4711	10,833	2782,5	8,4165	9,748	2782,3	8,3676	8,860	2782,2	8,3233
160	12,479	2801,9	8,5161	11,091	2801,8	8,4615	9,980	2801,6	8,4127	9,071	2801,5	8,3684
170	12,768	2821,2	8,5602	11,348	2821,1	8,5056	10,212	2821,0	8,4568	9,282	2820,8	8,4126
180	13,058	2840,5	8,6033	11,606	2840,4	8,5488	10,444	2840,3	8,5000	9,493	2840,2	8,4558
190	13,347	2859,9	8,6457	11,863	2859,8	8,5911	10,675	2859,7	8,5423	9,704	2859,6	8,4982
200	13,636	2879,4	8,6872	12,120	2879,3	8,6326	10,907	2879,2	8,5839	9,914	2879,1	8,5397
210	13,926	2898,8	8,7279	12,377	2898,7	8,6734	11,138	2898,7	8,6246	10,125	2898,6	8,5805
220	14,215	2918,4	8,7679	12,634	2918,3	8,7134	11,370	2918,2	8,6647	10,335	2918,1	8,6206
230	14,504	2937,9	8,8072	12,891	2937,9	8,7527	11,601	2937,8	8,7040	10,546	2937,7	8,6599
240	14,793	2957,6	8,8458	13,148	2957,5	8,7914	11,832	2957,4	8,7426	10,756	2957,4	8,6985
250	15,082	2977,2	8,8838	13,405	2977,2	8,8293	12,064	2977,1	8,7806	10,966	2977,0	8,7366
260	15,371	2997,0	8,9212	13,662	2996,9	8,8667	12,295	2996,9	8,8180	11,176	2996,8	8,7739
270	15,659	3016,8	8,9579	13,919	3016,7	8,9035	12,526	3016,7	8,8548	11,387	3016,6	8,8107
280	15,948	3036,6	8,9941	14,175	3036,6	8,9397	12,757	3036,5	8,8910	11,597	3036,5	8,8470
290	16,237	3056,5	9,0298	14,432	3056,5	8,9754	12,988	3056,4	8,9267	11,807	3056,4	8,8826
300	16,526	3076,5	9,0649	14,689	3076,4	9,0105	13,219	3076,4	8,9618	12,017	3076,3	8,9178
310	16,815	3096,5	9,0996	14,946	3096,5	9,0452	13,450	3096,4	8,9965	12,227	3096,4	8,9524
320	17,103	3116,6	9,1337	15,202	3116,5	9,0793	13,681	3116,5	9,0306	12,437	3116,5	8,9866
330	17,392	3136,7	9,1674	15,459	3136,7	9,1130	13,912	3136,6	9,0643	12,647	3136,6	9,0203
340	17,681	3156,9	9,2006	15,715	3156,9	9,1462	14,143	3156,9	9,0975	12,857	3156,8	9,0535
350	17,969	3177,2	9,2334	15,972	3177,2	9,1790	14,374	3177,1	9,1303	13,067	3177,1	9,0863
360	18,258	3197,5	9,2658	16,229	3197,5	9,2114	14,605	3197,5	9,1627	13,277	3197,4	9,1187
370	18,547	3217,9	9,2977	16,485	3217,9	9,2433	14,836	3217,9	9,1947	13,487	3217,8	9,1506
380	18,835	3238,4	9,3293	16,742	3238,3	9,2749	15,067	3238,3	9,2262	13,697	3238,3	9,1822
390	19,124	3258,9	9,3605	16,998	3258,9	9,3061	15,298	3258,8	9,2574	13,907	3258,8	9,2134
400	19,412	3279,5	9,3913	17,255	3279,5	9,3369	15,529	3279,4	9,2882	14,117	3279,4	9,2442
410	19,701	3200,1	9,4218	17,512	3300,1	9,3674	15,760	3300,1	9,3187	14,327	3300,1	9,2747
420	19,990	3320,9	9,4519	17,768	3320,8	9,3975	15,991	3320,8	9,3488	14,537	3320,8	9,3048
430	20,278	3341,7	9,4816	18,025	3341,6	9,4273	16,222	3341,6	9,3786	14,747	3341,6	9,3346
440	20,567	3362,5	9,5111	18,281	3362,5	9,4567	16,453	3362,5	9,4081	14,957	3362,4	9,3640
450	20,855	3383,4	9,5402	18,538	3383,4	9,4858	16,684	3383,4	9,4372	15,167	3383,4	9,3932
460	21,144	3404,4	9,5690	18,794	3404,4	9,5147	16,914	3404,4	9,4660	15,377	3404,3	9,4220
470	21,433	3425,5	9,5976	19,051	3425,5	9,5432	17,145	3425,4	9,4945	15,586	3425,4	9,4505
480	21,721	3446,6	9,6258	19,307	3446,6	9,5714	17,376	3446,6	9,5228	15,796	3446,5	9,4788
490	22,010	3467,8	9,6538	19,564	3467,8	9,5994	17,607	3467,8	9,5507	16,006	3467,7	9,5067
500	22,298	3489,1	9,6814	19,820	3489,0	9,6271	17,838	3489,0	9,5784	16,216	3489,0	9,5344
510	22,587	3510,4	9,7089	20,077	3510,4	9,6545	18,069	3510,3	9,6058	16,426	3510,3	9,5618
520	22,875	3531,8	9,7360	20,333	3531,8	9,6816	18,300	3531,8	9,6330	16,636	3531,7	9,5890
530	23,164	3553,3	9,7629	20,590	3553,2	9,7085	18,531	3553,2	9,6599	16,846	3553,2	9,6159
540	23,452	3574,8	9,7896	20,846	3574,8	9,7352	18,761	3574,8	9,6865	17,056	3574,7	9,6425
550	23,741	3596,4	9,8160	21,103	3596,4	9,7616	18,992	3596,4	9,7130	17,265	3596,3	9,6690

Table 3. Water and Superheated Steam (Continuation) Wasser und überhitzter Dampf (Fortsetzung)

t	0,16 bar t_s = 55,34 °C			0,18 bar t_s = 57,83 °C			0,20 bar t_s = 60,09 °C			0,22 bar t_s = 62,16 °C		
	v'' 9,433	h'' 2601,6	s'' 7,9869	v'' 8,445	h'' 2605,9	s'' 7,9460	v'' 7,650	h'' 2609,9	s'' 7,9094	v'' 6,995	h'' 2613,5	s'' 7,8764
°C	v	h	s	v	h	s	v	h	s	v	h	s
550	23,741	3596,4	9,8160	21,103	3596,4	9,7616	18,992	3596,4	9,7130	17,265	3596,3	9,6690
560	24,029	3618,1	9,8421	21,359	3618,1	9,7878	19,223	3618,0	9,7391	17,475	3618,0	9,6951
570	24,318	3639,8	9,8681	21,616	3639,8	9,8137	19,454	3639,8	9,7651	17,685	3639,8	9,7211
580	24,606	3661,6	9,8938	21,872	3661,6	9,8394	19,685	3661,6	9,7908	17,895	3661,6	9,7468
590	24,895	3683,5	9,9193	22,129	3683,5	9,8649	19,916	3683,5	9,8163	18,105	3683,5	9,7723
600	25,183	3705,5	9,9446	22,385	3705,5	9,8902	20,146	3705,4	9,8416	18,315	3705,4	9,7976
610	25,472	3727,5	9,9697	22,642	3727,5	9,9153	20,377	3727,5	9,8667	18,525	3727,5	9,8227
620	25,761	3749,6	9,9946	22,898	3749,6	9,9402	20,608	3749,6	9,8915	18,734	3749,5	9,8475
630	26,049	3771,7	10,0192	23,154	3771,7	9,9649	20,839	3771,7	9,9162	18,944	3771,7	9,8722
640	26,338	3794,0	10,0437	23,411	3794,0	9,9893	21,070	3793,9	9,9407	19,154	3793,9	9,8967
650	26,626	3816,3	10,0680	23,667	3816,3	10,0136	21,300	3816,2	9,9650	19,364	3816,2	9,9210
660	26,915	3838,6	10,0921	23,924	3838,6	10,0377	21,531	3838,6	9,9891	19,574	3838,6	9,9451
670	27,203	3861,1	10,1160	24,180	3861,1	10,0616	21,762	3861,0	10,0130	19,784	3861,0	9,9690
680	27,492	3883,6	10,1397	24,437	3883,6	10,0853	21,993	3883,5	10,0367	19,993	3883,5	9,9927
690	27,780	3906,1	10,1633	24,693	3906,1	10,1089	22,224	3906,1	10,0603	20,203	3906,1	10,0163
700	28,069	3928,8	10,1866	24,950	3928,7	10,1323	22,455	3928,7	10,0836	20,413	3928,7	10,0396
710	28,357	3951,5	10,2098	25,206	3951,4	10,1555	22,685	3951,4	10,1068	20,623	3951,4	10,0629
720	28,645	3974,2	10,2329	25,463	3974,2	10,1785	22,916	3974,2	10,1299	20,833	3974,2	10,0859
730	28,934	3997,1	10,2558	25,719	3997,0	10,2014	23,147	3997,0	10,1528	21,043	3997,0	10,1088
740	29,222	4020,0	10,2785	25,975	4019,9	10,2241	23,378	4019,9	10,1755	21,252	4019,9	10,1315
750	29,511	4042,9	10,3010	26,232	4042,9	10,2467	23,609	4042,9	10,1980	21,462	4042,9	10,1540
760	29,799	4065,9	10,3234	26,488	4065,9	10,2691	23,839	4065,9	10,2204	21,672	4065,9	10,1764
770	30,088	4089,0	10,3457	26,745	4089,0	10,2913	24,070	4089,0	10,2427	21,882	4089,0	10,1987
780	30,376	4112,2	10,3677	27,001	4112,2	10,3134	24,301	4112,2	10,2647	22,092	4112,2	10,2208
790	30,665	4135,4	10,3897	27,258	4135,4	10,3353	24,532	4135,4	10,2867	22,301	4135,4	10,2427
800	30,953	4158,7	10,4115	27,514	4158,7	10,3571	24,762	4158,7	10,3085	22,511	4158,6	10,2645

Table 3. Water and Superheated Steam (Continuation) Wasser und überhitzter Dampf (Fortsetzung)

t	0,24 bar $t_s = 64,08$ °C			0,26 bar $t_s = 65,87$ °C			0,28 bar $t_s = 67,55$ °C			0,30 bar $t_s = 69,12$ °C		
	v'' 6,447	h'' 2616,8	s'' 7,8464	v'' 5,980	h'' 2619,9	s'' 7,8188	v'' 5,579	h'' 2622,7	s'' 7,7933	v'' 5,229	h'' 2625,4	s'' 7,7695
°C	v	h	s	v	h	s	v	h	s	v	h	s
0	0,0010002	− 0,0	−0,0002	0,0010002	− 0,0	−0,0002	0,0010002	− 0,0	−0,0002	0,0010002	− 0,0	−0,0002
10	0,0010002	42,0	0,1510	0,0010002	42,0	0,1510	0,0010002	42,0	0,1510	0,0010002	42,0	0,1510
20	0,0010017	83,9	0,2963	0,0010017	83,9	0,2963	0,0010017	83,9	0,2963	0,0010017	83,9	0,2963
30	0,0010043	125,7	0,4365	0,0010043	125,7	0,4365	0,0010043	125,7	0,4365	0,0010043	125,7	0,4365
40	0,0010078	167,5	0,5721	0,0010078	167,5	0,5721	0,0010078	167,5	0,5721	0,0010078	167,5	0,5721
50	0,0010121	209,3	0,7035	0,0010121	209,3	0,7035	0,0010121	209,3	0,7035	0,0010121	209,3	0,7035
60	0,0010171	251,1	0,8310	0,0010171	251,1	0,8310	0,0010171	251,1	0,8310	0,0010171	251,1	0,8310
70	6,563	2628,2	7,8798	6,055	2627,8	7,8421	5,620	2627,5	7,8071	5,243	2627,1	7,7745
80	6,759	2647,4	7,9350	6,237	2647,1	7,8974	5,789	2646,8	7,8625	5,401	2646,5	7,8300
90	6,954	2666,6	7,9887	6,417	2666,3	7,9511	5,957	2666,0	7,9163	5,558	2665,8	7,8839
100	7,150	2685,8	8,0409	6,598	2685,6	8,0034	6,124	2685,3	7,9687	5,714	2685,1	7,9363
110	7,344	2705,0	8,0917	6,778	2704,8	8,0543	6,292	2704,6	8,0196	5,871	2704,3	7,9873
120	7,539	2724,2	8,1412	6,957	2724,0	8,1039	6,459	2723,8	8,0693	6,027	2723,6	8,0370
130	7,733	2743,5	8,1896	7,137	2743,3	8,1523	6,625	2743,1	8,1177	6,182	2742,9	8,0855
140	7,927	2762,8	8,2368	7,316	2762,6	8,1995	6,792	2762,4	8,1650	6,338	2762,3	8,1329
150	8,121	2782,0	8,2829	7,495	2781,9	8,2457	6,958	2781,7	8,2112	6,493	2781,6	8,1791
160	8,314	2801,4	8,3280	7,673	2801,2	8,2908	7,124	2801,1	8,2564	6,648	2801,0	8,2243
170	8,508	2820,7	8,3722	7,852	2820,6	8,3350	7,290	2820,5	8,3006	6,803	2820,3	8,2686
180	8,701	2840,1	8,4154	8,031	2840,0	8,3783	7,456	2839,9	8,3439	6,958	2839,8	8,3119
190	8,894	2859,5	8,4578	8,209	2859,4	8,4207	7,622	2859,3	8,3864	7,113	2859,2	8,3544
200	9,087	2879,0	8,4994	8,387	2878,9	8,4623	7,787	2878,8	8,4280	7,268	2878,7	8,3960
210	9,280	2898,5	8,5402	8,566	2898,4	8,5031	7,953	2898,3	8,4688	7,422	2898,2	8,4368
220	9,473	2918,0	8,5803	8,744	2918,0	8,5432	8,118	2917,9	8,5089	7,577	2917,8	8,4769
230	9,666	2937,6	8,6196	8,922	2937,6	8,5826	8,284	2937,5	8,5483	7,731	2937,4	8,5163
240	9,859	2957,3	8,6583	9,100	2957,2	8,6212	8,449	2957,1	8,5869	7,885	2957,1	8,5550
250	10,052	2977,0	8,6963	9,278	2976,9	8,6593	8,615	2976,9	8,6250	8,040	2976,8	8,5930
260	10,244	2996,7	8,7337	9,456	2996,7	8,6967	8,780	2996,6	8,6624	8,194	2996,6	8,6305
270	10,437	3016,5	8,7705	9,634	3016,5	8,7335	8,945	3016,4	8,6992	8,348	3016,4	8,6673
280	10,630	3036,4	8,8067	9,812	3036,3	8,7697	9,110	3036,3	8,7355	8,502	3036,2	8,7035
290	10,822	3056,3	8,8424	9,989	3056,3	8,8054	9,275	3056,2	8,7711	8,657	3056,2	8,7392
300	11,015	3076,3	8,8776	10,167	3076,2	8,8406	9,441	3076,2	8,8063	8,811	3076,1	8,7744
310	11,208	3096,3	8,9122	10,345	3096,3	8,8752	9,606	3096,2	8,8410	8,965	3096,2	8,8091
320	11,400	3116,4	8,9464	10,523	3116,4	8,9094	9,771	3116,3	8,8751	9,119	3116,3	8,8432
330	11,593	3136,6	8,9801	10,701	3136,5	8,9431	9,936	3136,5	8,9088	9,273	3136,4	8,8769
340	11,785	3156,8	9,0133	10,878	3156,7	8,9763	10,101	3156,7	8,9421	9,427	3156,7	8,9102
350	11,978	3177,0	9,0461	11,056	3177,0	9,0091	10,266	3177,0	8,9749	9,581	3176,9	8,9430
360	12,170	3197,4	9,0785	11,234	3197,4	9,0415	10,431	3197,3	9,0073	9,735	3197,3	8,9754
370	12,363	3217,8	9,1105	11,411	3217,8	9,0735	10,596	3217,7	9,0392	9,889	3217,7	9,0074
380	12,555	3238,2	9,1420	11,589	3238,2	9,1050	10,761	3238,2	9,0708	10,043	3238,2	9,0389
390	12,748	3258,8	9,1732	11,767	3258,7	9,1362	10,926	3258,7	9,1020	10,197	3258,7	9,0701
400	12,940	3279,4	9,2040	11,945	3279,3	9,1671	11,091	3279,3	9,1328	10,351	3279,3	9,1010
410	13,133	3300,0	9,2345	12,122	3300,0	9,1975	11,256	3300,0	9,1633	10,505	3299,9	9,1314
420	13,325	3320,8	9,2646	12,300	3320,7	9,2276	11,421	3320,7	9,1934	10,659	3320,7	9,1615
430	13,518	3341,5	9,2944	12,477	3341,5	9,2574	11,586	3341,5	9,2232	10,813	3341,5	9,1913
440	13,710	3362,4	9,3239	12,655	3362,4	9,2869	11,751	3362,4	9,2527	10,967	3362,3	9,2208
450	13,902	3383,3	9,3530	12,833	3383,3	9,3160	11,916	3383,3	9,2818	11,121	3383,3	9,2499
460	14,095	3404,3	9,3818	13,010	3404,3	9,3449	12,081	3404,3	9,3106	11,275	3404,2	9,2788
470	14,287	3425,4	9,4104	13,188	3425,4	9,3734	12,246	3425,3	9,3392	11,429	3425,3	9,3073
480	14,480	3446,5	9,4386	13,366	3446,5	9,4016	12,411	3446,5	9,3674	11,583	3446,4	9,3355
490	14,672	3467,7	9,4666	13,543	3467,7	9,4296	12,576	3467,7	9,3954	11,737	3467,6	9,3635
500	14,865	3489,0	9,4942	13,721	3489,0	9,4573	12,741	3488,9	9,4231	11,891	3488,9	9,3912
510	15,057	3510,3	9,5217	13,898	3510,3	9,4847	12,906	3510,3	9,4505	12,045	3510,2	9,4186
520	15,249	3531,7	9,5488	14,076	3531,7	9,5119	13,070	3531,7	9,4776	12,199	3531,6	9,4458
530	15,442	3553,2	9,5757	14,254	3553,2	9,5388	13,235	3553,1	9,5045	12,353	3553,1	9,4727
540	15,634	3574,7	9,6024	14,431	3574,7	9,5654	13,400	3574,7	9,5312	12,507	3574,7	9,4993
550	15,826	3596,3	9,6288	14,609	3596,3	9,5918	13,565	3596,3	9,5576	12,661	3596,3	9,5257

Table 3. Water and Superheated Steam (Continuation) Wasser und überhitzter Dampf (Fortsetzung)

	0,24 bar $t_s = 64{,}08\,°C$			0,26 bar $t_s = 65{,}87\,°C$			0,28 bar $t_s = 67{,}55\,°C$			0,30 bar $t_s = 69{,}12\,°C$		
t	v'' 6,447	h'' 2616,8	s'' 7,8464	v'' 5,980	h'' 2619,9	s'' 7,8188	v'' 5,579	h'' 2622,7	s'' 7,7933	v'' 5,229	h'' 2625,4	s'' 7,7695
°C	v	h	s	v	h	s	v	h	s	v	h	s
550	15,826	3596,3	9,6288	14,609	3596,3	9,5918	13,565	3596,3	9,5576	12,661	3596,3	9,5257
560	16,019	3618,0	9,6550	14,786	3618,0	9,6180	13,730	3618,0	9,5838	12,815	3618,0	9,5519
570	16,211	3639,8	9,6809	14,964	3639,7	9,6439	13,895	3639,7	9,6097	12,969	3639,7	9,5779
580	16,404	3661,6	9,7066	15,142	3661,6	9,6697	14,060	3661,5	9,6355	13,122	3661,5	9,6036
590	16,596	3683,5	9,7321	15,319	3683,4	9,6952	14,225	3683,4	9,6610	13,276	3683,4	9,6291
600	16,788	3705,4	9,7574	15,497	3705,4	9,7205	14,390	3705,4	9,6862	13,430	3705,4	9,6544
610	16,981	3727,4	9,7825	15,674	3727,4	9,7455	14,555	3727,4	9,7113	13,584	3727,4	9,6795
620	17,173	3749,5	9,8074	15,852	3749,5	9,7704	14,719	3749,5	9,7362	13,738	3749,5	9,7043
630	17,365	3771,7	9,8320	16,029	3771,7	9,7951	14,884	3771,7	9,7609	13,892	3771,6	9,7290
640	17,558	3793,9	9,8565	16,207	3793,9	9,8196	15,049	3793,9	9,7854	14,046	3793,9	9,7535
650	17,750	3816,2	9,8808	16,385	3816,2	9,8439	15,214	3816,2	9,8096	14,200	3816,2	9,7778
660	17,942	3838,6	9,9049	16,562	3838,6	9,8680	15,379	3838,6	9,8337	14,354	3838,5	9,8019
670	18,135	3861,0	9,9288	16,740	3861,0	9,8919	15,544	3861,0	9,8577	14,508	3861,0	9,8258
680	18,327	3883,5	9,9525	16,917	3883,5	9,9156	15,709	3883,5	9,8814	14,661	3883,5	9,8495
690	18,520	3906,1	9,9761	17,095	3906,1	9,9391	15,874	3906,1	9,9049	14,815	3906,0	9,8731
700	18,712	3928,7	9,9995	17,272	3928,7	9,9625	16,039	3928,7	9,9283	14,969	3928,7	9,8965
710	18,904	3951,4	10,0227	17,450	3951,4	9,9857	16,203	3951,4	9,9515	15,123	3951,4	9,9197
720	19,097	3974,2	10,0457	17,627	3974,2	10,0088	16,368	3974,2	9,9746	15,277	3974,1	9,9427
730	19,289	3997,0	10,0686	17,805	3997,0	10,0316	16,533	3997,0	9,9974	15,431	3997,0	9,9656
740	19,481	4019,9	10,0913	17,983	4019,9	10,0544	16,698	4019,9	10,0202	15,585	4019,9	9,9883
750	19,674	4042,9	10,1139	18,160	4042,9	10,0769	16,863	4042,8	10,0427	15,739	4042,8	10,0109
760	19,866	4065,9	10,1363	18,338	4065,9	10,0993	17,028	4065,9	10,0651	15,893	4065,9	10,0333
770	20,058	4089,0	10,1585	18,515	4089,0	10,1216	17,193	4089,0	10,0873	16,046	4089,0	10,0555
780	20,251	4112,1	10,1806	18,693	4112,1	10,1436	17,357	4112,1	10,1094	16,200	4112,1	10,0776
790	20,443	4135,4	10,2025	18,870	4135,3	10,1656	17,522	4135,3	10,1314	16,354	4135,3	10,0995
800	20,635	4158,6	10,2243	19,048	4158,6	10,1874	17,687	4158,6	10,1532	16,508	4158,6	10,1213

Table 3. Water and Superheated Steam (Continuation) Wasser und überhitzter Dampf (Fortsetzung)

t	0,32 bar t_s = 70,62 °C			0,34 bar t_s = 72,03 °C			0,36 bar t_s = 73,37 °C			0,38 bar t_s = 74,66 °C		
	v'' 4,922	h'' 2628,0	s'' 7,7474	v'' 4,650	h'' 2630,4	s'' 7,7266	v'' 4,408	h'' 2632,6	s'' 7,7070	v'' 4,190	h'' 2634,8	s'' 7,6884
°C	v	h	s	v	h	s	v	h	s	v	h	s
0	0,0010002	—0,0	—0,0002	0,0010002	—0,0	—0,0002	0,0010002	—0,0	—0,0002	0,0010002	—0,0	—0,0002
10	0,0010002	42,0	0,1510	0,0010002	42,0	0,1510	0,0010002	42,0	0,1510	0,0010002	42,0	0,1510
20	0,0010017	83,9	0,2963	0,0010017	83,9	0,2963	0,0010017	83,9	0,2963	0,0010017	83,9	0,2963
30	0,0010043	125,7	0,4365	0,0010043	125,7	0,4365	0,0010043	125,7	0,4365	0,0010043	125,7	0,4365
40	0,0010078	167,5	0,5721	0,0010078	167,5	0,5721	0,0010078	167,5	0,5721	0,0010078	167,5	0,5721
50	0,0010121	209,3	0,7035	0,0010121	209,3	0,7035	0,0010121	209,3	0,7035	0,0010121	209,3	0,7035
60	0,0010171	251,1	0,8310	0,0010171	251,1	0,8310	0,0010171	251,1	0,8310	0,0010171	251,1	0,8310
70	0,0010228	293,0	0,9548	0,0010228	293,0	0,9548	0,0010228	293,0	0,9548	0,0010228	293,0	0,9548
80	5,061	2646,1	7,7995	4,761	2645,8	7,7708	4,495	2645,5	7,7438	4,257	2645,2	7,7181
90	5,209	2665,5	7,8535	4,900	2665,2	7,8249	4,627	2664,9	7,7979	4,382	2664,6	7,7724
100	5,356	2684,8	7,9060	5,039	2684,6	7,8775	4,758	2684,3	7,8506	4,506	2684,1	7,8251
110	5,502	2704,1	7,9571	5,177	2703,9	7,9287	4,888	2703,7	7,9018	4,630	2703,5	7,8764
120	5,649	2723,4	8,0069	5,315	2723,2	7,9785	5,019	2723,0	7,9517	4,753	2722,8	7,9263
130	5,795	2742,8	8,0554	5,453	2742,6	8,0271	5,149	2742,4	8,0003	4,877	2742,2	7,9750
140	5,941	2762,1	8,1028	5,590	2761,9	8,0745	5,279	2761,8	8,0478	5,000	2761,6	8,0225
150	6,086	2781,4	8,1490	5,727	2781,3	8,1208	5,408	2781,1	8,0941	5,123	2781,0	8,0689
160	6,232	2800,8	8,1943	5,864	2800,7	8,1661	5,538	2800,5	8,1394	5,245	2800,4	8,1143
170	6,377	2820,2	8,2386	6,001	2820,1	8,2104	5,667	2820,0	8,1838	5,368	2819,8	8,1586
180	6,522	2839,6	8,2819	6,138	2839,5	8,2537	5,796	2839,4	8,2272	5,491	2839,3	8,2020
190	6,668	2859,1	8,3244	6,275	2859,0	8,2962	5,925	2858,9	8,2697	5,613	2858,8	8,2446
200	6,813	2878,6	8,3660	6,411	2878,5	8,3379	6,054	2878,4	8,3114	5,735	2878,3	8,2863
210	6,958	2898,1	8,4069	6,548	2898,1	8,3788	6,183	2898,0	8,3523	5,857	2897,9	8,3272
220	7,102	2917,7	8,4470	6,684	2917,6	8,4189	6,312	2917,6	8,3924	5,979	2917,5	8,3673
230	7,247	2937,3	8,4864	6,820	2937,3	8,4583	6,441	2937,2	8,4318	6,101	2937,1	8,4068
240	7,392	2957,0	8,5251	6,957	2956,9	8,4970	6,570	2956,9	8,4706	6,223	2956,8	8,4455
250	7,537	2976,7	8,5632	7,093	2976,7	8,5351	6,698	2976,6	8,5086	6,345	2976,5	8,4836
260	7,681	2996,5	8,6006	7,229	2996,4	8,5725	6,827	2996,4	8,5461	6,467	2996,3	8,5210
270	7,826	3016,3	8,6374	7,365	3016,3	8,6094	6,956	3016,2	8,5829	6,589	3016,1	8,5579
280	7,971	3036,2	8,6737	7,501	3036,1	8,6456	7,084	3036,1	8,6192	6,711	3036,0	8,5942
290	8,115	3056,1	8,7094	7,637	3056,1	8,6813	7,213	3056,0	8,6549	6,833	3056,0	8,6299
300	8,260	3076,1	8,7446	7,773	3076,1	8,7165	7,341	3076,0	8,6901	6,955	3076,0	8,6651
310	8,404	3096,1	8,7792	7,909	3096,1	8,7512	7,470	3096,0	8,7248	7,076	3096,0	8,6997
320	8,549	3116,2	8,8134	8,045	3116,2	8,7854	7,598	3116,2	8,7589	7,198	3116,1	8,7339
330	8,693	3136,4	8,8471	8,181	3136,4	8,8191	7,727	3136,3	8,7926	7,320	3136,3	8,7676
340	8,838	3156,6	8,8804	8,317	3156,6	8,8523	7,855	3156,5	8,8259	7,441	3156,5	8,8009
350	8,982	3176,9	8,9132	8,453	3176,9	8,8851	7,984	3176,8	8,8587	7,563	3176,8	8,8337
360	9,127	3197,2	8,9456	8,589	3197,2	8,9175	8,112	3197,2	8,8911	7,685	3197,1	8,8661
370	9,271	3217,6	8,9775	8,725	3217,6	8,9495	8,240	3217,6	8,9231	7,806	3217,5	8,8981
380	9,415	3238,1	9,0091	8,861	3238,1	8,9811	8,369	3238,1	8,9547	7,928	3238,0	8,9297
390	9,560	3258,7	9,0403	8,997	3258,6	9,0123	8,497	3258,6	8,9859	8,050	3258,6	8,9609
400	9,704	3279,2	9,0711	9,133	3279,2	9,0431	8,625	3279,2	9,0167	8,171	3279,2	8,9917
410	9,849	3299,9	9,1016	9,269	3299,9	9,0736	8,754	3299,9	9,0472	8,293	3299,8	9,0222
420	9,993	3320,6	9,1317	9,405	3320,6	9,1037	8,882	3320,6	9,0773	8,414	3320,6	9,0523
430	10,137	3341,4	9,1615	9,541	3341,4	9,1335	9,010	3341,4	9,1071	8,536	3341,4	9,0821
440	10,282	3362,3	9,1910	9,677	3362,3	9,1630	9,139	3362,2	9,1366	8,658	3362,2	9,1116
450	10,426	3383,2	9,2201	9,812	3383,2	9,1921	9,267	3383,2	9,1657	8,779	3383,2	9,1407
460	10,570	3404,2	9,2490	9,948	3404,2	9,2210	9,395	3404,2	9,1945	8,901	3404,1	9,1696
470	10,715	3425,3	9,2775	10,084	3425,3	9,2495	9,524	3425,2	9,2231	9,022	3425,2	9,1981
480	10,859	3446,4	9,3057	10,220	3446,4	9,2777	9,652	3446,4	9,2513	9,144	3446,3	9,2264
490	11,003	3467,6	9,3337	10,356	3467,6	9,3057	9,780	3467,6	9,2793	9,265	3467,6	9,2543
500	11,148	3488,9	9,3614	10,492	3488,9	9,3334	9,909	3488,8	9,3070	9,387	3488,8	9,2820
510	11,292	3510,2	9,3888	10,628	3510,2	9,3608	10,037	3510,2	9,3344	9,509	3510,2	9,3094
520	11,436	3531,6	9,4160	10,763	3531,6	9,3880	10,165	3531,6	9,3616	9,630	3531,6	9,3366
530	11,581	3553,1	9,4429	10,899	3553,1	9,4149	10,294	3553,1	9,3885	9,752	3553,0	9,3635
540	11,725	3574,6	9,4695	11,035	3574,6	9,4415	10,422	3574,6	9,4151	9,873	3574,6	9,3902
550	11,869	3596,3	9,4959	11,171	3596,2	9,4679	10,550	3596,2	9,4415	9,995	3596,2	9,4166

Table 3. Water and Superheated Steam (Continuation) **Wasser und überhitzter Dampf** (Fortsetzung)

t	0,32 bar $t_s = 70,62$ °C			0,34 bar $t_s = 72,03$ °C			0,36 bar $t_s = 73,37$ °C			0,38 bar $t_s = 74,66$ °C		
	v'' 4,922	h'' 2628,0	s'' 7,7474	v'' 4,650	h'' 2630,4	s'' 7,7266	v'' 4,408	h'' 2632,6	s'' 7,7070	v'' 4,190	h'' 2634,8	s'' 7,6884
°C	v	h	s	v	h	s	v	h	s	v	h	s
550	11,869	3596,3	9,4959	11,171	3596,2	9,4679	10,550	3596,2	9,4415	9,995	3596,2	9,4166
560	12,014	3617,9	9,5221	11,307	3617,9	9,4941	10,678	3617,9	9,4677	10,116	3617,9	9,4428
570	12,158	3639,7	9,5481	11,443	3639,7	9,5201	10,807	3639,6	9,4937	10,238	3639,6	9,4687
580	12,302	3661,5	9,5738	11,578	3661,5	9,5458	10,935	3661,5	9,5194	10,359	3661,5	9,4944
590	12,446	3683,4	9,5993	11,714	3683,4	9,5713	11,063	3683,4	9,5449	10,481	3683,3	9,5199
600	12,591	3705,3	9,6246	11,850	3705,3	9,5966	11,192	3705,3	9,5702	10,602	3705,3	9,5452
610	12,735	3727,4	9,6497	11,986	3727,4	9,6217	11,320	3727,3	9,5953	10,724	3727,3	9,5703
620	12,879	3749,5	9,6746	12,122	3749,5	9,6466	11,448	3749,4	9,6202	10,845	3749,4	9,5952
630	13,024	3771,6	9,6992	12,257	3771,6	9,6712	11,576	3771,6	9,6448	10,967	3771,6	9,6199
640	13,168	3793,9	9,7237	12,393	3793,8	9,6957	11,705	3793,8	9,6693	11,088	3793,8	9,6444
650	13,312	3816,2	9,7480	12,529	3816,1	9,7200	11,833	3816,1	9,6936	11,210	3816,1	9,6686
660	13,456	3838,5	9,7721	12,665	3838,5	9,7441	11,961	3838,5	9,7177	11,332	3838,5	9,6927
670	13,601	3861,0	9,7960	12,801	3860,9	9,7680	12,089	3860,9	9,7416	11,453	3860,9	9,7167
680	13,745	3883,5	9,8197	12,936	3883,4	9,7917	12,218	3883,4	9,7654	11,575	3883,4	9,7404
690	13,889	3906,0	9,8433	13,072	3906,0	9,8153	12,346	3906,0	9,7889	11,696	3906,0	9,7639
700	14,034	3928,7	9,8667	13,208	3928,7	9,8387	12,474	3928,6	9,8123	11,818	3928,6	9,7873
710	14,178	3951,4	9,8899	13,344	3951,4	9,8619	12,602	3951,3	9,8355	11,939	3951,3	9,8105
720	14,322	3974,1	9,9129	13,480	3974,1	9,8849	12,731	3974,1	9,8585	12,061	3974,1	9,8336
730	14,466	3997,0	9,9358	13,615	3997,0	9,9078	12,859	3996,9	9,8814	12,182	3996,9	9,8565
740	14,611	4019,9	9,9585	13,751	4019,9	9,9305	12,987	4019,8	9,9041	12,303	4019,8	9,8792
750	14,755	4042,8	9,9811	13,887	4042,8	9,9531	13,115	4042,8	9,9267	12,425	4042,8	9,9017
760	14,899	4065,9	10,0035	14,023	4065,8	9,9755	13,244	4065,8	9,9491	12,546	4065,8	9,9241
770	15,043	4088,9	10,0257	14,158	4088,9	9,9977	13,372	4088,9	9,9713	12,668	4088,9	9,9464
780	15,188	4112,1	10,0478	14,294	4112,1	10,0198	13,500	4112,1	9,9934	12,789	4112,1	9,9685
790	15,332	4135,3	10,0697	14,430	4135,3	10,0417	13,628	4135,3	10,0154	12,911	4135,3	9,9904
800	15,476	4158,6	10,0915	14,566	4158,6	10,0635	13,757	4158,6	10,0372	13,032	4158,6	10,0122

Table 3. Water and Superheated Steam (Continuation) Wasser und überhitzter Dampf (Fortsetzung)

t	0,40 bar $t_s = 75,89$ °C			0,42 bar $t_s = 77,06$ °C			0,44 bar $t_s = 78,19$ °C			0,46 bar $t_s = 79,28$ °C		
	v'' 3,993	h'' 2636,9	s'' 7,6709	v'' 3,815	h'' 2638,9	s'' 7,6542	v'' 3,652	h'' 2640,7	s'' 7,6383	v'' 3,503	h'' 2642,6	s'' 7,6232
°C	v	h	s	v	h	s	v	h	s	v	h	s
0	0,0010002	0,0	−0,0002	0,0010002	0,0	−0,0002	0,0010002	0,0	−0,0002	0,0010002	0,0	−0,0002
10	0,0010002	42,0	0,1510	0,0010002	42,0	0,1510	0,0010002	42,0	0,1510	0,0010002	42,0	0,1510
20	0,0010017	83,9	0,2963	0,0010017	83,9	0,2963	0,0010017	83,9	0,2963	0,0010017	83,9	0,2963
30	0,0010043	125,7	0,4365	0,0010043	125,7	0,4365	0,0010043	125,7	0,4365	0,0010043	125,7	0,4365
40	0,0010078	167,5	0,5721	0,0010078	167,5	0,5721	0,0010078	167,5	0,5721	0,0010078	167,5	0,5721
50	0,0010121	209,3	0,7035	0,0010121	209,3	0,7035	0,0010121	209,3	0,7035	0,0010121	209,3	0,7035
60	0,0010171	251,1	0,8310	0,0010171	251,1	0,8310	0,0010171	251,1	0,8310	0,0010171	251,1	0,8310
70	0,0010228	293,0	0,9548	0,0010228	293,0	0,9548	0,0010228	293,0	0,9548	0,0010228	293,0	0,9548
80	4,0424	2644,9	7,6937	3,8483	2644,6	7,6705	3,6719	2644,3	7,6484	3,5108	2644,0	7,6271
90	4,1610	2664,4	7,7481	3,9615	2664,1	7,7250	3,7801	2663,8	7,7029	3,6144	2663,5	7,6818
100	4,2792	2683,8	7,8009	4,0742	2683,6	7,7779	3,8878	2683,3	7,7559	3,7176	2683,1	7,7348
110	4,3971	2703,2	7,8523	4,1866	2703,0	7,8293	3,9952	2702,8	7,8074	3,8204	2702,6	7,7864
120	4,5146	2722,6	7,9023	4,2986	2722,4	7,8793	4,1022	2722,2	7,8575	3,9229	2722,0	7,8365
130	4,6319	2742,0	7,9510	4,4104	2741,9	7,9281	4,2090	2741,7	7,9063	4,0252	2741,5	7,8854
140	4,7489	2761,4	7,9985	4,5219	2761,3	7,9757	4,3156	2761,1	7,9539	4,1272	2760,9	7,9331
150	4,8658	2780,9	8,0450	4,6333	2780,7	8,0222	4,4219	2780,6	8,0004	4,2290	2780,4	7,9796
160	4,9825	2800,3	8,0903	4,7445	2800,1	8,0676	4,5281	2800,0	8,0459	4,3306	2799,9	8,0251
170	5,0990	2819,7	8,1347	4,8555	2819,6	8,1120	4,6342	2819,5	8,0903	4,4321	2819,4	8,0696
180	5,2154	2839,2	8,1782	4,9664	2839,1	8,1555	4,7401	2839,0	8,1338	4,5334	2838,9	8,1131
190	5,3317	2858,7	8,2207	5,0772	2858,6	8,1980	4,8459	2858,5	8,1764	4,6346	2858,4	8,1557
200	5,4478	2878,2	8,2625	5,1879	2878,1	8,2398	4,9516	2878,0	8,2182	4,7358	2877,9	8,1975
210	5,5639	2897,8	8,3034	5,2985	2897,7	8,2807	5,0572	2897,6	8,2591	4,8368	2897,5	8,2385
220	5,6800	2917,4	8,3435	5,4090	2917,3	8,3209	5,1627	2917,2	8,2993	4,9378	2917,2	8,2787
230	5,7959	2937,0	8,3830	5,5195	2937,0	8,3603	5,2682	2936,9	8,3388	5,0387	2936,8	8,3181
240	5,9118	2956,7	8,4217	5,6299	2956,7	8,3991	5,3736	2956,6	8,3775	5,1396	2956,5	8,3569
250	6,0277	2976,5	8,4598	5,7403	2976,4	8,4372	5,4790	2976,3	8,4156	5,2404	2976,3	8,3950
260	6,1435	2996,3	8,4973	5,8506	2996,2	8,4747	5,5843	2996,1	8,4531	5,3412	2996,1	8,4325
270	6,2593	3016,1	8,5341	5,9609	3016,0	8,5115	5,6896	3016,0	8,4900	5,4419	3015,9	8,4694
280	6,3751	3036,0	8,5704	6,0712	3035,9	8,5478	5,7949	3035,9	8,5263	5,5426	3035,8	8,5057
290	6,4908	3055,9	8,6061	6,1814	3055,9	8,5836	5,9001	3055,8	8,5620	5,6433	3055,8	8,5414
300	6,6065	3075,9	8,6413	6,2916	3075,9	8,6188	6,0053	3075,8	8,5972	5,7439	3075,8	8,5767
310	6,7221	3096,0	8,6760	6,4018	3095,9	8,6534	6,1105	3095,9	8,6319	5,8445	3095,8	8,6113
320	6,8378	3116,1	8,7102	6,5119	3116,0	8,6876	6,2156	3116,0	8,6661	5,9451	3115,9	8,6455
330	6,9534	3136,2	8,7439	6,6220	3136,2	8,7214	6,3208	3136,2	8,6998	6,0457	3136,1	8,6793
340	7,0690	3156,5	8,7772	6,7322	3156,4	8,7546	6,4259	3156,4	8,7331	6,1463	3156,3	8,7125
350	7,1846	3176,8	8,8100	6,8423	3176,7	8,7874	6,5310	3176,7	8,7659	6,2469	3176,6	8,7454
360	7,3002	3197,1	8,8424	6,9524	3197,1	8,8198	6,6361	3197,0	8,7983	6,3474	3197,0	8,7778
370	7,4158	3217,5	8,8744	7,0625	3217,5	8,8518	6,7412	3217,4	8,8303	6,4479	3217,4	8,8098
380	7,5314	3238,0	8,9060	7,1725	3238,0	8,8834	6,8463	3237,9	8,8619	6,5484	3237,9	8,8414
390	7,6469	3258,5	8,9372	7,2826	3258,5	8,9146	6,9514	3258,5	8,8931	6,6489	3258,4	8,8726
400	7,7625	3279,1	8,9680	7,3926	3279,1	8,9455	7,0564	3279,1	8,9240	6,7494	3279,0	8,9034
410	7,8780	3299,8	8,9985	7,5027	3299,8	8,9760	7,1615	3299,7	8,9545	6,8499	3299,7	8,9339
420	7,9935	3320,5	9,0286	7,6127	3320,5	9,0061	7,2665	3320,5	8,9846	6,9504	3320,4	8,9640
430	8,1091	3341,3	9,0584	7,7227	3341,3	9,0359	7,3715	3341,3	9,0144	7,0509	3341,2	8,9938
440	8,2246	3362,2	9,0879	7,8328	3362,2	9,0653	7,4766	3362,1	9,0438	7,1513	3362,1	9,0233
450	8,3401	3383,1	9,1170	7,9428	3383,1	9,0945	7,5816	3383,1	9,0730	7,2518	3383,0	9,0525
460	8,4556	3404,1	9,1459	8,0528	3404,1	9,1233	7,6866	3404,1	9,1018	7,3523	3404,1	9,0813
470	8,5711	3425,2	9,1744	8,1628	3425,2	9,1519	7,7916	3425,1	9,1304	7,4527	3425,1	9,1098
480	8,6866	3446,3	9,2027	8,2728	3446,3	9,1801	7,8966	3446,3	9,1586	7,5532	3446,3	9,1381
490	8,8021	3467,5	9,2306	8,3828	3467,5	9,2081	8,0016	3467,5	9,1866	7,6536	3467,5	9,1661
500	8,9176	3488,8	9,2583	8,4928	3488,8	9,2358	8,1066	3488,8	9,2143	7,7540	3488,7	9,1938
510	9,0330	3510,1	9,2857	8,6028	3510,1	9,2632	8,2116	3510,1	9,2417	7,8545	3510,1	9,2212
520	9,1485	3531,5	9,3129	8,7127	3531,5	9,2904	8,3166	3531,5	9,2689	7,9549	3531,5	9,2483
530	9,2640	3553,0	9,3398	8,8227	3553,0	9,3173	8,4216	3553,0	9,2958	8,0553	3553,0	9,2753
540	9,3795	3574,6	9,3665	8,9327	3574,5	9,3439	8,5266	3574,5	9,3225	8,1557	3574,5	9,3019
550	9,4949	3596,2	9,3929	9,0427	3596,2	9,3704	8,6315	3596,1	9,3489	8,2561	3596,1	9,3283

Table 3. Water and Superheated Steam (Continuation) Wasser und überhitzter Dampf (Fortsetzung)

t	0,40 bar $t_s = 75,89$ °C			0,42 bar $t_s = 77,06$ °C			0,44 bar $t_s = 78,19$ °C			0,46 bar $t_s = 79,28$ °C		
	v'' 3,993	h'' 2636,7	s'' 7,6709	v'' 3,815	h'' 2638,9	s'' 7,6542	v'' 3,652	h'' 2640,7	s'' 7,6383	v'' 3,503	h'' 2642,6	s'' 7,6232
°C	v	h	s	v	h	s	v	h	s	v	h	s
550	9,4949	3596,2	9,3929	9,0427	3596,2	9,3704	8,6315	3596,1	9,3489	8,2561	3596,1	9,3283
560	9,6104	3617,9	9,4191	9,1526	3617,8	9,3965	8,7365	3617,8	9,3751	8,3566	3617,8	9,3545
570	9,7258	3639,6	9,4450	9,2626	3639,6	9,4225	8,8415	3639,6	9,4010	8,4570	3639,6	9,3805
580	9,8413	3661,4	9,4708	9,3726	3661,4	9,4482	8,9464	3661,4	9,4267	8,5574	3661,4	9,4062
590	9,9567	3683,3	9,4963	9,4825	3683,3	9,4737	9,0514	3683,3	9,4522	8,6578	3683,3	9,4317
600	10,072	3705,3	9,5216	9,5925	3705,3	9,4990	9,1564	3705,2	9,4775	8,7582	3705,2	9,4570
610	10,188	3727,3	9,5466	9,7024	3727,3	9,5241	9,2613	3727,3	9,5026	8,8586	3727,3	9,4821
620	10,303	3749,4	9,5715	9,8124	3749,4	9,5490	9,3663	3749,4	9,5275	8,9590	3749,4	9,5070
630	10,419	3771,6	9,5962	9,9223	3771,6	9,5737	9,4712	3771,5	9,5522	9,0594	3771,5	9,5317
640	10,534	3793,8	9,6207	10,032	3793,8	9,5981	9,5762	3793,8	9,5767	9,1598	3793,8	9,5561
650	10,649	3816,1	9,6450	10,142	3816,1	9,6224	9,6811	3816,1	9,6010	9,2601	3816,1	9,5804
660	10,765	3838,5	9,6691	10,252	3838,5	9,6465	9,7861	3838,4	9,6251	9,3605	3838,4	9,6045
670	10,880	3860,9	9,6930	10,362	3860,9	9,6704	9,891	3860,9	9,6490	9,4609	3860,9	9,6284
680	10,996	3883,4	9,7167	10,472	3883,4	9,6942	9,996	3883,4	9,6727	9,5613	3883,4	9,6522
690	11,111	3906,0	9,7403	10,582	3906,0	9,7177	10,101	3906,0	9,6963	9,6617	3905,9	9,6757
700	11,227	3928,6	9,7636	10,692	3928,6	9,7411	10,206	3928,6	9,7196	9,7621	3928,6	9,6991
710	11,342	3951,3	9,7869	10,802	3951,3	9,7643	10,311	3951,3	9,7429	9,8624	3951,3	9,7223
720	11,457	3974,1	9,8099	10,912	3974,1	9,7874	10,416	3974,1	9,7659	9,9628	3974,1	9,7454
730	11,573	3996,9	9,8328	11,022	3996,9	9,8103	10,521	3996,9	9,7888	10,063	3996,9	9,7682
740	11,688	4019,8	9,8555	11,132	4019,8	9,8330	10,626	4019,8	9,8115	10,164	4019,8	9,7910
750	11,804	4042,8	9,8780	11,242	4042,8	9,8555	10,731	4042,8	9,8340	10,264	4042,8	9,8135
760	11,919	4065,8	9,9004	11,351	4065,8	9,8779	10,835	4065,8	9,8564	10,364	4065,8	9,8359
770	12,035	4088,9	9,9227	11,461	4088,9	9,9002	10,940	4088,9	9,8787	10,465	4088,9	9,8582
780	12,150	4112,1	9,9448	11,571	4112,1	9,9223	11,045	4112,0	9,9008	10,565	4112,0	9,8803
790	12,265	4135,3	9,9667	11,681	4135,3	9,9442	11,150	4135,3	9,9227	10,665	4135,3	9,9022
800	12,381	4158,6	9,9885	11,791	4158,6	9,9660	11,255	4158,5	9,9445	10,766	4158,5	9,9240

Table 3. Water and Superheated Steam (Continuation) Wasser und überhitzter Dampf (Fortsetzung)

t	0,48 bar $t_s = 80,33$ °C			0,50 bar $t_s = 81,35$ °C			0,55 bar $t_s = 83,74$ °C			0,60 bar $t_s = 85,95$ °C		
	v'' 3,366	h'' 2644,3	s'' 7,6086	v'' 3,240	h'' 2646,0	s'' 7,5947	v'' 2,964	h'' 2649,9	s'' 7,5623	v'' 2,732	h'' 2653,6	s'' 7,5327
°C	v	h	s	v	h	s	v	h	s	v	h	s
0	0,0010002	0,0	−0,0002	0,0010002	0,0	−0,0002	0,0010002	0,0	−0,0002	0,0010002	0,0	−0,0002
10	0,0010002	42,0	0,1510	0,0010002	42,0	0,1510	0,0010002	42,0	0,1510	0,0010002	42,1	0,1510
20	0,0010017	83,9	0,2963	0,0010017	83,9	0,2963	0,0010017	83,9	0,2963	0,0010017	83,9	0,2963
30	0,0010043	125,7	0,4365	0,0010043	125,7	0,4365	0,0010043	125,7	0,4365	0,0010043	125,7	0,4365
40	0,0010078	167,5	0,5721	0,0010078	167,5	0,5721	0,0010078	167,5	0,5721	0,0010078	167,5	0,5721
50	0,0010121	209,3	0,7035	0,0010121	209,3	0,7035	0,0010121	209,3	0,7035	0,0010121	209,3	0,7035
60	0,0010171	251,1	0,8310	0,0010171	251,1	0,8310	0,0010171	251,1	0,8310	0,0010171	251,1	0,8310
70	0,0010228	293,0	0,9548	0,0010228	293,0	0,9548	0,0010228	293,0	0,9548	0,0010228	293,0	0,9548
80	0,0010292	334,9	1,0753	0,0010292	334,9	1,0753	0,0010292	334,9	1,0753	0,0010292	334,9	1,0752
90	3,4626	2663,3	7,6615	3,323	2663,0	7,6421	3,018	2662,3	7,5966	2,764	2661,6	7,5549
100	3,5616	2682,8	7,7146	3,418	2682,6	7,6953	3,105	2681,9	7,6499	2,844	2681,3	7,6085
110	3,6602	2702,3	7,7663	3,513	2702,1	7,7470	3,191	2701,5	7,7018	2,923	2701,0	7,6605
120	3,7586	2721,8	7,8165	3,607	2721,6	7,7972	3,277	2721,1	7,7522	3,002	2720,6	7,7111
130	3,8566	2741,3	7,8654	3,702	2741,1	7,8462	3,363	2740,7	7,8013	3,081	2740,2	7,7603
140	3,9544	2760,8	7,9131	3,796	2760,6	7,8940	3,449	2760,2	7,8492	3,160	2759,8	7,8083
150	4,0521	2780,3	7,9597	3,889	2780,1	7,9406	3,534	2779,7	7,8959	3,238	2779,4	7,8551
160	4,1495	2799,7	8,0052	3,983	2799,6	7,9861	3,619	2799,3	7,9415	3,317	2798,9	7,9008
170	4,2468	2819,2	8,0497	4,076	2819,1	8,0307	3,705	2818,8	7,9861	3,395	2818,5	7,9454
180	4,3440	2838,7	8,0933	4,170	2838,6	8,0742	3,789	2838,3	8,0298	3,473	2838,1	7,9891
190	4,4410	2858,3	8,1359	4,263	2858,2	8,1169	3,874	2857,9	8,0725	3,550	2857,7	8,0319
200	4,5380	2877,8	8,1777	4,356	2877,7	8,1587	3,959	2877,5	8,1143	3,628	2877,3	8,0738
210	4,6349	2897,4	8,2187	4,449	2897,4	8,1997	4,044	2897,1	8,1554	3,706	2896,9	8,1149
220	4,7317	2917,1	8,2589	4,542	2917,0	8,2399	4,128	2916,8	8,1956	3,783	2916,6	8,1552
230	4,8284	2936,7	8,2984	4,635	2936,7	8,2794	4,213	2936,5	8,2352	3,861	2936,3	8,1947
240	4,9251	2956,5	8,3372	4,728	2956,4	8,3182	4,297	2956,2	8,2740	3,938	2956,0	8,2336
250	5,0217	2976,2	8,3753	4,821	2976,1	8,3564	4,382	2976,0	8,3122	4,016	2975,8	8,2718
260	5,1183	2996,0	8,4128	4,913	2995,9	8,3939	4,466	2995,8	8,3497	4,093	2995,6	8,3093
270	5,2149	3015,9	8,4497	5,006	3015,8	8,4308	4,550	3015,7	8,3866	4,170	3015,5	8,3462
280	5,3114	3035,8	8,4860	5,099	3035,7	8,4671	4,634	3035,6	8,4229	4,248	3035,4	8,3826
290	5,4079	3055,7	8,5217	5,191	3055,7	8,5028	4,719	3055,5	8,4587	4,325	3055,4	8,4184
300	5,5043	3075,7	8,5569	5,284	3075,7	8,5380	4,803	3075,6	8,4939	4,402	3075,4	8,4536
310	5,6008	3095,8	8,5916	5,376	3095,7	8,5728	4,887	3095,6	8,5286	4,479	3095,5	8,4883
320	5,6972	3115,9	8,6259	5,469	3115,9	8,6070	4,971	3115,7	8,5628	4,557	3115,6	8,5226
330	5,7936	3136,1	8,6596	5,562	3136,0	8,6407	5,056	3135,9	8,5966	4,634	3135,8	8,5563
340	5,8900	3156,3	8,6929	5,654	3156,3	8,6740	5,140	3156,2	8,6299	4,711	3156,1	8,5896
350	5,9864	3176,6	8,7257	5,747	3176,6	8,7068	5,224	3176,5	8,6627	4,788	3176,4	8,6224
360	6,0827	3197,0	8,7581	5,839	3196,9	8,7392	5,308	3196,8	8,6951	4,865	3196,7	8,6549
370	6,1791	3217,4	8,7901	5,932	3217,3	8,7712	5,392	3217,3	8,7271	4,942	3217,2	8,6869
380	6,2754	3237,9	8,8217	6,024	3237,8	8,8028	5,476	3237,7	8,7587	5,019	3237,7	8,7185
390	6,3717	3258,4	8,8529	6,117	3258,4	8,8340	5,560	3258,3	8,7900	5,097	3258,2	8,7497
400	6,4680	3279,0	8,8837	6,209	3279,0	8,8649	5,644	3278,9	8,8208	5,174	3278,8	8,7806
410	6,5643	3299,7	8,9142	6,302	3299,6	8,8954	5,728	3299,6	8,8513	5,251	3299,5	8,8111
420	6,6606	3320,4	8,9444	6,394	3320,4	8,9255	5,812	3320,3	8,8814	5,328	3320,2	8,8412
430	6,7569	3341,2	8,9742	6,487	3341,2	8,9553	5,896	3341,1	8,9112	5,405	3341,1	8,8710
440	6,8532	3362,1	9,0036	6,579	3362,1	8,9848	5,981	3362,0	8,9407	5,482	3361,9	8,9005
450	6,9495	3383,0	9,0328	6,671	3383,0	9,0139	6,065	3382,9	8,9699	5,559	3382,9	8,9296
460	7,0458	3404,0	9,0616	6,764	3404,0	9,0428	6,149	3403,9	8,9987	5,636	3403,9	8,9585
470	7,1420	3425,1	9,0902	6,856	3425,1	9,0713	6,233	3425,0	9,0273	5,713	3425,0	8,9871
480	7,2383	3446,2	9,1184	6,949	3446,2	9,0996	6,317	3446,2	9,0555	5,790	3446,1	9,0153
490	7,3346	3467,4	9,1464	7,041	3467,4	9,1275	6,401	3467,4	9,0835	5,867	3467,3	9,0433
500	7,4308	3488,7	9,1741	7,133	3488,7	9,1552	6,485	3488,6	9,1112	5,944	3488,6	9,0710
510	7,5271	3510,1	9,2015	7,226	3510,0	9,1827	6,569	3510,0	9,1386	6,021	3509,9	9,0984
520	7,6233	3531,5	9,2287	7,318	3531,4	9,2098	6,653	3531,4	9,1658	6,098	3531,3	9,1256
530	7,7196	3552,9	9,2556	7,411	3552,9	9,2367	6,737	3552,9	9,1927	6,175	3552,8	9,1525
540	7,8158	3574,5	9,2823	7,503	3574,5	9,2634	6,821	3574,4	9,2194	6,252	3574,4	9,1792
550	7,9120	3596,1	9,3087	7,595	3596,1	9,2898	6,905	3596,1	9,2458	6,329	3596,0	9,2056

Table 3. Water and Superheated Steam (Continuation) Wasser und überhitzter Dampf (Fortsetzung)

t	0,48 bar $t_s = 80,33\,°C$			0,50 bar $t_s = 81,35\,°C$			0,55 bar $t_s = 83,74\,°C$			0,60 bar $t_s = 85,95\,°C$		
	v'' 3,366	h'' 2644,3	s'' 7,6086	v'' 3,240	h'' 2646,0	s'' 7,5947	v'' 2,964	h'' 2649,9	s'' 7,5623	v'' 2,732	h'' 2653,6	s'' 7,5327
°C	v	h	s	v	h	s	v	h	s	v	h	s
550	7,9120	3596,1	9,3087	7,595	3596,1	9,2898	6,905	3596,0	9,2458	6,329	3596,0	9,2056
560	8,0083	3617,8	9,3349	7,688	3617,8	9,3160	6,989	3617,7	9,2720	6,406	3617,7	9,2318
570	8,1045	3639,5	9,3608	7,780	3639,5	9,3420	7,073	3639,5	9,2979	6,483	3639,4	9,2577
580	8,2007	3661,4	9,3865	7,873	3661,3	9,3677	7,157	3661,3	9,3237	6,560	3661,3	9,2835
590	8,2969	3683,3	9,4121	7,965	3683,2	9,3932	7,241	3683,2	9,3492	6,637	3683,2	9,3090
600	8,3932	3705,2	9,4374	8,057	3705,2	9,4185	7,325	3705,2	9,3745	6,714	3705,1	9,3343
610	8,4894	3727,2	9,4624	8,150	3727,2	9,4436	7,409	3727,2	9,3996	6,791	3727,1	9,3594
620	8,5856	3749,3	9,4873	8,242	3749,3	9,4685	7,493	3749,3	9,4244	6,868	3749,2	9,3843
630	8,6818	3771,5	9,5120	8,334	3771,5	9,4931	7,577	3771,5	9,4491	6,945	3771,4	9,4089
640	8,7780	3793,7	9,5365	8,427	3793,7	9,5176	7,661	3793,7	9,4736	7,022	3793,7	9,4334
650	8,8742	3816,0	9,5608	8,519	3816,0	9,5419	7,745	3816,0	9,4979	7,099	3816,0	9,4577
660	8,9704	3838,4	9,5849	8,612	3838,4	9,5660	7,829	3838,4	9,5220	7,176	3838,3	9,4818
670	9,0666	3860,8	9,6088	8,704	3860,8	9,5899	7,913	3860,8	9,5459	7,253	3860,8	9,5057
680	9,1628	3883,4	9,6325	8,796	3883,3	9,6137	7,996	3883,3	9,5697	7,330	3883,3	9,5295
690	9,2590	3905,9	9,6561	8,889	3905,9	9,6372	8,080	3905,9	9,5932	7,407	3905,8	9,5530
700	9,3552	3928,6	9,6795	8,981	3928,6	9,6606	8,164	3928,5	9,6166	7,484	3928,5	9,5764
710	9,4514	3951,3	9,7027	9,073	3951,3	9,6838	8,248	3951,2	9,6398	7,561	3951,2	9,5996
720	9,5476	3974,0	9,7257	9,166	3974,0	9,7069	8,332	3974,0	9,6629	7,638	3974,0	9,6227
730	9,6438	3996,9	9,7486	9,258	3996,9	9,7297	8,416	3996,8	9,6857	7,715	3996,8	9,6456
740	9,7400	4019,8	9,7713	9,350	4019,8	9,7525	8,500	4019,7	9,7085	7,792	4019,7	9,6683
750	9,8362	4042,7	9,7939	9,443	4042,7	9,7750	8,584	4042,7	9,7310	7,869	4042,7	9,6908
760	9,9324	4065,8	9,8163	9,535	4065,8	9,7974	8,668	4065,7	9,7534	7,946	4065,7	9,7132
770	10,029	4088,9	9,8385	9,627	4088,9	9,8197	8,752	4088,8	9,7757	8,023	4088,8	9,7355
780	10,125	4112,0	9,8606	9,720	4112,0	9,8418	8,836	4112,0	9,7978	8,100	4112,0	9,7576
790	10,221	4135,2	9,8826	9,812	4135,2	9,8637	8,920	4135,2	9,8197	8,177	4135,2	9,7795
800	10,317	4158,5	9,9043	9,904	4158,5	9,8855	9,004	4158,5	9,8415	8,254	4158,5	9,8013

Table 3. Water and Superheated Steam (Continuation) Wasser und überhitzter Dampf (Fortsetzung)

t	0,65 bar $t_s = 88,02$ °C			0,70 bar $t_s = 89,96$ °C			0,75 bar $t_s = 91,79$ °C			0,80 bar $t_s = 93,51$ °C		
	v''	h''	s''	v''	h''	s''	v''	h''	s''	v''	h''	s''
	2,535	2656,9	7,5055	2,365	2660,1	7,4804	2,217	2663,0	7,4570	2,087	2665,8	7,4352
°C	v	h	s	v	h	s	v	h	s	v	h	s
0	0,0010002	0,0	−0,0002	0,0010002	0,0	−0,0002	0,0010002	0,0	−0,0001	0,0010002	0,0	−0,0001
10	0,0010002	42,1	0,1510	0,0010002	42,1	0,1510	0,0010002	42,1	0,1510	0,0010002	42,1	0,1510
20	0,0010017	83,9	0,2963	0,0010017	83,9	0,2963	0,0010017	83,9	0,2963	0,0010017	83,9	0,2963
30	0,0010043	125,7	0,4365	0,0010043	125,7	0,4365	0,0010043	125,7	0,4365	0,0010043	125,7	0,4365
40	0,0010078	167,5	0,5721	0,0010078	167,5	0,5721	0,0010078	167,5	0,5721	0,0010078	167,5	0,5721
50	0,0010121	209,3	0,7035	0,0010121	209,3	0,7035	0,0010121	209,3	0,7035	0,0010121	209,3	0,7035
60	0,0010171	251,1	0,8310	0,0010171	251,1	0,8310	0,0010171	251,1	0,8310	0,0010171	251,1	0,8310
70	0,0010228	293,0	0,9548	0,0010228	293,0	0,9548	0,0010228	293,0	0,9548	0,0010228	293,0	0,9548
80	0,0010292	334,9	1,0752	0,0010292	334,9	1,0752	0,0010292	334,9	1,0752	0,0010292	334,9	1,0752
90	2,549	2660,9	7,5164	2,365	2660,2	7,4806	0,0010361	376,9	1,1925	0,0010361	376,9	1,1925
100	2,623	2680,7	7,5702	2,434	2680,0	7,5346	2,270	2679,4	7,5014	2,126	2678,8	7,4703
110	2,697	2700,4	7,6224	2,502	2699,8	7,5870	2,334	2699,3	7,5540	2,186	2698,7	7,5230
120	2,770	2720,1	7,6731	2,570	2719,6	7,6379	2,398	2719,1	7,6050	2,246	2718,6	7,5742
130	2,843	2739,7	7,7225	2,638	2739,3	7,6874	2,461	2738,8	7,6546	2,306	2738,4	7,6239
140	2,915	2759,4	7,7705	2,706	2759,0	7,7356	2,524	2758,5	7,7029	2,365	2758,1	7,6723
150	2,988	2779,0	7,8174	2,773	2778,6	7,7825	2,587	2778,2	7,7500	2,425	2777,8	7,7195
160	3,060	2798,6	7,8632	2,841	2798,2	7,8284	2,650	2797,9	7,7959	2,484	2797,5	7,7655
170	3,132	2818,2	7,9079	2,908	2817,9	7,8732	2,713	2817,6	7,8408	2,542	2817,2	7,8105
180	3,204	2837,8	7,9517	2,975	2837,5	7,9170	2,775	2837,2	7,8847	2,601	2836,9	7,8544
190	3,276	2857,4	7,9945	3,041	2857,1	7,9599	2,838	2856,9	7,9276	2,660	2856,6	7,8974
200	3,348	2877,0	8,0365	3,108	2876,8	8,0019	2,900	2876,6	7,9697	2,718	2876,3	7,9395
210	3,420	2896,7	8,0776	3,175	2896,5	8,0430	2,962	2896,3	8,0108	2,777	2896,0	7,9807
220	3,492	2916,4	8,1179	3,241	2916,2	8,0834	3,025	2916,0	8,0513	2,835	2915,8	8,0212
230	3,563	2936,1	8,1575	3,308	2935,9	8,1230	3,087	2935,7	8,0909	2,893	2935,5	8,0680
240	3,635	2955,9	8,1964	3,374	2955,7	8,1619	3,149	2955,5	8,1298	2,952	2955,3	8,0998
250	3,706	2975,7	8,2346	3,441	2975,5	8,2002	3,211	2975,3	8,1681	3,010	2975,2	8,1381
260	3,778	2995,5	8,2722	3,507	2995,3	8,2377	3,273	2995,2	8,2057	3,068	2995,0	8,1757
270	3,849	3015,4	8,3091	3,574	3015,2	8,2747	3,335	3015,1	8,2427	3,126	3014,9	8,2127
280	3,920	3035,3	8,3455	3,640	3035,2	8,3111	3,397	3035,0	8,2791	3,184	3034,9	8,2491
290	3,992	3055,3	8,3813	3,706	3055,2	8,3469	3,459	3055,0	8,3149	3,242	3054,9	8,2849
300	4,063	3075,3	8,4165	3,772	3075,2	8,3822	3,520	3075,1	8,3502	3,300	3075,0	8,3202
310	4,134	3095,4	8,4512	3,839	3095,3	8,4169	3,582	3095,2	8,3849	3,358	3095,1	8,3550
320	4,206	3115,5	8,4855	3,905	3115,4	8,4511	3,644	3115,3	8,4192	3,416	3115,2	8,3893
330	4,277	3135,7	8,5192	3,971	3135,6	8,4849	3,706	3135,5	8,4530	3,474	3135,4	8,4230
340	4,348	3156,0	8,5525	4,037	3155,9	8,5182	3,768	3155,8	8,4863	3,532	3155,7	8,4564
350	4,419	3176,3	8,5854	4,103	3176,2	8,5511	3,829	3176,1	8,5191	3,590	3176,0	8,4892
360	4,491	3196,7	8,6178	4,169	3196,6	8,5835	3,891	3196,5	8,5516	3,648	3196,4	8,5217
370	4,562	3217,1	8,6498	4,236	3217,0	8,6155	3,953	3216,9	8,5836	3,706	3216,8	8,5537
380	4,633	3237,6	8,6815	4,302	3237,5	8,6472	4,015	3237,4	8,6152	3,763	3237,3	8,5854
390	4,704	3258,1	8,7127	4,368	3258,1	8,6784	4,076	3258,0	8,6465	3,821	3257,9	8,6166
400	4,775	3278,7	8,7435	4,434	3278,7	8,7093	4,138	3278,6	8,6773	3,879	3278,5	8,6475
410	4,846	3299,4	8,7740	4,500	3299,4	8,7398	4,200	3299,3	8,7078	3,937	3299,2	8,6780
420	4,918	3320,2	8,8042	4,566	3320,1	8,7699	4,261	3320,0	8,7380	3,995	3320,0	8,7081
430	4,989	3341,0	8,8340	4,632	3340,9	8,7997	4,323	3340,8	8,7678	4,053	3340,8	8,7380
440	5,060	3361,9	8,8635	4,698	3361,8	8,8292	4,385	3361,7	8,7973	4,111	3361,7	8,7675
450	5,131	3382,8	8,8926	4,764	3382,7	8,8584	4,446	3382,7	8,8265	4,168	3382,6	8,7966
460	5,202	3403,8	8,9215	4,830	3403,8	8,8872	4,508	3403,7	8,8553	4,226	3403,6	8,8255
470	5,273	3424,9	8,9501	4,896	3424,8	8,9158	4,570	3424,8	8,8839	4,284	3424,7	8,8540
480	5,344	3446,0	8,9783	4,962	3446,0	8,9441	4,631	3445,9	8,9122	4,342	3445,9	8,8823
490	5,416	3467,2	9,0063	5,028	3467,2	8,9720	4,693	3467,1	8,9401	4,400	3467,1	8,9103
500	5,487	3488,5	9,0340	5,095	3488,5	8,9997	4,755	3488,4	8,9678	4,457	3488,4	8,9380
510	5,558	3509,9	9,0614	5,161	3509,8	9,0272	4,816	3509,8	8,9953	4,515	3509,7	8,9654
520	5,629	3531,3	9,0886	5,227	3531,2	9,0543	4,878	3531,2	9,0225	4,573	3531,1	8,9926
530	5,700	3552,8	9,1155	5,293	3552,7	9,0813	4,940	3552,7	9,0494	4,631	3552,6	9,0195
540	5,771	3574,3	9,1422	5,359	3574,3	9,1079	5,001	3574,2	9,0761	4,688	3574,2	9,0462
550	5,842	3595,9	9,1686	5,425	3595,9	9,1344	5,063	3595,8	9,1025	4,746	3595,8	9,0727

Table 3. Water and Superheated Steam (Continuation) **Wasser und überhitzter Dampf** (Fortsetzung)

t	0,65 bar $t_s = 88,02$ °C			0,70 bar $t_s = 89,96$ °C			0,75 bar $t_s = 91,79$ °C			0,80 bar $t_s = 93,51$ °C		
	v'' 2,535	h'' 2656,9	s'' 7,5055	v'' 2,365	h'' 2660,1	s'' 7,4804	v'' 2,217	h'' 2663,0	s'' 7,4570	v'' 2,087	h'' 2665,8	s'' 7,4352
°C	v	h	s	v	h	s	v	h	s	v	h	s
550	5,842	3595,9	9,1686	5,425	3595,9	9,1344	5,063	3595,8	9,1025	4,746	3595,8	9,0727
560	5,913	3617,6	9,1948	5,491	3617,6	9,1606	5,124	3617,5	9,1287	4,804	3617,5	9,0988
570	5,984	3639,4	9,2208	5,557	3639,3	9,1865	5,186	3639,3	9,1546	4,862	3639,3	9,1248
580	6,055	3661,2	9,2465	5,623	3661,2	9,2123	5,248	3661,1	9,1804	4,920	3661,1	9,1506
590	6,126	3683,1	9,2720	5,689	3683,1	9,2378	5,309	3683,0	9,2059	4,977	3683,0	9,1761
600	6,198	3705,1	9,2973	5,755	3705,0	9,2631	5,371	3705,0	9,2312	5,035	3705,0	9,2014
610	6,269	3727,1	9,3224	5,821	3727,1	9,2882	5,433	3727,0	9,2563	5,093	3727,0	9,2265
620	6,340	3749,2	9,3473	5,887	3749,2	9,3130	5,494	3749,1	9,2812	5,151	3749,1	9,2514
630	6,411	3771,4	9,3720	5,953	3771,3	9,3377	5,556	3771,3	9,3059	5,208	3771,3	9,2760
640	6,482	3793,6	9,3965	6,019	3793,6	9,3622	5,617	3793,5	9,3303	5,266	3793,5	9,3005
650	6,553	3815,9	9,4207	6,085	3815,9	9,3865	5,679	3815,9	9,3546	5,324	3815,8	9,3248
660	6,624	3838,3	9,4449	6,151	3838,3	9,4106	5,741	3838,2	9,3788	5,382	3838,2	9,3489
670	6,695	3860,7	9,4688	6,217	3860,7	9,4345	5,802	3860,7	9,4027	5,439	3860,6	9,3729
680	6,766	3883,2	9,4925	6,283	3883,2	9,4583	5,864	3883,2	9,4264	5,497	3883,1	9,3966
690	6,837	3905,8	9,5161	6,349	3905,8	9,4818	5,925	3905,8	9,4500	5,555	3905,7	9,4202
700	6,908	3928,5	9,5395	6,415	3928,4	9,5052	5,987	3928,4	9,4734	5,613	3928,4	9,4436
710	6,979	3951,2	9,5627	6,481	3951,1	9,5284	6,048	3951,1	9,4966	5,670	3951,1	9,4668
720	7,050	3973,9	9,5857	6,547	3973,9	9,5515	6,110	3973,9	9,5196	5,728	3973,9	9,4898
730	7,121	3996,8	9,6086	6,613	3996,7	9,5744	6,172	3996,7	9,5425	5,786	3996,7	9,5127
740	7,192	4019,7	9,6313	6,678	4019,7	9,5971	6,233	4019,6	9,5652	5,844	4019,6	9,5354
750	7,263	4042,6	9,6539	6,744	4042,6	9,6197	6,295	4042,6	9,5878	5,901	4042,6	9,5580
760	7,334	4065,7	9,6763	6,810	4065,7	9,6421	6,356	4065,6	9,6102	5,959	4065,6	9,5804
770	7,405	4088,8	9,6985	6,876	4088,8	9,6643	6,418	4088,7	9,6324	6,017	4088,7	9,6026
780	7,476	4111,9	9,7206	6,942	4111,9	9,6864	6,479	4111,9	9,6545	6,074	4111,9	9,6247
790	7,548	4135,2	9,7426	7,008	4135,1	9,7083	6,541	4135,1	9,6765	6,132	4135,1	9,6467
800	7,619	4158,4	9,7644	7,074	4158,4	9,7301	6,603	4158,4	9,6983	6,190	4158,4	9,6685

Table 3. Water and Superheated Steam (Continuation) Wasser und überhitzter Dampf (Fortsetzung)

t	0,85 bar t_s = 95,15 °C v'' 1,972	h'' 2668,4	s'' 7,4147	0,90 bar t_s = 96,71 °C v'' 1,869	h'' 2670,9	s'' 7,3954	0,95 bar t_s = 98,20 °C v'' 1,777	h'' 2673,2	s'' 7,3771	1,0 bar t_s = 99,63 °C v'' 1,694	h'' 2675,4	s'' 7,3598
°C	v	h	s	v	h	s	v	h	s	v	h	s
0	0,0010002	0,0	−0,0001	0,0010002	0,1	−0,0001	0,0010002	0,1	−0,0001	0,0010002	0,1	−0,0001
10	0,0010002	42,1	0,1510	0,0010002	42,1	0,1510	0,0010002	42,1	0,1510	0,0010002	42,1	0,1510
20	0,0010017	83,9	0,2963	0,0010017	83,9	0,2963	0,0010017	83,9	0,2963	0,0010017	84,0	0,2963
30	0,0010043	125,7	0,4365	0,0010043	125,7	0,4365	0,0010043	125,7	0,4365	0,0010043	125,8	0,4365
40	0,0010078	167,5	0,5721	0,0010078	167,5	0,5721	0,0010078	167,5	0,5721	0,0010078	167,5	0,5721
50	0,0010121	209,3	0,7035	0,0010121	209,3	0,7035	0,0010121	209,3	0,7035	0,0010121	209,3	0,7035
60	0,0010171	251,1	0,8310	0,0010171	251,1	0,8309	0,0010171	251,2	0,8309	0,0010171	251,2	0,8309
70	0,0010228	293,0	0,9548	0,0010228	293,0	0,9548	0,0010228	293,0	0,9548	0,0010228	293,0	0,9548
80	0,0010292	334,9	1,0752	0,0010292	335,0	1,0752	0,0010292	335,0	1,0752	0,0010292	335,0	1,0752
90	0,0010361	377,0	1,1925	0,0010361	377,0	1,1925	0,0010361	377,0	1,1925	0,0010361	377,0	1,1925
100	2,000	2678,1	7,4409	1,887	2677,5	7,4132	1,786	2676,8	7,3869	1,696	2676,2	7,3618
110	2,056	2698,1	7,4938	1,941	2697,5	7,4663	1,837	2697,0	7,4401	1,744	2696,4	7,4152
120	2,113	2718,0	7,5452	1,994	2717,5	7,5177	1,888	2717,0	7,4917	1,793	2716,5	7,4670
130	2,169	2737,9	7,5950	2,048	2737,4	7,5677	1,939	2737,0	7,5419	1,841	2736,5	7,5173
140	2,225	2757,7	7,6436	2,101	2757,3	7,6164	1,989	2756,9	7,5906	1,889	2756,4	7,5662
150	2,281	2777,5	7,6908	2,153	2777,1	7,6638	2,039	2776,7	7,6381	1,936	2776,3	7,6137
160	2,337	2797,2	7,7369	2,206	2796,9	7,7099	2,089	2796,5	7,6844	1,984	2796,2	7,6601
170	2,392	2816,9	7,7820	2,258	2816,6	7,7550	2,139	2816,3	7,7295	2,031	2816,0	7,7053
180	2,447	2836,6	7,8259	2,311	2836,4	7,7991	2,188	2836,1	7,7736	2,078	2835,8	7,7495
190	2,503	2856,4	7,8690	2,363	2856,1	7,8422	2,238	2855,8	7,8168	2,125	2855,6	7,7927
200	2,558	2876,1	7,9111	2,415	2875,8	7,8843	2,287	2875,6	7,8590	2,172	2875,4	7,8349
210	2,613	2895,8	7,9524	2,467	2895,6	7,9257	2,337	2895,4	7,9004	2,219	2895,2	7,8763
220	2,668	2915,6	7,9929	2,519	2915,4	7,9662	2,386	2915,2	7,9409	2,266	2915,0	7,9169
230	2,723	2935,4	8,0326	2,571	2935,2	8,0059	2,435	2935,0	7,9807	2,313	2934,8	7,9567
240	2,777	2955,2	8,0716	2,623	2955,0	8,0449	2,484	2954,8	8,0197	2,359	2954,6	7,9958
250	2,832	2975,0	8,1099	2,674	2974,8	8,0832	2,533	2974,7	8,0581	2,406	2974,5	8,0342
260	2,887	2994,9	8,1475	2,726	2994,7	8,1209	2,582	2994,6	8,0957	2,453	2994,4	8,0719
270	2,942	3014,8	8,1845	2,778	3014,7	8,1579	2,631	3014,5	8,1328	2,499	3014,4	8,1089
280	2,996	3034,8	8,2209	2,829	3034,6	8,1944	2,680	3034,5	8,1692	2,546	3034,4	8,1454
290	3,051	3054,8	8,2568	2,881	3054,6	8,2302	2,729	3054,5	8,2051	2,592	3054,4	8,1813
300	3,106	3074,8	8,2921	2,933	3074,7	8,2656	2,778	3074,6	8,2405	2,639	3074,5	8,2166
310	3,160	3094,9	8,3269	2,984	3094,8	8,3003	2,827	3094,7	8,2753	2,685	3094,6	8,2514
320	3,215	3115,1	8,3611	3,036	3115,0	8,3346	2,876	3114,9	8,3096	2,732	3114,8	8,2857
330	3,269	3135,3	8,3949	3,087	3135,2	8,3684	2,925	3135,1	8,3434	2,778	3135,0	8,3196
340	3,324	3155,6	8,4283	3,139	3155,5	8,4018	2,973	3155,4	8,3767	2,824	3155,3	8,3529
350	3,378	3175,9	8,4612	3,190	3175,8	8,4347	3,022	3175,7	8,4095	2,871	3175,6	8,3858
360	3,433	3196,3	8,4936	3,242	3196,2	8,4671	3,071	3196,1	8,4421	2,917	3196,0	8,4183
370	3,487	3216,7	8,5256	3,293	3216,7	8,4992	3,120	3216,6	8,4741	2,964	3216,5	8,4504
380	3,542	3237,2	8,5573	3,345	3237,2	8,5308	3,168	3237,1	8,5058	3,010	3237,0	8,4820
390	3,596	3257,8	8,5885	3,396	3257,7	8,5621	3,217	3257,7	8,5370	3,056	3257,6	8,5133
400	3,651	3278,4	8,6194	3,448	3278,4	8,5929	3,266	3278,3	8,5679	3,102	3278,2	8,5442
410	3,705	3299,1	8,6499	3,499	3299,1	8,6235	3,315	3299,0	8,5984	3,149	3298,9	8,5747
420	3,760	3319,9	8,6801	3,551	3319,8	8,6536	3,363	3319,8	8,6286	3,195	3319,7	8,6049
430	3,814	3340,7	8,7099	3,602	3340,6	8,6835	3,412	3340,6	8,6584	3,241	3340,5	8,6347
440	3,869	3361,6	8,7394	3,653	3361,5	8,7130	3,461	3361,5	8,6879	3,288	3361,4	8,6642
450	3,923	3382,6	8,7686	3,705	3382,5	8,7421	3,510	3382,4	8,7171	3,334	3382,4	8,6934
460	3,977	3403,6	8,7974	3,756	3403,5	8,7710	3,558	3403,4	8,7460	3,380	3403,4	8,7223
470	4,032	3424,7	8,8260	3,808	3424,6	8,7996	3,607	3424,5	8,7746	3,427	3424,5	8,7508
480	4,086	3445,8	8,8543	3,859	3445,7	8,8278	3,656	3445,7	8,8028	3,473	3445,6	8,7791
490	4,141	3467,0	8,8823	3,910	3467,0	8,8558	3,704	3466,9	8,8308	3,519	3466,9	8,8071
500	4,195	3488,3	8,9100	3,962	3488,3	8,8835	3,753	3488,2	8,8585	3,565	3488,1	8,8348
510	4,249	3509,7	8,9374	4,013	3509,6	8,9110	3,802	3509,6	8,8860	3,612	3509,5	8,8623
520	4,304	3531,1	8,9646	4,065	3531,0	8,9382	3,850	3531,0	8,9132	3,658	3530,9	8,8894
530	4,358	3552,6	8,9915	4,116	3552,5	8,9651	3,899	3552,5	8,9401	3,704	3552,4	8,9164
540	4,413	3574,1	9,0182	4,167	3574,1	8,9918	3,948	3574,0	8,9668	3,750	3574,0	8,9431
550	4,467	3595,8	9,0446	4,219	3595,7	9,0182	3,996	3595,7	8,9932	3,797	3595,6	8,9695

Table 3. Water and Superheated Steam (Continuation) Wasser und überhitzter Dampf (Fortsetzung)

t	0,85 bar $t_s = 95,15$ °C			0,90 bar $t_s = 96,71$ °C			0,95 bar $t_s = 98,20$ °C			1,0 bar $t_s = 99,63$ °C		
	v'' 1,972	h'' 2668,4	s'' 7,4147	v'' 1,869	h'' 2670,9	s'' 7,3954	v'' 1,777	h'' 2673,2	s'' 7,3771	v'' 1,694	h'' 2675,4	s'' 7,3598
°C	v	h	s	v	h	s	v	h	s	v	h	s
550	4,467	3595,8	9,0446	4,219	3595,7	9,0182	3,996	3595,7	8,9932	3,797	3595,6	8,9695
560	4,521	3617,4	9,0708	4,270	3617,4	9,0444	4,045	3617,4	9,0194	3,843	3617,3	8,9957
570	4,576	3639,2	9,0968	4,321	3639,2	9,0704	4,094	3639,1	9,0454	3,889	3639,1	9,0217
580	4,630	3661,0	9,1225	4,373	3661,0	9,0961	4,142	3661,0	9,0711	3,935	3660,9	9,0474
590	4,684	3682,9	9,1481	4,424	3682,9	9,1216	4,191	3682,9	9,0966	3,981	3682,8	9,0729
600	4,739	3704,9	9,1734	4,475	3704,9	9,1469	4,240	3704,8	9,1220	4,028	3704,8	9,0982
610	4,793	3727,0	9,1985	4,527	3726,9	9,1720	4,288	3726,9	9,1471	4,074	3726,8	9,1233
620	4,848	3749,1	9,2233	4,578	3749,0	9,1969	4,337	3749,0	9,1719	4,120	3748,9	9,1482
630	4,902	3771,2	9,2480	4,629	3771,2	9,2216	4,386	3771,2	9,1966	4,166	3771,1	9,1729
640	4,956	3793,5	9,2725	4,681	3793,4	9,2461	4,434	3793,4	9,2211	4,213	3793,4	9,1974
650	5,011	3815,8	9,2968	4,732	3815,7	9,2704	4,483	3815,7	9,2454	4,259	3815,7	9,2217
660	5,065	3838,2	9,3209	4,783	3838,1	9,2945	4,532	3838,1	9,2695	4,305	3838,1	9,2458
670	5,119	3860,6	9,3449	4,835	3860,6	9,3184	4,580	3860,5	9,2935	4,351	3860,5	9,2698
680	5,174	3883,1	9,3686	4,886	3883,1	9,3422	4,629	3883,0	9,3172	4,397	3883,0	9,2935
690	5,228	3905,7	9,3922	4,937	3905,7	9,3658	4,678	3905,6	9,3408	4,444	3905,6	9,3171
700	5,282	3928,3	9,4156	4,989	3928,3	9,3891	4,726	3928,3	9,3642	4,490	3928,2	9,3405
710	5,337	3951,0	9,4388	5,040	3951,0	9,4124	4,775	3951,0	9,3874	4,536	3951,0	9,3637
720	5,391	3973,8	9,4618	5,091	3973,8	9,4354	4,823	3973,8	9,4104	4,582	3973,7	9,3867
730	5,445	3996,7	9,4847	5,143	3996,6	9,4583	4,872	3996,6	9,4333	4,628	3996,6	9,4096
740	5,500	4019,6	9,5074	5,194	4019,5	9,4810	4,921	4019,5	9,4561	4,675	4019,5	9,4324
750	5,554	4042,5	9,5300	5,245	4042,5	9,5036	4,969	4042,5	9,4786	4,721	4042,5	9,4549
760	5,608	4065,6	9,5524	5,297	4065,6	9,5260	5,018	4065,5	9,5010	4,767	4065,5	9,4773
770	5,663	4088,7	9,5746	5,348	4088,7	9,5482	5,067	4088,6	9,5233	4,813	4088,6	9,4996
780	5,717	4111,8	9,5967	5,399	4111,8	9,5703	5,115	4111,8	9,5454	4,859	4111,8	9,5217
790	5,771	4135,1	9,6187	5,451	4135,0	9,5923	5,164	4135,0	9,5673	4,906	4135,0	9,5436
800	5,826	4158,3	9,6405	5,502	4158,3	9,6141	5,212	4158,3	9,5891	4,952	4158,3	9,5654

Table 3. Water and Superheated Steam (Continuation) Wasser und überhitzter Dampf (Fortsetzung)

t	1,1 bar $t_s = 102,32$ °C			1,2 bar $t_s = 104,81$ °C			1,3 bar $t_s = 107,13$ °C			1,4 bar $t_s = 109,32$ °C		
	v'' 1,549	h'' 2679,6	s'' 7,3277	v'' 1,428	h'' 2683,4	s'' 7,2984	v'' 1,325	h'' 2687,0	s'' 7,2715	v'' 1,236	h'' 2690,3	s'' 7,2465
°C	v	h	s	v	h	s	v	h	s	v	h	s
0	0,0010002	0,1	−0,0001	0,0010002	0,1	−0,0001	0,0010002	0,1	−0,0001	0,0010002	0,1	−0,0001
10	0,0010002	42,1	0,1510	0,0010002	42,1	0,1510	0,0010002	42,1	0,1510	0,0010002	42,1	0,1510
20	0,0010017	84,0	0,2963	0,0010017	84,0	0,2963	0,0010017	84,0	0,2963	0,0010017	84,0	0,2963
30	0,0010043	125,8	0,4365	0,0010043	125,8	0,4365	0,0010043	125,8	0,4365	0,0010042	125,8	0,4365
40	0,0010078	167,5	0,5721	0,0010078	167,6	0,5721	0,0010078	167,6	0,5721	0,0010078	167,6	0,5721
50	0,0010121	209,3	0,7035	0,0010121	209,3	0,7035	0,0010121	209,4	0,7035	0,0010121	209,4	0,7034
60	0,0010171	251,2	0,8309	0,0010171	251,2	0,8309	0,0010171	251,2	0,8309	0,0010171	251,2	0,8309
70	0,0010228	293,0	0,9548	0,0010228	293,0	0,9548	0,0010228	293,1	0,9548	0,0010228	293,1	0,9548
80	0,0010292	335,0	1,0752	0,0010292	335,0	1,0752	0,0010292	335,0	1,0752	0,0010291	335,0	1,0752
90	0,0010361	377,0	1,1925	0,0010361	377,0	1,1925	0,0010361	377,0	1,1925	0,0010361	377,0	1,1925
100	0,0010437	419,1	1,3069	0,0010437	419,1	1,3069	0,0010437	419,1	1,3068	0,0010437	419,1	1,3068
110	1,583	2695,2	7,3689	1,449	2694,1	7,3263	1,336	2692,9	7,2869	1,239	2691,7	7,2502
120	1,628	2715,4	7,4209	1,490	2714,4	7,3787	1,374	2713,3	7,3396	1,274	2712,3	7,3033
130	1,672	2735,6	7,4715	1,531	2734,6	7,4295	1,411	2733,7	7,3907	1,309	2732,7	7,3546
140	1,715	2755,6	7,5206	1,571	2754,7	7,4788	1,449	2753,9	7,4402	1,344	2753,0	7,4044
150	1,759	2775,6	7,5683	1,611	2774,8	7,5267	1,486	2774,0	7,4884	1,378	2773,2	7,4527
160	1,802	2795,5	7,6148	1,651	2794,8	7,5734	1,522	2794,1	7,5352	1,413	2793,4	7,4998
170	1,845	2815,4	7,6602	1,690	2814,7	7,6189	1,559	2814,1	7,5809	1,447	2813,4	7,5456
180	1,888	2835,2	7,7045	1,730	2834,6	7,6634	1,596	2834,1	7,6254	1,481	2833,5	7,5903
190	1,931	2855,0	7,7478	1,769	2854,5	7,7068	1.632	2854,0	7,6690	1,515	2853,5	7,6339
200	1,974	2874,9	7,7902	1,808	2874,4	7,7492	1,668	2873,9	7,7115	1,548	2873,4	7,6765
210	2,016	2894,7	7,8317	1,848	2894,3	7,7908	1,705	2893,8	7,7532	1,582	2893,4	7,7183
220	2,059	2914,5	7,8723	1,887	2914,1	7,8315	1,741	2913,7	7,7939	1,616	2913,3	7,7591
230	2,102	2934,4	7,9122	1,926	2934,0	7,8714	1,777	2933,6	7,8339	1,649	2933,3	7,7992
240	2,144	2954,3	7,9513	1,965	2953,9	7,9106	1,813	2953,6	7,8732	1,683	2953,2	7,8384
250	2,187	2974,2	7,9897	2,004	2973,9	7,9491	1,849	2973,5	7,9117	1,716	2973,2	7,8770
260	2,229	2994,1	8,0274	2,043	2993,8	7,9869	1,885	2993,5	7,9495	1,750	2993,2	7,9149
270	2,271	3014,1	8,0645	2,081	3013,8	8,0240	1,921	3013,5	7,9867	1,783	3013,2	7,9521
280	2,314	3034,1	8,1010	2,120	3033,8	8,0605	1,957	3033,6	8,0232	1,816	3033,3	7,9887
290	2,356	3054,1	8,1370	2,159	3053,9	8,0965	1,993	3053,6	8,0592	1,850	3053,4	8,0247
300	2,398	3074,2	8,1723	2,198	3074,0	8,1319	2,028	3073,8	8,0946	1,883	3073,5	8,0601
310	2,441	3094,4	8,2072	2,237	3094,1	8,1667	2,064	3093,9	8,1295	1,916	3093,7	8,0950
320	2,483	3114,6	8,2415	2,275	3114,3	8,2011	2,100	3114,1	8,1639	1,949	3113,9	8,1294
330	2,525	3134,8	8,2753	2,314	3134,6	8,2349	2,136	3134,4	8,1977	1,983	3134,2	8,1633
340	2,567	3155,1	8,3087	2,353	3154,9	8,2683	2,171	3154,7	8,2311	2,016	3154,5	8,1967
350	2,609	3175,4	8,3416	2,391	3175,3	8,3012	2,207	3175,1	8,2641	2,049	3174,9	8,2297
360	2,652	3195,9	8,3741	2,430	3195,7	8,3337	2,243	3195,5	8,2966	2,082	3195,3	8,2622
370	2,694	3216,3	8,4062	2,469	3216,1	8,3658	2,279	3216,0	8,3287	2,115	3215,8	8,2943
380	2,736	3236,8	8,4378	2,507	3236,7	8,3975	2,314	3236,5	8,3604	2,149	3236,3	8,3260
390	2,778	3257,4	8,4691	2,546	3257,3	8,4288	2,350	3257,1	8,3917	2,182	3256,9	8,3573
400	2,820	3278,1	8,5000	2,585	3277,9	8,4597	2,386	3277,8	8,4226	2,215	3277,6	8,3882
410	2,862	3298,8	8,5305	2,623	3298,6	8,4902	2,421	3298,5	8,4531	2,248	3298,3	8,4188
420	2,904	3319,5	8,5607	2,662	3319,4	8,5204	2,457	3319,3	8,4833	2,281	3319,1	8,4490
430	2,946	3340,4	8,5906	2,701	3340,2	8,5503	2,493	3340,1	8,5132	2,314	3340,0	8,4788
440	2,988	3361,3	8,6201	2,739	3361,1	8,5798	2,528	3361,0	8,5427	2,347	3360,9	8,5084
450	3,031	3382,2	8,6493	2,778	3382,1	8,6090	2,564	3382,0	8,5719	2,380	3381,8	8,5376
460	3,073	3403,3	8,6781	2,816	3403,1	8,6379	2,599	3403,0	8,6008	2,414	3402,9	8,5665
470	3,115	3424,4	8,7067	2,855	3424,2	8,6664	2,635	3424,1	8,6294	2,447	3424,0	8,5951
480	3,157	3445,5	8,7350	2,893	3445,4	8,6947	2,671	3445,3	8,6577	2,480	3445,2	8,6234
490	3,199	3466,7	8,7630	2,932	3466,6	8,7227	2,706	3466,5	8,6857	2,513	3466,4	8,6514
500	3,241	3488,0	8,7907	2,971	3487,9	8,7505	2,742	3487,8	8,7134	2,546	3487,7	8,6791
510	3,283	3509,4	8,8182	3,009	3509,3	8,7779	2,777	3509,2	8,7409	2,579	3509,1	8,7066
520	3,325	3530,8	8,8454	3,048	3530,7	8,8051	2,813	3530,6	8,7681	2,612	3530,5	8,7338
530	3,367	3552,3	8,8723	3,086	3552,2	8,8321	2,849	3552,1	8,7950	2,645	3552,0	8,7607
540	3,409	3573,9	8,8990	3,125	3573,8	8,8587	2,884	3573,7	8,8217	2,678	3573,6	8,7874
550	3,451	3595,5	8,9254	3,163	3595,4	8,8852	2,920	3595,3	8,8482	2,711	3595,2	8,8139

57

Table 3. Water and Superheated Steam (Continuation) Wasser und überhitzter Dampf (Fortsetzung)

t	1,1 bar $t_s = 102,32$ °C			1,2 bar $t_s = 104,81$ °C			1,3 bar $t_s = 107,13$ °C			1,4 bar $t_s = 109,32$ °C		
	v'' 1,549	h'' 2679,6	s'' 7,3277	v'' 1,428	h'' 2683,4	s'' 7,2984	v'' 1,325	h'' 2687,0	s'' 7,2715	v'' 1,236	h'' 2690,3	s'' 7,2465
°C	v	h	s	v	h	s	v	h	s	v	h	s
550	3,451	3595,5	8,9254	3,163	3595,4	8,8852	2,920	3595,3	8,8482	2,711	3595,2	8,8139
560	3,493	3617,2	8,9516	3,202	3617,1	8,9114	2,955	3617,0	8,8744	2,744	3616,9	8,8401
570	3,535	3639,0	8,9776	3,240	3638,9	8,9374	2,991	3638,8	8,9003	2,777	3638,7	8,8661
580	3,577	3660,8	9,0034	3,279	3660,7	8,9631	3,027	3660,7	8,9261	2,810	3660,6	8,8918
590	3,619	3682,7	9,0289	3,318	3682,6	8,9886	3,062	3682,6	8,9516	2,843	3682,5	8,9174
600	3,661	3704,7	9,0542	3,356	3704,6	9,0140	3,098	3704,5	8,9770	2,876	3704,5	8,9427
610	3,703	3726,8	9,0793	3,395	3726,7	9,0391	3,133	3726,6	9,0021	2,909	3726,5	8,9678
620	3,745	3748,9	9,1042	3,433	3748,8	9,0640	3,169	3748,7	9,0270	2,942	3748,6	8,9927
630	3,787	3771,0	9,1289	3,472	3771,0	9,0887	3,204	3770,9	9,0517	2,975	3770,8	9,0174
640	3,829	3793,3	9,1534	3,510	3793,2	9,1132	3,240	3793,1	9,0762	3,008	3793,1	9,0419
650	3,871	3815,6	9,1777	3,549	3815,5	9,1375	3,276	3815,5	9,1005	3,042	3815,4	9,0662
660	3,913	3838,0	9,2018	3,587	3837,9	9,1616	3,311	3837,8	9,1246	3,075	3837,8	9,0903
670	3,955	3860,4	9,2257	3,626	3860,4	9,1855	3,347	3860,3	9,1485	3,108	3860,2	9,1143
680	3,997	3882,9	9,2495	3,664	3882,9	9,2093	3,382	3882,8	9,1723	3,141	3882,8	9,1380
690	4,039	3905,5	9,2730	3,703	3905,5	9,2328	3,418	3905,4	9,1958	3,174	3905,3	9,1616
700	4,082	3928,2	9,2964	3,741	3928,1	9,2562	3,453	3928,1	9,2192	3,207	3928,0	9,1850
710	4,124	3950,9	9,3197	3,780	3950,8	9,2795	3,489	3950,8	9,2425	3,240	3950,7	9,2082
720	4,166	3973,7	9,3427	3,818	3973,6	9,3025	3,524	3973,6	9,2655	3,273	3973,5	9,2313
730	4,208	3996,5	9,3656	3,857	3996,5	9,3254	3,560	3996,4	9,2884	3,306	3996,3	9,2542
740	4,250	4019,4	9,3883	3,895	4019,4	9,3481	3,596	4019,3	9,3111	3,339	4019,3	9,2769
750	4,292	4042,4	9,4109	3,934	4042,4	9,3707	3,631	4042,3	9,3337	3,372	4042,2	9,2995
760	4,334	4065,4	9,4333	3,972	4065,4	9,3931	3,667	4065,3	9,3561	3,405	4065,3	9,3219
770	4,375	4088,5	9,4556	4,011	4088,5	9,4154	3,702	4088,4	9,3784	3,438	4088,4	9,3441
780	4,417	4111,7	9,4777	4,049	4111,7	9,4375	3,738	4111,6	9,4005	3,471	4111,6	9,3662
790	4,459	4134,9	9,4996	4,088	4134,9	9,4594	3,773	4134,8	9,4224	3,504	4134,8	9,3882
800	4,501	4158,2	9,5214	4,126	4158,2	9,4812	3,809	4158,1	9,4442	3,537	4158,1	9,4100

Table 3. Water and Superheated Steam (Continuation) Wasser und überhitzter Dampf (Fortsetzung)

t	1,5 bar $t_s = 111,37\ °C$			1,6 bar $t_s = 113,32\ °C$			1,7 bar $t_s = 115,17\ °C$			1,8 bar $t_s = 116,93\ °C$		
	v'' 1,159	h'' 2693,4	s'' 7,2234	v'' 1,091	h'' 2696,2	s'' 7,2017	v'' 1,031	h'' 2699,0	s'' 7,1813	v'' 0,9772	h'' 2701,5	s'' 7,1622
°C	v	h	s	v	h	s	v	h	s	v	h	s
0	0,0010001	0,1	−0,0001	0,0010001	0,1	−0,0001	0,0010001	0,1	−0,0001	0,0010001	0,1	−0,0001
10	0,0010002	42,1	0,1510	0,0010002	42,1	0,1510	0,0010002	42,2	0,1510	0,0010002	42,2	0,1510
20	0,0010017	84,0	0,2963	0,0010017	84,0	0,2963	0,0010016	84,0	0,2963	0,0010016	84,0	0,2963
30	0,0010042	125,8	0,4365	0,0010042	125,8	0,4365	0,0010042	125,8	0,4365	0,0010042	125,8	0,4365
40	0,0010077	167,6	0,5721	0,0010077	167,6	0,5721	0,0010077	167,6	0,5721	0,0010077	167,6	0,5721
50	0,0010121	209,4	0,7034	0,0010120	209,4	0,7034	0,0010120	209,4	0,7034	0,0010120	209,4	0,7034
60	0,0010171	251,2	0,8309	0,0010171	251,2	0,8309	0,0010171	251,2	0,8309	0,0010171	251,2	0,8309
70	0,0010228	293,1	0,9547	0,0010228	293,1	0,9547	0,0010228	293,1	0,9547	0,0010228	293,1	0,9547
80	0,0010291	335,0	1,0752	0,0010291	335,0	1,0752	0,0010291	335,0	1,0752	0,0010291	335,0	1,0752
90	0,0010361	377,0	1,1925	0,0010361	377,0	1,1925	0,0010361	377,0	1,1925	0,0010361	377,0	1,1925
100	0,0010437	419,1	1,3068	0,0010437	419,1	1,3068	0,0010437	419,1	1,3068	0,0010437	419;1	1,3068
110	0,0010519	461,3	1,4185	0,0010519	461,3	1,4185	0,0010519	461,3	1,4185	0,0010518	461,3	1,4185
120	1,188	2711,2	7,2693	1,112	2710,1	7,2373	1,045	2709,1	7,2072	0,9858	2708,0	7,1786
130	1,220	2731,8	7,3209	1,143	2730,8	7,2892	1,074	2729,8	7,2593	1,013	2728,9	7,2311
140	1,253	2752,2	7,3709	1,173	2751,3	7,3395	1,103	2750,4	7,3098	1,041	2749,6	7,2818
150	1,285	2772,5	7,4194	1,204	2771,7	7,3882	1,132	2770,9	7,3588	1,068	2770,1	7,3309
160	1,317	2792,7	7,4667	1,234	2792,0	7,4356	1,160	2791,3	7,4063	1,095	2790,5	7,3787
170	1,349	2812,8	7,5126	1,264	2812,2	7,4817	1,189	2811,5	7,4526	1,122	2810,9	7,4251
180	1,381	2832,9	7,5574	1,294	2832,3	7,5266	1,217	2831,7	7,4977	1,149	2831,1	7,4703
190	1,413	2852,9	7,6012	1,324	2852,4	7,5705	1,245	2851,9	7,5416	1,175	2851,3	7,5144
200	1,444	2872,9	7,6439	1,353	2872,5	7,6133	1,273	2872,0	7,5846	1,202	2871,5	7,5574
210	1,476	2892,9	7,6857	1,383	2892,5	7,6552	1,301	2892,0	7,6265	1,228	2891,6	7,5994
220	1,507	2912,9	7,7266	1,413	2912,5	7,6962	1,329	2912,1	7,6676	1,254	2911,7	7,6406
230	1,539	2932,9	7,7667	1,442	2932,5	7,7364	1,357	2932,1	7,7078	1,281	2931,7	7,6809
240	1,570	2952,9	7,8061	1,471	2952,5	7,7758	1,384	2952,2	7,7473	1,307	2951,8	7,7204
250	1,601	2972,9	7,8447	1,501	2972,5	7,8144	1,412	2972,2	7,7860	1,333	2971,9	7,7591
260	1,633	2992,9	7,8826	1,530	2992,6	7,8524	1,440	2992,3	7,8240	1,359	2992,0	7,7972
270	1,664	3012,9	7,9198	1,559	3012,7	7,8897	1,467	3012,4	7,8613	1,385	3012,1	7,8345
280	1,695	3033,0	7,9565	1,588	3032,7	7,9263	1,495	3032,5	7,8980	1,411	3032,2	7,8712
290	1,726	3053,1	7,9925	1,618	3052,9	7,9624	1,522	3052,6	7,9341	1,437	3052,4	7,9074
300	1,757	3073,3	8,0280	1,647	3073,0	7,9979	1,550	3072,8	7,9696	1,463	3072,6	7,9429
310	1,788	3093,5	8,0629	1,676	3093,2	8,0328	1,577	3093,0	8,0046	1,489	3092,8	7,9779
320	1,819	3113,7	8,0973	1,705	3113,5	8,0672	1,604	3113,3	8,0390	1,515	3113,1	8,0124
330	1,850	3134,0	8,1312	1,734	3133,8	8,1012	1,632	3133,6	8,0729	1,541	3133,4	8,0463
340	1,881	3154,3	8,1646	1,763	3154,1	8,1346	1,659	3153,9	8,1064	1,567	3153,7	8,0798
350	1,912	3174,7	8,1976	1,792	3174,5	8,1676	1,687	3174,3	8,1394	1,593	3174,1	8,1128
360	1,943	3195,1	8,2301	1,821	3195,0	8,2002	1,714	3194,8	8,1720	1,618	3194,6	8,1454
370	1,974	3215,6	8,2623	1,850	3215,5	8,2323	1,741	3215,3	8,2041	1,644	3215,1	8,1775
380	2,005	3236,2	8,2940	1,879	3236,0	8,2640	1,769	3235,8	8,2358	1,670	3235,7	8,2093
390	2,036	3256,8	8,3253	1,908	3256,6	8,2953	1,796	3256,5	8,2672	1,696	3256,3	8,2406
400	2,067	3277,5	8,3562	1,937	3277,3	8,3263	1,823	3277,1	8,2981	1,722	3277,0	8,2716
410	2,098	3298,2	8,3868	1,967	3298,0	8,3568	1,851	3297,9	8,3287	1,748	3297,7	8,3022
420	2,129	3319,0	8,4170	1,995	3318,8	8,3871	1,878	3318,7	8,3589	1,773	3318,5	8,3324
430	2,160	3339,8	8,4469	2,024	3339,7	8,4169	1,905	3339,6	8,3888	1,799	3339,4	8,3623
440	2,191	3360,7	8,4764	2,053	3360,6	8,4465	1,932	3360,5	8,4184	1,825	3360,3	8,3918
450	2,222	3381,7	8,5056	2,082	3381,6	8,4757	1,960	3381,5	8,4476	1,851	3381,3	8,4211
460	2,252	3402,8	8,5345	2,111	3402,6	8,5046	1,987	3402,5	8,4765	1,876	3402,4	8,4500
470	2,283	3423,9	8,5631	2,140	3423,8	8,5332	2,014	3423,6	8,5051	1,902	3423,5	8,4786
480	2,314	3445,0	8,5914	2,169	3444,9	8,5615	2,042	3444,8	8,5334	1,928	3444,7	8,5069
490	2,345	3466,3	8,6194	2,198	3466,2	8,5895	2,069	3466,1	8,5614	1,954	3466,0	8,5350
500	2,376	3487,6	8,6472	2,227	3487,5	8,6173	2,096	3487,4	8,5892	1,979	3487,3	8,5627
510	2,407	3509,0	8,6746	2,256	3508,9	8,6447	2,123	3508,8	8,6167	2,005	3508,7	8,5902
520	2,438	3530,4	8,7018	2,285	3530,3	8,6720	2,151	3530,2	8,6439	2,031	3530,1	8,6174
530	2,469	3551,9	8,7288	2,314	3551,8	8,6989	2,178	3551,7	8,6708	2,057	3551,6	8,6444
540	2,499	3573,5	8,7555	2,343	3573,4	8,7256	2,205	3573,3	8,6975	2,082	3573,2	8,6711
550	2,530	3595,1	8,7819	2,372	3595,0	8,7521	2,232	3595,0	8,7240	2,108	3594,9	8,6975

Table 3. Water and Superheated Steam (Continuation) Wasser und überhitzter Dampf (Fortsetzung)

t	1,5 bar $t_s = 111,37\,°C$			1,6 bar $t_s = 113,32\,°C$			1,7 bar $t_s = 115,17\,°C$			1,8 bar $t_s = 116,93\,°C$		
	v'' 1,159	h'' 2693,4	s'' 7,2234	v'' 1,091	h'' 2696,2	s'' 7,2017	v'' 1,031	h'' 2699,0	s'' 7,1813	v'' 0,9772	h'' 2701,5	s'' 7,1622
°C	v	h	s	v	h	s	v	h	s	v	h	s
550	2,530	3595,1	8,7819	2,372	3595,0	8,7521	2,232	3595,0	8,7240	2,108	3594,9	8,6975
560	2,561	3616,9	8,8082	2,401	3616,8	8,7783	2,260	3616,7	8,7502	2,134	3616,6	8,7238
570	2,592	3638,6	8,8341	2,430	3638,5	8,8043	2,287	3638,5	8,7762	2,160	3638,4	8,7498
580	2,623	3660,5	8,8599	2,459	3660,4	8,8300	2,314	3660,3	8,8020	2,185	3660,2	8,7755
590	2,654	3682,4	8,8854	2,488	3682,3	8,8556	2,341	3682,2	8,8275	2,211	3682,1	8,8011
600	2,684	3704,4	8,9108	2,517	3704,3	8,8809	2,368	3704,2	8,8529	2,237	3704,1	8,8264
610	2,715	3726,4	8,9359	2,545	3726,4	8,9060	2,396	3726,3	8,8780	2,262	3726,2	8,8515
620	2,746	3748,6	8,9608	2,574	3748,5	8,9309	2,423	3748,4	8,9029	2,288	3748,3	8,8764
630	2,777	3770,7	8,9855	2,603	3770,7	8,9556	2,450	3770,6	8,9276	2,314	3770,5	8,9012
640	2,808	3793,0	9,0100	2,632	3792,9	8,9801	2,477	3792,8	8,9521	2,340	3792,8	8,9257
650	2,839	3815,3	9,0343	2,661	3815,2	9,0045	2,504	3815,2	8,9764	2,365	3815,1	8,9500
660	2,869	3837,7	9,0584	2,690	3837,6	9,0286	2,532	3837,6	9,0006	2,391	3837,5	8,9741
670	2,900	3860,2	9,0824	2,719	3860,1	9,0525	2,559	3860,0	9,0245	2,417	3860,0	8,9981
680	2,931	3882,7	9,1061	2,748	3882,6	9,0763	2,586	3882,6	9,0483	2,442	3882,5	9,0218
690	2,962	3905,3	9,1297	2,777	3905,2	9,0999	2,613	3905,1	9,0718	2,468	3905,1	9,0454
700	2,993	3927,9	9,1531	2,806	3927,9	9,1233	2,640	3927,8	9,0952	2,494	3927,7	9,0688
710	3,024	3950,7	9,1763	2,834	3950,6	9,1465	2,668	3950,5	9,1185	2,519	3950,5	9,0920
720	3,054	3973,4	9,1994	2,863	3973,4	9,1696	2,695	3973,3	9,1415	2,545	3973,3	9,1151
730	3,085	3996,3	9,2223	2,892	3996,2	9,1925	2,722	3996,2	9,1644	2,571	3996,1	9,1380
740	3,116	4019,2	9,2450	2,921	4019,2	9,2152	2,749	4019,1	9,1872	2,596	4019,0	9,1607
750	3,147	4042,2	9,2676	2,950	4042,1	9,2378	2,776	4042,1	9,2097	2,622	4042,0	9,1833
760	3,178	4065,2	9,2900	2,979	4065,2	9,2602	2,804	4065,1	9,2322	2,648	4065,1	9,2057
770	3,208	4088,3	9,3123	3,008	4088,3	9,2824	2,831	4088,2	9,2544	2,673	4088,2	9,2280
780	3,239	4111,5	9,3344	3,037	4111,5	9,3045	2,858	4111,4	9,2765	2,699	4111,4	9,2501
790	3,270	4134,7	9,3563	3,066	4134,7	9,3265	2,885	4134,7	9,2985	2,725	4134,6	9,2721
800	3,301	4158,0	9,3781	3,094	4158,0	9,3483	2,912	4158,0	9,3203	2,751	4157,9	9,2939

60

Table 3. Water and Superheated Steam (Continuation) Wasser und überhitzter Dampf (Fortsetzung)

t	**1,9 bar** $t_s = 118{,}62$ °C			**2,0 bar** $t_s = 120{,}23$ °C			**2,1 bar** $t_s = 121{,}78$ °C			**2,2 bar** $t_s = 123{,}27$ °C		
	v'' 0,9290	h'' 2704,0	s'' 7,1440	v'' 0,8854	h'' 2706,3	s'' 7,1268	v'' 0,8459	h'' 2708,5	s'' 7,1105	v'' 0,8098	h'' 2710,6	s'' 7,0949
°C	v	h	s	v	h	s	v	h	s	v	h	s
0	0,0010001	0,2	−0,0001	0,0010001	0,2	−0,0001	0,0010001	0,2	−0,0001	0,0010001	0,2	−0,0001
10	0,0010002	42,2	0,1510	0,0010002	42,2	0,1510	0,0010002	42,2	0,1510	0,0010001	42,2	0,1510
20	0,0010016	84,0	0,2963	0,0010016	84,0	0,2963	0,0010016	84,1	0,2963	0,0010016	84,1	0,2963
30	0,0010042	125,8	0,4364	0,0010042	125,8	0,4364	0,0010042	125,9	0,4364	0,0010042	125,9	0,4364
40	0,0010077	167,6	0,5721	0,0010077	167,6	0,5720	0,0010077	167,6	0,5720	0,0010077	167,6	0,5720
50	0,0010120	209,4	0,7034	0,0010120	209,4	0,7034	0,0010120	209,4	0,7034	0,0010120	209,4	0,7034
60	0,0010171	251,2	0,8309	0,0010171	251,2	0,8309	0,0010171	251,2	0,8309	0,0010171	251,3	0,8309
70	0,0010228	293,1	0,9547	0,0010228	293,1	0,9547	0,0010228	293,1	0,9547	0,0010228	293,1	0,9547
80	0,0010291	335,0	1,0752	0,0010291	335,0	1,0752	0,0010291	335,0	1,0751	0,0010291	335,1	1,0751
90	0,0010361	377,0	1,1924	0,0010361	377,0	1,1924	0,0010361	377,0	1,1924	0,0010361	377,1	1,1924
100	0,0010437	419,1	1,3068	0,0010437	419,1	1,3068	0,0010437	419,1	1,3068	0,0010436	419,2	1,3068
110	0,0010518	461,4	1,4185	0,0010518	461,4	1,4184	0,0010518	461,4	1,4184	0,0010518	461,4	1,4184
120	0,9327	2706,9	7,1515	0,0010606	503,7	1,5276	0,0010606	503,7	1,5276	0,0010606	503,7	1,5276
130	0,9590	2727,9	7,2042	0,9100	2726,9	7,1786	0,8657	2725,9	7,1541	0,8253	2724,9	7,1307
140	0,9851	2748,7	7,2551	0,9349	2747,8	7,2298	0,8894	2746,9	7,2056	0,8481	2746,0	7,1824
150	1,0110	2769,3	7,3045	0,9595	2768,5	7,2794	0,9130	2767,7	7,2554	0,8707	2766,9	7,2324
160	1,0366	2789,8	7,3524	0,9840	2789,1	7,3275	0,9364	2788,4	7,3036	0,8931	2787,7	7,2809
170	1,0621	2810,2	7,3990	1,0083	2809,6	7,3742	0,9596	2808,9	7,3505	0,9153	2808,3	7,3279
180	1,0875	2830,5	7,4443	1,0325	2830,0	7,4196	0,9826	2829,4	7,3961	0,9374	2828,8	7,3736
190	1,1127	2850,8	7,4885	1,0565	2850,3	7,4639	1,0056	2849,7	7,4405	0,9593	2849,2	7,4181
200	1,1379	2871,0	7,5316	1,0804	2870,5	7,5072	1,0284	2870,0	7,4838	0,9811	2869,5	7,4616
210	1,1629	2891,1	7,5738	1,1042	2890,7	7,5494	1,0511	2890,2	7,5262	1,0029	2889,8	7,5040
220	1,1879	2911,3	7,6150	1,1280	2910,8	7,5907	1,0738	2910,4	7,5675	1,0245	2910,0	7,5454
230	1,2128	2931,4	7,6554	1,1517	2931,0	7,6311	1,0964	2930,6	7,6080	1,0461	2930,2	7,5860
240	1,2376	2951,5	7,6949	1,1753	2951,1	7,6707	1,1189	2950,7	7,6477	1,0676	2950,4	7,6257
250	1,2624	2971,6	7,7337	1,1989	2971,2	7,7096	1,1414	2970,9	7,6866	1,0891	2970,6	7,6646
260	1,2872	2991,7	7,7718	1,2224	2991,4	7,7477	1,1638	2991,0	7,7247	1,1106	2990,7	7,7028
270	1,3119	3011,8	7,8092	1,2459	3011,5	7,7851	1,1862	3011,2	7,7622	1,1319	3010,9	7,7403
280	1,3365	3031,9	7,8459	1,2693	3031,7	7,8219	1,2086	3031,4	7,7990	1,1533	3031,1	7,7772
290	1,3612	3052,1	7,8821	1,2928	3051,9	7,8581	1,2309	3051,6	7,8352	1,1746	3051,3	7,8134
300	1,3858	3072,3	7,9176	1,3162	3072,1	7,8937	1,2532	3071,8	7,8708	1,1959	3071,6	7,8491
310	1,4104	3092,6	7,9527	1,3395	3092,3	7,9287	1,2755	3092,1	7,9059	1,2172	3091,9	7,8841
320	1,4349	3112,8	7,9871	1,3629	3112,6	7,9632	1,2977	3112,4	7,9404	1,2385	3112,2	7,9187
330	1,4595	3133,2	8,0211	1,3862	3133,0	7,9972	1,3199	3132,7	7,9744	1,2597	3132,5	7,9527
340	1,4840	3153,5	8,0546	1,4095	3153,3	8,0307	1,3422	3153,1	8,0080	1,2809	3152,9	7,9863
350	1,5085	3174,0	8,0876	1,4328	3173,8	8,0638	1,3644	3173,6	8,0410	1,3021	3173,4	8,0193
360	1,5330	3194,4	8,1202	1,4561	3194,2	8,0964	1,3866	3194,1	8,0736	1,3233	3193,9	8,0520
370	1,5575	3214,9	8,1524	1,4794	3214,8	8,1285	1,4087	3214,6	8,1058	1,3445	3214,4	8,0842
380	1,5820	3235,5	8,1841	1,5027	3235,4	8,1603	1,4309	3235,2	8,1376	1,3656	3235,0	8,1159
390	1,6065	3256,2	8,2155	1,5259	3256,0	8,1916	1,4531	3255,8	8,1690	1,3868	3255,7	8,1473
400	1,6309	3276,8	8,2465	1,5492	3276,7	8,2226	1,4752	3276,5	8,1999	1,4080	3276,4	8,1783
410	1,6554	3297,6	8,2771	1,5724	3297,4	8,2532	1,4973	3297,3	8,2306	1,4291	3297,2	8,2089
420	1,6798	3318,4	8,3073	1,5956	3318,3	8,2835	1,5195	3318,1	8,2608	1,4502	3318,0	8,2392
430	1,7042	3339,3	8,3372	1,6188	3339,1	8,3134	1,5416	3339,0	8,2907	1,4713	3338,9	8,2691
440	1,7287	3360,2	8,3668	1,6421	3360,1	8,3429	1,5637	3359,9	8,3203	1,4925	3359,8	8,2987
450	1,7531	3381,2	8,3960	1,6653	3381,1	8,3722	1,5858	3381,0	8,3495	1,5136	3380,8	8,3279
460	1,7775	3402,3	8,4249	1,6885	3402,1	8,4011	1,6079	3402,0	8,3785	1,5347	3401,9	8,3569
470	1,8019	3423,4	8,4535	1,7117	3423,3	8,4297	1,6300	3423,2	8,4071	1,5558	3423,0	8,3855
480	1,8263	3444,6	8,4819	1,7349	3444,5	8,4581	1,6521	3444,3	8,4354	1,5769	3444,2	8,4139
490	1,8507	3465,8	8,5099	1,7580	3465,7	8,4861	1,6742	3465,6	8,4635	1,5980	3465,5	8,4419
500	1,8751	3487,2	8,5377	1,7812	3487,0	8,5139	1,6963	3486,9	8,4913	1,6190	3486,8	8,4697
510	1,8995	3508,5	8,5651	1,8044	3508,4	8,5414	1,7184	3508,3	8,5187	1,6401	3508,2	8,4972
520	1,9239	3530,0	8,5924	1,8276	3529,9	8,5686	1,7404	3529,8	8,5460	1,6612	3529,7	8,5244
530	1,9483	3551,5	8,6193	1,8508	3551,4	8,5956	1,7625	3551,3	8,5729	1,6823	3551,2	8,5514
540	1,9727	3573,1	8,6460	1,8739	3573,0	8,6223	1,7846	3572,9	8,5997	1,7033	3572,8	8,5781
550	1,9971	3594,8	8,6725	1,8971	3594,7	8,6487	1,8066	3594,6	8,6261	1,7244	3594,5	8,6046

Table 3. Water and Superheated Steam (Continuation) Wasser und überhitzter Dampf (Fortsetzung)

t	1,9 bar $t_s = 118,62$ °C			2,0 bar $t_s = 120,23$ °C			2,1 bar $t_s = 121,78$ °C			2,2 bar $t_s = 123,27$ °C		
	v'' 0,9290	h'' 2704,0	s'' 7,1440	v'' 0,8854	h'' 2706,3	s'' 7,1268	v'' 0,8459	h'' 2708,5	s'' 7,1105	v'' 0,8098	h'' 2710,6	s'' 7,0949
°C	v	h	s	v	h	s	v	h	s	v	h	s
550	1,9971	3594,8	8,6725	1,8971	3594,7	8,6487	1,8066	3594,6	8,6261	1,7244	3594,5	8,6046
560	2,0214	3616,5	8,6987	1,9202	3616,4	8,6750	1,8287	3616,3	8,6524	1,7455	3616,2	8,6308
570	2,0458	3638,3	8,7247	1,9434	3638,2	8,7010	1,8508	3638,1	8,6784	1,7665	3638,0	8,6568
580	2,0702	3660,1	8,7505	1,9666	3660,0	8,7268	1,8728	3660,0	8,7042	1,7876	3659,9	8,6826
590	2,0945	3682,1	8,7761	1,9897	3682,0	8,7523	1,8949	3681,9	8,7297	1,8086	3681,8	8,7082
600	2,1189	3704,1	8,8014	2,0129	3704,0	8,7776	1,9169	3703,9	8,7551	1,8297	3703,8	8,7335
610	2,1433	3726,1	8,8265	2,0360	3726,0	8,8028	1,9390	3726,0	8,7802	1,8507	3725,9	8,7586
620	2,1676	3748,2	8,8514	2,0591	3748,2	8,8277	1,9610	3748,1	8,8051	1,8718	3748,0	8,7836
630	2,1920	3770,4	8,8761	2,0823	3770,4	8,8524	1,9830	3770,3	8,8298	1,8928	3770,2	8,8083
640	2,2163	3792,7	8,9007	2,1054	3792,6	8,8769	2,0051	3792,6	8,8543	1,9139	3792,5	8,8328
650	2,2407	3815,0	8,9250	2,1286	3815,0	8,9012	2,0271	3814,9	8,8787	1,9349	3814,8	8,8571
660	2,2650	3837,4	8,9491	2,1517	3837,4	8,9254	2,0491	3837,3	8,9028	1,9559	3837,2	8,8813
670	2,2894	3859,9	8,9731	2,1748	3859,8	8,9493	2,0712	3859,8	8,9268	1,9770	3859,7	8,9052
680	2,3137	3882,4	8,9968	2,1979	3882,4	8,9731	2,0932	3882,3	8,9505	1,9980	3882,2	8,9290
690	2,3380	3905,0	9,0204	2,2211	3905,0	8,9967	2,1152	3904,9	8,9741	2,0190	3904,8	8,9526
700	2,3624	3927,7	9,0438	2,2442	3927,6	9,0201	2,1373	3927,6	8,9975	2,0401	3927,5	8,9760
710	2,3867	3950,4	9,0670	2,2673	3950,4	9,0433	2,1593	3950,3	9,0208	2,0611	3950,2	8,9992
720	2,4111	3973,2	9,0901	2,2904	3973,1	9,0664	2,1813	3973,1	9,0438	2,0821	3973,0	9,0223
730	2,4354	3996,1	9,1130	2,3136	3996,0	9,0893	2,2033	3996,0	9,0667	2,1031	3995,9	9,0452
740	2,4597	4019,0	9,1357	2,3367	4018,9	9,1120	2,2254	4018,9	9,0895	2,1242	4018,8	9,0680
750	2,4841	4042,0	9,1583	2,3598	4041,9	9,1346	2,2474	4041,9	9,1121	2,1452	4041,8	9,0905
760	2,5084	4065,0	9,1807	2,3829	4065,0	9,1570	2,2694	4064,9	9,1345	2,1662	4064,9	9,1130
770	2,5327	4088,1	9,2030	2,4060	4088,1	9,1793	2,2914	4088,0	9,1567	2,1872	4088,0	9,1352
780	2,5571	4111,3	9,2251	2,4291	4111,3	9,2014	2,3134	4111,2	9,1789	2,2082	4111,2	9,1574
790	2,5814	4134,6	9,2471	2,4523	4134,5	9,2234	2,3354	4134,5	9,2008	2,2292	4134,4	9,1793
800	2,6057	4157,9	9,2689	2,4754	4157,8	9,2452	2,3575	4157,8	9,2226	2,2502	4157,7	9,2011

Table 3. Water and Superheated Steam (Continuation) Wasser und überhitzter Dampf (Fortsetzung)

t	2,3 bar $t_s = 124{,}71\ °C$			2,4 bar $t_s = 126{,}09\ °C$			2,5 bar $t_s = 127{,}43\ °C$			2,6 bar $t_s = 128{,}73\ °C$		
	v'' 0,7768	h'' 2712,6	s'' 7,0800	v'' 0,7465	h'' 2714,5	s'' 7,0657	v'' 0,7184	h'' 2716,4	s'' 7,0520	v'' 0,6925	h'' 2718,2	s'' 7,0389
°C	v	h	s	v	h	s	v	h	s	v	h	s
0	0,0010001	0,2	−0,0001	0,0010001	0,2	−0,0001	0,0010001	0,2	−0,0001	0,0010001	0,2	−0,0001
10	0,0010001	42,2	0,1510	0,0010001	42,2	0,1510	0,0010001	42,2	0,1510	0,0010001	42,2	0,1510
20	0,0010016	84,1	0,2963	0,0010016	84,1	0,2963	0,0010016	84,1	0,2962	0,0010016	84,1	0,2962
30	0,0010042	125,9	0,4364	0,0010042	125,9	0,4364	0,0010042	125,9	0,4364	0,0010042	125,9	0,4364
40	0,0010077	167,6	0,5720	0,0010077	167,7	0,5720	0,0010077	167,7	0,5720	0,0010077	167,7	0,5720
50	0,0010120	209,4	0,7034	0,0010120	209,5	0,7034	0,0010120	209,5	0,7034	0,0010120	209,5	0,7034
60	0,0010170	251,3	0,8309	0,0010170	251,3	0,8309	0,0010170	251,3	0,8309	0,0010170	251,3	0,8309
70	0,0010228	293,1	0,9547	0,0010228	293,1	0,9547	0,0010227	293,2	0,9547	0,0010227	293,2	0,9547
80	0,0010291	335,1	1,0751	0,0010291	335,1	1,0751	0,0010291	335,1	1,0751	0,0010291	335,1	1,0751
90	0,0010361	377,1	1,1924	0,0010361	377,1	1,1924	0,0010361	377,1	1,1924	0,0010361	377,1	1,1924
100	0,0010436	419,2	1,3068	0,0010436	419,2	1,3068	0,0010436	419,2	1,3068	0,0010436	419,2	1,3067
110	0,0010518	461,4	1,4184	0,0010518	461,4	1,4184	0,0010518	461,4	1,4184	0,0010518	461,4	1,4184
120	0,0010606	503,7	1,5276	0,0010606	503,7	1,5276	0,0010606	503,8	1,5275	0,0010606	503,8	1,5275
130	0,7885	2723,9	7,1082	0,7548	2722,9	7,0866	0,7237	2721,9	7,0658	0,6950	2720,9	7,0457
140	0,8104	2745,1	7,1602	0,7758	2744,2	7,1388	0,7440	2743,3	7,1183	0,7146	2742,4	7,0984
150	0,8321	2766,1	7,2104	0,7967	2765,3	7,1893	0,7641	2764,5	7,1689	0,7340	2763,7	7,1493
160	0,8536	2787,0	7,2590	0,8173	2786,2	7,2381	0,7840	2785,5	7,2179	0,7532	2784,8	7,1985
170	0,8749	2807,6	7,3062	0,8378	2807,0	7,2854	0,8037	2806,3	7,2654	0,7722	2805,6	7,2461
180	0,8960	2828,2	7,3521	0,8581	2827,6	7,3314	0,8232	2827,0	7,3115	0,7910	2826,4	7,2924
190	0,9170	2848,6	7,3967	0,8783	2848,1	7,3762	0,8427	2847,5	7,3564	0,8098	2847,0	7,3374
200	0,9380	2869,0	7,4402	0,8984	2868,5	7,4198	0,8620	2868,0	7,4001	0,8284	2867,5	7,3812
210	0,9588	2889,3	7,4827	0,9184	2888,9	7,4624	0,8812	2888,4	7,4428	0,8469	2887,9	7,4240
220	0.9795	2909,6	7,5242	0,9383	2909,2	7,5040	0,9004	2908,7	7,4845	0,8653	2908,3	7,4657
230	1,0002	2929,8	7,5649	0,9581	2929,4	7,5446	0,9194	2929,0	7,5252	0,8837	2928,7	7,5065
240	1,0208	2950,0	7,6046	0,9779	2949,7	7,5845	0,9385	2949,3	7,5651	0,9020	2948,9	7,5465
250	1,0414	2970,2	7,6436	0,9977	2969,9	7,6235	0,9574	2969,6	7,6042	0,9203	2969,2	7,5856
260	1,0619	2990,4	7,6819	1,0173	2990,1	7,6618	0,9763	2989,8	7,6425	0,9385	2989,5	7,6240
270	1,0824	3010,6	7,7194	1,0370	3010,3	7,6994	0,9952	3010,0	7,6801	0,9566	3009,8	7,6616
280	1,1028	3030,8	7,7563	1,0566	3030,6	7,7363	1,0140	3030,3	7,7171	0,9748	3030,0	7,6986
290	1,1233	3051,1	7,7926	1,0762	3050,8	7,7726	1,0329	3050,6	7,7534	0,9929	3050,3	7,7350
300	1,1437	3071,3	7,8282	1,0957	3071,1	7,8083	1,0516	3070,9	7,7891	1,0109	3070,6	7,7707
310	1,1640	3091,6	7,8633	1,1153	3091,4	7,8434	1,0704	3091,2	7,8243	1,0290	3091,0	7,8059
320	1,1844	3112,0	7,8979	1,1348	3111,8	7,8780	1,0891	3111,5	7,8589	1,0470	3111,3	7,8405
330	1,2047	3132,3	7,9319	1,1543	3132,1	7,9121	1,1079	3131,9	7,8930	1,0650	3131,7	7,8746
340	1,2250	3152,7	7,9655	1,1737	3152,6	7,9456	1,1266	3152,4	7,9266	1,0830	3152,2	7,9082
350	1,2453	3173,2	7,9986	1,1932	3173,0	7,9787	1,1452	3172,8	7,9597	1,1010	3172,6	7,9414
360	1,2656	3193,7	8,0312	1,2126	3193,5	8,0114	1,1639	3193,3	7,9923	1,1190	3193,2	7,9740
370	1,2858	3214,3	8,0634	1,2321	3214,1	8,0436	1,1826	3213,9	8,0246	1,1369	3213,7	8,0063
380	1,3061	3234,9	8,0952	1,2515	3234,7	8,0754	1,2012	3234,5	8,0564	1,1549	3234,4	8,0381
390	1,3263	3255,5	8,1266	1,2709	3255,4	8,1068	1,2199	3255,2	8,0878	1,1728	3255,0	8,0695
400	1,3466	3276,2	8,1576	1,2903	3276,1	8,1378	1,2385	3275,9	8,1188	1,1907	3275,8	8,1005
410	1,3668	3297,0	8,1883	1,3097	3296,9	8,1685	1,2571	3296,7	8,1495	1,2086	3296,6	8,1312
420	1,3870	3317,8	8,2185	1,3290	3317,7	8,1987	1,2757	3317,6	8,1797	1,2265	3317,4	8,1615
430	1,4072	3338,7	8,2485	1,3484	3338,6	8,2287	1,2943	3338,5	8,2097	1,2444	3338,3	8,1914
440	1,4274	3359,7	8,2780	1,3678	3359,5	8,2583	1,3129	3359,4	8,2393	1,2623	3359,3	8,2211
450	1,4476	3380,7	8,3073	1,3872	3380,6	8,2875	1,3315	3380,4	8,2686	1,2802	3380,3	8,2503
460	1,4678	3401,8	8,3362	1,4065	3401,6	8,3165	1,3501	3401,5	8,2975	1,2981	3401,4	8,2793
470	1,4880	3422,9	8,3649	1,4259	3422,8	8,3451	1,3687	3422,7	8,3262	1,3159	3422,6	8,3080
480	1,5082	3444,1	8,3932	1,4452	3444,0	8,3735	1,3873	3443,9	8,3545	1,3338	3443,8	8,3363
490	1,5284	3465,4	8,4213	1,4645	3465,3	8,4015	1,4058	3465,2	8,3826	1,3517	3465,0	8,3644
500	1,5485	3486,7	8,4491	1,4839	3486,6	8,4293	1,4244	3486,5	8,4104	1,3695	3486,4	8,3922
510	1,5687	3508,1	8,4766	1,5032	3508,0	8,4568	1,4430	3507,9	8,4379	1,3874	3507,8	8,4197
520	1,5889	3529,6	8,5038	1,5225	3529,5	8,4841	1,4615	3529,4	8,4651	1,4052	3529,3	8,4469
530	1,6090	3551,1	8,5308	1,5419	3551,0	8,5110	1,4801	3550,9	8,4921	1,4231	3550,8	8,4739
540	1,6292	3572,7	8,5575	1,5612	3572,6	8,5378	1,4986	3572,5	8,5188	1,4409	3572,4	8,5007
550	1,6493	3594,4	8,5840	1,5805	3594,3	8,5643	1,5172	3594,2	8,5453	1,4587	3594,1	8,5271

Table 3. Water and Superheated Steam (Continuation) **Wasser und überhitzter Dampf** (Fortsetzung)

t	2,3 bar $t_s = 124{,}71$ °C			2,4 bar $t_s = 126{,}09$ °C			2,5 bar $t_s = 127{,}43$ °C			2,6 bar $t_s = 128{,}73$ °C		
	v'' 0,7768	h'' 2712,6	s'' 7,0800	v'' 0,7465	h'' 2714,5	s'' 7,0657	v'' 0,7184	h'' 2716,4	s'' 7,0520	v'' 0,6925	h'' 2718,2	s'' 7,0389
°C	v	h	s	v	h	s	v	h	s	v	h	s
550	1,6493	3594,4	8,5840	1,5805	3594,3	8,5643	1,5172	3594,2	8,5453	1,4587	3594,1	8,5271
560	1,6695	3616,1	8,6102	1,5998	3616,0	8,5905	1,5357	3615,9	8,5716	1,4766	3615,8	8,5534
570	1,6896	3637,9	8,6362	1,6191	3637,8	8,6165	1,5543	3637,7	8,5976	1,4944	3637,7	8,5794
580	1,7098	3659,8	8,6620	1,6384	3659,7	8,6423	1,5728	3659,6	8,6234	1,5122	3659,5	8,6052
590	1,7299	3681,7	8,6876	1,6577	3681,6	8,6679	1,5914	3681,6	8,6490	1,5301	3681,5	8,6308
600	1,7500	3703,7	8,7129	1,6770	3703,6	8,6932	1,6099	3703,6	8,6743	1,5479	3703,5	8,6561
610	1,7702	3725,8	8,7381	1,6963	3725,7	8,7184	1,6284	3725,6	8,6994	1,5657	3725,6	8,6813
620	1,7903	3747,9	8,7630	1,7156	3747,9	8,7433	1,6469	3747,8	8,7244	1,5835	3747,7	8,7062
630	1,8104	3770,1	8,7877	1,7349	3770,1	8,7680	1,6655	3770,0	8,7491	1,6013	3769,9	8,7309
640	1,8306	3792,4	8,8122	1,7542	3792,3	8,7925	1,6840	3792,3	8,7736	1,6192	3792,2	8,7555
650	1,8507	3814,7	8,8366	1,7735	3814,7	8,8169	1,7025	3814,6	8,7980	1,6370	3814,5	8,7798
660	1,8708	3837,2	8,8607	1,7928	3837,1	8,8410	1,7210	3837,0	8,8221	1,6548	3836,9	8,8039
670	1,8909	3859,6	8,8847	1,8121	3859,6	8,8650	1,7395	3859,5	8,8461	1,6726	3859,4	8,8279
680	1,9111	3882,2	8,9084	1,8314	3882,1	8,8887	1,7581	3882,0	8,8698	1,6904	3882,0	8,8517
690	1,9312	3904,8	8,9320	1,8507	3904,7	8,9123	1,7766	3904,6	8,8934	1,7082	3904,6	8,8753
700	1,9513	3927,4	8,9554	1,8699	3927,4	8,9357	1,7951	3927,3	8,9169	1,7260	3927,2	8,8987
710	1,9714	3950,2	8,9787	1,8892	3950,1	8,9590	1,8136	3950,1	8,9401	1,7438	3950,0	8,9220
720	1,9915	3973,0	9,0018	1,9085	3972,9	8,9821	1,8321	3972,9	8,9632	1,7616	3972,8	8,9450
730	2,0116	3995,8	9,0247	1,9278	3995,8	9,0050	1,8506	3995,7	8,9861	1,7794	3995,7	8,9679
740	2,0317	4018,8	9,0474	1,9470	4018,7	9,0277	1,8691	4018,7	9,0088	1,7972	4018,6	8,9907
750	2,0519	4041,8	9,0700	1,9663	4041,7	9,0503	1,8876	4041,7	9,0314	1,8150	4041,6	9,0133
760	2,0720	4064,8	9,0924	1,9856	4064,8	9,0727	1,9061	4064,7	9,0539	1,8328	4064,7	9,0357
770	2,0921	4087,9	9,1147	2,0048	4087,9	9,0950	1,9246	4087,8	9,0761	1,8505	4087,8	9,0580
780	2,1122	4111,1	9,1368	2,0241	4111,1	9,1171	1,9431	4111,0	9,0982	1,8683	4111,0	9,0801
790	2,1323	4134,4	9,1588	2,0434	4134,3	9,1391	1,9616	4134,3	9,1202	1,8861	4134,2	9,1021
800	2,1524	4157,7	9,1806	2,0626	4157,6	9,1609	1,9801	4157,6	9,1420	1,9039	4157,5	9,1239

Table 3. Water and Superheated Steam (Continuation) Wasser und überhitzter Dampf (Fortsetzung)

t	2,7 bar t_s = 129,98 °C			2,8 bar t_s = 131,20 °C			2,9 bar t_s = 132,39 °C			3,0 bar t_s = 133,54 °C		
	v'' 0,6684	h'' 2719,9	s'' 7,0262	v'' 0,6460	h'' 2721,5	s'' 7,0140	v'' 0,6251	h'' 2723,1	s'' 7,0023	v'' 0,6056	h'' 2724,7	s'' 6,9909
°C	v	h	s	v	h	s	v	h	s	v	h	s
0	0,0010001	0,2	−0,0001	0,0010001	0,2	−0,0001	0,0010001	0,3	−0,0001	0,0010001	0,3	−0,0001
10	0,0010001	42,3	0,1510	0,0010001	42,3	0,1510	0,0010001	42,3	0,1510	0,0010001	42,3	0,1510
20	0,0010016	84,1	0,2962	0,0010016	84,1	0,2962	0,0010016	84,1	0,2962	0,0010016	84,1	0,2962
30	0,0010042	125,9	0,4364	0,0010042	125,9	0,4364	0,0010042	125,9	0,4364	0,0010042	125,9	0,4364
40	0,0010077	167,7	0,5720	0,0010077	167,7	0,5720	0,0010077	167,7	0,5720	0,0010077	167,7	0,5720
50	0,0010120	209,5	0,7034	0,0010120	209,5	0,7034	0,0010120	209,5	0,7034	0,0010120	209,5	0,7034
60	0,0010170	251,3	0,8309	0,0010170	251,3	0,8308	0,0010170	251,3	0,8308	0,0010170	251,3	0,8308
70	0,0010227	293,2	0,9547	0,0010227	293,2	0,9547	0,0010227	293,2	0,9547	0,0010227	293,2	0,9547
80	0,0010291	335,1	1,0751	0,0010291	335,1	1,0751	0,0010291	335,1	1,0751	0,0010291	335,1	1,0751
90	0,0010360	377,1	1,1924	0,0010360	377,1	1,1924	0,0010360	377,1	1,1924	0,0010360	377,1	1,1924
100	0,0010436	419,2	1,3067	0,0010436	419,2	1,3067	0,0010436	419,2	1,3067	0,0010436	419,2	1,3067
110	0,0010518	461,4	1,4184	0,0010518	461,4	1,4184	0,0010518	461,4	1,4184	0,0010518	461,4	1,4184
120	0,0010606	503,8	1,5275	0,0010606	503,8	1,5275	0,0010606	503,8	1,5275	0,0010606	503,8	1,5275
130	0,6685	2719,9	7,0263	0,0010700	546,3	1,6344	0,0010700	546,3	1,6343	0,0010700	546,3	1,6343
140	0,6874	2741,5	7,0793	0,6622	2740,6	7,0607	0,6387	2739,7	7,0428	0,6167	2738,8	7,0254
150	0,7062	2762,9	7,1304	0,6803	2762,1	7,1120	0,6562	2761,3	7,0943	0,6337	2760,4	7,0771
160	0,7247	2784,0	7,1797	0,6982	2783,3	7,1616	0,6736	2782,6	7,1441	0,6506	2781,8	7,1271
170	0,7430	2805,0	7,2276	0,7160	2804,3	7,2096	0,6908	2803,6	7,1922	0,6672	2803,0	7,1754
180	0,7612	2825,8	7,2740	0,7336	2825,2	7,2561	0,7078	2824,6	7,2389	0,6837	2824,0	7,2222
190	0,7793	2846,4	7,3191	0,7510	2845,9	7,3014	0,7247	2845,3	7,2842	0,7001	2844,8	7,2677
200	0,7973	2867,0	7,3630	0,7684	2866,5	7,3454	0,7415	2866,0	7,3284	0,7164	2865,5	7,3119
210	0,8151	2887,5	7,4058	0,7856	2887,0	7,3883	0,7582	2886,6	7,3714	0,7325	2886,1	7,3550
220	0,8329	2907,9	7,4476	0,8028	2907,5	7,4302	0,7748	2907,1	7,4134	0,7486	2906,6	7,3971
230	0,8506	2928,3	7,4885	0,8199	2927,9	7,4712	0,7913	2927,5	7,4544	0,7646	2927,1	7,4381
240	0,8683	2948,6	7,5285	0,8369	2948,2	7,5112	0,8078	2947,9	7,4945	0,7805	2947,5	7,4783
250	0,8859	2968,9	7,5677	0,8539	2968,6	7,5504	0,8242	2968,2	7,5338	0,7964	2967,9	7,5176
260	0,9034	2989,2	7,6061	0,8709	2988,9	7,5889	0,8406	2988,6	7,5723	0,8123	2988,2	7,5562
270	0,9209	3009,5	7,6438	0,8878	3009,2	7,6266	0,8569	3008,9	7,6100	0,8281	3008,6	7,5940
280	0,9384	3029,8	7,6808	0,9046	3029,5	7,6637	0,8732	3029,2	7,6471	0,8438	3028,9	7,6311
290	0,9558	3050,1	7,7172	0,9215	3049,8	7,7001	0,8895	3049,5	7,6836	0,8596	3049,3	7,6676
300	0,9733	3070,4	7,7530	0,9383	3070,1	7,7359	0,9057	3069,9	7,7194	0,8753	3069,7	7,7034
310	0,9907	3090,7	7,7882	0,9551	3090,5	7,7711	0,9219	3090,3	7,7546	0,8910	3090,0	7,7387
320	1,0080	3111,1	7,8228	0,9718	3110,9	7,8058	0,9381	3110,7	7,7893	0,9066	3110,5	7,7734
330	1,0254	3131,5	7,8569	0,9886	3131,3	7,8399	0,9543	3131,1	7,8235	0,9223	3130,9	7,8076
340	1,0427	3152,0	7,8906	1,0053	3151,8	7,8736	0,9704	3151,6	7,8571	0,9379	3151,4	7,8412
350	1,0600	3172,5	7,9237	1,0220	3172,3	7,9067	0,9866	3172,1	7,8903	0,9535	3171,9	7,8744
360	1,0773	3193,0	7,9564	1,0387	3192,8	7,9394	1,0027	3192,6	7,9230	0,9691	3192,4	7,9072
370	1,0946	3213,6	7,9887	1,0554	3213,4	7,9717	1,0188	3213,2	7,9553	0,9847	3213,1	7,9395
380	1,1119	3234,2	8,0205	1,0720	3234,0	8,0035	1,0349	3233,9	7,9872	1,0003	3233,7	7,9713
390	1,1292	3254,9	8,0519	1,0887	3254,7	8,0350	1,0510	3254,6	8,0186	1,0158	3254,4	8,0028
400	1,1464	3275,6	8,0830	1,1054	3275,5	8,0660	1,0671	3275,3	8,0497	1,0314	3275,2	8,0338
410	1,1637	3296,4	8,1136	1,1220	3296,3	8,0967	1,0832	3296,1	8,0803	1,0469	3296,0	8,0645
420	1,1809	3317,3	8,1439	1,1386	3317,1	8,1270	1,0992	3317,0	8,1107	1,0625	3316,8	8,0949
430	1,1982	3338,2	8,1739	1,1553	3338,0	8,1570	1,1153	3337,9	8,1406	1,0780	3337,8	8,1248
440	1,2154	3359,2	8,2035	1,1719	3359,0	8,1866	1,1314	3358,9	8,1702	1,0935	3358,8	8,1545
450	1,2326	3380,2	8,2328	1,1885	3380,1	8,2159	1,1474	3379,9	8,1995	1,1090	3379,8	8,1838
460	1,2499	3401,3	8,2618	1,2051	3401,2	8,2448	1,1634	3401,0	8,2285	1,1245	3400,9	8,2128
470	1,2671	3422,4	8,2904	1,2217	3422,3	8,2735	1,1795	3422,2	8,2572	1,1401	3422,1	8,2414
480	1,2843	3443,7	8,3188	1,2383	3443,5	8,3019	1,1955	3443,4	8,2856	1,1556	3443,3	8,2698
490	1,3015	3464,9	8,3469	1,2549	3464,8	8,3300	1,2115	3464,7	8,3137	1,1710	3464,6	8,2979
500	1,3187	3486,3	8,3746	1,2715	3486,2	8,3578	1,2276	3486,1	8,3415	1,1865	3486,0	8,3257
510	1,3359	3507,7	8,4022	1,2881	3507,6	8,3853	1,2436	3507,5	8,3690	1,2020	3507,4	8,3532
520	1,3531	3529,2	8,4294	1,3047	3529,1	8,4125	1,2596	3529,0	8,3962	1,2175	3528,9	8,3805
530	1,3703	3550,7	8,4564	1,3212	3550,6	8,4395	1,2756	3550,5	8,4232	1,2330	3550,4	8,4075
540	1,3874	3572,3	8,4831	1,3378	3572,2	8,4663	1,2916	3572,1	8,4500	1,2485	3572,0	8,4343
550	1,4046	3594,0	8,5096	1,3544	3593,9	8,4928	1,3076	3593,8	8,4765	1,2639	3593,7	8,4608

Table 3. Water and Superheated Steam (Continuation) Wasser und überhitzter Dampf (Fortsetzung)

t	2,7 bar $t_s = 129,98\ °C$			2,8 bar $t_s = 131,20\ °C$			2,9 bar $t_s = 132,39\ °C$			3,0 bar $t_s = 133,54\ °C$		
	v'' 0,6684	h'' 2719,9	s'' 7,0262	v'' 0,6460	h'' 2721,5	s'' 7,0140	v'' 0,6251	h'' 2723,1	s'' 7,0023	v'' 0,6056	h'' 2724,7	s'' 6,9909
°C	v	h	s	v	h	s	v	h	s	v	h	s
550	1,4046	3594,0	8,5096	1,3544	3593,9	8,4928	1,3076	3593,8	8,4765	1,2639	3593,7	8,4608
560	1,4218	3615,8	8,5359	1,3709	3615,7	8,5190	1,3236	3615,6	8,5028	1,2794	3615,5	8,4870
570	1,4390	3637,6	8,5619	1,3875	3637,5	8,5451	1,3396	3637,4	8,5288	1,2949	3637,3	8,5131
580	1,4562	3659,4	8,5877	1,4041	3659,4	8,5709	1,3556	3659,3	8,5546	1,3103	3659,2	8,5389
590	1,4733	3681,4	8,6133	1,4206	3681,3	8,5964	1,3716	3681,2	8,5802	1,3258	3681,1	8,5644
600	1,4905	3703,4	8,6386	1,4372	3703,3	8,6218	1,3876	3703,2	8,6055	1,3412	3703,2	8,5898
610	1,5076	3725,5	8,6638	1,4537	3725,4	8,6469	1,4035	3725,3	8,6307	1,3567	3725,2	8,6150
620	1,5248	3747,6	8,6887	1,4703	3747,5	8,6719	1,4195	3747,5	8,6556	1,3721	3747,4	8,6399
630	1,5420	3769,8	8,7135	1,4868	3769,8	8,6966	1,4355	3769,7	8,6803	1,3876	3769,6	8,6646
640	1,5591	3792,1	8,7380	1,5034	3792,0	8,7211	1,4515	3792,0	8,7049	1,4030	3791,9	8,6892
650	1,5763	3814,5	8,7623	1,5199	3814,4	8,7455	1,4675	3814,3	8,7292	1,4185	3814,2	8,7135
660	1,5934	3836,9	8,7865	1,5365	3836,8	8,7696	1,4834	3836,7	8,7534	1,4339	3836,7	8,7377
670	1,6106	3859,4	8,8104	1,5530	3859,3	8,7936	1,4994	3859,2	8,7773	1,4494	3859,2	8,7616
680	1,6277	3881,9	8,8342	1,5695	3881,8	8,8174	1,5154	3881,8	8,8011	1,4648	3881,7	8,7854
690	1,6449	3904,5	8,8578	1,5861	3904,4	8,8410	1,5313	3904,4	8,8247	1,4802	3904,3	8,8090
700	1,6620	3927,2	8,8812	1,6026	3927,1	8,8644	1,5473	3927,1	8,8482	1,4957	3927,0	8,8325
710	1,6792	3949,9	8,9045	1,6191	3949,9	8,8877	1,5633	3949,8	8,8714	1,5111	3949,7	8,8557
720	1,6963	3972,7	8,9276	1,6357	3972,7	8,9107	1,5792	3972,6	8,8945	1,5265	3972,6	8,8788
730	1,7134	3995,6	8,9505	1,6522	3995,6	8,9337	1,5952	3995,5	8,9174	1,5420	3995,4	8,9017
740	1,7306	4018,5	8,9732	1,6687	4018,5	8,9564	1,6111	4018,4	8,9402	1,5574	4018,4	8,9245
750	1,7477	4041,5	8,9958	1,6852	4041,5	8,9790	1,6271	4041,4	8,9628	1,5728	4041,4	8,9471
760	1,7648	4064,6	9,0183	1,7018	4064,6	9,0014	1,6430	4064,5	8,9852	1,5882	4064,4	8,9695
770	1,7820	4087,7	9,0405	1,7183	4087,7	9,0237	1,6590	4087,6	9,0075	1,6037	4087,6	8,9918
780	1,7991	4110,9	9,0627	1,7348	4110,9	9,0458	1,6750	4110,8	9,0296	1,6191	4110,8	9,0139
790	1,8162	4134,2	9,0846	1,7513	4134,1	9,0678	1,6909	4134,1	9,0516	1,6345	4134,0	9,0359
800	1,8334	4157,5	9,1065	1,7678	4157,4	9,0896	1,7069	4157,4	9,0734	1,6499	4157,3	9,0577

Table 3. Water and Superheated Steam (Continuation) Wasser und überhitzter Dampf (Fortsetzung)

t	3,2 bar t_s = 135,75 °C			3,4 bar t_s = 137,86 °C			3,6 bar t_s = 139,86 °C			3,8 bar t_s = 141,78 °C		
	v'' 0,5700	h'' 2727,6	s'' 6,9693	v'' 0,5385	h'' 2730,3	s'' 6,9489	v'' 0,5103	h'' 2732,9	s'' 6,9297	v'' 0,4851	h'' 2735,3	s'' 6,9116
°C	v	h	s	v	h	s	v	h	s	v	h	s
0	0,0010001	0,3	−0,0001	0,0010001	0,3	−0,0001	0,0010000	0,3	−0,0001	0,0010000	0,3	−0,0001
10	0,0010001	42,3	0,1510	0,0010001	42,3	0,1510	0,0010001	42,3	0,1510	0,0010001	42,4	0,1510
20	0,0010016	84,2	0,2962	0,0010016	84,2	0,2962	0,0010016	84,2	0,2962	0,0010016	84,2	0,2962
30	0,0010042	126,0	0,4364	0,0010042	126,0	0,4364	0,0010041	126,0	0,4364	0,0010041	126,0	0,4364
40	0,0010077	167,7	0,5720	0,0010077	167,7	0,5720	0,0010077	167,8	0,5720	0,0010076	167,8	0,5720
50	0,0010120	209,5	0,7034	0,0010120	209,5	0,7034	0,0010120	209,6	0,7033	0,0010119	209,6	0,7033
60	0,0010170	251,3	0,8308	0,0010170	251,4	0,8308	0,0010170	251,4	0,8308	0,0010170	251,4	0,8308
70	0,0010227	293,2	0,9546	0,0010227	293,2	0,9546	0,0010227	293,2	0,9546	0,0010227	293,3	0,9546
80	0,0010291	335,1	1,0751	0,0010291	335,1	1,0751	0,0010290	335,2	1,0750	0,0010290	335,2	1,0750
90	0,0010360	377,1	1,1924	0,0010360	377,1	1,1923	0,0010360	377,2	1,1923	0,0010360	377,2	1,1923
100	0,0010436	419,2	1,3067	0,0010436	419,2	1,3067	0,0010436	419,3	1,3067	0,0010436	419,3	1,3067
110	0,0010518	461,4	1,4183	0,0010518	461,5	1,4183	0,0010517	461,5	1,4183	0,0010517	461,5	1,4183
120	0,0010606	503,8	1,5275	0,0010605	503,8	1,5275	0,0010605	503,8	1,5274	0,0010605	503,8	1,5274
130	0,0010700	546,3	1,6343	0,0010700	546,4	1,6343	0,0010700	546,4	1,6343	0,0010700	546,4	1,6343
140	0,5769	2736,9	6,9920	0,5418	2735,1	6,9605	0,5105	2733,2	6,9305	0,0010800	589,1	1,7390
150	0,5930	2758,8	7,0442	0,5570	2757,1	7,0131	0,5250	2755,4	6,9836	0,4964	2753,7	6,9554
160	0,6089	2780,3	7,0946	0,5721	2778,8	7,0638	0,5393	2777,3	7,0347	0,5101	2775,8	7,0070
170	0,6246	2801,6	7,1432	0,5869	2800,3	7,1128	0,5535	2798,9	7,0840	0,5235	2797,5	7,0566
180	0,6401	2822,7	7,1903	0,6016	2821,5	7,1602	0,5674	2820,3	7,1317	0,5368	2819,0	7,1046
190	0,6555	2843,7	7,2360	0,6162	2842,6	7,2062	0,5813	2841,4	7,1779	0,5500	2840,3	7,1511
200	0,6708	2864,5	7,2805	0,6307	2863,5	7,2508	0,5950	2862,4	7,2228	0,5630	2861,4	7,1961
210	0,6860	2885,2	7,3238	0,6450	2884,2	7,2943	0,6086	2883,3	7,2664	0,5760	2882,4	7,2400
220	0,7012	2905,8	7,3660	0,6593	2904,9	7,3367	0,6221	2904,1	7,3089	0,5888	2903,2	7,2826
230	0,7162	2926,3	7,4072	0,6735	2925,5	7,3780	0,6356	2924,7	7,3504	0,6016	2923,9	7,3243
240	0,7312	2946,8	7,4475	0,6877	2946,1	7,4184	0,6489	2945,3	7,3909	0,6143	2944,6	7,3649
250	0,7461	2967,2	7,4869	0,7017	2966,5	7,4579	0,6623	2965,9	7,4306	0,6270	2965,2	7,4047
260	0,7610	2987,6	7,5255	0,7158	2987,0	7,4967	0,6756	2986,4	7,4694	0,6396	2985,7	7,4436
270	0,7759	3008,0	7,5634	0,7298	3007,4	7,5346	0,6888	3006,8	7,5074	0,6522	3006,2	7,4817
280	0,7907	3028,4	7,6006	0,7437	3027,8	7,5719	0,7020	3027,3	7,5448	0,6647	3026,7	7,5191
290	0,8054	3048,8	7,6371	0,7577	3048,3	7,6085	0,7152	3047,7	7,5814	0,6772	3047,2	7,5558
300	0,8202	3069,2	7,6730	0,7716	3068,7	7,6444	0,7283	3068,2	7,6174	0,6897	3067,7	7,5918
310	0,8349	3089,6	7,7083	0,7854	3089,1	7,6798	0,7415	3088,7	7,6528	0,7021	3088,2	7,6273
320	0,8496	3110,0	7,7431	0,7993	3109,6	7,7146	0,7546	3109,1	7,6876	0,7145	3108,7	7,6621
330	0,8643	3130,5	7,7773	0,8131	3130,1	7,7488	0,7676	3129,7	7,7219	0,7269	3129,2	7,6965
340	0,8790	3151,0	7,8110	0,8269	3150,6	7,7825	0,7807	3150,2	7,7557	0,7393	3149,8	7,7303
350	0,8936	3171,5	7,8442	0,8407	3171,1	7,8158	0,7937	3170,8	7,7890	0,7517	3170,4	7,7636
360	0,9082	3192,1	7,8770	0,8545	3191,7	7,8486	0,8068	3191,4	7,8218	0,7641	3191,0	7,7964
370	0,9229	3212,7	7,9093	0,8683	3212,4	7,8809	0,8198	3212,0	7,8541	0,7764	3211,7	7,8288
380	0,9375	3233,4	7,9412	0,8821	3233,0	7,9128	0,8328	3232,7	7,8861	0,7887	3232,4	7,8608
390	0,9521	3254,1	7,9726	0,8958	3253,8	7,9443	0,8458	3253,5	7,9176	0,8011	3253,1	7,8923
400	0,9667	3274,9	8,0037	0,9096	3274,6	7,9754	0,8588	3274,2	7,9487	0,8134	3273,9	7,9234
410	0,9812	3295,7	8,0344	0,9233	3295,4	8,0061	0,8718	3295,1	7,9795	0,8257	3294,8	7,9542
420	0,9958	3316,6	8,0648	0,9370	3316,3	8,0365	0,8847	3316,0	8,0098	0,8380	3315,7	7,9846
430	1,0104	3337,5	8,0948	0,9507	3337,2	8,0665	0,8977	3336,9	8,0398	0,8503	3336,7	8,0146
440	1,0250	3358,5	8,1244	0,9645	3358,2	8,0962	0,9107	3358,0	8,0695	0,8626	3357,7	8,0443
450	1,0395	3379,5	8,1537	0,9782	3379,3	8,1255	0,9236	3379,0	8,0988	0,8748	3378,8	8,0736
460	1,0541	3400,7	8,1827	0,9919	3400,4	8,1545	0,9366	3400,2	8,1279	0,8871	3399,9	8,1027
470	1,0686	3421,8	8,2114	1,0056	3421,6	8,1832	0,9495	3421,4	8,1566	0,8994	3421,1	8,1314
480	1,0831	3443,1	8,2398	1,0192	3442,8	8,2116	0,9624	3442,6	8,1850	0,9116	3442,4	8,1598
490	1,0977	3464,4	8,2679	1,0329	3464,1	8,2397	0,9754	3463,9	8,2131	0,9239	3463,7	8,1879
500	1,1122	3485,7	8,2957	1,0466	3485,5	8,2675	0,9883	3485,3	8,2409	0,9361	3485,1	8,2158
510	1,1267	3507,2	8,3232	1,0603	3507,0	8,2951	1,0012	3506,7	8,2685	0,9484	3506,5	8,2433
520	1,1412	3528,7	8,3505	1,0740	3528,5	8,3223	1,0141	3528,2	8,2958	0,9606	3528,0	8,2706
530	1,1558	3550,2	8,3775	1,0876	3550,0	8,3494	1,0271	3549,8	8,3228	0,9729	3549,6	8,2977
540	1,1703	3571,8	8,4043	1,1013	3571,7	8,3761	1,0400	3571,5	8,3496	0,9851	3571,3	8,3244
550	1,1848	3593,5	8,4308	1,1150	3593,3	8,4027	1,0529	3593,2	8,3761	0,9973	3593,0	8,3510

Table 3. Water and Superheated Steam (Continuation) Wasser und überhitzter Dampf (Fortsetzung)

t	3,2 bar $t_s = 135{,}75$ °C			3,4 bar $t_s = 137{,}86$ °C			3,6 bar $t_s = 139{,}86$ °C			3,8 bar $t_s = 141{,}78$ °C		
	v'' 0,5700	h'' 2727,6	s'' 6,9693	v'' 0,5385	h'' 2730,3	s'' 6,9489	v'' 0,5103	h'' 2732,9	s'' 6,9297	v'' 0,4851	h'' 2735,3	s'' 6,9116
°C	v	h	s	v	h	s	v	h	s	v	h	s
550	1,1848	3593,5	8,4308	1,1150	3593,3	8,4027	1,0529	3593,2	8,3761	0,9973	3593,0	8,3510
560	1,1993	3615,3	8,4571	1,1286	3615,1	8,4289	1,0658	3614,9	8,4024	1,0096	3614,7	8,3773
570	1,2138	3637,1	8,4831	1,1423	3636,9	8,4550	1,0787	3636,8	8,4284	1,0218	3636,6	8,4033
580	1,2283	3659,0	8,5089	1,1559	3658,8	8,4808	1,0916	3658,7	8,4543	1,0340	3658,5	8,4292
590	1,2428	3681,0	8,5345	1,1696	3680,8	8,5064	1,1045	3680,6	8,4799	1,0462	3680,5	8,4548
600	1,2573	3703,0	8,5599	1,1832	3702,8	8,5318	1,1174	3702,7	8,5052	1,0585	3702,5	8,4801
610	1,2718	3725,1	8,5850	1,1969	3724,9	8,5569	1,1303	3724,8	8,5304	1,0707	3724,6	8,5053
620	1,2863	3747,2	8,6100	1,2105	3747,1	8,5819	1,1431	3746,9	8,5554	1,0829	3746,8	8,5303
630	1,3008	3769,5	8,6347	1,2241	3769,3	8,6066	1,1560	3769,2	8,5801	1,0951	3769,0	8,5550
640	1,3152	3791,7	8,6593	1,2378	3791,6	8,6312	1,1689	3791,5	8,6047	1,1073	3791,3	8,5796
650	1,3297	3814,1	8,6836	1,2514	3814,0	8,6555	1,1818	3813,8	8,6290	1,1195	3813,7	8,6040
660	1,3442	3836,5	8,7078	1,2650	3836,4	8,6797	1,1947	3836,3	8,6532	1,1317	3836,1	8,6281
670	1,3587	3859,0	8,7318	1,2787	3858,9	8,7037	1,2075	3858,7	8,6772	1,1439	3858,6	8,6521
680	1,3732	3881,6	8,7555	1,2923	3881,4	8,7275	1,2204	3881,3	8,7010	1,1561	3881,2	8,6759
690	1,3876	3904,2	8,7791	1,3059	3904,1	8,7511	1,2333	3903,9	8,7246	1,1683	3903,8	8,6995
700	1,4021	3926,9	8,8026	1,3195	3926,8	8,7745	1,2462	3926,6	8,7480	1,1805	3926,5	8,7230
710	1,4166	3949,6	8,8258	1,3332	3949,5	8,7978	1,2590	3949,4	8,7713	1,1927	3949,3	8,7462
720	1,4310	3972,4	8,8489	1,3468	3972,3	8,8209	1,2719	3972,2	8,7944	1,2049	3972,1	8,7693
730	1,4455	3995,3	8,8719	1,3604	3995,2	8,8438	1,2848	3995,1	8,8173	1,2171	3995,0	8,7923
740	1,4600	4018,3	8,8946	1,3740	4018,2	8,8665	1,2976	4018,0	8,8401	1,2293	4017,9	8,8150
750	1,4744	4041,3	8,9172	1,3876	4041,2	8,8891	1,3105	4041,1	8,8627	1,2415	4041,0	8,8376
760	1,4889	4064,3	8,9396	1,4013	4064,2	8,9116	1,3233	4064,1	8,8851	1,2536	4064,0	8,8601
770	1,5034	4087,5	8,9619	1,4149	4087,4	8,9339	1,3362	4087,3	8,9074	1,2658	4087,2	8,8824
780	1,5178	4110,7	8,9841	1,4285	4110,6	8,9560	1,3491	4110,5	8,9296	1,2780	4110,4	8,9045
790	1,5323	4133,9	9,0060	1,4421	4133,8	8,9780	1,3619	4133,7	8,9515	1,2902	4133,6	8,9265
800	1,5467	4157,2	9,0279	1,4557	4157,2	8,9998	1,3748	4157,1	8,9734	1,3024	4157,0	8,9483

Table 3. Water and Superheated Steam (Continuation) Wasser und überhitzter Dampf (Fortsetzung)

	4,0 bar $t_s = 143,62$ °C			4,2 bar $t_s = 145,39$ °C			4,4 bar $t_s = 147,09$ °C			4,6 bar $t_s = 148,73$ °C		
t	v'' 0,4622	h'' 2737,6	s'' 6,8943	v'' 0,4415	h'' 2739,8	s'' 6,8779	v'' 0,4226	h'' 2741,9	s'' 6,8623	v'' 0,4053	h'' 2743,9	s'' 6,8473
°C	v	h	s	v	h	s	v	h	s	v	h	s
0	0,0010000	0,4	−0,0001	0,0010000	0,4	−0,0001	0,0010000	0,4	−0,0001	0,0010000	0,4	−0,0001
10	0,0010001	42,4	0,1510	0,0010001	42,4	0,1510	0,0010000	42,4	0,1510	0,0010000	42,4	0,1510
20	0,0010015	84,2	0,2962	0,0010015	84,3	0,2962	0,0010015	84,3	0,2962	0,0010015	84,3	0,2962
30	0,0010041	126,0	0,4364	0,0010041	126,0	0,4364	0,0010041	126,1	0,4364	0,0010041	126,1	0,4364
40	0,0010076	167,8	0,5720	0,0010076	167,8	0,5720	0,0010076	167,8	0,5720	0,0010076	167,9	0,5719
50	0,0010119	209,6	0,7033	0,0010119	209,6	0,7033	0,0010119	209,6	0,7033	0,0010119	209,6	0,7033
60	0,0010170	251,4	0,8308	0,0010170	251,4	0,8308	0,0010170	251,4	9,8308	0,0010169	251,5	0,8307
70	0,0010227	293,3	0,9546	0,0010227	293,3	0,9546	0,0010227	293,3	0,9546	0,0010226	293,3	0,9546
80	0,0010290	335,2	1,0750	0,0010290	335,2	1,0750	0,0010290	335,2	1,0750	0,0010290	335,2	1,0750
90	0,0010360	377,2	1,1923	0,0010360	377,2	1,1923	0,0010360	377,2	1,1923	0,0010360	377,2	1,1922
100	0,0010436	419,3	1,3066	0,0010435	419,3	1,3066	0,0010435	419,3	1,3066	0,0010435	419,3	1,3066
110	0,0010517	461,5	1,4183	0,0010517	461,5	1,4183	0,0010517	461,5	1,4182	0,0010517	461,5	1,4182
120	0,0010605	503,9	1,5274	0,0010605	503,9	1,5274	0,0010605	503,9	1,5274	0,0010605	503,9	1,5274
130	0,0010699	546,4	1,6342	0,0010699	546,4	1,6342	0,0010699	546,4	1,6342	0,0010699	546,4	1,6342
140	0,0010800	589,1	1,7390	0,0010800	589,1	1,7389	0,0010800	589,2	1,7389	0,0010800	589,2	1,7389
150	0,4707	2752,0	6,9285	0,4473	2750,3	6,9028	0,4261	2748,5	6,8780	0,4068	2746,8	6,8542
160	0,4837	2774,2	6,9805	0,4599	2772,7	6,9551	0,4382	2771,1	6,9308	0,4184	2769,5	6,9074
170	0,4966	2796,1	7,0305	0,4722	2794,7	7,0055	0,4500	2793,3	6,9815	0,4297	2791,9	6,9585
180	0,5093	2817,8	7,0788	0,4843	2816,5	7,0541	0,4617	2815,2	7,0304	0,4410	2814,0	7,0077
190	0,5218	2839,2	7,1255	0,4963	2838,0	7,1010	0,4732	2836,9	7,0776	0,4520	2835,7	7,0552
200	0,5343	2860,4	7,1708	0,5082	2859,3	7,1466	0,4846	2858,3	7,1234	0,4630	2857,2	7,1011
210	0,5466	2881,4	7,2148	0,5200	2880,5	7,1908	0,4959	2879,5	7,1678	0,4738	2878,6	7,1457
220	0,5589	2902,3	7,2576	0,5317	2901,5	7,2338	0,5071	2900,6	7,2109	0,4846	2899,7	7,1891
230	0,5710	2923,1	7,2994	0,5434	2922,3	7,2757	0,5182	2921,5	7,2530	0,4953	2920,7	7,2312
240	0,5831	2943,9	7,3402	0,5549	2943,1	7,3166	0,5293	2942,4	7,2940	0,5059	2941,6	7,2724
250	0,5952	2964,5	7,3800	0,5664	2963,8	7,3565	0,5403	2963,1	7,3340	0,5164	2962,4	7,3125
260	0,6072	2985,1	7,4190	0,5779	2984,5	7,3956	0,5513	2983,8	7,3732	0,5269	2983,2	7,3518
270	0,6192	3005,6	7,4572	0,5893	3005,1	7,4339	0,5622	3004,5	7,4116	0,5374	3003,9	7,3902
280	0,6311	3026,2	7,4947	0,6007	3025,6	7,4714	0,5731	3025,1	7,4492	0,5478	3024,5	7,4279
290	0,6430	3046,7	7,5314	0,6120	3046,2	7,5082	0,5839	3045,7	7,4861	0,5582	3045,1	7,4649
300	0,6549	3067,2	7,5675	0,6234	3066,7	7,5444	0,5947	3066,2	7,5223	0,5686	3065,7	7,5011
310	0,6667	3087,7	7,6030	0,6346	3087,3	7,5799	0,6055	3086,8	7,5579	0,5789	3086,3	7,5368
320	0,6785	3108,3	7,6379	0,6459	3107,8	7,6149	0,6163	3107,4	7,5929	0,5892	3107,0	7,5718
330	0,6903	3128,8	7,6723	0,6572	3128,4	7,6493	0,6270	3128,0	7,6273	0,5995	3127,6	7,6063
340	0,7021	3149,4	7,7061	0,6684	3149,0	7,6831	0,6378	3148,6	7,6612	0,6098	3148,2	7,6402
350	0,7139	3170,0	7,7395	0,6796	3169,6	7,7165	0,6485	3169,2	7,6946	0,6201	3168,9	7,6736
360	0,7256	3190,6	7,7723	0,6908	3190,3	7,7494	0,6592	3189,9	7,7275	0,6303	3189,6	7,7066
370	0,7373	3211,3	7,8047	0,7020	3211,0	7,7818	0,6699	3210,6	7,7600	0,6406	3210,3	7,7391
380	0,7491	3232,1	7,8367	0,7132	3231,7	7,8138	0,6806	3231,4	7,7920	0,6508	3231,1	7,7711
390	0,7608	3252,8	7,8683	0,7244	3252,5	7,8454	0,6912	3252,2	7,8236	0,6610	3251,9	7,8027
400	0,7725	3273,6	7,8994	0,7355	3273,3	7,8766	0,7019	3273,0	7,8548	0,6712	3272,7	7,8339
410	0,7842	3294,5	7,9302	0,7467	3294,2	7,9074	0,7125	3293,9	7,8856	0,6814	3293,6	7,8648
420	0,7959	3315,4	7,9606	0,7578	3315,1	7,9378	0,7232	3314,9	7,9160	0,6916	3314,6	7,8952
430	0,8076	3336,4	7,9906	0,7689	3336,1	7,9678	0,7338	3335,9	7,9461	0,7018	3335,6	7,9253
440	0,8192	3357,4	8,0203	0,7801	3357,2	7,9976	0,7444	3356,9	7,9758	0,7119	3356,6	7,9550
450	0,8309	3378,5	8,0497	0,7912	3378,3	8,0269	0,7551	3378,0	8,0052	0,7221	3377,7	7,9844
460	0,8426	3399,7	8,0787	0,8023	3399,4	8,0560	0,7657	3399,2	8,0343	0,7322	3398,9	8,0135
470	0,8542	3420,9	8,1075	0,8134	3420,6	8,0847	0,7763	3420,4	8,0630	0,7424	3420,2	8,0423
480	0,8659	3442,1	8,1359	0,8245	3441,9	8,1132	0,7869	3441,7	8,0915	0,7526	3441,4	8,0707
490	0,8775	3463,5	8,1640	0,8356	3463,2	8,1413	0,7975	3463,0	8,1196	0,7627	3462,8	8,0989
500	0,8892	3484,9	8,1919	0,8467	3484,6	8,1692	0,8081	3484,4	8,1475	0,7728	3484,2	8,1268
510	0,9008	3506,3	8,2195	0,8578	3506,1	8,1967	0,8187	3505,9	8,1751	0,7830	3505,7	8,1544
520	0,9125	3527,8	8,2468	0,8689	3527,6	8,2241	0,8293	3527,4	8,2024	0,7931	3527,2	8,1817
530	0,9241	3549,4	8,2738	0,8800	3549,2	8,2511	0,8399	3549,0	8,2294	0,8032	3548,8	8,2087
540	0,9357	3571,1	8,3006	0,8911	3570,9	8,2779	0,8504	3570,7	8,2563	0,8134	3570,5	8,2356
550	0,9474	3592,8	8,3271	0,9021	3592,6	8,3044	0,8610	3592,4	8,2828	0,8235	3592,2	8,2621

Table 3. Water and Superheated Steam (Continuation) **Wasser und überhitzter Dampf** (Fortsetzung)

t	4,0 bar $t_s = 143{,}62$ °C			4,2 bar $t_s = 145{,}39$ °C			4,4 bar $t_s = 147{,}09$ °C			4,6 bar $t_s = 148{,}73$ °C		
	v'' 0,4622	h'' 2737,6	s'' 6,8943	v'' 0,4415	h'' 2739,8	s'' 6,8779	v'' 0,4226	h'' 2741,9	s'' 6,8623	v'' 0,4053	h'' 2743,9	s'' 6,8473
°C	v	h	s	v	h	s	v	h	s	v	h	s
550	0,9474	3592,8	8,3271	0,9021	3592,6	8,3044	0,8610	3592,4	8,2828	0,8235	3592,2	8,2621
560	0,9590	3614,6	8,3534	0,9132	3614,4	8,3307	0,8716	3614,2	8,3091	0,8336	3614,0	8,2884
570	0,9706	3636,4	8,3795	0,9243	3636,2	8,3568	0,8822	3636,0	8,3352	0,8437	3635,9	8,3145
580	0,9822	3658,3	8,4053	0,9353	3658,1	8,3827	0,8927	3658,0	8,3610	0,8538	3657,8	8,3404
590	0,9938	3680,3	8,4309	0,9464	3680,1	8,4083	0,9033	3680,0	8,3867	0,8639	3679,8	8,3660
600	1,0054	3702,3	8,4563	0,9575	3702,2	8,4337	0,9138	3702,0	8,4121	0,8740	3701,8	8,3914
610	1,0170	3724,4	8,4815	0,9685	3724,3	8,4588	0,9244	3724,1	8,4372	0,8841	3724,0	8,4166
620	1,0286	3746,6	8,5065	0,9796	3746,5	8,4838	0,9350	3746,3	8,4622	0,8942	3746,2	8,4416
630	1,0402	3768,9	8,5312	0,9906	3768,7	8,5086	0,9455	3768,6	8,4870	0,9043	3768,4	8,4664
640	1,0518	3791,2	8,5558	1,0017	3791,0	8,5332	0,9561	3790,9	8,5116	0,9144	3790,7	8,4909
650	1,0634	3813,5	8,5802	1,0127	3813,4	8,5575	0,9666	3813,3	8,5359	0,9245	3813,1	8,5153
660	1,0750	3836,0	8,6043	1,0238	3835,8	8,5817	0,9772	3835,7	8,5601	0,9346	3835,6	8,5395
670	1,0866	3858,5	8,6283	1,0348	3858,3	8,6057	0,9877	3858,2	8,5841	0,9447	3858,1	8,5635
680	1,0982	3881,0	8,6521	1,0459	3880,9	8,6295	0,9983	3880,8	8,6079	0,9548	3880,7	8,5873
690	1,1098	3903,7	8,6758	1,0569	3903,6	8,6531	1,0088	3903,4	8,6316	0,9649	3903,3	8,6109
700	1,1214	3926,4	8,6992	1,0679	3926,3	8,6766	1,0193	3926,1	8,6550	0,9750	3926,0	8,6344
710	1,1330	3949,1	8,7225	1,0790	3949,0	8,6999	1,0299	3948,9	8,6783	0,9850	3948,8	8,6577
720	1,1446	3972,0	8,7456	1,0900	3971,9	8,7230	1,0404	3971,7	8,7014	0,9951	3971,6	8,6808
730	1,1562	3994,9	8,7685	1,1010	3994,8	8,7459	1,0509	3994,6	8,7243	1,0052	3994,5	8,7037
740	1,1677	4017,8	8,7913	1,1121	4017,7	8,7687	1,0615	4017,6	8,7471	1,0153	4017,5	8,7265
750	1,1793	4040,8	8,8139	1,1231	4040,7	8,7913	1,0720	4040,6	8,7697	1,0253	4040,5	8,7491
760	1,1909	4063,9	8,8363	1,1341	4063,8	8,8137	1,0825	4063,7	8,7922	1,0354	4063,6	8,7716
770	1,2025	4087,1	8,8586	1,1452	4087,0	8,8360	1,0931	4086,9	8,8145	1,0455	4086,8	8,7939
780	1,2141	4110,3	8,8808	1,1562	4110,2	8,8582	1,1036	4110,1	8,8366	1,0556	4110,0	8,8161
790	1,2256	4133,5	8,9028	1,1672	4133,4	8,8802	1,1141	4133,4	8,8586	1,0656	4133,3	8,8381
800	1,2372	4156,9	8,9246	1,1782	4156,8	8,9020	1,1246	4156,7	8,8805	1,0757	4156,6	8,8599

Table 3. Water and Superheated Steam (Continuation) Wasser und überhitzter Dampf (Fortsetzung)

t	4,8 bar t_s = 150,31 °C			5,0 bar t_s = 151,84 °C			5,5 bar t_s = 155,47 °C			6,0 bar t_s = 158,84 °C		
	v'' 0,3894	h'' 2745,7	s'' 6,8330	v'' 0,3747	h'' 2747,5	s'' 6,8192	v'' 0,3425	h'' 2751,7	s'' 6,7870	v'' 0,3155	h'' 2755,5	s'' 6,7575
°C	v	h	s	v	h	s	v	h	s	v	h	s
0	0,0010000	0,4	−0,0001	0,0010000	0,5	−0,0001	0,0009999	0,5	−0,0001	0,0009999	0,6	−0,0001
10	0,0010000	42,5	0,1509	0,0010000	42,5	0,1509	0,0010000	42,5	0,1509	0,0010000	42,6	0,1509
20	0,0010015	84,3	0,2962	0,0010015	84,3	0,2962	0,0010015	84,4	0,2962	0,0010015	84,4	0,2962
30	0,0010041	126,1	0,4364	0,0010041	126,1	0,4364	0,0010041	126,2	0,4363	0,0010040	126,2	0,4363
40	0,0010076	167,9	0,5719	0,0010076	167,9	0,5719	0,0010076	167,9	0,5719	0,0010075	168,0	0,5719
50	0,0010119	209,7	0,7033	0,0010119	209,7	0,7033	0,0010119	209,7	0,7033	0,0010119	209,8	0,7032
60	0,0010169	251,5	0,8307	0,0010169	251,5	0,8307	0,0010169	251,5	0,8307	0,0010169	251,6	0,8307
70	0,0010226	293,3	0,9545	0,0010226	293,4	0,9545	0,0010226	293,4	0,9545	0,0010226	293,4	0,9545
80	0,0010290	335,3	1,0750	0,0010290	335,3	1,0750	0,0010289	335,3	1,0749	0,0010289	335,4	1,0749
90	0,0010359	377,3	1,1922	0,0010359	377,3	1,1922	0,0010359	377,3	1,1922	0,0010359	377,3	1,1921
100	0,0010435	419,3	1,3066	0,0010435	419,4	1,3066	0,0010435	419,4	1,3065	0,0010434	419,4	1,3065
110	0,0010517	461,6	1,4182	0,0010517	461,6	1,4182	0,0010516	461,6	1,4181	0,0010516	461,6	1,4181
120	0,0010605	503,9	1,5273	0,0010605	503,9	1,5273	0,0010604	504,0	1,5273	0,0010604	504,0	1,5272
130	0,0010699	546,4	1,6342	0,0010699	546,5	1,6341	0,0010699	546,5	1,6341	0,0010698	546,5	1,6340
140	0,0010800	589,2	1,7389	0,0010800	589,2	1,7388	0,0010799	589,2	1,7388	0,0010799	589,3	1,7387
150	0,0010908	632,2	1,8416	0,0010908	632,2	1,8416	0,0010907	632,2	1,8415	0,0010907	632,2	1,8415
160	0,4002	2768,0	6,8849	0,3835	2766,4	6,8631	0,3470	2762,3	6,8117	0,3165	2758,2	6,7640
170	0,4112	2790,5	6,9363	0,3941	2789,1	6,9149	0,3568	2785,4	6,8644	0,3257	2781,8	6,8177
180	0,4220	2812,7	6,9858	0,4045	2811,4	6,9647	0,3664	2808,1	6,9151	0,3346	2804,8	6,8691
190	0,4326	2834,6	7,0336	0,4148	2833,4	7,0127	0,3759	2830,5	6,9638	0,3434	2827,5	6,9185
200	0,4432	2856,2	7,0798	0,4250	2855,1	7,0592	0,3852	2852,5	7,0108	0,3520	2849,7	6,9662
210	0,4536	2877,5	7,1245	0,4350	2876,6	7,1042	0,3944	2874,2	7,0563	0,3606	2871,7	7,0121
220	0,4640	2898,8	7,1680	0,4450	2898,0	7,1478	0,4036	2895,7	7,1004	0,3690	2893,5	7,0567
230	0,4742	2919,9	7,2104	0,4549	2919,1	7,1903	0,4126	2917,1	7,1432	0,3774	2915,0	7,0999
240	0,4844	2940,9	7,2516	0,4647	2940,1	7,2317	0,4216	2938,3	7,1849	0,3857	2936,4	7,1419
250	0,4946	2961,8	7,2919	0,4744	2961,1	7,2721	0,4305	2959,3	7,2256	0,3939	2957,6	7,1829
260	0,5046	2982,5	7,3313	0,4841	2981,9	7,3115	0,4394	2980,3	7,2653	0,4021	2978,7	7,2228
270	0,5147	3003,3	7,3698	0,4938	3002,7	7,3501	0,4482	3001,2	7,3041	0,4102	2999,7	7,2618
280	0,5247	3024,0	7,4075	0,5034	3023,4	7,3879	0,4570	3022,0	7,3421	0,4183	3020,6	7,3000
290	0,5347	3044,6	7,4445	0,5130	3044,1	7,4250	0,4658	3042,8	7,3793	0,4264	3041,5	7,3374
300	0,5446	3065,2	7,4809	0,5226	3064,8	7,4614	0,4745	3063,5	7,4158	0,4344	3062,3	7,3740
310	0,5545	3085,9	7,5165	0,5321	3085,4	7,4971	0,4832	3084,2	7,4517	0,4424	3083,1	7,4100
320	0,5644	3106,5	7,5516	0,5416	3106,1	7,5322	0,4919	3105,0	7,4869	0,4504	3103,9	7,4454
330	0,5743	3127,2	7,5861	0,5511	3126,7	7,5668	0,5005	3125,7	7,5215	0,4583	3124,6	7,4801
340	0,5842	3147,8	7,6201	0,5606	3147,4	7,6008	0,5091	3146,4	7,5556	0,4663	3145,4	7,5143
350	0,5940	3168,5	7,6536	0,5701	3168,1	7,6343	0,5178	3167,2	7,5892	0,4742	3166,2	7,5479
360	0,6039	3189,2	7,6865	0,5795	3188,8	7,6673	0,5264	3187,9	7,6222	0,4821	3187,0	7,5810
370	0,6137	3209,9	7,7190	0,5889	3209,6	7,6998	0,5350	3208,7	7,6548	0,4900	3207,9	7,6137
380	0,6235	3230,7	7,7511	0,5984	3230,4	7,7319	0,5435	3229,6	7,6870	0,4979	3228,7	7,6459
390	0,6333	3251,5	7,7827	0,6078	3251,2	7,7635	0,5521	3250,4	7,7187	0,5057	3249,6	7,6776
400	0,6431	3272,4	7,8140	0,6172	3272,1	7,7948	0,5607	3271,3	7,7500	0,5136	3270,6	7,7090
410	0,6528	3293,3	7,8448	0,6266	3293,0	7,8256	0,5692	3292,3	7,7809	0,5215	3291,6	7,7399
420	0,6626	3314,3	7,8753	0,6359	3314,0	7,8561	0,5778	3313,3	7,8114	0,5293	3312,6	7,7705
430	0,6724	3335,3	7,9054	0,6453	3335,0	7,8862	0,5863	3334,3	7,8415	0,5371	3333,7	7,8007
440	0,6821	3356,4	7,9351	0,6547	3356,1	7,9160	0,5948	3355,4	7,8713	0,5450	3354,8	7,8305
450	0,6919	3377,5	7,9645	0,6640	3377,2	7,9454	0,6034	3376,6	7,9008	0,5528	3376,0	7,8600
460	0,7016	3398,7	7,9936	0,6734	3398,4	7,9745	0,6119	3397,8	7,9299	0,5606	3397,2	7,8891
470	0,7113	3419,9	8,0224	0,6828	3419,7	8,0033	0,6204	3419,1	7,9587	0,5684	3418,5	7,9180
480	0,7211	3441,2	8,0509	0,6921	3441,0	8,0318	0,6289	3440,4	7,9872	0,5762	3439,8	7,9465
490	0,7308	3462,6	8,0790	0,7014	3462,3	8,0600	0,6374	3461,8	8,0154	0,5840	3461,2	7,9747
500	0,7405	3484,0	8,1069	0,7108	3483,8	8,0879	0,6459	3483,2	8,0433	0,5918	3482,7	8,0027
510	0,7502	3505,5	8,1345	0,7201	3505,3	8,1155	0,6544	3504,7	8,0710	0,5996	3504,2	8,0303
520	0,7599	3527,0	8,1618	0,7294	3526,8	8,1428	0,6629	3526,3	8,0983	0,6074	3525,8	8,0577
530	0,7697	3548,6	8,1889	0,7388	3548,4	8,1699	0,6714	3547,9	8,1254	0,6152	3547,4	8,0848
540	0,7794	3570,3	8,2157	0,7481	3570,1	8,1967	0,6799	3569,6	8,1523	0,6230	3569,1	8,1117
550	0,7891	3592,0	8,2423	0,7574	3591,8	8,2233	0,6883	3591,4	8,1789	0,6308	3590,9	8,1383

Table 3. Water and Superheated Steam (Continuation) **Wasser und überhitzter Dampf** (Fortsetzung)

t	4,8 bar $t_s = 150,31$ °C			5,0 bar $t_s = 151,84$ °C			5,5 bar $t_s = 155,47$ °C			6,0 bar $t_s = 158,84$ °C		
	v'' 0,3894	h'' 2745,7	s'' 6,8330	v'' 0,3747	h'' 2747,5	s'' 6,8192	v'' 0,3425	h'' 2751,7	s'' 6,7870	v'' 0,3155	h'' 2755,5	s'' 6,7575
°C	v	h	s	v	h	s	v	h	s	v	h	s
550	0,7891	3592,0	8,2423	0,7574	3591,8	8,2233	0,6883	3591,4	8,1789	0,6308	3590,9	8,1383
560	0,7988	3613,8	8,2686	0,7667	3613,6	8,2496	0,6968	3613,2	8,2052	0,6386	3612,7	8,1647
570	0,8085	3635,7	8,2947	0,7760	3635,5	8,2757	0,7053	3635,1	8,2313	0,6463	3634,6	8,1908
580	0,8182	3657,6	8,3206	0,7853	3657,4	8,3016	0,7138	3657,0	8,2572	0,6541	3656,6	8,2167
590	0,8278	3679,6	8,3462	0,7946	3679,4	8,3272	0,7222	3679,0	8,2829	0,6619	3678,6	8,2423
600	0,8375	3701,7	8,3716	0,8039	3701,5	8,3526	0,7307	3701,1	8,3083	0,6696	3700,7	8,2678
610	0,8472	3723,8	8,3968	0,8132	3723,6	8,3778	0,7391	3723,2	8,3335	0,6774	3722,8	8,2930
620	0,8569	3746,0	8,4218	0,8225	3745,8	8,4028	0,7476	3745,5	8,3585	0,6851	3745,1	8,3180
630	0,8666	3768,3	8,4466	0,8318	3768,1	8,4276	0,7561	3767,7	8,3833	0,6929	3767,3	8,3428
640	0,8763	3790,6	8,4712	0,8411	3790,4	8,4522	0,7645	3790,1	8,4079	0,7007	3789,7	8,3674
650	0,8859	3813,0	8,4955	0,8504	3812,8	8,4766	0,7730	3812,5	8,4323	0,7084	3812,1	8,3919
660	0,8956	3835,4	8,5197	0,8597	3835,3	8,5008	0,7814	3834,9	8,4565	0,7162	3834,6	8,4161
670	0,9053	3857,9	8,5437	0,8690	3857,8	8,5248	0,7899	3857,5	8,4805	0,7239	3857,1	8,4401
680	0,9149	3880,5	8,5676	0,8783	3880,4	8,5486	0,7983	3880,1	8,5044	0,7317	3879,7	8,4639
690	0,9246	3903,2	8,5912	0,8876	3903,0	8,5723	0,8067	3902,7	8,5280	0,7394	3902,4	8,4876
700	0,9343	3925,9	8,6147	0,8968	3925,8	8,5957	0,8152	3925,5	8,5515	0,7471	3925,1	8,5111
710	0,9439	3948,7	8,6380	0,9061	3948,5	8,6190	0,8236	3948,2	8,5748	0,7549	3947,9	8,5344
720	0,9536	3971,5	8,6611	0,9154	3971,4	8,6421	0,8321	3971,1	8,5979	0,7626	3970,8	8,5575
730	0,9633	3994,4	8,6840	0,9247	3994,3	8,6651	0,8405	3994,0	8,6209	0,7704	3993,7	8,5805
740	0,9729	4017,4	8,7068	0,9340	4017,3	8,6879	0,8489	4017,0	8,6437	0,7781	4016,7	8,6033
750	0,9826	4040,4	8,7294	0,9432	4040,3	8,7105	0,8574	4040,0	8,6663	0,7858	4039,8	8,6259
760	0,9922	4063,5	8,7519	0,9525	4063,4	8,7330	0,8658	4063,1	8,6888	0,7936	4062,9	8,6484
770	1,0019	4086,7	8,7742	0,9618	4086,6	8,7553	0,8742	4086,3	8,7111	0,8013	4086,1	8,6707
780	1,0115	4109,9	8,7963	0,9710	4109,8	8,7774	0,8827	4109,5	8,7333	0,8090	4109,3	8,6929
790	1,0212	4133,2	8,8183	0,9803	4133,1	8,7994	0,8911	4132,8	8,7553	0,8167	4132,6	8,7149
800	1,0308	4156,5	8,8402	0,9896	4156,4	8,8213	0,8995	4156,2	8,7771	0,8245	4155,9	8,7368

Table 3. Water and Superheated Steam (Continuation) Wasser und überhitzter Dampf (Fortsetzung)

	6,5 bar $t_s = 161,99$ °C			7,0 bar $t_s = 164,96$ °C			7,5 bar $t_s = 167,76$ °C			8,0 bar $t_s = 170,41$ °C		
t	v''	h''	s''	v''	h''	s''	v''	h''	s''	v''	h''	s''
	0,2925	2758,9	6,7304	0,2727	2762,0	6,7052	0,2554	2764,8	6,6817	0,2403	2767,5	6,6596
°C	v	h	s	v	h	s	v	h	s	v	h	s
0	0,0009999	0,6	−0,0001	0,0009999	0,7	−0,0001	0,0009998	0,7	−0,0001	0,0009998	0,8	−0,0001
10	0,0009999	42,6	0,1509	0,0009999	42,7	0,1509	0,0009999	42,7	0,1509	0,0009999	42,8	0,1509
20	0,0010014	84,5	0,2962	0,0010014	84,5	0,2962	0,0010014	84,6	0,2961	0,0010014	84,6	0,2961
30	0,0010040	126,3	0,4363	0,0010040	126,3	0,4363	0,0010040	126,3	0,4363	0,0010040	126,4	0,4363
40	0,0010075	168,0	0,5719	0,0010075	168,1	0,5719	0,0010075	168,1	0,5718	0,0010075	168,2	0,5718
50	0,0010118	209,8	0,7032	0,0010118	209,8	0,7032	0,0010118	209,9	0,7032	0,0010118	209,9	0,7031
60	0,0010169	251,6	0,8306	0,0010168	251,7	0,8306	0,0010168	251,7	0,8306	0,0010168	251,7	0,8306
70	0,0010226	293,5	0,9544	0,0010225	293,5	0,9544	0,0010225	293,6	0,9544	0,0010225	293,6	0,9544
80	0,0010289	335,4	1,0749	0,0010289	335,4	1,0748	0,0010289	335,5	1,0748	0,0010288	335,5	1,0748
90	0,0010359	377,4	1,1921	0,0010358	377,4	1,1921	0,0010358	377,5	1,1920	0,0010358	377,5	1,1920
100	0,0010434	419,5	1,3064	0,0010434	419,5	1,3064	0,0010434	419,6	1,3064	0,0010433	419,6	1,3063
110	0,0010516	461,7	1,4181	0,0010516	461,7	1,4180	0,0010515	461,8	1,4180	0,0010515	461,8	1,4179
120	0,0010604	504,0	1,5272	0,0010603	504,1	1,5271	0,0010603	504,1	1,5271	0,0010603	504,1	1,5270
130	0,0010698	546,6	1,6340	0,0010698	546,6	1,6339	0,0010697	546,6	1,6339	0,0010697	546,7	1,6338
140	0,0010799	589,3	1,7387	0,0010798	589,3	1,7386	0,0010798	589,4	1,7386	0,0010798	589,4	1,7385
150	0,0010907	632,3	1,8414	0,0010906	632,3	1,8414	0,0010906	632,3	1,8413	0,0010906	632,4	1,8413
160	0,0011022	675,5	1,9424	0,0011022	675,5	1,9424	0,0011021	675,6	1,9423	0,0011021	675,6	1,9423
170	0,2993	2778,0	6,7740	0,2767	2774,2	6,7330	0,2571	2770,4	6,6941	0,0011144	719,1	2,0416
180	0,3077	2801,5	6,8263	0,2846	2798,0	6,7861	0,2646	2794,6	6,7482	0,2471	2791,1	6,7122
190	0,3159	2824,4	6,8764	0,2923	2821,4	6,8370	0,2719	2818,2	6,7998	0,2540	2815,1	6,7647
200	0,3240	2847,0	6,9247	0,2999	2844,2	6,8859	0,2791	2841,4	6,8494	0,2608	2838,6	6,8148
210	0,3319	2869,2	6,9712	0,3074	2866,7	6,9329	0,2861	2864,2	6,8970	0,2675	2861,6	6,8630
220	0,3398	2891,2	7,0162	0,3147	2888,9	6,9784	0,2930	2886,6	6,9429	0,2740	2884,2	6,9094
230	0,3476	2912,9	7,0598	0,3220	2910,8	7,0224	0,2999	2908,7	6,9873	0,2805	2906,6	6,9542
240	0,3553	2934,4	7,1021	0,3292	2932,5	7,0651	0,3066	2930,6	7,0303	0,2869	2928,6	6,9976
250	0,3629	2955,8	7,1434	0,3364	2954,0	7,1066	0,3134	2952,2	7,0721	0,2932	2950,4	7,0397
260	0,3705	2977,0	7,1835	0,3435	2975,4	7,1470	0,3200	2973,7	7,1128	0,2995	2972,1	7,0807
270	0,3781	2998,1	7,2228	0,3505	2996,6	7,1865	0,3266	2995,1	7,1525	0,3057	2993,5	7,1205
280	0,3856	3019,2	7,2611	0,3575	3017,7	7,2250	0,3332	3016,3	7,1912	0,3119	3014,9	7,1595
290	0,3930	3040,1	7,2987	0,3645	3038,8	7,2627	0,3397	3037,4	7,2291	0,3180	3036,1	7,1975
300	0,4005	3061,0	7,3355	0,3714	3059,8	7,2997	0,3462	3058,5	7,2662	0,3241	3057,3	7,2348
310	0,4079	3081,9	7,3716	0,3783	3080,7	7,3359	0,3527	3079,5	7,3026	0,3302	3078,3	7,2713
320	0,4153	3102,8	7,4070	0,3852	3101,6	7,3715	0,3591	3100,5	7,3382	0,3363	3099,4	7,3070
330	0,4226	3123,6	7,4419	0,3920	3122,5	7,4064	0,3655	3121,5	7,3733	0,3423	3120,4	7,3422
340	0,4300	3144,4	7,4761	0,3989	3143,4	7,4407	0,3719	3142,4	7,4077	0,3483	3141,4	7,3767
350	0,4373	3165,3	7,5099	0,4057	3164,3	7,4745	0,3783	3163,4	7,4416	0,3543	3162,4	7,4107
360	0,4446	3186,1	7,5431	0,4125	3185,2	7,5078	0,3847	3184,3	7,4749	0,3603	3183,4	7,4441
370	0,4519	3207,0	7,5758	0,4193	3206,1	7,5406	0,3910	3205,3	7,5078	0,3663	3204,4	7,4770
380	0,4592	3227,9	7,6080	0,4261	3227,1	7,5729	0,3974	3226,2	7,5401	0,3723	3225,4	7,5094
390	0,4665	3248,8	7,6398	0,4329	3248,0	7,6047	0,4037	3247,2	7,5720	0,3782	3246,4	7,5414
400	0,4738	3269,8	7,6712	0,4396	3269,0	7,6362	0,4100	3268,3	7,6035	0,3842	3267,5	7,5729
410	0,4810	3290,8	7,7022	0,4464	3290,1	7,6672	0,4164	3289,3	7,6346	0,3901	3288,6	7,6040
420	0,4883	3311,9	7,7328	0,4531	3311,2	7,6978	0,4227	3310,4	7,6652	0,3960	3309,7	7,6347
430	0,4955	3333,0	7,7630	0,4599	3332,3	7,7281	0,4290	3331,6	7,6955	0,4019	3330,9	7,6650
440	0,5028	3354,1	7,7929	0,4666	3353,4	7,7580	0,4353	3352,8	7,7255	0,4078	3352,1	7,6950
450	0,5100	3375,3	7,8224	0,4733	3374,7	7,7875	0,4416	3374,0	7,7550	0,4137	3373,4	7,7246
460	0,5172	3396,6	7,8516	0,4801	3395,9	7,8168	0,4478	3395,3	7,7843	0,4196	3394,7	7,7539
470	0,5245	3417,9	7,8804	0,4868	3417,3	7,8456	0,4541	3416,7	7,8132	0,4255	3416,1	7,7828
480	0,5317	3439,2	7,9090	0,4935	3438,6	7,8742	0,4604	3438,1	7,8418	0,4314	3437,5	7,8115
490	0,5389	3460,6	7,9372	0,5002	3460,1	7,9025	0,4666	3459,5	7,8701	0,4373	3459,0	7,8398
500	0,5461	3482,1	7,9652	0,5069	3481,6	7,9305	0,4729	3481,0	7,8981	0,4432	3480,5	7,8678
510	0,5533	3503,7	7,9929	0,5136	3503,1	7,9582	0,4792	3502,6	7,9258	0,4490	3502,1	7,8955
520	0,5605	3525,3	8,0203	0,5203	3524,7	7,9856	0,4854	3524,2	7,9533	0,4549	3523,7	7,9230
530	0,5677	3546,9	8,0474	0,5270	3546,4	8,0128	0,4917	3545,9	7,9804	0,4608	3545,4	7,9502
540	0,5749	3568,6	8,0743	0,5336	3568,1	8,0396	0,4979	3567,7	8,0074	0,4666	3567,2	7,9771
550	0,5821	3590,4	8,1009	0,5403	3589,9	8,0663	0,5041	3589,5	8,0340	0,4725	3589,0	8,0038

Table 3. Water and Superheated Steam (Continuation) Wasser und überhitzter Dampf (Fortsetzung)

t	6,5 bar $t_s = 161,99$ °C			7,0 bar $t_s = 164,96$ °C			7,5 bar $t_s = 167,76$ °C			8,0 bar $t_s = 170,41$ °C		
	v'' 0,2925	h'' 2758,9	s'' 6,7304	v'' 0,2727	h'' 2762,0	s'' 6,7052	v'' 0,2554	h'' 2764,8	s'' 6,6817	v'' 0,2403	h'' 2767,5	s'' 6,6596
°C	v	h	s	v	h	s	v	h	s	v	h	s
550	0,5821	3590,4	8,1009	0,5403	3589,9	8,0663	0,5041	3589,5	8,0340	0,4725	3589,0	8,0038
560	0,5893	3612,3	8,1273	0,5470	3611,8	8,0927	0,5104	3611,3	8,0604	0,4783	3610,9	8,0302
570	0,5964	3634,2	8,1534	0,5537	3633,7	8,1188	0,5166	3633,3	8,0866	0,4842	3632,8	8,0564
580	0,6036	3656,1	8,1793	0,5603	3655,7	8,1448	0,5229	3655,3	8,1125	0,4900	3654,8	8,0824
590	0,6108	3678,2	8,2050	0,5670	3677,8	8,1705	0,5291	3677,3	8,1382	0,4959	3676,9	8,1081
600	0,6180	3700,3	8,2305	0,5737	3699,9	8,1959	0,5353	3699,5	8,1637	0,5017	3699,1	8,1336
610	0,6251	3722,4	8,2557	0,5803	3722,1	8,2212	0,5415	3721,7	8,1890	0,5076	3721,3	8,1589
620	0,6323	3744,7	8,2808	0,5870	3744,3	8,2462	0,5477	3743,9	8,2141	0,5134	3743,5	8,1839
630	0,6395	3767,0	8,3056	0,5937	3766,6	8,2711	0,5540	3766,2	8,2389	0,5192	3765,8	8,2088
640	0,6466	3789,3	8,3302	0,6003	3789,0	8,2957	0,5602	3788,6	8,2635	0,5251	3788,2	8,2334
650	0,6538	3811,8	8,3546	0,6070	3811,4	8,3201	0,5664	3811,0	8,2880	0,5309	3810,7	8,2579
660	0,6609	3834,2	8,3788	0,6136	3833,9	8,3444	0,5726	3833,5	8,3122	0,5367	3833,2	8,2822
670	0,6681	3856,8	8,4029	0,6203	3856,5	8,3684	0,5788	3856,1	8,3363	0,5425	3855,8	8,3062
680	0,6753	3879,4	8,4267	0,6269	3879,1	8,3923	0,5850	3878,8	8,3602	0,5484	3878,4	8,3301
690	0,6824	3902,1	8,4504	0,6336	3901,8	8,4160	0,5912	3901,5	8,3839	0,5542	3901,1	8,3538
700	0,6896	3924,8	8,4739	0,6402	3924,5	8,4395	0,5974	3924,2	8,4074	0,5600	3923,9	8,3773
710	0,6967	3947,6	8,4972	0,6468	3947,3	8,4628	0,6036	3947,0	8,4307	0,5658	3946,7	8,4007
720	0,7039	3970,5	8,5204	0,6535	3970,2	8,4859	0,6098	3969,9	8,4539	0,5716	3969,6	8,4238
730	0,7110	3993,4	8,5433	0,6601	3993,2	8,5089	0,6160	3992,9	8,4769	0,5775	3992,6	8,4468
740	0,7181	4016,4	8,5661	0,6668	4016,2	8,5317	0,6222	4015,9	8,4997	0,5833	4015,6	8,4697
750	0,7253	4039,5	8,5888	0,6734	4039,2	8,5544	0,6284	4039,0	8,5223	0,5891	4038,7	8,4923
760	0,7324	4062,6	8,6113	0,6800	4062,4	8,5769	0,6346	4062,1	8,5448	0,5949	4061,8	8,5149
770	0,7396	4085,8	8,6336	0,6867	4085,5	8,5992	0,6408	4085,3	8,5672	0,6007	4085,0	8,5372
780	0,7467	4109,0	8,6558	0,6933	4108,8	8,6214	0,6470	4108,5	8,5894	0,6065	4108,3	8,5594
790	0,7538	4132,3	8,6778	0,6999	4132,1	8,6434	0,6532	4131,9	8,6114	0,6123	4131,6	8,5814
800	0,7610	4155,7	8,6997	0,7066	4155,5	8,6653	0,6594	4155,2	8,6333	0,6181	4155,0	8,6033

Table 3. Water and Superheated Steam (Continuation) Wasser und überhitzter Dampf (Fortsetzung)

	8,5 bar $t_s = 172,94\,°C$			9,0 bar $t_s = 175,36\,°C$			9,5 bar $t_s = 177,67\,°C$			10,0 bar $t_s = 179,88\,°C$		
t	v'' 0,2268	h'' 2769,9	s'' 6,6388	v'' 0,2148	h'' 2772,1	s'' 6,6192	v'' 0,2040	h'' 2774,2	s'' 6,6005	v'' 0,1943	h'' 2776,2	s'' 6,5828
°C	v	h	s	v	h	s	v	h	s	v	h	s
0	0,0009998	0,8	−0,0001	0,0009998	0,9	−0,0001	0,0009997	0,9	−0,0001	0,0009997	1,0	−0,0001
10	0,0009999	42,8	0,1509	0,0009999	42,9	0,1509	0,0009998	42,9	0,1509	0,0009998	43,0	0,1509
20	0,0010013	84,7	0,2961	0,0010013	84,7	0,2961	0,0010013	84,8	0,2961	0,0010013	84,8	0,2961
30	0,0010039	126,4	0,4362	0,0010039	126,5	0,4362	0,0010039	126,5	0,4362	0,0010039	126,6	0,4362
40	0,0010074	168,2	0,5718	0,0010074	168,2	0,5718	0,0010074	168,3	0,5718	0,0010074	168,3	0,5717
50	0,0010117	210,0	0,7031	0,0010117	210,0	0,7031	0,0010117	210,1	0,7031	0,0010117	210,1	0,7030
60	0,0010168	251,8	0,8305	0,0010167	251,8	0,8305	0,0010167	251,9	0,8305	0,0010167	251,9	0,8305
70	0,0010225	293,6	0,9543	0,0010224	293,7	0,9543	0,0010224	293,7	0,9543	0,0010224	293,8	0,9542
80	0,0010288	335,6	1,0747	0,0010288	335,6	1,0747	0,0010288	335,6	1,0747	0,0010287	335,7	1,0746
90	0,0010358	377,5	1,1920	0,0010357	377,6	1,1919	0,0010357	377,6	1,1919	0,0010357	377,7	1,1919
100	0,0010433	419,6	1,3063	0,0010433	419,7	1,3062	0,0010433	419,7	1,3062	0,0010432	419,7	1,3062
110	0,0010515	461,8	1,4179	0,0010514	461,9	1,4179	0,0010514	461,9	1,4178	0,0010514	461,9	1,4178
120	0,0010603	504,2	1,5270	0,0010602	504,2	1,5270	0,0010602	504,2	1,5269	0,0010602	504,3	1,5269
130	0,0010697	546,7	1,6338	0,0010696	546,7	1,6338	0,0010696	546,8	1,6337	0,0010696	546,8	1,6337
140	0,0010797	589,4	1,7385	0,0010797	589,5	1,7384	0,0010797	589,5	1,7384	0,0010796	589,5	1,7383
150	0,0010905	632,4	1,8412	0,0010905	632,4	1,8412	0,0010905	632,4	1,8411	0,0010904	632,5	1,8410
160	0,0011021	675,6	1,9422	0,0011020	675,6	1,9421	0,0011020	675,7	1,9421	0,0011019	675,7	1,9420
170	0,0011144	719,1	2,0416	0,0011144	719,2	2,0415	0,0011143	719,2	2,0414	0,0011143	719,2	2,0414
180	0,2316	2787,5	6,6780	0,2178	2783,9	6,6452	0,2055	2780,2	6,6137	0,1944	2776,5	6,5835
190	0,2382	2811,9	6,7312	0,2241	2808,6	6,6992	0,2116	2805,3	6,6686	0,2002	2802,0	6,6392
200	0,2447	2835,7	6,7820	0,2303	2832,7	6,7508	0,2175	2829,8	6,7209	0,2059	2826,8	6,6922
210	0,2510	2859,0	6,8308	0,2364	2856,3	6,8001	0,2233	2853,7	6,7708	0,2115	2851,0	6,7427
220	0,2572	2881,9	6,8777	0,2423	2879,5	6,8475	0,2290	2877,0	6,8187	0,2169	2874,6	6,7911
230	0,2634	2904,4	6,9229	0,2482	2902,2	6,8931	0,2345	2900,0	6,8648	0,2223	2897,8	6,8377
240	0,2694	2926,6	6,9666	0,2539	2924,6	6,9373	0,2400	2922,6	6,9093	0,2276	2920,6	6,8825
250	0,2754	2948,6	7,0091	0,2596	2946,8	6,9800	0,2455	2944,9	6,9523	0,2327	2943,0	6,9259
260	0,2814	2970,4	7,0503	0,2653	2968,7	7,0215	0,2509	2967,0	6,9941	0,2379	2965,2	6,9680
270	0,2873	2992,0	7,0904	0,2709	2990,4	7,0618	0,2562	2988,8	7,0347	0,2430	2987,2	7,0088
280	0,2931	3013,4	7,1295	0,2764	3012,0	7,1012	0,2615	3010,5	7,0742	0,2480	3009,0	7,0485
290	0,2989	3034,7	7,1678	0,2819	3033,4	7,1396	0,2667	3032,0	7,1128	0,2530	3030,6	7,0873
300	0,3047	3056,0	7,2051	0,2874	3054,7	7,1771	0,2719	3053,4	7,1505	0,2580	3052,1	7,1251
310	0,3104	3077,2	7,2418	0,2928	3076,0	7,2139	0,2771	3074,8	7,1874	0,2629	3073,5	7,1622
320	0,3162	3098,3	7,2777	0,2983	3097,1	7,2499	0,2822	3096,0	7,2235	0,2678	3094,9	7,1984
330	0,3219	3119,3	7,3129	0,3037	3118,3	7,2852	0,2874	3117,2	7,2590	0,2727	3116,1	7,2340
340	0,3275	3140,4	7,3475	0,3090	3139,4	7,3199	0,2925	3138,4	7,2938	0,2776	3137,4	7,2689
350	0,3332	3161,4	7,3815	0,3144	3160,5	7,3540	0,2976	3159,5	7,3279	0,2824	3158,5	7,3031
360	0,3388	3182,5	7,4150	0,3197	3181,6	7,3876	0,3027	3180,6	7,3616	0,2873	3179,7	7,3368
370	0,3445	3203,5	7,4480	0,3251	3202,6	7,4206	0,3077	3201,8	7,3947	0,2921	3200,9	7,3700
380	0,3501	3224,6	7,4805	0,3304	3223,7	7,4532	0,3128	3222,9	7,4273	0,2969	3222,0	7,4027
390	0,3557	3245,6	7,5125	0,3357	3244,8	7,4852	0,3178	3244,0	7,4594	0,3017	3243,2	7,4348
400	0,3613	3266,7	7,5441	0,3410	3266,0	7,5169	0,3228	3265,2	7,4911	0,3065	3264,4	7,4665
410	0,3669	3287,8	7,5752	0,3463	3287,1	7,5480	0,3279	3286,4	7,5223	0,3113	3285,6	7,4978
420	0,3725	3309,0	7,6060	0,3516	3308,3	7,5788	0,3329	3307,6	7,5531	0,3160	3306,9	7,5287
430	0,3781	3330,2	7,6363	0,3569	3329,5	7,6092	0,3379	3328,8	7,5836	0,3208	3328,1	7,5592
440	0,3836	3351,5	7,6663	0,3621	3350,8	7,6393	0,3429	3350,1	7,6136	0,3256	3349,5	7,5893
450	0,3892	3372,7	7,6960	0,3674	3372,1	7,6689	0,3479	3371,5	7,6433	0,3303	3370,8	7,6190
460	0,3948	3394,1	7,7253	0,3726	3393,5	7,6983	0,3529	3392,8	7,6727	0,3350	3392,2	7,6484
470	0,4003	3415,5	7,7542	0,3779	3414,9	7,7273	0,3578	3414,3	7,7017	0,3398	3413,6	7,6775
480	0,4059	3436,9	7,7829	0,3831	3436,3	7,7560	0,3628	3435,7	7,7304	0,3445	3435,1	7,7062
490	0,4114	3458,4	7,8112	0,3884	3457,8	7,7843	0,3678	3457,3	7,7588	0,3492	3456,7	7,7346
500	0,4169	3479,9	7,8393	0,3936	3479,4	7,8124	0,3727	3478,8	7,7869	0,3540	3478,3	7,7627
510	0,4225	3501,5	7,8671	0,3988	3501,0	7,8402	0,3777	3500,5	7,8147	0,3587	3499,9	7,7905
520	0,4280	3523,2	7,8945	0,4041	3522,7	7,8677	0,3827	3522,2	7,8422	0,3634	3521,6	7,8181
530	0,4335	3544,9	7,9218	0,4093	3544,4	7,8949	0,3876	3543,9	7,8695	0,3681	3543,4	7,8454
540	0,4390	3566,7	7,9487	0,4145	3566,2	7,9219	0,3926	3565,7	7,8965	0,3728	3565,2	7,8724
550	0,4446	3588,5	7,9754	0,4197	3588,1	7,9486	0,3975	3587,6	7,9232	0,3775	3587,1	7,8991

Table 3. Water and Superheated Steam (Continuation) Wasser und überhitzter Dampf (Fortsetzung)

t	8,5 bar $t_s = 172,94$ °C			9,0 bar $t_s = 175,36$ °C			9,5 bar $t_s = 177,67$ °C			10,0 bar $t_s = 179,88$ °C		
	v'' 0,2268	h'' 2769,9	s'' 6,6388	v'' 0,2148	h'' 2772,1	s'' 6,6192	v'' 0,2040	h'' 2774,2	s'' 6,6005	v'' 0,1943	h'' 2776,2	s'' 6,5828
°C	v	h	s	v	h	s	v	h	s	v	h	s
550	0,4446	3588,5	7,9754	0,4197	3588,1	7,9486	0,3975	3587,6	7,9232	0,3775	3587,1	7,8991
560	0,4501	3610,4	8,0018	0,4249	3610,0	7,9750	0,4025	3609,5	7,9497	0,3822	3609,0	7,9256
570	0,4556	3632,4	8,0280	0,4301	3631,9	8,0013	0,4074	3631,5	7,9759	0,3869	3631,0	7,9518
580	0,4611	3654,4	8,0540	0,4354	3654,0	8,0272	0,4123	3653,5	8,0019	0,3916	3653,1	7,9779
590	0,4666	3676,5	8,0797	0,4406	3676,1	8,0530	0,4173	3675,7	8,0277	0,3963	3675,2	8,0036
600	0,4721	3698,6	8,1053	0,4458	3698,2	8,0785	0,4222	3697,8	8,0532	0,4010	3697,4	8,0292
610	0,4776	3720,9	8,1305	0,4510	3720,5	8,1038	0,4271	3720,1	8,0785	0,4057	3719,7	8,0545
620	0,4831	3743,1	8,1556	0,4561	3742,7	8,1289	0,4320	3742,4	8,1036	0,4104	3742,0	8,0796
630	0,4886	3765,5	8,1805	0,4613	3765,1	8,1538	0,4370	3764,7	8,1285	0,4150	3764,3	8,1045
640	0,4941	3787,9	8,2052	0,4665	3787,5	8,1785	0,4419	3787,1	8,1532	0,4197	3786,8	8,1292
650	0,4996	3810,3	8,2296	0,4717	3810,0	8,2030	0,4468	3809,6	8,1777	0,4244	3809,3	8,1537
660	0,5051	3832,9	8,2539	0,4769	3832,5	8,2272	0,4517	3832,2	8,2020	0,4291	3831,8	8,1780
670	0,5105	3855,4	8,2780	0,4821	3855,1	8,2513	0,4566	3854,8	8,2261	0,4337	3854,4	8,2021
680	0,5160	3878,1	8,3019	0,4873	3877,8	8,2752	0,4615	3877,4	8,2500	0,4384	3877,1	8,2261
690	0,5215	3900,8	8,3256	0,4925	3900,5	8,2989	0,4665	3900,2	8,2737	0,4431	3899,9	8,2498
700	0,5270	3923,6	8,3491	0,4976	3923,3	8,3225	0,4714	3923,0	8,2973	0,4477	3922,7	8,2734
710	0,5325	3946,4	8,3725	0,5028	3946,1	8,3458	0,4763	3945,8	8,3206	0,4524	3945,5	8,2967
720	0,5379	3969,3	8,3956	0,5080	3969,0	8,3690	0,4812	3968,8	8,3438	0,4571	3968,5	8,3199
730	0,5434	3992,3	8,4186	0,5132	3992,0	8,3920	0,4861	3991,7	8,3669	0,4617	3991,4	8,3430
740	0,5489	4015,3	8,4415	0,5183	4015,1	8,4149	0,4910	4014,8	8,3897	0,4664	4014,5	8,3658
750	0,5544	4038,4	8,4642	0,5235	4038,1	8,4376	0,4959	4037,9	8,4124	0,4710	4037,6	8,3885
760	0,5598	4061,6	8,4867	0,5287	4061,3	8,4601	0,5008	4061,0	8,4349	0,4757	4060,8	8,4111
770	0,5653	4084,8	8,5090	0,5338	4084,5	8,4825	0,5057	4084,3	8,4573	0,4803	4084,0	8,4335
780	0,5708	4108,0	8,5312	0,5390	4107,8	8,5047	0,5106	4107,6	8,4795	0,4850	4107,3	8,4557
790	0,5762	4131,4	8,5533	0,5442	4131,1	8,5267	0,5155	4130,9	8,5016	0,4896	4130,7	8,4777
800	0,5817	4154,8	8,5752	0,5493	4154,5	8,5486	0,5204	4154,3	8,5235	0,4943	4154,1	8,4997

Table 3. Water and Superheated Steam (Continuation) Wasser und überhitzter Dampf (Fortsetzung)

t	10,5 bar t_s = 182,02 °C			11,0 bar t_s = 184,07 °C			11,5 bar t_s = 186,05 °C			12,0 bar t_s = 187,96 °C		
	v'' 0,1855	h'' 2778,0	s'' 6,5659	v'' 0,1774	h'' 2779,7	s'' 6,5497	v'' 0,1700	h'' 2781,3	s'' 6,5342	v'' 0,1632	h'' 2782,7	s'' 6,5194
°C	v	h	s	v	h	s	v	h	s	v	h	s
0	0,0009997	1,0	−0,0001	0,0009997	1,1	−0,0001	0,0009996	1,1	−0,0001	0,0009996	1,2	−0,0001
10	0,0009998	43,0	0,1509	0,0009997	43,1	0,1509	0,0009997	43,1	0,1509	0,0009997	43,2	0,1509
20	0,0010012	84,8	0,2961	0,0010012	84,9	0,2961	0,0010012	84,9	0,2961	0,0010012	85,0	0,2961
30	0,0010038	126,6	0,4362	0,0010038	126,7	0,4362	0,0010038	126,7	0,4362	0,0010038	126,8	0,4361
40	0,0010073	168,4	0,5717	0,0010073	168,4	0,5717	0,0010073	168,5	0,5717	0,0010073	168,5	0,5717
50	0,0010116	210,1	0,7030	0,0010116	210,2	0,7030	0,0010116	210,2	0,7030	0,0010116	210,3	0,7030
60	0,0010167	252,0	0,8304	0,0010167	252,0	0,8304	0,0010166	252,0	0,8304	0,0010166	252,1	0,8304
70	0,0010224	293,8	0,9542	0,0010223	293,8	0,9542	0,0010223	293,9	0,9541	0,0010223	293,9	0,9541
80	0,0010287	335,7	1,0746	0,0010287	335,8	1,0746	0,0010287	335,8	1,0745	0,0010286	335,8	1,0745
90	0,0010357	377,7	1,1918	0,0010356	377,7	1,1918	0,0010356	377,8	1,1917	0,0010356	377,8	1,1917
100	0,0010432	419,8	1,3061	0,0010432	419,8	1,3061	0,0010432	419,9	1,3060	0,0010431	419,9	1,3060
110	0,0010514	462,0	1,4177	0,0010513	462,0	1,4177	0,0010513	462,0	1,4176	0,0010513	462,1	1,4176
120	0,0010601	504,3	1,5268	0,0010601	504,4	1,5268	0,0010601	504,4	1,5267	0,0010601	504,4	1,5267
130	0,0010695	546,8	1,6336	0,0010695	546,9	1,6336	0,0010695	546,9	1,6335	0,0010695	546,9	1,6335
140	0,0010796	589,6	1,7383	0,0010796	589,6	1,7382	0,0010795	589,6	1,7382	0,0010795	589,6	1,7381
150	0,0010904	632,5	1,8410	0,0010903	632,5	1,8409	0,0010903	632,6	1,8409	0,0010903	632,6	1,8408
160	0,0011019	675,7	1,9420	0,0011019	675,8	1,9419	0,0011018	675,8	1,9418	0,0011018	675,8	1,9418
170	0,0011142	719,3	2,0413	0,0011142	719,3	2,0412	0,0011142	719,3	2,0412	0,0011141	719,3	2,0411
180	0,0011275	763,1	2,1392	0,0011274	763,2	2,1392	0,0011274	763,2	2,1391	0,0011274	763,2	2,1390
190	0,1900	2798,6	6,6108	0,1806	2795,2	6,5834	0,1721	2791,7	6,5569	0,1642	2788,2	6,5312
200	0,1954	2823,8	6,6615	0,1859	2820,7	6,6379	0,1772	2817,6	6,6122	0,1692	2814,4	6,5872
210	0,2008	2848,2	6,7157	0,1911	2845,5	6,6897	0,1822	2842,7	6,6646	0,1741	2839,8	6,6403
220	0,2060	2872,1	6,7647	0,1961	2869,6	6,7392	0,1871	2867,1	6,7147	0,1788	2864,5	6,6909
230	0,2112	2895,5	6,8117	0,2011	2893,2	6,7867	0,1919	2891,0	6,7626	0,1834	2888,6	6,7394
240	0,2162	2918,5	6,8569	0,2060	2916,4	6,8323	0,1966	2914,4	6,8086	0,1879	2912,2	6,7858
250	0,2212	2941,2	6,9006	0,2107	2939,3	6,8764	0,2012	2937,4	6,8530	0,1924	2935,4	6,8305
260	0,2261	2963,5	6,9430	0,2155	2961,8	6,9190	0,2057	2960,0	6,8959	0,1968	2958,2	6,8738
270	0,2310	2985,6	6,9840	0,2201	2984,0	6,9603	0,2102	2982,4	6,9375	0,2011	2980,8	6,9156
280	0,2358	3007,5	7,0240	0,2248	3006,0	7,0005	0,2147	3004,5	6,9779	0,2054	3003,0	6,9562
290	0,2406	3029,3	7,0629	0,2294	3027,9	7,0396	0,2191	3026,5	7,0172	0,2096	3025,1	6,9957
300	0,2454	3050,8	7,1009	0,2339	3049,6	7,0778	0,2234	3048,2	7,0556	0,2139	3046,9	7,0342
310	0,2501	3072,3	7,1381	0,2384	3071,1	7,1151	0,2278	3069,9	7,0930	0,2180	3068,7	7,0718
320	0,2548	3093,7	7,1745	0,2429	3092,6	7,1516	0,2321	3091,4	7,1296	0,2222	3090,3	7,1085
330	0,2595	3115,1	7,2101	0,2474	3114,0	7,1873	0,2364	3112,9	7,1655	0,2263	3111,8	7,1445
340	0,2641	3136,3	7,2451	0,2518	3135,3	7,2224	0,2407	3134,3	7,2007	0,2304	3133,2	7,1798
350	0,2687	3157,6	7,2795	0,2563	3156,6	7,2569	0,2449	3155,6	7,2352	0,2345	3154,6	7,2144
360	0,2734	3178,8	7,3133	0,2607	3177,9	7,2907	0,2491	3176,9	7,2691	0,2386	3176,0	7,2484
370	0,2780	3200,0	7,3465	0,2651	3199,1	7,3240	0,2534	3198,2	7,3025	0,2426	3197,3	7,2818
380	0,2825	3221,2	7,3792	0,2695	3220,3	7,3568	0,2576	3219,5	7,3353	0,2467	3218,7	7,3147
390	0,2871	3242,4	7,4114	0,2739	3241,6	7,3891	0,2618	3240,8	7,3677	0,2507	3240,0	7,3471
400	0,2917	3263,6	7,4432	0,2782	3262,9	7,4209	0,2660	3262,1	7,3995	0,2547	3261,3	7,3790
410	0,2963	3284,9	7,4745	0,2826	3284,1	7,4523	0,2701	3283,4	7,4309	0,2587	3282,6	7,4105
420	0,3008	3306,1	7,5054	0,2870	3305,4	7,4832	0,2743	3304,7	7,4619	0,2627	3304,0	7,4415
430	0,3053	3327,4	7,5359	0,2913	3326,8	7,5138	0,2785	3326,1	7,4925	0,2667	3325,4	7,4722
440	0,3099	3348,8	7,5661	0,2956	3348,1	7,5439	0,2826	3347,4	7,5227	0,2707	3346,8	7,5024
450	0,3144	3370,2	7,5958	0,3000	3369,5	7,5737	0,2868	3368,9	7,5526	0,2747	3368,2	7,5323
460	0,3189	3391,6	7,6253	0,3043	3391,0	7,6032	0,2909	3390,3	7,5820	0,2787	3389,7	7,5618
470	0,3235	3413,0	7,6543	0,3086	3412,4	7,6323	0,2950	3411,8	7,6112	0,2826	3411,2	7,5909
480	0,3280	3434,6	7,6831	0,3129	3434,0	7,6611	0,2992	3433,4	7,6400	0,2866	3432,8	7,6198
490	0,3325	3456,1	7,7115	0,3172	3455,6	7,6895	0,3033	3455,0	7,6685	0,2905	3454,4	7,6483
500	0,3370	3477,7	7,7397	0,3215	3477,2	7,7177	0,3074	3476,6	7,6966	0,2945	3476,1	7,6765
510	0,3415	3499,4	7,7675	0,3258	3498,9	7,7455	0,3115	3498,3	7,7245	0,2984	3497,8	7,7044
520	0,3460	3521,1	7,7951	0,3301	3520,6	7,7731	0,3157	3520,1	7,7521	0,3024	3519,6	7,7320
530	0,3505	3542,9	7,8224	0,3344	3542,4	7,8004	0,3198	3541,9	7,7795	0,3063	3541,4	7,7593
540	0,3549	3564,7	7,8494	0,3387	3564,3	7,8275	0,3239	3563,8	7,8065	0,3103	3563,3	7,7864
550	0,3594	3586,6	7,8762	0,3430	3586,2	7,8543	0,3280	3585,7	7,8333	0,3142	3585,2	7,8132

Table 3. Water and Superheated Steam (Continuation) Wasser und überhitzter Dampf (Fortsetzung)

t	10,5 bar $t_s = 182,02\,°C$			11,0 bar $t_s = 184,07\,°C$			11,5 bar $t_s = 186,05\,°C$			12,0 bar $t_s = 187,96\,°C$		
	v'' 0,1855	h'' 2778,0	s'' 6,5659	v'' 0,1774	h'' 2779,7	s'' 6,5497	v'' 0,1700	h'' 2781,3	s'' 6,5342	v'' 0,1632	h'' 2782,7	s'' 6,5194
°C	v	h	s	v	h	s	v	h	s	v	h	s
550	0,3594	3586,6	7,8762	0,3430	3586,2	7,8543	0,3280	3585,7	7,8333	0,3142	3585,2	7,8132
560	0,3639	3608,6	7,9027	0,3473	3608,1	7,8808	0,3321	3607,7	7,8598	0,3181	3607,2	7,8398
570	0,3684	3630,6	7,9289	0,3515	3630,2	7,9071	0,3362	3629,7	7,8861	0,3221	3629,3	7,8661
580	0,3729	3652,7	7,9550	0,3558	3652,2	7,9331	0,3402	3651,8	7,9122	0,3260	3651,4	7,8922
590	0,3773	3674,8	7,9807	0,3601	3674,4	7,9589	0,3443	3674,0	7,9380	0,3299	3673,5	7,9180
600	0,3818	3697,0	8,0063	0,3643	3696,6	7,9845	0,3484	3696,2	7,9636	0,3338	3695,8	7,9436
610	0,3863	3719,3	8,0317	0,3686	3718,9	8,0098	0,3525	3718,5	7,9890	0,3377	3718,1	7,9690
620	0,3907	3741,6	8,0568	0,3729	3741,2	8,0350	0,3566	3740,8	8,0141	0,3417	3740,4	7,9942
630	0,3952	3764,0	8,0817	0,3771	3763,6	8,0599	0,3607	3763,2	8,0391	0,3456	3762,8	8,0191
640	0,3996	3786,4	8,1064	0,3814	3786,0	8,0846	0,3647	3785,7	8,0638	0,3495	3785,3	8,0439
650	0,4041	3808,9	8,1309	0,3857	3808,5	8,1092	0,3688	3808,2	8,0883	0,3534	3807,8	8,0684
660	0,4085	3831,5	8,1552	0,3899	3831,1	8,1335	0,3729	3830,8	8,1127	0,3573	3830,4	8,0928
670	0,4130	3854,1	8,1794	0,3942	3853,8	8,1576	0,3770	3853,4	8,1368	0,3612	3853,1	8,1169
680	0,4175	3876,8	8,2033	0,3984	3876,5	8,1816	0,3810	3876,1	8,1608	0,3651	3875,8	8,1409
690	0,4219	3899,5	8,2270	0,4027	3899,2	8,2053	0,3851	3898,9	8,1845	0,3690	3898,6	8,1646
700	0,4263	3922,4	8,2506	0,4069	3922,0	8,2289	0,3891	3921,7	8,2081	0,3729	3921,4	8,1882
710	0,4308	3945,2	8,2740	0,4111	3944,9	8,2523	0,3932	3944,6	8,2315	0,3768	3944,3	8,2116
720	0,4352	3968,2	8,2972	0,4154	3967,9	8,2755	0,3973	3967,6	8,2547	0,3807	3967,3	8,2349
730	0,4397	3991,2	8,3202	0,4196	3990,9	8,2985	0,4013	3990,6	8,2778	0,3846	3990,3	8,2579
740	0,4441	4014,2	8,3431	0,4239	4013,9	8,3214	0,4054	4013,7	8,3007	0,3884	4013,4	8,2808
750	0,4485	4037,3	8,3658	0,4281	4037,1	8,3441	0,4094	4036,8	8,3234	0,3923	4036,5	8,3036
760	0,4530	4060,5	8,3884	0,4323	4060,3	8,3667	0,4135	4060,0	8,3460	0,3962	4059,7	8,3261
770	0,4574	4083,8	8,4107	0,4366	4083,5	8,3891	0,4175	4083,2	8,3684	0,4001	4083,0	8,3485
780	0,4618	4107,1	8,4330	0,4408	4106,8	8,4113	0,4216	4106,6	8,3906	0,4040	4106,3	8,3708
790	0,4663	4130,4	8,4550	0,4450	4130,2	8,4334	0,4256	4129,9	8,4127	0,4079	4129,7	8,3929
800	0,4707	4153,8	8,4770	0,4493	4153,6	8,4553	0,4297	4153,4	8,4346	0,4118	4153,1	8,4148

Table 3. Water and Superheated Steam (Continuation) Wasser und überhitzter Dampf (Fortsetzung)

t	12,5 bar t_s = 189,81 °C			13,0 bar t_s = 191,61 °C			13,5 bar t_s = 193,35 °C			14,0 bar t_s = 195,04 °C		
	v''	h''	s''	v''	h''	s''	v''	h''	s''	v''	h''	s''
	0,1569	2784,1	6,5051	0,1511	2785,4	6,4913	0,1457	2786,6	6,4780	0,1407	2787,8	6,4651
°C	v	h	s	v	h	s	v	h	s	v	h	s
0	0,0009996	1,2	−0,0001	0,0009996	1,3	−0,0000	0,0009995	1,3	−0,0000	0,0009995	1,4	−0,0000
10	0,0009997	43,2	0,1509	0,0009996	43,3	0,1509	0,0009996	43,3	0,1509	0,0009996	43,4	0,1509
20	0,0010012	85,0	0,2960	0,0010011	85,1	0,2960	0,0010011	85,1	0,2960	0,0010011	85,2	0,2960
30	0,0010038	126,8	0,4361	0,0010037	126,8	0,4361	0,0010037	126,9	0,4361	0,0010037	126,9	0,4361
40	0,0010073	168,6	0,5716	0,0010072	168,6	0,5716	0,0010072	168,6	0,5716	0,0010072	168,7	0,5716
50	0,0010116	210,3	0,7029	0,0010115	210,4	0,7029	0,0010115	210,4	0,7029	0,0010115	210,5	0,7029
60	0,0010166	252,1	0,8303	0,0010166	252,2	0,8303	0,0010165	252,2	0,8303	0,0010165	252,2	0,8302
70	0,0010223	294,0	0,9541	0,0010223	294,0	0,9541	0,0010222	294,0	0,9540	0,0010222	294,1	0,9540
80	0,0010286	335,9	1,0745	0,0010286	335,9	1,0744	0,0010286	336,0	1,0744	0,0010285	336,0	1,0744
90	0,0010356	377,9	1,1917	0,0010355	377,9	1,1916	0,0010355	377,9	1,1916	0,0010355	378,0	1,1916
100	0,0010431	419,9	1,3060	0,0010431	420,0	1,3059	0,0010431	420,0	1,3059	0,0010430	420,0	1,3059
110	0,0010513	462,1	1,4176	0,0010512	462,2	1,4175	0,0010512	462,2	1,4175	0,0010512	462,2	1,4174
120	0,0010600	504,5	1,5266	0,0010600	504,5	1,5266	0,0010600	504,5	1,5265	0,0010599	504,6	1,5265
130	0,0010694	547,0	1,6334	0,0010694	547,0	1,6334	0,0010694	547,0	1,6333	0,0010693	547,1	1,6333
140	0,0010795	589,7	1,7381	0,0010794	589,7	1,7380	0,0010794	589,7	1,7380	0,0010794	589,8	1,7379
150	0,0010902	632,6	1,8408	0,0010902	632,7	1,8407	0,0010902	632,7	1,8407	0,0010901	632,7	1,8406
160	0,0011018	675,8	1,9417	0,0011017	675,9	1,9417	0,0011017	675,9	1,9416	0,0011016	675,9	1,9415
170	0,0011141	719,4	2,0410	0,0011140	719,4	2,0410	0,0011140	719,4	2,0409	0,0011140	719,5	2,0409
180	0,0011273	763,2	2,1390	0,0011273	763,3	2,1389	0,0011272	763,3	2,1388	0,0011272	763,3	2,1388
190	0,1570	2784,6	6,5061	0,0011415	807,5	2,2355	0,0011414	807,6	2,2354	0,0011414	807,6	2,2354
200	0,1619	2811,2	6,5630	0,1551	2808,0	6,5394	0,1488	2804,7	6,5165	0,1429	2801,4	6,4941
210	0,1666	2837,0	6,6168	0,1597	2834,1	6,5939	0,1532	2831,1	6,5717	0,1473	2828,2	6,5500
220	0,1712	2861,9	6,6680	0,1641	2859,3	6,6457	0,1576	2856,7	6,6240	0,1515	2854,0	6,6030
230	0,1756	2886,3	6,7169	0,1685	2883,9	6,6951	0,1618	2881,6	6,6739	0,1556	2879,1	6,6534
240	0,1800	2910,1	6,7637	0,1727	2908,0	6,7424	0,1659	2905,8	6,7217	0,1596	2903,6	6,7016
250	0,1843	2933,5	6,8088	0,1769	2931,5	6,7878	0,1700	2929,5	6,7675	0,1635	2927,6	6,7477
260	0,1886	2956,5	6,8523	0,1810	2954,7	6,8316	0,1739	2952,9	6,8116	0,1674	2951,0	6,7922
270	0,1927	2979,1	6,8944	0,1850	2977,5	6,8740	0,1779	2975,8	6,8542	0,1712	2974,1	6,8351
280	0,1969	3001,5	6,9353	0,1890	3000,0	6,9151	0,1817	2998,4	6,8955	0,1749	2996,9	6,8766
290	0,2010	3023,6	6,9750	0,1930	3022,2	6,9550	0,1855	3020,8	6,9356	0,1787	3019,4	6,9169
300	0,2050	3045,6	7,0136	0,1969	3044,3	6,9938	0,1893	3043,0	6,9746	0,1823	3041,6	6,9561
310	0,2090	3067,4	7,0514	0,2008	3066,2	7,0317	0,1931	3064,9	7,0127	0,1860	3063,7	6,9943
320	0,2130	3089,1	7,0882	0,2046	3088,0	7,0687	0,1968	3086,8	7,0498	0,1896	3085,6	7,0315
330	0,2170	3110,7	7,1243	0,2084	3109,6	7,1049	0,2005	3108,5	7,0861	0,1931	3107,4	7,0680
340	0,2210	3132,2	7,1597	0,2123	3131,2	7,1404	0,2042	3130,1	7,1217	0,1967	3129,1	7,1036
350	0,2249	3153,7	7,1944	0,2160	3152,7	7,1751	0,2078	3151,7	7,1566	0,2002	3150,7	7,1386
360	0,2288	3175,1	7,2285	0,2198	3174,1	7,2093	0,2115	3173,2	7,1908	0,2038	3172,3	7,1729
370	0,2327	3196,4	7,2620	0,2236	3195,6	7,2429	0,2151	3194,7	7,2244	0,2073	3193,8	7,2066
380	0,2366	3217,8	7,2949	0,2273	3217,0	7,2759	0,2187	3216,1	7,2575	0,2108	3215,3	7,2398
390	0,2405	3239,2	7,3274	0,2311	3238,3	7,3084	0,2223	3237,5	7,2901	0,2142	3236,7	7,2724
400	0,2443	3260,5	7,3593	0,2348	3259,7	7,3404	0,2259	3259,0	7,3221	0,2177	3258,2	7,3045
410	0,2482	3281,9	7,3909	0,2385	3281,1	7,3720	0,2295	3280,4	7,3537	0,2212	3279,6	7,3361
420	0,2521	3303,3	7,4219	0,2422	3302,5	7,4031	0,2331	3301,8	7,3849	0,2246	3301,1	7,3673
430	0,2559	3324,7	7,4526	0,2459	3324,0	7,4338	0,2367	3323,3	7,4156	0,2281	3322,6	7,3981
440	0,2597	3346,1	7,4829	0,2496	3345,4	7,4641	0,2402	3344,8	7,4460	0,2315	3344,1	7,4285
450	0,2636	3367,6	7,5128	0,2533	3366,9	7,4940	0,2438	3366,3	7,4759	0,2349	3365,6	7,4585
460	0,2674	3389,1	7,5423	0,2570	3388,5	7,5236	0,2473	3387,8	7,5055	0,2384	3387,2	7,4881
470	0,2712	3410,6	7,5715	0,2606	3410,0	7,5528	0,2509	3409,4	7,5348	0,2418	3408,8	7,5174
480	0,2750	3432,2	7,6003	0,2643	3431,6	7,5817	0,2544	3431,0	7,5637	0,2452	3430,5	7,5463
490	0,2788	3453,9	7,6289	0,2680	3453,3	7,6102	0,2579	3452,7	7,5923	0,2486	3452,2	7,5749
500	0,2826	3475,5	7,6571	0,2716	3475,0	7,6385	0,2615	3474,4	7,6205	0,2520	3473,9	7,6032
510	0,2864	3497,3	7,6850	0,2753	3496,7	7,6664	0,2650	3496,2	7,6485	0,2554	3495,7	7,6312
520	0,2902	3519,1	7,7127	0,2789	3518,5	7,6941	0,2685	3518,0	7,6762	0,2588	3517,5	7,6589
530	0,2940	3540,9	7,7400	0,2826	3540,4	7,7215	0,2720	3539,9	7,7036	0,2622	3539,4	7,6863
540	0,2978	3562,8	7,7671	0,2862	3562,3	7,7486	0,2755	3561,8	7,7307	0,2656	3561,3	7,7135
550	0,3015	3584,7	7,7940	0,2898	3584,3	7,7754	0,2790	3583,8	7,7576	0,2690	3583,3	7,7404

Table 3. Water and Superheated Steam (Continuation) **Wasser und überhitzter Dampf** (Fortsetzung)

t	12,5 bar $t_s = 189,81$ °C			13,0 bar $t_s = 191,61$ °C			13,5 bar $t_s = 193,35$ °C			14,0 bar $t_s = 195,04$ °C		
	v'' 0,1569	h'' 2784,1	s'' 6,5051	v'' 0,1511	h'' 2785,4	s'' 6,4913	v'' 0,1457	h'' 2786,6	s'' 6,4780	v'' 0,1407	h'' 2787,8	s'' 6,4651
°C	v	h	s	v	h	s	v	h	s	v	h	s
550	0,3015	3584,7	7,7940	0,2898	3584,3	7,7754	0,2790	3583,8	7,7576	0,2690	3583,3	7,7404
560	0,3053	3606,8	7,8205	0,2935	3606,3	7,8020	0,2825	3605,8	7,7842	0,2724	3605,4	7,7670
570	0,3091	3628,8	7,8469	0,2971	3628,4	7,8284	0,2860	3627,9	7,8106	0,2757	3627,5	7,7934
580	0,3129	3650,9	7,8730	0,3007	3650,5	7,8545	0,2895	3650,1	7,8367	0,2791	3649,6	7,8195
590	0,3166	3673,1	7,8988	0,3044	3672,7	7,8803	0,2930	3672,3	7,8625	0,2825	3671,9	7,8454
600	0,3204	3695,4	7,9244	0,3080	3695,0	7,9060	0,2965	3694,5	7,8882	0,2859	3694,1	7,8710
610	0,3242	3717,7	7,9498	0,3116	3717,3	7,9314	0,3000	3716,9	7,9136	0,2892	3716,5	7,8965
620	0,3279	3740,0	7,9750	0,3152	3739,6	7,9566	0,3035	3739,3	7,9388	0,2926	3738,9	7,9217
630	0,3317	3762,5	8,0000	0,3188	3762,1	7,9815	0,3070	3761,7	7,9638	0,2959	3761,3	7,9467
640	0,3354	3784,9	8,0247	0,3225	3784,6	8,0063	0,3104	3784,2	7,9886	0,2993	3783,8	7,9715
650	0,3392	3807,5	8,0493	0,3261	3807,1	8,0309	0,3139	3806,8	8,0132	0,3027	3806,4	7,9961
660	0,3429	3830,1	8,0736	0,3297	3829,7	8,0552	0,3174	3829,4	8,0375	0,3060	3829,0	8,0205
670	0,3467	3852,7	8,0978	0,3333	3852,4	8,0794	0,3209	3852,1	8,0617	0,3094	3851,7	8,0447
680	0,3504	3875,5	8,1218	0,3369	3875,1	8,1034	0,3244	3874,8	8,0857	0,3127	3874,5	8,0687
690	0,3542	3898,3	8,1455	0,3405	3897,9	8,1272	0,3278	3897,6	8,1095	0,3161	3897,3	8,0925
700	0,3579	3921,1	8,1691	0,3441	3920,8	8,1508	0,3313	3920,5	8,1331	0,3194	3920,2	8,1161
710	0,3616	3944,0	8,1926	0,3477	3943,7	8,1742	0,3348	3943,4	8,1566	0,3228	3943,1	8,1395
720	0,3654	3967,0	8,2158	0,3513	3966,7	8,1975	0,3382	3966,4	8,1798	0,3261	3966,1	8,1628
730	0,3691	3990,0	8,2389	0,3549	3989,7	8,2206	0,3417	3989,5	8,2029	0,3294	3989,2	8,1859
740	0,3729	4013,1	8,2618	0,3585	4012,8	8,2435	0,3452	4012,6	8,2258	0,3328	4012,3	8,2088
750	0,3766	4036,3	8,2845	0,3621	4036,0	8,2662	0,3486	4035,7	8,2486	0,3361	4035,5	8,2316
760	0,3803	4059,5	8,3071	0,3657	4059,2	8,2888	0,3521	4058,9	8,2712	0,3395	4058,7	8,2542
770	0,3841	4082,7	8,3295	0,3692	4082,5	8,3112	0,3555	4082,2	8,2936	0,3428	4082,0	8,2766
780	0,3878	4106,1	8,3518	0,3728	4105,8	8,3335	0,3590	4105,6	8,3159	0,3461	4105,3	8,2989
790	0,3915	4129,4	8,3739	0,3764	4129,2	8,3556	0,3624	4129,0	8,3380	0,3495	4128,7	8,3210
800	0,3952	4152,9	8,3958	0,3800	4152,7	8,3775	0,3659	4152,4	8,3599	0,3528	4152,2	8,3430

Table 3. Water and Superheated Steam (Continuation) Wasser und überhitzter Dampf (Fortsetzung)

t	14,5 bar t_s = 196,69 °C			15,0 bar t_s = 198,29 °C			15,5 bar t_s = 199,85 °C			16,0 bar t_s = 201,37 °C		
	v'' 0,1360	h'' 2788,9	s'' 6,4526	v'' 0,1317	h'' 2789,9	s'' 6,4406	v'' 0,1275	h'' 2790,8	s'' 6,4289	v'' 0,1237	h'' 2791,7	s'' 6,4175
°C	v	h	s	v	h	s	v	h	s	v	h	s
0	0,0009995	1,4	—0,0000	0,0009995	1,5	—0,0000	0,0009994	1,5	—0,0000	0,0009994	1,6	—0,0000
10	0,0009996	43,4	0,1509	0,0009995	43,5	0,1509	0,0009995	43,5	0,1509	0,0009995	43,6	0,1509
20	0,0010011	85,2	0,2960	0,0010010	85,3	0,2960	0,0010010	85,3	0,2960	0,0010010	85,4	0,2960
30	0,0010037	127,0	0,4361	0,0010036	127,0	0,4360	0,0010036	127,1	0,4360	0,0010036	127,1	0,4360
40	0,0010072	168,7	0,5716	0,0010071	168,8	0,5715	0,0010071	168,8	0,5715	0,0010071	168,9	0,5715
50	0,0010115	210,5	0,7028	0,0010114	210,5	0,7028	0,0010114	210,6	0,7028	0,0010114	210,6	0,7028
60	0,0010165	252,3	0,8302	0,0010165	252,3	0,8302	0,0010164	252,4	0,8302	0,0010164	252,4	0,8301
70	0,0010222	294,1	0,9540	0,0010222	294,2	0,9539	0,0010221	294,2	0,9539	0,0010221	294,3	0,9539
80	0,0010285	336,0	1,0743	0,0010285	336,1	1,0743	0,0010285	336,1	1,0743	0,0010284	336,1	1,0742
90	0,0010355	378,0	1,1915	0,0010354	378,0	1,1915	0,0010354	378,1	1,1915	0,0010354	378,1	1,1914
100	0,0010430	420,1	1,3058	0,0010430	420,1	1,3058	0,0010429	420,2	1,3057	0,0010429	420,2	1,3057
110	0,0010511	462,3	1,4174	0,0010511	462,3	1,4173	0,0010511	462,3	1,4173	0,0010511	462,4	1,4173
120	0,0010599	504,6	1,5265	0,0010599	504,6	1,5264	0,0010598	504,7	1,5264	0,0010598	504,7	1,5263
130	0,0010693	547,1	1,6332	0,0010693	547,1	1,6332	0,0010692	547,2	1,6331	0,0010692	547,2	1,6331
140	0,0010794	589,8	1,7379	0,0010793	589,8	1,7378	0,0010793	589,9	1,7378	0,0010793	589,9	1,7377
150	0,0010901	632,8	1,8405	0,0010901	632,8	1,8405	0,0010900	632,8	1,8404	0,0010900	632,8	1,8404
160	0,0011016	676,0	1,9415	0,0011016	676,0	1,9414	0,0011015	676,0	1,9414	0,0011015	676,0	1,9413
170	0,0011139	719,5	2,0408	0,0011139	719,5	2,0407	0,0011138	719,5	2,0407	0,0011138	719,6	2,0406
180	0,0011271	763,3	2,1387	0,0011271	763,4	2,1386	0,0011270	763,4	2,1385	0,0011270	763,4	2,1385
190	0,0011413	807,6	2,2353	0,0011413	807,6	2,2352	0,0011412	807,7	2,2351	0,0011412	807,7	2,2351
200	0,1375	2798,1	6,4722	0,1324	2794,7	6,4508	0,1276	2791,3	6,4298	0,0011564	852,4	2,3306
210	0,1417	2825,2	6,5288	0,1365	2822,2	6,5082	0,1317	2819,1	6,4880	0,1271	2816,0	6,4682
220	0,1458	2851,3	6,5824	0,1406	2848,6	6,5624	0,1356	2845,9	6,5428	0,1310	2843,1	6,5237
230	0,1498	2876,7	6,6334	0,1445	2874,3	6,6139	0,1394	2871,8	6,5949	0,1347	2869,3	6,5763
240	0,1538	2901,4	6,6820	0,1483	2899,2	6,6630	0,1431	2897,0	6,6444	0,1383	2894,7	6,6263
250	0,1576	2925,5	6,7286	0,1520	2923,5	6,7099	0,1468	2921,5	6,6917	0,1419	2919,4	6,6740
260	0,1613	2949,2	6,7733	0,1556	2947,3	6,7550	0,1503	2945,5	6,7372	0,1453	2943,6	6,7198
270	0,1650	2972,4	6,8165	0,1592	2970,7	6,7985	0,1538	2969,0	6,7809	0,1487	2967,3	6,7638
280	0,1686	2995,3	6,8583	0,1628	2993,7	6,8405	0,1573	2992,2	6,8232	0,1521	2990,6	6,8063
290	0,1722	3017,9	6,8988	0,1663	3016,5	6,8812	0,1606	3015,0	6,8641	0,1554	3013,5	6,8474
300	0,1758	3040,3	6,9381	0,1697	3038,9	6,9207	0,1640	3037,6	6,9038	0,1587	3036,2	6,8873
310	0,1793	3062,4	6,9765	0,1731	3061,2	6,9592	0,1673	3059,9	6,9424	0,1619	3058,6	6,9261
320	0,1828	3084,4	7,0139	0,1765	3083,3	6,9967	0,1706	3082,1	6,9801	0,1651	3080,9	6,9639
330	0,1863	3106,3	7,0504	0,1799	3105,2	7,0334	0,1739	3104,1	7,0169	0,1683	3102,9	7,0008
340	0,1897	3128,0	7,0862	0,1832	3127,0	7,0693	0,1771	3125,9	7,0529	0,1714	3124,9	7,0369
350	0,1931	3149,7	7,1212	0,1865	3148,7	7,1044	0,1803	3147,7	7,0881	0,1745	3146,7	7,0723
360	0,1966	3171,3	7,1556	0,1898	3170,4	7,1389	0,1836	3169,4	7,1227	0,1777	3168,5	7,1069
370	0,2000	3192,9	7,1894	0,1931	3192,0	7,1727	0,1867	3191,1	7,1566	0,1808	3190,2	7,1409
380	0,2033	3214,4	7,2226	0,1964	3213,5	7,2060	0,1899	3212,7	7,1899	0,1838	3211,8	7,1743
390	0,2067	3235,9	7,2553	0,1997	3235,1	7,2387	0,1931	3234,3	7,2227	0,1869	3233,4	7,2071
400	0,2101	3257,4	7,2874	0,2029	3256,6	7,2709	0,1962	3255,8	7,2550	0,1900	3255,0	7,2394
410	0,2134	3278,9	7,3191	0,2062	3278,1	7,3027	0,1994	3277,4	7,2867	0,1930	3276,6	7,2713
420	0,2168	3300,4	7,3504	0,2094	3299,7	7,3340	0,2025	3298,9	7,3181	0,1961	3298,2	7,3026
430	0,2201	3321,9	7,3812	0,2126	3321,2	7,3648	0,2056	3320,5	7,3490	0,1991	3319,8	7,3336
440	0,2234	3343,4	7,4116	0,2158	3342,8	7,3953	0,2088	3342,1	7,3794	0,2021	3341,4	7,3641
450	0,2267	3365,0	7,4416	0,2191	3364,3	7,4253	0,2119	3363,7	7,4095	0,2051	3363,0	7,3942
460	0,2300	3386,6	7,4713	0,2223	3386,0	7,4550	0,2150	3385,3	7,4392	0,2082	3384,7	7,4240
470	0,2333	3408,2	7,5006	0,2255	3407,6	7,4843	0,2181	3407,0	7,4686	0,2112	3406,4	7,4534
480	0,2366	3429,9	7,5295	0,2287	3429,3	7,5133	0,2212	3428,7	7,4976	0,2142	3428,1	7,4824
490	0,2399	3451,6	7,5582	0,2318	3451,0	7,5420	0,2243	3450,5	7,5263	0,2172	3449,9	7,5111
500	0,2432	3473,3	7,5865	0,2350	3472,8	7,5703	0,2274	3472,2	7,5547	0,2202	3471,7	7,5395
510	0,2465	3495,1	7,6145	0,2382	3494,6	7,5984	0,2304	3494,1	7,5827	0,2231	3493,5	7,5676
520	0,2498	3517,0	7,6422	0,2414	3516,5	7,6261	0,2335	3516,0	7,6105	0,2261	3515,4	7,5954
530	0,2531	3538,9	7,6697	0,2446	3538,4	7,6536	0,2366	3537,9	7,6380	0,2291	3537,4	7,6229
540	0,2563	3560,8	7,6968	0,2477	3560,4	7,6808	0,2397	3559,9	7,6652	0,2321	3559,4	7,6501
550	0,2596	3582,9	7,7237	0,2509	3582,4	7,7077	0,2427	3581,9	7,6921	0,2351	3581,4	7,6770

Table 3. Water and Superheated Steam (Continuation) Wasser und überhitzter Dampf (Fortsetzung)

t	14,5 bar $t_s = 196,69\ °C$			15,0 bar $t_s = 198,29\ °C$			15,5 bar $t_s = 199,85\ °C$			16,0 bar $t_s = 201,37\ °C$		
	v'' 0,1360	h'' 2788,9	s'' 6,4526	v'' 0,1317	h'' 2789,9	s'' 6,4406	v'' 0,1275	h'' 2790,8	s'' 6,4289	v'' 0,1237	h'' 2791,7	s'' 6,4175
°C	v	h	s	v	h	s	v	h	s	v	h	s
550	0,2596	3582,9	7,7237	0,2509	3582,4	7,7077	0,2427	3581,9	7,6921	0,2351	3581,4	7,6770
560	0,2629	3604,9	7,7504	0,2540	3604,5	7,7343	0,2458	3604,0	7,7188	0,2380	3603,5	7,7037
570	0,2662	3627,0	7,7768	0,2572	3626,6	7,7607	0,2488	3626,1	7,7452	0,2410	3625,7	7,7301
580	0,2694	3649,2	7,8029	0,2604	3648,8	7,7869	0,2519	3648,3	7,7714	0,2440	3647,9	7,7563
590	0,2727	3671,4	7,8288	0,2635	3671,0	7,8128	0,2549	3670,6	7,7973	0,2469	3670,2	7,7823
600	0,2759	3693,7	7,8545	0,2667	3693,3	7,8385	0,2580	3692,9	7,8230	0,2499	3692,5	7,8080
610	0,2792	3716,1	7,8799	0,2698	3715,7	7,8640	0,2610	3715,3	7,8485	0,2528	3714,9	7,8335
620	0,2824	3738,5	7,9052	0,2730	3738,1	7,8892	0,2641	3737,7	7,8737	0,2558	3737,3	7,8588
630	0,2857	3760,9	7,9302	0,2761	3760,6	7,9142	0,2671	3760,2	7,8988	0,2587	3759,8	7,8838
640	0,2889	3783,5	7,9550	0,2792	3783,1	7,9390	0,2702	3782,7	7,9236	0,2617	3782,4	7,9086
650	0,2922	3806,1	7,9796	0,2824	3805,7	7,9636	0,2732	3805,3	7,9482	0,2646	3805,0	7,9333
660	0,2954	3828,7	8,0040	0,2855	3828,4	7,9881	0,2762	3828,0	7,9726	0,2676	3827,7	7,9577
670	0,2986	3851,4	8,0282	0,2886	3851,1	8,0123	0,2793	3850,7	7,9969	0,2705	3850,4	7,9819
680	0,3019	3874,2	8,0522	0,2918	3873,8	8,0363	0,2823	3873,5	8,0209	0,2734	3873,2	8,0060
690	0,3051	3897,0	8,0760	0,2949	3896,7	8,0601	0,2853	3896,4	8,0447	0,2764	3896,0	8,0298
700	0,3083	3919,9	8,0997	0,2980	3919,6	8,0838	0,2884	3919,3	8,0684	0,2793	3918,9	8,0535
710	0,3116	3942,8	8,1231	0,3012	3942,5	8,1072	0,2914	3942,2	8,0919	0,2822	3941,9	8,0770
720	0,3148	3965,8	8,1464	0,3043	3965,5	8,1305	0,2944	3965,2	8,1152	0,2852	3964,9	8,1003
730	0,3180	3988,9	8,1695	0,3074	3988,6	8,1536	0,2974	3988,3	8,1383	0,2881	3988,0	8,1234
740	0,3213	4012,0	8,1924	0,3105	4011,7	8,1766	0,3005	4011,4	8,1612	0,2910	4011,2	8,1464
750	0,3245	4035,2	8,2152	0,3136	4034,9	8,1993	0,3035	4034,6	8,1840	0,2940	4034,4	8,1691
760	0,3277	4058,4	8,2378	0,3168	4058,2	8,2220	0,3065	4057,9	8,2066	0,2969	4057,6	8,1918
770	0,3309	4081,7	8,2602	0,3199	4081,5	8,2444	0,3095	4081,2	8,2291	0,2998	4081,0	8,2142
780	0,3342	4105,1	8,2825	0,3230	4104,8	8,2667	0,3125	4104,6	8,2514	0,3027	4104,3	8,2365
790	0,3374	4128,5	8,3046	0,3261	4128,2	8,2888	0,3155	4128,0	8,2735	0,3057	4127,8	8,2587
800	0,3406	4152,0	8,3266	0,3292	4151,7	8,3108	0,3186	4151,5	8,2955	0,3086	4151,3	8,2807

Table 3. Water and Superheated Steam (Continuation) Wasser und überhitzter Dampf (Fortsetzung)

t	16,5 bar $t_s=202,86$ °C			17,0 bar $t_s=204,31$ °C			17,5 bar $t_s=205,72$ °C			18,0 bar $t_s=207,11$ °C		
	v'' 0,1201	h'' 2792,6	s'' 6,4065	v'' 0,1166	h'' 2793,4	s'' 6,3957	v'' 0,1134	h'' 2794,1	s'' 6,3853	v'' 0,1103	h'' 2794,8	s'' 6,3751
°C	v	h	s	v	h	s	v	h	s	v	h	s
0	0,0009994	1,6	−0,0000	0,0009994	1,7	−0,0000	0,0009993	1 7	−0,0000	0,0009993	1,8	−0,0000
10	0,0009995	43,6	0,1508	0,0009995	43,7	0,1508	0,0009994	43,7	0,1508	0,0009994	43,7	0,1508
20	0,0010010	85,4	0,2960	0,0010010	85,5	0,2959	0,0010009	85,5	0,2959	0,0010009	85,6	0,2959
30	0,0010036	127,2	0,4360	0,0010036	127,2	0,4360	0,0010035	127,3	0,4360	0,0010035	127,3	0,4360
40	0,0010071	168,9	0,5715	0,0010071	169,0	0,5715	0,0010070	169,0	0,5714	0,0010070	169,0	0,5714
50	0,0010114	210,7	0,7027	0,0010114	210,7	0,7027	0,0010113	210,8	0,7027	0,0010113	210,8	0,7027
60	0,0010164	252,5	0,8301	0,0010164	252,5	0,8301	0,0010164	252,5	0,8301	0,0010163	252,6	0,8300
70	0,0010221	294,3	0,9538	0,0010221	294,3	0,9538	0,0010220	294,4	0,9538	0,0010220	294,4	0,9538
80	0,0010284	336,2	1,0742	0,0010284	336,2	1,0742	0,0010284	336,3	1,0741	0,0010283	336,3	1,0741
90	0,0010354	378,2	1,1914	0,0010353	378,2	1,1913	0,0010353	378,2	1,1913	0,0010353	378,3	1,1913
100	0,0010429	420,2	1,3057	0,0010429	420,3	1,3056	0,0010428	420,3	1,3056	0,0010428	420,3	1,3055
110	0,0010510	462,4	1,4172	0,0010510	462,4	1,4172	0,0010510	462,5	1,4171	0,0010510	462,5	1,4171
120	0,0010598	504,7	1,5263	0,0010598	504,8	1,5262	0,0010597	504,8	1,5262	0,0010597	504,8	1,5261
130	0,0010692	547,2	1,6330	0,0010691	547,3	1,6330	0,0010691	547,3	1,6329	0,0010691	547,3	1,6329
140	0,0010792	589,9	1,7377	0,0010792	590,0	1,7376	0,0010792	590,0	1,7375	0,0010791	590,0	1,7375
150	0,0010900	632,9	1,8403	0,0010899	632,9	1,8403	0,0010899	632,9	1,8402	0,0010899	633,0	1,8402
160	0,0011015	676,1	1,9412	0,0011014	676,1	1,9412	0,0011014	676,1	1,9411	0,0011013	676,2	1,9411
170	0,0011138	719,6	2,0405	0,0011137	719,6	2,0405	0,0011137	719,6	2,0404	0,0011136	719,7	2,0403
180	0,0011270	763,4	2,1384	0,0011269	763,5	2,1383	0,0011269	763,5	2,1383	0,0011268	763,5	2,1382
190	0,0011411	807,7	2,2350	0,0011411	807,7	2,2349	0,0011410	807,7	2,2349	0,0011410	807,8	2,2348
200	0,0011564	852,4	2,3305	0,0011563	852,4	2,3305	0,0011563	852,4	2,3304	0,0011562	852,5	2,3303
210	0,1228	2812,9	6,4488	0,1188	2809,7	6,4297	0,1150	2806,5	6,4110	0,1114	2803,3	6,3926
220	0,1266	2840,3	6,5050	0,1225	2837,5	6,4866	0,1186	2834,6	6,4686	0,1150	2831,7	6,4509
230	0,1303	2866,8	6,5581	0,1261	2864,2	6,5403	0,1221	2861,6	6,5229	0,1184	2859,1	6,5058
240	0,1338	2892,4	6,6086	0,1296	2890,1	6,5912	0,1255	2887,8	6,5743	0,1217	2885,4	6,5577
250	0,1373	2917,4	6,6567	0,1329	2915,3	6,6398	0,1288	2913,2	6,6233	0,1250	2911,0	6,6071
260	0,1407	2941,7	6,7028	0,1362	2939,8	6,6863	0,1321	2937,9	6,6701	0,1282	2935,9	6,6543
270	0,1440	2965,6	6,7471	0,1395	2963,8	6,7309	0,1353	2962,1	6,7150	0,1313	2960,3	6,6995
280	0,1472	2989,0	6,7899	0,1427	2987,4	6,7739	0,1384	2985,8	6,7583	0,1343	2984,1	6,7430
290	0,1505	3012,1	6,8312	0,1458	3010,6	6,8154	0,1414	3009,1	6,8000	0,1373	3007,6	6,7850
300	0,1536	3034,8	6,8713	0,1489	3033,5	6,8557	0,1445	3032,1	6,8405	0,1402	3030,7	6,8257
310	0,1568	3057,4	6,9103	0,1520	3056,1	6,8948	0,1474	3054,8	6,8798	0,1432	3053,5	6,8651
320	0,1599	3079,7	6,9482	0,1550	3078,5	6,9329	0,1504	3077,3	6,9180	0,1460	3076,1	6,9035
330	0,1630	3101,8	6,9853	0,1580	3100,7	6,9701	0,1533	3099,6	6,9553	0,1489	3098,4	6,9409
340	0,1660	3123,8	7,0214	0,1610	3122,8	7,0064	0,1562	3121,7	6,9917	0,1517	3120,6	6,9774
350	0,1691	3145,7	7,0569	0,1640	3144,7	7,0419	0,1591	3143,7	7,0273	0,1546	3142,7	7,0131
360	0,1721	3167,5	7,0916	0,1669	3166,6	7,0767	0,1620	3165,6	7,0622	0,1573	3164,7	7,0481
370	0,1751	3189,3	7,1257	0,1698	3188,4	7,1108	0,1648	3187,5	7,0964	0,1601	3186,6	7,0824
380	0,1781	3211,0	7,1591	0,1728	3210,1	7,1444	0,1677	3209,2	7,1300	0,1629	3208,4	7,1160
390	0,1811	3232,6	7,1920	0,1757	3231,8	7,1773	0,1705	3231,0	7,1630	0,1656	3230,1	7,1491
400	0,1841	3254,2	7,2244	0,1785	3253,5	7,2098	0,1733	3252,7	7,1955	0,1684	3251,9	7,1816
410	0,1871	3275,9	7,2563	0,1814	3275,1	7,2417	0,1761	3274,3	7,2275	0,1711	3273,6	7,2137
420	0,1900	3297,5	7,2877	0,1843	3296,8	7,2731	0,1789	3296,0	7,2590	0,1738	3295,3	7,2452
430	0,1930	3319,1	7,3187	0,1872	3318,4	7,3041	0,1817	3317,7	7,2900	0,1766	3317,0	7,2763
440	0,1959	3340,7	7,3492	0,1900	3340,1	7,3347	0,1845	3339,4	7,3207	0,1793	3338,7	7,3070
450	0,1988	3362,4	7,3794	0,1929	3361,7	7,3649	0,1873	3361,1	7,3509	0,1820	3360,4	7,3372
460	0,2018	3384,1	7,4091	0,1957	3383,4	7,3947	0,1900	3382,8	7,3807	0,1847	3382,2	7,3671
470	0,2047	3405,8	7,4385	0,1986	3405,2	7,4242	0,1928	3404,6	7,4102	0,1873	3404,0	7,3966
480	0,2076	3427,5	7,4676	0,2014	3426,9	7,4533	0,1955	3426,4	7,4393	0,1900	3425,8	7,4257
490	0,2105	3449,3	7,4963	0,2042	3448,7	7,4820	0,1983	3448,2	7,4681	0,1927	3447,6	7,4545
500	0,2134	3471,1	7,5248	0,2070	3470,6	7,5105	0,2010	3470,0	7,4965	0,1954	3469,5	7,4830
510	0,2163	3493,0	7,5529	0,2099	3492,5	7,5386	0,2038	3491,9	7,5247	0,1980	3491,4	7,5112
520	0,2192	3514,9	7,5807	0,2127	3514,4	7,5664	0,2065	3513,9	7,5525	0,2007	3513,4	7,5391
530	0,2221	3536,9	7,6082	0,2155	3536,4	7,5939	0,2093	3535,9	7,5801	0,2034	3535,4	7,5666
540	0,2250	3558,9	7,6354	0,2183	3558,4	7,6212	0,2120	3557,9	7,6074	0,2060	3557,4	7,5939
550	0,2279	3581,0	7,6624	0,2211	3580,5	7,6482	0,2147	3580,0	7,6344	0,2087	3579,5	7,6209

Table 3. Water and Superheated Steam (Continuation) Wasser und überhitzter Dampf (Fortsetzung)

t	16,5 bar $t_s = 202,86$ °C			17,0 bar $t_s = 204,31$ °C			17,5 bar $t_s = 205,72$ °C			18,0 bar $t_s = 207,11$ °C		
	v'' 0,1201	h'' 2792,6	s'' 6,4065	v'' 0,1166	h'' 2793,4	s'' 6,3957	v'' 0,1134	h'' 2794,1	s'' 6,3853	v'' 0,1103	h'' 2794,8	s'' 6,3751
°C	v	h	s	v	h	s	v	h	s	v	h	s
550	0,2279	3581,0	7,6624	0,2211	3580,5	7,6482	0,2147	3580,0	7,6344	0,2087	3579,5	7,6209
560	0,2307	3603,1	7,6891	0,2239	3602,6	7,6749	0,2174	3602,2	7,6611	0,2113	3601,7	7,6477
570	0,2336	3625,2	7,7155	0,2267	3624,8	7,7014	0,2201	3624,3	7,6876	0,2140	3623,9	7,6742
580	0,2365	3647,5	7,7417	0,2295	3647,0	7,7276	0,2229	3646,6	7,7138	0,2166	3646,2	7,7004
590	0,2394	3669,7	7,7677	0,2323	3669,3	7,7536	0,2256	3668,9	7,7398	0,2193	3668,5	7,7264
600	0,2423	3692,1	7,7934	0,2351	3691,7	7,7793	0,2283	3691,3	7,7656	0,2219	3690,9	7,7522
610	0,2451	3714,5	7,8189	0,2378	3714,1	7,8048	0,2310	3713,7	7,7911	0,2245	3713,3	7,7778
620	0,2480	3736,9	7,8442	0,2406	3736,5	7,8301	0,2337	3736,2	7,8164	0,2272	3735,8	7,8031
630	0,2508	3759,4	7,8693	0,2434	3759,1	7,8552	0,2364	3758,7	7,8415	0,2298	3758,3	7,8282
640	0,2537	3782,0	7,8941	0,2462	3781,6	7,8800	0,2391	3781,3	7,8664	0,2324	3780,9	7,8531
650	0,2566	3804,6	7,9188	0,2490	3804,3	7,9047	0,2418	3803,9	7,8910	0,2350	3803,6	7,8777
660	0,2594	3827,3	7,9432	0,2517	3827,0	7,9291	0,2445	3826,6	7,9155	0,2377	3826,3	7,9022
670	0,2633	3850,1	7,9675	0,2545	3849,7	7,9534	0,2472	3849,4	7,9397	0,2403	3849,0	7,9265
680	0,2651	3872,9	7,9915	0,2573	3872,5	7,9775	0,2499	3872,2	7,9638	0,2429	3871,9	7,9505
690	0,2680	3895,7	8,0154	0,2600	3895,4	8,0013	0,2526	3895,1	7,9877	0,2455	3894,8	7,9744
700	0,2708	3918,6	8,0390	0,2628	3918,3	8,0250	0,2553	3918,0	8,0114	0,2481	3917,7	7,9981
710	0,2737	3941,6	8,0625	0,2656	3941,3	8,0485	0,2579	3941,0	8,0349	0,2507	3940,7	8,0217
720	0,2765	3964,6	8,0858	0,2683	3964,4	8,0718	0,2606	3964,1	8,0582	0,2533	3963,8	8,0450
730	0,2793	3987,7	8,1090	0,2711	3987,5	8,0950	0,2633	3987,2	8,0814	0,2560	3986,9	8,0682
740	0,2822	4010,9	8,1319	0,2738	4010,6	8,1179	0,2660	4010,3	8,1044	0,2586	4010,1	8,0911
750	0,2850	4034,1	8,1547	0,2766	4033,8	8,1408	0,2687	4033,6	8,1272	0,2612	4033,3	8,1140
760	0,2879	4057,4	8,1774	0,2794	4057,1	8,1634	0,2713	4056,9	8,1498	0,2638	4056,6	8,1366
770	0,2907	4080,7	8,1998	0,2821	4080,4	8,1859	0,2740	4080,2	8,1723	0,2664	4079,9	8,1591
780	0,2935	4104,1	8,2221	0,2849	4103,8	8,2082	0,2767	4103,6	8,1946	0,2690	4103,3	8,1814
790	0,2964	4127,5	8,2443	0,2876	4127,3	8,2303	0,2794	4127,0	8,2168	0,2716	4126,8	8,2036
800	0,2992	4151,0	8,2663	0,2904	4150,8	8,2523	0,2820	4150,6	8,2388	0,2742	4150,3	8,2256

Table 3. Water and Superheated Steam (Continuation) Wasser und überhitzter Dampf (Fortsetzung)

	18,5 bar $t_s = 208,47$ °C			19,0 bar $t_s = 209,80$ °C			19,5 bar $t_s = 211,10$ °C			20 bar $t_s = 212,37$ °C		
t	v'' 0,1074	h'' 2795,5	s'' 6,3651	v'' 0,1047	h'' 2796,1	s'' 6,3554	v'' 0,1020	h'' 2796,7	s'' 6,3459	v'' 0,09954	h'' 2797,2	s'' 6,3366
°C	v	h	s	v	h	s	v	h	s	v	h	s
0	0,0009993	1,8	—0,0000	0,0009993	1,9	—0,0000	0,0009992	2,0	—0,0000	0,0009992	2,0	—0,0000
10	0,0009994	43,8	0,1508	0,0009994	43,8	0,1508	0,0009993	43,9	0,1508	0,0009993	43,9	0,1508
20	0,0010009	85,6	0,2959	0,0010009	85,6	0,2959	0,0010008	85,7	0,2959	0,0010008	85,7	0,2959
30	0,0010035	127,3	0,4359	0,0010035	127,4	0,4359	0,0010034	127,4	0,4359	0,0010034	127,5	0,4359
40	0,0010070	169,1	0,5714	0,0010070	169,1	0,5714	0,0010069	169,2	0,5714	0,0010069	169,2	0,5713
50	0,0010113	210,8	0,7027	0,0010113	210,9	0,7026	0,0010112	210,9	0,7026	0,0010112	211,0	0,7026
60	0,0010163	252,6	0,8300	0,0010163	252,7	0,8300	0,0010163	252,7	0,8300	0,0010162	252,7	0,8299
70	0,0010220	294,5	0,9537	0,0010220	294,5	0,9537	0,0010220	294,5	0,9537	0,0010219	294,6	0,9536
80	0,0010283	336,3	1,0741	0,0010283	336,4	1,0740	0,0010283	336,4	1,0740	0,0010282	336,5	1,0740
90	0,0010353	378,3	1,1912	0,0010352	378,4	1,1912	0,0010352	378,4	1,1912	0,0010352	378,4	1,1911
100	0,0010428	420,4	1,3055	0,0010428	420,4	1,3055	0,0010427	420,5	1,3054	0,0010427	420,5	1,3054
110	0,0010509	462,6	1,4171	0,0010509	462,6	1,4170	0,0010509	462,6	1,4170	0,0010508	462,7	1,4169
120	0,0010597	504,9	1,5261	0,0010596	504,9	1,5260	0,0010596	504,9	1,5260	0,0010596	505,0	1,5260
130	0,0010691	547,4	1,6328	0,0010690	547,4	1,6328	0,0010690	547,4	1,6327	0,0010690	547,5	1,6327
140	0,0010791	590,1	1,7374	0,0010791	590,1	1,7374	0,0010790	590,1	1,7373	0,0010790	590,2	1,7373
150	0,0010898	633,0	1,8401	0,0010898	633,0	1,8400	0,0010898	633,1	1,8400	0,0010897	633,1	1,8399
160	0,0011013	676,2	1,9410	0,0011013	676,2	1,9409	0,0011012	676,3	1,9409	0,0011012	676,3	1,9408
170	0,0011136	719,7	2,0403	0,0011136	719,7	2,0402	0,0011135	719,8	2,0402	0,0011135	719,8	2,0401
180	0,0011268	763,5	2,1381	0,0011267	763,6	2,1381	0,0011267	763,6	2,1380	0,0011267	763,6	2,1379
190	0,0011409	807,8	2,2347	0,0011409	807,8	2,2346	0,0011408	807,8	2,2346	0,0011408	807,9	2,2345
200	0,0011562	852,5	2,3302	0,0011561	852,5	2,3301	0,0011561	852,5	2,3301	0,0011560	852,6	2,3300
210	0,1080	2800,0	6,3745	0,1047	2796,7	6,3567	0,0011726	897,7	2,4246	0,0011725	897,8	2,4245
220	0,1115	2828,8	6,4335	0,1082	2825,9	6,4164	0,1051	2822,9	6,3995	0,1021	2819,9	6,3829
230	0,1149	2856,4	6,4890	0,1115	2853,8	6,4725	0,1083	2851,1	6,4562	0,1053	2848,4	6,4403
240	0,1181	2883,1	6,5414	0,1147	2880,7	6,5254	0,1115	2878,3	6,5097	0,1084	2875,9	6,4943
250	0,1213	2908,9	6,5912	0,1179	2906,7	6,5757	0,1146	2904,6	6,5604	0,1114	2902,4	6,5454
260	0,1244	2934,0	6,6388	0,1209	2932,0	6,6236	0,1176	2930,1	6,6087	0,1144	2928,1	6,5941
270	0,1275	2958,5	6,6843	0,1239	2956,7	6,6694	0,1205	2954,9	6,6549	0,1172	2953,1	6,6406
280	0,1304	2982,5	6,7281	0,1268	2980,9	6,7135	0,1233	2979,2	6,6992	0,1200	2977,5	6,6852
290	0,1334	3006,1	6,7703	0,1297	3004,6	6,7559	0,1261	3003,0	6,7419	0,1228	3001,5	6,7281
300	0,1363	3029,3	6,8112	0,1325	3027,9	6,7970	0,1289	3026,5	6,7831	0,1255	3025,0	6,7696
310	0,1391	3052,2	6,8508	0,1353	3050,9	6,8368	0,1316	3049,6	6,8231	0,1282	3048,2	6,8097
320	0,1419	3074,8	6,8893	0,1380	3073,6	6,8755	0,1343	3072,4	6,8619	0,1308	3071,2	6,8487
330	0,1447	3097,3	6,9268	0,1408	3096,1	6,9131	0,1370	3095,0	6,8997	0,1334	3093,8	6,8866
340	0,1475	3119,6	6,9635	0,1435	3118,5	6,9498	0,1396	3117,4	6,9365	0,1360	3116,3	6,9235
350	0,1502	3141,7	6,9993	0,1461	3140,7	6,9857	0,1422	3139,7	6,9725	0,1386	3138,6	6,9596
360	0,1530	3163,7	7,0343	0,1488	3162,8	7,0209	0,1449	3161,8	7,0078	0,1411	3160,8	6,9950
370	0,1557	3185,6	7,0687	0,1514	3184,7	7,0553	0,1474	3183,8	7,0423	0,1436	3182,9	7,0296
380	0,1584	3207,5	7,1024	0,1541	3206,6	7,0891	0,1500	3205,8	7,0762	0,1461	3204,9	7,0635
390	0,1611	3229,3	7,1356	0,1567	3228,5	7,1223	0,1526	3227,6	7,1094	0,1486	3226,8	7,0968
400	0,1637	3251,1	7,1681	0,1593	3250,3	7,1550	0,1551	3249,5	7,1421	0,1511	3248,7	7,1296
410	0,1664	3272,8	7,2002	0,1619	3272,1	7,1871	0,1576	3271,3	7,1743	0,1536	3270,5	7,1618
420	0,1690	3294,6	7,2318	0,1645	3293,8	7,2187	0,1602	3293,1	7,2059	0,1561	3292,4	7,1935
430	0,1717	3316,3	7,2629	0,1671	3315,6	7,2499	0,1627	3314,9	7,2372	0,1585	3314,2	7,2247
440	0,1743	3338,0	7,2936	0,1696	3337,4	7,2806	0,1652	3336,7	7,2679	0,1610	3336,0	7,2555
450	0,1770	3359,8	7,3239	0,1722	3359,1	7,3109	0,1677	3358,5	7,2983	0,1634	3357,8	7,2859
460	0,1796	3381,6	7,3538	0,1748	3380,9	7,3409	0,1702	3380,3	7,3282	0,1659	3379,7	7,3159
470	0,1822	3403,4	7,3833	0,1773	3402,7	7,3704	0,1727	3402,1	7,3578	0,1683	3401,5	7,3455
480	0,1848	3425,2	7,4125	0,1799	3424,6	7,3996	0,1752	3424,0	7,3871	0,1707	3423,4	7,3748
490	0,1874	3447,0	7,4413	0,1824	3446,5	7,4285	0,1777	3445,9	7,4159	0,1731	3445,3	7,4037
500	0,1900	3468,9	7,4698	0,1849	3468,4	7,4570	0,1801	3467,8	7,4445	0,1756	3467,3	7,4323
510	0,1926	3490,9	7,4980	0,1875	3490,3	7,4852	0,1826	3489,8	7,4727	0,1780	3489,3	7,4605
520	0,1952	3512,8	7,5259	0,1900	3512,3	7,5131	0,1851	3511,8	7,5007	0,1804	3511,3	7,4885
530	0,1978	3534,9	7,5535	0,1925	3534,4	7,5407	0,1875	3533,9	7,5283	0,1828	3533,4	7,5161
540	0,2004	3556,9	7,5808	0,1951	3556,5	7,5681	0,1900	3556,0	7,5556	0,1852	3555,5	7,5435
550	0,2030	3579,1	7,6079	0,1976	3578,6	7,5951	0,1924	3578,1	7,5827	0,1876	3577,6	7,5706

Table 3. Water and Superheated Steam (Continuation) Wasser und überhitzter Dampf (Fortsetzung)

t	18,5 bar $t_s = 208,47\,°C$			19,0 bar $t_s = 209,80\,°C$			19,5 bar $t_s = 211,10\,°C$			20 bar $t_s = 212,37\,°C$		
	v'' 0,1074	h'' 2795,5	s'' 6,3651	v'' 0,1047	h'' 2796,1	s'' 6,3554	v'' 0,1020	h'' 2796,7	s'' 6,3459	v'' 0,09954	h'' 2797,2	s'' 6,3366
°C	v	h	s	v	h	s	v	h	s	v	h	s
550	0,2030	3579,1	7,6079	0,1976	3578,6	7,5951	0,1924	3578,1	7,5827	0,1876	3577,6	7,5706
560	0,2056	3601,2	7,6346	0,2001	3600,8	7,6219	0,1949	3600,3	7,6095	0,1900	3599,9	7,5974
570	0,2081	3623,5	7,6611	0,2026	3623,0	7,6484	0,1973	3622,6	7,6361	0,1924	3622,1	7,6240
580	0,2107	3645,7	7,6874	0,2051	3645,3	7,6747	0,1998	3644,9	7,6623	0,1947	3644,4	7,6503
590	0,2133	3668,1	7,7134	0,2076	3667,6	7,7008	0,2022	3667,2	7,6884	0,1971	3666,8	7,6763
600	0,2158	3690,4	7,7392	0,2101	3690,0	7,7265	0,2047	3689,6	7,7142	0,1995	3689,2	7,7022
610	0,2184	3712,9	7,7648	0,2126	3712,5	7,7521	0,2071	3712,1	7,7398	0,2019	3711,7	7,7278
620	0,2210	3735,4	7,7901	0,2151	3735,0	7,7775	0,2095	3734,6	7,7651	0,2043	3734,2	7,7531
630	0,2235	3757,9	7,8152	0,2176	3757,6	7,8026	0,2120	3757,2	7,7903	0,2066	3756,8	7,7783
640	0,2261	3780,5	7,8401	0,2201	3780,2	7,8275	0,2144	3779,8	7,8152	0,2090	3779,4	7,8032
650	0,2286	3803,2	7,8648	0,2226	3802,8	7,8522	0,2168	3802,5	7,8399	0,2114	3802,1	7,8279
660	0,2312	3825,9	7,8893	0,2251	3825,6	7,8767	0,2193	3825,2	7,8644	0,2137	3824,9	7,8524
670	0,2337	3848,7	7,9136	0,2275	3848,4	7,9010	0,2217	3848,0	7,8887	0,2161	3847,7	7,8767
680	0,2363	3871,5	7,9376	0,2300	3871,2	7,9251	0,2241	3870,9	7,9128	0,2185	3870,6	7,9009
690	0,2388	3894,4	7,9615	0,2325	3894,1	7,9490	0,2265	3893,8	7,9367	0,2208	3893,5	7,9248
700	0,2414	3917,4	7,9852	0,2350	3917,1	7,9727	0,2289	3916,8	7,9605	0,2232	3916,5	7,9485
710	0,2439	3940,4	8,0088	0,2375	3940,1	7,9962	0,2313	3939,8	7,9840	0,2255	3939,5	7,9721
720	0,2465	3963,5	8,0321	0,2399	3963,2	8,0196	0,2338	3962,9	8,0074	0,2279	3962,6	7,9954
730	0,2490	3986,6	8,0553	0,2424	3986,3	8,0428	0,2362	3986,0	8,0305	0,2302	3985,7	8,0186
740	0,2515	4009,8	8,0783	0,2449	4009,5	8,0658	0,2386	4009,2	8,0536	0,2326	4009,0	8,0417
750	0,2541	4033,0	8,1011	0,2474	4032,8	8,0886	0,2410	4032,5	8,0764	0,2349	4032,2	8,0645
760	0,2566	4056,3	8,1238	0,2498	4056,1	8,1113	0,2434	4055,8	8,0991	0,2373	4055,5	8,0872
770	0,2591	4079,7	8,1463	0,2523	4079,4	8,1338	0,2458	4079,2	8,1216	0,2396	4078,9	8,1097
780	0,2617	4103,1	8,1686	0,2548	4102,8	8,1561	0,2482	4102,6	8,1439	0,2420	4102,3	8,1321
790	0,2642	4126,6	8,1908	0,2572	4126,3	8,1783	0,2506	4126,1	8,1661	0,2443	4125,8	8,1543
800	0,2667	4150,1	8,2128	0,2597	4149,8	8,2003	0,2530	4149,6	8,1882	0,2467	4149,4	8,1763

Table 3. Water and Superheated Steam (Continuation) Wasser und überhitzter Dampf (Fortsetzung)

t	21 bar $t_s = 214,85$ °C			22 bar $t_s = 217,24$ °C			23 bar $t_s = 219,55$ °C			24 bar $t_s = 221,78$ °C		
	v'' 0,09489	h'' 2798,2	s'' 6,3187	v'' 0,09065	h'' 2799,1	s'' 6,3015	v'' 0,08677	h'' 2799,8	s'' 6,2849	v'' 0,08320	h'' 2800,4	s'' 6,2690
°C	v	h	s	v	h	s	v	h	s	v	h	s
0	0,0009992	2,1	0,0000	0,0009991	2,2	0,0000	0,0009991	2,3	0,0000	0,0009990	2,4	0,0000
10	0,0009993	44,0	0,1508	0,0009992	44,1	0,1508	0,0009992	44,2	0,1508	0,0009991	44,3	0,1508
20	0,0010008	85,8	0,2959	0,0010007	85,9	0,2958	0,0010007	86,0	0,2958	0,0010006	86,1	0,2958
30	0,0010034	127,6	0,4359	0,0010033	127,7	0,4358	0,0010033	127,8	0,4358	0,0010032	127,8	0,4358
40	0,0010069	169,3	0,5713	0,0010068	169,4	0,5713	0,0010068	169,5	0,5712	0,0010067	169,6	0,5712
50	0,0010112	211,1	0,7025	0,0010111	211,1	0,7025	0,0010111	211,2	0,7024	0,0010110	211,3	0,7024
60	0,0010162	252,8	0,8299	0,0010161	252,9	0,8298	0,0010161	253,0	0,8298	0,0010161	253,1	0,8297
70	0,0010219	294,7	0,9536	0,0010218	294,7	0,9535	0,0010218	294,8	0,9534	0,0010217	294,9	0,9534
80	0,0010282	336,5	1,0739	0,0010282	336,6	1,0738	0,0010281	336,7	1,0738	0,0010281	336,8	1,0737
90	0,0010351	378,5	1,1911	0,0010351	378,6	1,1910	0,0010350	378,7	1,1909	0,0010350	378,7	1,1908
100	0,0010427	420,6	1,3053	0,0010426	420,6	1,3052	0,0010426	420,7	1,3051	0,0010425	420,8	1,3051
110	0,0010508	462,7	1,4168	0,0010507	462,8	1,4168	0,0010507	462,9	1,4167	0,0010506	463,0	1,4166
120	0,0010595	505,1	1,5259	0,0010595	505,1	1,5258	0,0010594	505,2	1,5257	0,0010594	505,3	1,5256
130	0,0010689	547,5	1,6326	0,0010688	547,6	1,6325	0,0010688	547,7	1,6324	0,0010687	547,7	1,6323
140	0,0010789	590,2	1,7372	0,0010789	590,3	1,7371	0,0010788	590,4	1,7370	0,0010787	590,4	1,7369
150	0,0010897	633,2	1,8398	0,0010896	633,2	1,8397	0,0010895	633,3	1,8396	0,0010894	633,3	1,8395
160	0,0011011	676,3	1,9407	0,0011011	676,4	1,9406	0,0011010	676,5	1,9405	0,0011009	676,5	1,9404
170	0,0011134	719,8	2,0400	0,0011133	719,9	2,0398	0,0011132	719,9	2,0397	0,0011132	720,0	2,0396
180	0,0011266	763,7	2,1378	0,0011265	763,7	2,1377	0,0011264	763,8	2,1375	0,0011263	763,8	2,1374
190	0,0011407	807,9	2,2343	0,0011406	808,0	2,2342	0,0011405	808,0	2,2340	0,0011404	808,0	2,2339
200	0,0011559	852,6	2,3298	0,0011558	852,6	2,3297	0,0011557	852,7	2,3295	0,0011556	852,7	2,3293
210	0,0011724	897,8	2,4244	0,0011723	897,8	2,4242	0,0011722	897,9	2,4240	0,0011720	897,9	2,4238
220	0,09656	2813,8	6,3504	0,09152	2807,5	6,3187	0,08691	2801,2	6,2878	0,0011899	943,7	2,5176
230	0,09970	2843,0	6,4091	0,09458	2837,4	6,3787	0,08990	2831,8	6,3491	0,08559	2826,0	6,3202
240	0,10272	2871,0	6,4642	0,09752	2866,0	6,4349	0,09276	2860,9	6,4065	0,08839	2855,7	6,3788
250	0,10564	2897,9	6,5162	0,10035	2893,4	6,4879	0,09551	2888,9	6,4605	0,09107	2884,2	6,4338
260	0,10847	2924,0	6,5656	0,10309	2920,0	6,5382	0,09818	2915,8	6,5115	0,09367	2911,6	6,4857
270	0,11123	2949,4	6,6128	0,10576	2945,7	6,5860	0,10076	2941,9	6,5600	0,09618	2938,1	6,5349
280	0,11393	2974,2	6,6580	0,10837	2970,8	6,6317	0,10329	2967,3	6,6064	0,09863	2963,8	6,5818
290	0,11658	2998,4	6,7014	0,11092	2995,3	6,6756	0,10576	2992,1	6,6508	0,10102	2988,9	6,6267
300	0,11918	3022,2	6,7432	0,11343	3019,3	6,7179	0,10818	3016,4	6,6935	0,10336	3013,4	6,6699
310	0,12174	3045,6	6,7837	0,11590	3042,9	6,7588	0,11056	3040,2	6,7347	0,10566	3037,5	6,7115
320	0,12427	3068,7	6,8230	0,11833	3066,2	6,7983	0,11290	3063,7	6,7746	0,10793	3061,1	6,7517
330	0,12677	3091,5	6,8612	0,12073	3089,2	6,8368	0,11522	3086,8	6,8133	0,11016	3084,5	6,7907
340	0,12925	3114,1	6,8984	0,12311	3112,0	6,8742	0,11751	3109,7	6,8510	0,11237	3107,5	6,8286
350	0,13170	3136,6	6,9347	0,12547	3134,5	6,9107	0,11977	3132,4	6,8877	0,11455	3130,4	6,8656
360	0,13414	3158,9	6,9702	0,12780	3156,9	6,9464	0,12202	3155,0	6,9236	0,11672	3153,0	6,9016
370	0,13655	3181,1	7,0049	0,13012	3179,2	6,9813	0,12425	3177,3	6,9587	0,11887	3175,5	6,9369
380	0,13896	3203,1	7,0390	0,13243	3201,4	7,0155	0,12646	3199,6	6,9930	0,12100	3197,8	6,9714
390	0,14135	3225,1	7,0724	0,13472	3223,5	7,0491	0,12867	3221,8	7,0267	0,12312	3220,1	7,0052
400	0,14373	3247,1	7,1053	0,13700	3245,5	7,0821	0,13085	3243,9	7,0598	0,12522	3242,3	7,0384
410	0,14609	3269,0	7,1376	0,13927	3267,5	7,1145	0,13303	3265,9	7,0923	0,12732	3264,4	7,0710
420	0,14845	3290,9	7,1694	0,14152	3289,4	7,1464	0,13520	3288,0	7,1243	0,12940	3286,5	7,1031
430	0,15080	3312,8	7,2007	0,14377	3311,4	7,1778	0,13736	3310,0	7,1558	0,13148	3308,5	7,1347
440	0,15314	3334,6	7,2316	0,14602	3333,3	7,2088	0,13951	3331,9	7,1868	0,13355	3330,6	7,1658
450	0,15548	3356,5	7,2621	0,14825	3355,2	7,2393	0,14165	3353,9	7,2174	0,13561	3352,6	7,1964
460	0,15780	3378,4	7,2921	0,15048	3377,1	7,2694	0,14379	3375,9	7,2476	0,13766	3374,6	7,2267
470	0,16013	3400,3	7,3218	0,15270	3399,1	7,2991	0,14592	3397,9	7,2774	0,13971	3396,6	7,2565
480	0,16244	3422,2	7,3511	0,15492	3421,1	7,3285	0,14805	3419,9	7,3068	0,14175	3418,7	7,2860
490	0,16475	3444,2	7,3801	0,15713	3443,0	7,3575	0,15017	3441,9	7,3359	0,14379	3440,8	7,3151
500	0,16706	3466,2	7,4087	0,15934	3465,1	7,3862	0,15228	3464,0	7,3646	0,14582	3462,9	7,3439
510	0,16936	3488,2	7,4370	0,16154	3487,1	7,4145	0,15439	3486,0	7,3930	0,14785	3485,0	7,3723
520	0,17166	3510,3	7,4650	0,16373	3509,2	7,4425	0,15650	3508,2	7,4211	0,14987	3507,1	7,4004
530	0,17395	3532,4	7,4927	0,16593	3531,3	7,4703	0,15860	3530,3	7,4488	0,15189	3529,3	7,4282
540	0,17624	3554,5	7,5201	0,16812	3553,5	7,4977	0,16070	3552,5	7,4763	0,15390	3551,6	7,4558
550	0,17853	3576,7	7,5472	0,17030	3575,7	7,5249	0,16280	3574,8	7,5035	0,15591	3573,8	7,4830

Table 3. Water and Superheated Steam (Continuation) **Wasser und überhitzter Dampf** (Fortsetzung)

t	21 bar $t_s = 214,85$ °C			22 bar $t_s = 217,24$ °C			23 bar $t_s = 219,55$ °C			24 bar $t_s = 221,78$ °C		
	v'' 0,09489	h'' 2798,2	s'' 6,3187	v'' 0,09065	h'' 2799,1	s'' 6,3015	v'' 0,08677	h'' 2799,8	s'' 6,2849	v'' 0,08320	h'' 2800,4	s'' 6,2690
°C	v	h	s	v	h	s	v	h	s	v	h	s
550	0,17853	3576,7	7,5472	0,17030	3575,7	7,5249	0,16280	3574,8	7,5035	0,15591	3573,8	7,4830
560	0,18081	3598,9	7,5741	0,17249	3598,0	7,5518	0,16489	3597,1	7,5304	0,15792	3596,2	7,5099
570	0,18309	3621,2	7,6007	0,17467	3620,3	7,5784	0,16698	3619,4	7,5571	0,15993	3618,5	7,5366
580	0,18536	3643,6	7,6270	0,17684	3642,7	7,6048	0,16906	3641,8	7,5835	0,16193	3640,9	7,5631
590	0,18764	3665,9	7,6531	0,17901	3665,1	7,6309	0,17114	3664,3	7,6096	0,16393	3663,4	7,5892
600	0,18991	3688,4	7,6789	0,18119	3687,6	7,6568	0,17322	3686,7	7,6355	0,16592	3685,9	7,6152
610	0,19217	3710,9	7,7046	0,18335	3710,1	7,6824	0,17530	3709,3	7,6612	0,16791	3708,5	7,6409
620	0,19444	3733,4	7,7299	0,18552	3732,7	7,7078	0,17737	3731,9	7,6866	0,16991	3731,1	7,6663
630	0,19670	3756,0	7,7551	0,18768	3755,3	7,7330	0,17944	3754,5	7,7119	0,17189	3753,8	7,6916
640	0,19896	3778,7	7,7801	0,18984	3778,0	7,7580	0,18151	3777,2	7,7368	0,17388	3776,5	7,7166
650	0,20122	3801,4	7,8048	0,19200	3800,7	7,7827	0,18358	3800,0	7,7616	0,17586	3799,3	7,7414
660	0,20348	3824,2	7,8293	0,19416	3823,5	7,8073	0,18564	3822,8	7,7862	0,17784	3822,1	7,7660
670	0,20573	3847,0	7,8537	0,19631	3846,3	7,8317	0,18771	3845,7	7,8106	0,17982	3845,0	7,7904
680	0,20798	3869,9	7,8778	0,19846	3869,2	7,8558	0,18977	3868,6	7,8348	0,18180	3867,9	7,8146
690	0,21023	3892,8	7,9018	0,20061	3892,2	7,8798	0,19183	3891,6	7,8587	0,18377	3890,9	7,8386
700	0,21248	3915,8	7,9255	0,20276	3915,2	7,9035	0,19388	3914,6	7,8825	0,18575	3914,0	7,8624
710	0,21473	3938,9	7,9491	0,20491	3938,3	7,9271	0,19594	3937,7	7,9061	0,18772	3937,1	7,8860
720	0,21697	3962,0	7,9725	0,20705	3961,4	7,9505	0,19799	3960,8	7,9296	0,18969	3960,3	7,9095
730	0,21921	3985,2	7,9957	0,20919	3984,6	7,9738	0,20004	3984,0	7,9528	0,19166	3983,5	7,9327
740	0,22146	4008,4	8,0187	0,21133	4007,8	7,9968	0,20209	4007,3	7,9759	0,19362	4006,7	7,9558
750	0,22370	4031,7	8,0416	0,21347	4031,1	8,0197	0,20414	4030,6	7,9988	0,19559	4030,1	7,9787
760	0,22593	4055,0	8,0643	0,21561	4054,5	8,0424	0,20619	4054,0	8,0215	0,19755	4053,4	8,0015
770	0,22817	4078,4	8,0868	0,21775	4077,9	8,0650	0,20824	4077,4	8,0441	0,19951	4076,9	8,0240
780	0,23041	4101,9	8,1092	0,21989	4101,4	8,0873	0,21028	4100,9	8,0665	0,20148	4100,4	8,0464
790	0,23264	4125,4	8,1314	0,22202	4124,9	8,1096	0,21232	4124,4	8,0887	0,20344	4123,9	8,0687
800	0,23487	4148,9	8,1534	0,22415	4148,4	8,1316	0,21437	4148,0	8,1108	0,20539	4147,5	8,0908

Table 3. Water and Superheated Steam (Continuation) Wasser und überhitzter Dampf (Fortsetzung)

t	25 bar $t_s=223,94\,°C$			26 bar $t_s=226,04\,°C$			27 bar $t_s=228,07\,°C$			28 bar $t_s=230,05\,°C$		
	v'' 0,07991	h'' 2800,9	s'' 6,2536	v'' 0,07686	h'' 2801,4	s'' 6,2387	v'' 0,07402	h'' 2801,7	s'' 6,2244	v'' 0,07139	h'' 2802,0	s'' 6,2104
°C	v	h	s	v	h	s	v	h	s	v	h	s
0	0,0009990	2,5	0,0000	0,0009989	2,6	0,0001	0,0009989	2,7	0,0001	0,0009988	2,8	0,0001
10	0,0009991	44,4	0,1508	0,0009990	44,5	0,1508	0,0009990	44,6	0,1508	0,0009989	44,7	0,1507
20	0,0010006	86,2	0,2958	0,0010005	86,3	0,2958	0,0010005	86,4	0,2957	0,0010005	86,5	0,2957
30	0,0010032	127,9	0,4357	0,0010032	128,0	0,4357	0,0010031	128,1	0,4357	0,0010031	128,2	0,4356
40	0,0010067	169,7	0,5711	0,0010067	169,7	0,5711	0,0010066	169,8	0,5711	0,0010066	169,9	0,5710
50	0,0010110	211,4	0,7023	0,0010110	211,5	0,7023	0,0010109	211,6	0,7023	0,0010109	211,7	0,7022
60	0,0010160	253,2	0,8297	0,0010160	253,3	0,8296	0,0010159	253,3	0,8295	0,0010159	253,4	0,8295
70	0,0010217	295,0	0,9533	0,0010216	295,1	0,9533	0,0010216	295,1	0,9532	0,0010216	295,2	0,9531
80	0,0010280	336,9	1,0736	0,0010280	336,9	1,0736	0,0010279	337,0	1,0735	0,0010279	337,1	1,0734
90	0,0010349	378,8	1,1908	0,0010349	378,9	1,1907	0,0010348	379,0	1,1906	0,0010348	379,0	1,1906
100	0,0010425	420,9	1,3050	0,0010424	420,9	1,3049	0,0010423	421,0	1,3048	0,0010423	421,1	1,3048
110	0,0010506	463,0	1,4165	0,0010505	463,1	1,4164	0,0010505	463,2	1,4163	0,0010504	463,2	1,4162
120	0,0010593	505,3	1,5255	0,0010592	505,4	1,5254	0,0010592	505,5	1,5253	0,0010591	505,5	1,5252
130	0,0010687	547,8	1,6322	0,0010686	547,9	1,6321	0,0010685	548,0	1,6320	0,0010685	548,0	1,6319
140	0,0010787	590,5	1,7368	0,0010786	590,6	1,7367	0,0010785	590,6	1,7366	0,0010785	590,7	1,7365
150	0,0010894	633,4	1,8394	0,0010893	633,5	1,8393	0,0010892	633,5	1,8392	0,0010892	633,6	1,8390
160	0,0011008	676,6	1,9402	0,0011008	676,6	1,9401	0,0011007	676,7	1,9400	0,0011006	676,8	1,9399
170	0,0011131	720,1	2,0395	0,0011130	720,1	2,0393	0,0011129	720,2	2,0392	0,0011128	720,2	2,0391
180	0,0011262	763,9	2,1372	0,0011261	763,9	2,1371	0,0011260	764,0	2,1370	0,0011260	764,0	2,1368
190	0,0011403	808,1	2,2338	0,0011402	808,1	2,2336	0,0011401	808,2	2,2335	0,0011400	808,2	2,2333
200	0,0011555	852,8	2,3292	0,0011554	852,8	2,3290	0,0011553	852,8	2,3289	0,0011552	852,9	2,3287
210	0,0011719	897,9	2,4237	0,0011718	898,0	2,4235	0,0011717	898,0	2,4233	0,0011716	898,0	2,4232
220	0,0011897	943,7	2,5175	0,0011896	943,7	2,5173	0,0011895	943,8	2,5171	0,0011893	943,8	2,5169
230	0,08163	2820,1	6,2920	0,07796	2814,1	6,2642	0,07455	2808,0	6,2369	0,0012087	990,3	2,6102
240	0,08436	2850,5	6,3517	0,08064	2845,2	6,3253	0,07718	2839,7	6,2993	0,07397	2834,2	6,2738
250	0,08699	2879,5	6,4077	0,08320	2874,7	6,3823	0,07970	2869,9	6,3575	0,07644	2864,9	6,3331
260	0,08951	2907,4	6,4605	0,08567	2903,0	6,4360	0,08211	2898,7	6,4120	0,07880	2894,2	6,3886
270	0,09196	2934,2	6,5104	0,08806	2930,3	6,4867	0,08444	2926,4	6,4635	0,08108	2922,3	6,4408
280	0,09433	2960,3	6,5580	0,09037	2956,7	6,5348	0,08670	2953,1	6,5123	0,08328	2949,5	6,4903
290	0,09665	2985,7	6,6034	0,09262	2982,4	6,5808	0,08889	2979,1	6,5588	0,08542	2975,7	6,5374
300	0,09893	3010,4	6,6470	0,09483	3007,4	6,6249	0,09104	3004,4	6,6034	0,08751	3001,3	6,5824
310	0,10115	3034,7	6,6890	0,09699	3031,9	6,6673	0,09314	3029,1	6,6462	0,08955	3026,3	6,6256
320	0,10335	3058,6	6,7296	0,09912	3056,0	6,7082	0,09520	3053,4	6,6874	0,09156	3050,8	6,6672
330	0,10551	3082,1	6,7689	0,10121	3079,7	6,7478	0,09723	3077,2	6,7273	0,09353	3074,8	6,7074
340	0,10764	3105,3	6,8071	0,10328	3103,0	6,7862	0,09923	3100,8	6,7660	0,09548	3098,5	6,7464
350	0,10975	3128,2	6,8442	0,10532	3126,1	6,8236	0,10121	3124,0	6,8036	0,09740	3121,9	6,7842
360	0,11184	3151,0	6,8804	0,10734	3149,0	6,8600	0,10317	3147,0	6,8402	0,09929	3145,0	6,8210
370	0,11391	3173,6	6,9158	0,10934	3171,7	6,8956	0,10510	3169,8	6,8759	0,10117	3167,9	6,8570
380	0,11597	3196,1	6,9505	0,11133	3194,3	6,9304	0,10703	3192,5	6,9109	0,10303	3190,7	6,8921
390	0,11801	3218,4	6,9845	0,11330	3216,7	6,9644	0,10893	3215,0	6,9451	0,10488	3213,3	6,9264
400	0,12004	3240,7	7,0178	0,11526	3239,0	6,9979	0,11083	3237,4	6,9787	0,10671	3235,8	6,9601
410	0,12206	3262,9	7,0505	0,11721	3261,3	7,0307	0,11271	3259,8	7,0116	0,10854	3258,2	6,9931
420	0,12407	3285,0	7,0827	0,11914	3283,5	7,0630	0,11458	3282,0	7,0440	0,11035	3280,5	7,0256
430	0,12607	3307,1	7,1143	0,12107	3305,7	7,0947	0,11645	3304,3	7,0758	0,11215	3302,8	7,0575
440	0,12806	3329,2	7,1455	0,12299	3327,8	7,1260	0,11830	3326,5	7,1072	0,11395	3325,1	7,0890
450	0,13004	3351,3	7,1763	0,12491	3349,9	7,1568	0,12015	3348,6	7,1381	0,11574	3347,3	7,1199
460	0,13202	3373,3	7,2066	0,12681	3372,1	7,1872	0,12199	3370,8	7,1685	0,11752	3369,5	7,1504
470	0,13399	3395,4	7,2365	0,12871	3394,2	7,2172	0,12383	3393,0	7,1985	0,11929	3391,7	7,1805
480	0,13596	3417,5	7,2660	0,13061	3416,3	7,2467	0,12566	3415,1	7,2282	0,12106	3413,9	7,2102
490	0,13792	3439,6	7,2952	0,13250	3438,5	7,2760	0,12748	3437,3	7,2574	0,12282	3436,2	7,2395
500	0,13987	3461,7	7,3240	0,13438	3460,6	7,3048	0,12930	3459,5	7,2863	0,12458	3458,4	7,2685
510	0,14182	3483,9	7,3525	0,13626	3482,8	7,3333	0,13111	3481,8	7,3149	0,12633	3480,7	7,2971
520	0,14377	3506,1	7,3806	0,13814	3505,1	7,3615	0,13292	3504,0	7,3431	0,12808	3503,0	7,3254
530	0,14571	3528,3	7,4085	0,14001	3527,3	7,3894	0,13473	3526,3	7,3711	0,12982	3525,3	7,3533
540	0,14765	3550,6	7,4360	0,14187	3549,6	7,4170	0,13653	3548,6	7,3987	0,13156	3547,6	7,3810
550	0,14958	3572,9	7,4633	0,14374	3571,9	7,4443	0,13833	3571,0	7,4260	0,13330	3570,0	7,4084

Table 3. Water and Superheated Steam (Continuation) Wasser und überhitzter Dampf (Fortsetzung)

t	25 bar $t_s = 223{,}94$ °C			26 bar $t_s = 226{,}04$ °C			27 bar $t_s = 228{,}07$ °C			28 bar $t_s = 230{,}05$ °C		
	v'' 0,07991	h'' 2800,9	s'' 6,2536	v'' 0,7686	h'' 2801,4	s'' 6,2387	v'' 0,07402	h'' 2801,7	s'' 6,2244	v'' 0,07139	h'' 2802,0	s'' 6,2104
°C	v	h	s	v	h	s	v	h	s	v	h	s
550	0,14958	3572,9	7,4633	0,14374	3571,9	7,4443	0,13833	3571,0	7,4260	0,13330	3570,0	7,4084
560	0,15151	3595,2	7,4903	0,14560	3594,3	7,4713	0,14012	3593,4	7,4531	0,13503	3592,5	7,4355
570	0,15344	3617,6	7,5170	0,14745	3616,7	7,4981	0,14191	3615,8	7,4799	0,13676	3614,9	7,4623
580	0,15537	3640,1	7,5434	0,14931	3639,2	7,5246	0,14370	3638,3	7,5064	0,13849	3637,5	7,4888
590	0,15729	3662,6	7,5697	0,15116	3661,7	7,5508	0,14548	3660,9	7,5326	0,14022	3660,0	7,5151
600	0,15921	3685,1	7,5956	0,15301	3684,3	7,5768	0,14727	3683,5	7,5587	0,14194	3682,6	7,5412
610	0,16112	3707,7	7,6213	0,15485	3706,9	7,6025	0,14905	3706,1	7,5844	0,14365	3705,3	7,5670
620	0,16304	3730,3	7,6468	0,15669	3729,6	7,6281	0,15082	3728,8	7,6100	0,14537	3728,0	7,5925
630	0,16495	3753,0	7,6721	0,15853	3752,3	7,6534	0,15260	3751,5	7,6353	0,14708	3750,8	7,6179
640	0,16686	3775,8	7,6971	0,16037	3775,0	7,6784	0,15437	3774,3	7,6604	0,14879	3773,6	7,6430
650	0,16876	3798,6	7,7220	0,16221	3797,9	7,7033	0,15614	3797,1	7,6853	0,15050	3796,4	7,6679
660	0,17067	3821,4	7,7466	0,16404	3820,7	7,7279	0,15791	3820,0	7,7099	0,15221	3819,3	7,6926
670	0,17257	3844,3	7,7710	0,16587	3843,6	7,7524	0,15967	3843,0	7,7344	0,15539	3842,3	7,7170
680	0,17447	3867,3	7,7952	0,16770	3866,6	7,7766	0,16144	3866,0	7,7586	0,15562	3865,3	7,7413
690	0,17637	3890,3	7,8192	0,16953	3889,7	7,8006	0,16320	3889,0	7,7827	0,15732	3888,4	7,7654
700	0,17826	3913,4	7,8431	0,17135	3912,7	7,8245	0,16496	3912,1	7,8066	0,15902	3911,5	7,7893
710	0,18016	3936,5	7,8667	0,17318	3935,9	7,8481	0,16671	3935,3	7,8302	0,16071	3934,7	7,8130
720	0,18205	3959,7	7,8902	0,17500	3959,1	7,8716	0,16847	3958,5	7,8537	0,16241	3957,9	7,8365
730	0,18394	3982,9	7,9134	0,17682	3982,3	7,8949	0,17022	3981,8	7,8770	0,16410	3981,2	7,8598
740	0,18583	4006,2	7,9365	0,17864	4005,6	7,9180	0,17198	4005,1	7,9002	0,16579	4004,5	7,8829
750	0,18772	4029,5	7,9595	0,18046	4029,0	7,9409	0,17373	4028,4	7,9231	0,16748	4027,9	7,9059
760	0,18961	4052,9	7,9822	0,18227	4052,4	7,9637	0,17548	4051,9	7,9459	0,16917	4051,4	7,9287
770	0,19149	4076,4	8,0048	0,18409	4075,9	7,9863	0,17723	4075,4	7,9685	0,17086	4074,8	7,9513
780	0,19338	4099,9	8,0272	0,18590	4099,4	8,0088	0,17897	4098,9	7,9910	0,17255	4098,4	7,9738
790	0,19526	4123,4	8,0495	0,18771	4122,9	8,0310	0,18072	4122,5	8,0133	0,17423	4122,0	7,9961
800	0,19714	4147,0	8,0716	0,18952	4146,6	8,0531	0,18247	4146,1	8,0354	0,17591	4145,6	8,0183

Table 3. Water and Superheated Steam (Continuation) Wasser und überhitzter Dampf (Fortsetzung)

t	29 bar $t_s = 231,97$ °C			30 bar $t_s = 233,84$ °C			31 bar $t_s = 235,67$ °C			32 bar $t_s = 237,45$ °C		
	v'' 0,06893	h'' 2802,2	s'' 6,1969	v'' 0,06663	h'' 2802,3	s'' 6,1837	v'' 0,06447	h'' 2802,3	s'' 6,1709	v'' 0,06244	h'' 2802,3	s'' 6,1585
°C	v	h	s	v	h	s	v	h	s	v	h	s
0	0,0009988	2,9	0,0001	0,0009987	3,0	0,0001	0,0009987	3,1	0,0001	0,0009986	3,2	0,0001
10	0,0009989	44,8	0,1507	0,0009988	44,9	0,1507	0,0009988	45,0	0,1507	0,0009987	45,1	0,1507
20	0,0010004	86,6	0,2957	0,0010004	86,7	0,2957	0,0010003	86,8	0,2957	0,0010003	86,9	0,2956
30	0,0010030	128,3	0,4356	0,0010030	128,4	0,4356	0,0010029	128,5	0,4356	0,0010029	128,6	0,4355
40	0,0010065	170,0	0,5710	0,0010065	170,1	0,5710	0,0010064	170,2	0,5709	0,0010064	170,3	0,5709
50	0,0010108	211,7	0,7022	0,0010108	211,8	0,7021	0,0010107	211,9	0,7021	0,0010107	212,0	0,7020
60	0,0010158	253,5	0,8294	0,0010158	253,6	0,8294	0,0010157	253,7	0,8293	0,0010157	253,8	0,8293
70	0,0010215	295,3	0,9531	0,0010215	295,4	0,9530	0,0010214	295,5	0,9530	0,0010214	295,6	0,9529
80	0,0010278	337,2	1,0734	0,0010278	337,3	1,0733	0,0010277	337,3	1,0732	0,0010277	337,4	1,0732
90	0,0010347	379,1	1,1905	0,0010347	379,2	1,1904	0,0010346	379,3	1,1903	0,0010346	379,4	1,1903
100	0,0010422	421,2	1,3047	0,0010422	421,2	1,3046	0,0010421	421,3	1,3045	0,0010421	421;4	1,3044
110	0,0010504	463,3	1,4162	0,0010503	463,4	1,4161	0,0010502	463,5	1,4160	0,0010502	463,5	1,4159
120	0,0010591	505,6	1,5251	0,0010590	505,7	1,5251	0,0010590	505,8	1,5250	0,0010589	505,8	1,5249
130	0,0010684	548,1	1,6318	0,0010684	548,2	1,6317	0,0010683	548,2	1,6316	0,0010682	548,3	1,6315
140	0,0010784	590,8	1,7364	0,0010783	590,8	1,7363	0,0010783	590,9	1,7361	0,0010782	591,0	1,7360
150	0,0010891	633,7	1,8389	0,0010890	633,7	1,8388	0,0010890	633,8	1,8387	0,0010889	633,8	1,8386
160	0,0011005	676,8	1,9398	0,0011005	676,9	1,9396	0,0011004	676,9	1,9395	0,0011003	677,0	1,9394
170	0,0011128	720,3	2,0390	0,0011127	720,3	2,0388	0,0011126	720,4	2,0387	0,0011125	720,4	2,0386
180	0,0011259	764,1	2,1367	0,0011258	764,1	2,1366	0,0011257	764,2	2,1364	0,0011256	764,2	2,1363
190	0,0011400	808,3	2,2332	0,0011399	808,3	2,2330	0,0011398	808,4	2,2329	0,0011397	808,4	2,2327
200	0,0011551	852,9	2,3286	0,0011550	853,0	2,3284	0,0011549	853,0	2,3282	0,0011548	853,0	2,3281
210	0,0011715	898,1	2,4230	0,0011714	898,1	2,4228	0,0011712	898,1	2,4226	0,0011711	898,2	2,4225
220	0,0011892	943,8	2,5167	0,0011891	943,9	2,5165	0,0011890	943,9	2,5163	0,0011888	943,9	2,5162
230	0,0012086	990,3	2,6100	0,0012084	990,3	2,6098	0,0012083	990,3	2,6096	0,0012081	990,3	2,6094
240	0,07097	2828,6	6,2488	0,06816	2822,9	6,2241	0,06553	2817,1	6,1997	0,06305	2811,2	6,1757
250	0,07340	2859,9	6,3092	0,07055	2854,8	6,2857	0,06788	2849,6	6,2626	0,06538	2844,4	6,2398
260	0,07571	2889,7	6,3657	0,07283	2885,1	6,3432	0,07013	2880,5	6,3211	0,06759	2875,8	6,2993
270	0,07794	2918,3	6,4187	0,07501	2914,1	6,3970	0,07227	2910,0	6,3758	0,06970	2905,7	6,3549
280	0,08010	2945,8	6,4689	0,07712	2942,0	6,4479	0,07434	2938,2	6,4274	0,07173	2934,4	6,4072
290	0,08219	2972,4	6,5166	0,07917	2968,9	6,4962	0,07634	2965,5	6,4762	0,07369	2962,0	6,4567
300	0,08423	2998,2	6,5621	0,08116	2995,1	6,5422	0,07829	2991,9	6,5227	0,07559	2988,7	6,5037
310	0,08622	3023,4	6,6057	0,08310	3020,5	6,5862	0,08019	3017,6	6,5672	0,07745	3014,7	6,5487
320	0,08817	3048,1	6,6476	0,08500	3045,4	6,6285	0,08204	3042,7	6,6099	0,07926	3040,0	6,5917
330	0,09009	3072,3	6,6881	0,08687	3069,9	6,6694	0,08386	3067,4	6,6511	0,08104	3064,8	6,6332
340	0,09198	3096,2	6,7273	0,08871	3093,9	6,7088	0,08566	3091,5	6,6908	0,08279	3089,2	6,6733
350	0,09384	3119,7	6,7654	0,09053	3117,5	6,7471	0,08742	3115,4	6,7294	0,08451	3113,2	6,7120
360	0,09569	3143,0	6,8024	0,09232	3140,9	6,7844	0,08917	3138,9	6,7668	0,08621	3136,8	6,7497
370	0,09751	3166,0	6,8385	0,09409	3164,1	6,8207	0,09089	3162,1	6,8033	0,08789	3160,2	6,7863
380	0,09931	3188,9	6,8738	0,09584	3187,0	6,8561	0,09260	3185,2	6,8388	0,08955	3183,4	6,8221
390	0,10111	3211,6	6,9083	0,09758	3209,8	6,8907	0,09429	3208,1	6,8736	0,09120	3206,4	6,8570
400	0,10288	3234,1	6,9421	0,09931	3232,5	6,9246	0,09597	3230,8	6,9077	0,09283	3229,2	6,8912
410	0,10465	3256,6	6,9753	0,10102	3255,1	6,9579	0,09763	3253,5	6,9411	0,09445	3251,9	6,9247
420	0,10641	3279,0	7,0078	0,10273	3277,5	6,9906	0,09929	3276,0	6,9738	0,09606	3274,5	6,9576
430	0,10815	3301,4	7,0398	0,10442	3299,9	7,0227	0,10093	3298,5	7,0060	0,09766	3297,1	6,9898
440	0,10989	3323,7	7,0713	0,10611	3322,3	7,0543	0,10257	3320,9	7,0377	0,09925	3319,5	7,0216
450	0,11162	3346,0	7,1024	0,10779	3344,6	7,0854	0,10420	3343,3	7,0689	0,10083	3342,0	7,0528
460	0,11335	3368,2	7,1329	0,10946	3367,0	7,1160	0,10582	3365,7	7,0996	0,10241	3364,4	7,0836
470	0,11507	3390,5	7,1631	0,11112	3389,3	7,1462	0,10743	3388,0	7,1298	0,10397	3386,8	7,1140
480	0,11678	3412,8	7,1928	0,11278	3411,6	7,1760	0,10904	3410,4	7,1597	0,10554	3409,2	7,1439
490	0,11848	3435,0	7,2222	0,11443	3433,9	7,2054	0,11065	3432,7	7,1892	0,10709	3431,6	7,1734
500	0,12018	3457,3	7,2512	0,11608	3456,2	7,2345	0,11224	3455,1	7,2183	0,10865	3454,0	7,2026
510	0,12188	3479,6	7,2799	0,11772	3478,5	7,2632	0,11384	3477,4	7,2470	0,11019	3476,4	7,2314
520	0,12357	3501,9	7,3082	0,11936	3500,9	7,2916	0,11543	3499,8	7,2755	0,11174	3498,8	7,2598
530	0,12526	3524,3	7,3362	0,12100	3523,3	7,3196	0,11701	3522,3	7,3035	0,11328	3521,2	7,2879
540	0,12694	3546,7	7,3639	0,12263	3545,7	7,3474	0,11859	3544,7	7,3313	0,11481	3543,7	7,3158
550	0,12862	3569,1	7,3913	0,12426	3568,1	7,3748	0,12017	3567,2	7,3588	0,11634	3566,2	7,3433

Table 3. Water and Superheated Steam (Continuation) Wasser und überhitzter Dampf (Fortsetzung)

t	29 bar t_s = 231,97 °C			30 bar t_s = 233,84 °C			31 bar t_s = 235,67 °C			32 bar t_s = 237,45 °C		
	v'' 0,06893	h'' 2802,2	s'' 6,1969	v'' 0,06663	h'' 2802,3	s'' 6,1837	v'' 0,06447	h'' 2802,3	s'' 6,1709	v'' 0,06244	h'' 2802,3	s'' 6,1585
°C	v	h	s	v	h	s	v	h	s	v	h	s
550	0,12862	3569,1	7,3913	0,12426	3568,1	7,3748	0,12017	3567,2	7,3588	0,11634	3566,2	7,3433
560	0,13030	3591,5	7,4184	0,12588	3590,6	7,4020	0,12174	3589,7	7,3860	0,11787	3588,8	7,3705
570	0,13197	3614,0	7,4453	0,12750	3613,2	7,4288	0,12332	3612,3	7,4129	0,11939	3611,4	7,3974
580	0,13364	3636,6	7,4719	0,12912	3635,7	7,4554	0,12488	3634,8	7,4395	0,12091	3634,0	7,4241
590	0,13531	3659,2	7,4982	0,13073	3658,3	7,4818	0,12645	3657,5	7,4659	0,12243	3656,6	7,4505
600	0,13697	3681,8	7,5243	0,13234	3681,0	7,5079	0,12801	3680,2	7,4920	0,12395	3679,3	7,4767
610	0,13864	3704,5	7,5501	0,13395	3703,7	7,5338	0,12957	3702,9	7,5179	0,12546	3702,1	7,5026
620	0,14029	3727,2	7,5757	0,13556	3726,5	7,5594	0,13112	3725,7	7,5436	0,12697	3724,9	7,5283
630	0,14195	3750,0	7,6010	0,13716	3749,3	7,5848	0,13268	3748,5	7,5690	0,12848	3747,7	7,5537
640	0,14360	3772,8	7,6262	0,13876	3772,1	7,6099	0,13423	3771,4	7,5942	0,12998	3770,6	7,5789
650	0,14526	3795,7	7,6511	0,14036	3795,0	7,6349	0,13578	3794,3	7,6191	0,13148	3793,6	7,6039
660	0,14691	3818,6	7,6758	0,14196	3818,0	7,6596	0,13733	3817,3	7,6439	0,13299	3816,6	7,6287
670	0,14855	3841,6	7,7003	0,14355	3841,0	7,6841	0,13887	3840,3	7,6684	0,13448	3839,6	7,6532
680	0,15020	3864,7	7,7246	0,14514	3864,0	7,7084	0,14041	3863,3	7,6928	0,13598	3862,7	7,6776
690	0,15184	3887,7	7,7487	0,14673	3887,1	7,7325	0,14196	3886,5	7,7169	0,13747	3885,8	7,7017
700	0,15349	3910,9	7,7726	0,14832	3910,3	7,7564	0,14349	3909,6	7,7408	0,13897	3909,0	7,7257
710	0,15513	3934,1	7,7963	0,14991	3933,5	7,7802	0,14503	3932,9	7,7646	0,14046	3932,3	7,7494
720	0,15676	3957,3	7,8198	0,15150	3956,7	7,8037	0,14657	3956,1	7,7881	0,14195	3955,6	7,7730
730	0,15840	3980,6	7,8432	0,15308	3980,0	7,8271	0,14810	3979,5	7,8115	0,14343	3978,9	7,7964
740	0,16004	4004,0	7,8663	0,15466	4003,4	7,8502	0,14963	4002,9	7,8347	0,14492	4002,3	7,8196
750	0,16167	4027,4	7,8893	0,15624	4026,8	7,8733	0,15116	4026,3	7,8577	0,14640	4025,8	7,8426
760	0,16330	4050,8	7,9121	0,15782	4050,3	7,8961	0,15269	4049,8	7,8805	0,14789	4049,3	7,8655
770	0,16493	4074,3	7,9348	0,15940	4073,8	7,9187	0,15422	4073,3	7,9032	0,14937	4072,8	7,8882
780	0,16656	4097,9	7,9572	0,16097	4097,4	7,9412	0,15575	4096,9	7,9257	0,15085	4096,4	7,9107
790	0,16819	4121,5	7,9796	0,16255	4121,0	7,9636	0,15727	4120,5	7,9481	0,15233	4120,1	7,9331
800	0,16982	4145,2	8,0017	0,16412	4144,7	7,9857	0,15880	4144,2	7,9702	0,15381	4143,8	7,9552

Table 3. Water and Superheated Steam (Continuation) Wasser und überhitzter Dampf (Fortsetzung)

t	33 bar t_s = 239,18 °C			34 bar t_s = 240,88 °C			35 bar t_s = 242,54 °C			36 bar t_s = 244,16 °C		
	v'' 0,06053	h'' 2802,3	s'' 6,1463	v'' 0,05873	h'' 2802,1	s'' 6,1344	v'' 0,05703	h'' 2802,0	s'' 6,1228	v'' 0,05541	h'' 2801,7	s'' 6,1115
°C	v	h	s	v	h	s	v	h	s	v	h	s
0	0,0009986	3,3	0,0001	0,0009985	3,4	0,0001	0,0009985	3,5	0,0001	0,0009984	3,6	0,0001
10	0,0009987	45,2	0,1507	0,0009987	45,3	0,1507	0,0009986	45,4	0,1507	0,0009986	45,5	0,1507
20	0,0010002	87,0	0,2956	0,0010002	87,1	0,2956	0,0010001	87,1	0,2956	0,0010001	87,2	0,2955
30	0,0010028	128,7	0,4355	0,0010028	128,8	0,4355	0,0010028	128,8	0,4354	0,0010027	128,9	0,4354
40	0,0010063	170,4	0,5708	0,0010063	170,5	0,5708	0,0010063	170,5	0,5708	0,0010062	170,6	0,5707
50	0,0010106	212,1	0,7020	0,0010106	212,2	0,7019	0,0010106	212,3	0,7019	0,0010105	212,3	0,7018
60	0,0010157	253,8	0,8292	0,0010156	253,9	0,8292	0,0010156	254,0	0,8291	0,0010155	254,1	0,8291
70	0,0010213	295,6	0,9529	0,0010213	295,7	0,9528	0,0010212	295,8	0,9527	0,0010212	295,9	0,9527
80	0,0010276	337,5	1,0731	0,0010276	337,6	1,0730	0,0010275	337,7	1,0730	0,0010275	337,7	1,0729
90	0,0010345	379,4	1,1902	0,0010345	379,5	1,1901	0,0010344	379,6	1,1900	0,0010344	379,7	1,1900
100	0,0010420	421,5	1,3044	0,0010420	421,5	1,3043	0,0010419	421,6	1,3042	0,0010419	421,7	1,3041
110	0,0010501	463,6	1,4158	0,0010501	463,7	1,4157	0,0010500	463,8	1,4157	0,0010500	463,8	1,4156
120	0,0010588	505,9	1,5248	0,0010588	506,0	1,5247	0,0010587	506,0	1,5246	0,0010587	506,1	1,5245
130	0,0010682	548,4	1,6314	0,0010681	548,4	1,6313	0,0010681	548,5	1,6312	0,0010680	548,6	1,6311
140	0,0010782	591,0	1,7359	0,0010781	591,1	1,7358	0,0010780	591,1	1,7357	0,0010780	591,2	1,7356
150	0,0010888	633,9	1,8385	0,0010888	634,0	1,8384	0,0010887	634,0	1,8383	0,0010886	634,1	1,8382
160	0,0011002	677,0	1,9393	0,0011002	677,1	1,9392	0,0011001	677,2	1,9391	0,0011000	677,2	1,9389
170	0,0011125	720,5	2,0384	0,0011124	720,6	2,0383	0,0011123	720,6	2,0382	0,0011122	720,7	2,0381
180	0,0011255	764,3	2,1362	0,0011254	764,3	2,1360	0,0011254	764,4	2,1359	0,0011253	764,4	2,1358
190	0,0011396	808,5	2,2326	0,0011395	808,5	2,2324	0,0011394	808,6	2,2323	0,0011393	808,6	2,2322
200	0,0011547	853,1	2,3279	0,0011546	853,1	2,3278	0,0011545	853,2	2,3276	0,0011544	853,2	2,3275
210	0,0011710	898,2	2,4223	0,0011709	898,3	2,4221	0,0011708	898,3	2,4220	0,0011707	898,3	2,4218
220	0,0011887	943,9	2,5160	0,0011886	944,0	2,5158	0,0011885	944,0	2,5156	0,0011883	944,0	2,5154
230	0,0012080	990,4	2,6091	0,0012079	990,4	2,6089	0,0012077	990,4	2,6087	0,0012076	990,4	2,6085
240	0,06072	2805,1	6,1519	0,0012290	1037,6	2,7019	0,0012288	1037,6	2,7017	0,0012287	1037,6	2,7014
250	0,06302	2839,0	6,2173	0,06080	2833,6	6,1951	0,05869	2828,1	6,1732	0,05670	2822,5	6,1514
260	0,06520	2871,0	6,2779	0,06295	2866,2	6,2568	0,06082	2861,3	6,2360	0,05880	2856,3	6,2155
270	0,06728	2901,4	6,3344	0,06499	2897,1	6,3143	0,06284	2892,7	6,2944	0,06080	2888,2	6,2748
280	0,06927	2930,5	6,3875	0,06695	2926,6	6,3681	0,06477	2922,6	6,3490	0,06270	2918,6	6,3302
290	0,07119	2958,5	6,4376	0,06884	2954,9	6,4188	0,06663	2951,3	6,4004	0,06453	2947,7	6,3823
300	0,07306	2985,5	6,4851	0,07068	2982,2	6,4669	0,06842	2979,0	6,4491	0,06630	2975,6	6,4315
310	0,07488	3011,7	6,5305	0,07246	3008,7	6,5128	0,07017	3005,7	6,4954	0,06801	3002,7	6,4783
320	0,07665	3037,3	6,5740	0,07419	3034,5	6,5566	0,07187	3031,8	6,5396	0,06968	3028,9	6,5230
330	0,07839	3062,3	6,6158	0,07589	3059,7	6,5988	0,07354	3057,2	6,5821	0,07131	3054,6	6,5658
340	0,08010	3086,8	6,6561	0,07756	3084,4	6,6394	0,07517	3082,0	6,6230	0,07291	3079,6	6,6070
350	0,08178	3110,9	6,6952	0,07920	3108,7	6,6787	0,07678	3106,5	6,6626	0,07448	3104,2	6,6468
360	0,08344	3134,7	6,7330	0,08082	3132,7	6,7168	0,07836	3130,6	6,7009	0,07603	3128,4	6,6854
370	0,08507	3158,2	6,7699	0,08242	3156,3	6,7538	0,07992	3154,3	6,7381	0,07755	3152,3	6,7228
380	0,08669	3181,5	6,8058	0,08400	3179,7	6,7899	0,08146	3177,8	6,7744	0,07906	3175,9	6,7592
390	0,08829	3204,6	6,8408	0,08556	3202,8	6,8251	0,08298	3201,1	6,8097	0,08055	3199,3	6,7947
400	0,08988	3227,5	6,8752	0,08711	3225,9	6,8595	0,08449	3224,2	6,8443	0,08202	3222,5	6,8294
410	0,09146	3250,3	6,9088	0,08865	3248,7	6,8933	0,08599	3247,1	6,8781	0,08349	3245,5	6,8634
420	0,09303	3273,0	6,9417	0,09017	3271,5	6,9263	0,08748	3270,0	6,9113	0,08494	3268,4	6,8967
430	0,09458	3295,6	6,9741	0,09169	3294,2	6,9588	0,08896	3292,7	6,9439	0,08638	3291,2	6,9293
440	0,09613	3318,2	7,0060	0,09319	3316,8	6,9907	0,09043	3315,4	6,9759	0,08781	3314,0	6,9614
450	0,09767	3340,6	7,0373	0,09469	3339,3	7,0221	0,09189	3338,0	7,0074	0,08924	3336,6	6,9930
460	0,09920	3363,1	7,0681	0,09618	3361,8	7,0530	0,09334	3360,5	7,0384	0,09065	3359,2	7,0241
470	0,10073	3385,6	7,0985	0,09767	3384,3	7,0835	0,09478	3383,1	7,0689	0,09206	3381,8	7,0546
480	0,10225	3408,0	7,1285	0,09915	3406,8	7,1135	0,09622	3405,6	7,0990	0,09347	3404,4	7,0848
490	0,10376	3430,4	7,1581	0,10062	3429,2	7,1432	0,09766	3428,1	7,1287	0,09486	3426,9	7,1145
500	0,10527	3452,8	7,1873	0,10209	3451,7	7,1724	0,09909	3450,6	7,1580	0,09626	3449,5	7,1439
510	0,10677	3475,3	7,2161	0,10355	3474,2	7,2013	0,10051	3473,1	7,1869	0,09764	3472,0	7,1729
520	0,10827	3497,7	7,2446	0,10501	3496,7	7,2299	0,10193	3495,6	7,2155	0,09903	3494,6	7,2015
530	0,10977	3520,2	7,2728	0,10646	3519,2	7,2581	0,10335	3518,2	7,2438	0,10040	3517,2	7,2298
540	0,11126	3542,7	7,3007	0,10791	3541,8	7,2860	0,10476	3540,8	7,2717	0,10178	3539,8	7,2578
550	0,11274	3565,3	7,3282	0,10936	3564,3	7,3136	0,10617	3563,4	7,2993	0,10315	3562,4	7,2854

Table 3. Water and Superheated Steam (Continuation) Wasser und überhitzter Dampf (Fortsetzung)

	33 bar $t_s = 239{,}18\ °C$			34 bar $t_s = 240{,}88\ °C$			35 bar $t_s = 242{,}54\ °C$			36 bar $t_s = 244{,}16\ °C$		
t	v'' 0,06053	h'' 2802,3	s'' 6,1463	v'' 0,05873	h'' 2802,1	s'' 6,1344	v'' 0,05703	h'' 2802,0	s'' 6,1228	v'' 0,05541	h'' 2801,7	s'' 6,1115
°C	v	h	s	v	h	s	v	h	s	v	h	s
550	0,11274	3565,3	7,3282	0,10936	3564,3	7,3136	0,10617	3563,4	7,2993	0,10315	3562,4	7,2854
560	0,11423	3587,8	7,3555	0,11080	3586,9	7,3408	0,10757	3586,0	7,3266	0,10452	3585,1	7,3128
570	0,11571	3610,5	7,3824	0,11224	3609,6	7,3679	0,10897	3608,7	7,3537	0,10588	3607,8	7,3399
580	0,11719	3633,1	7,4091	0,11368	3632,2	7,3946	0,11037	3631,4	7,3804	0,10724	3630,5	7,3667
590	0,11866	3655,8	7,4356	0,11511	3654,9	7,4211	0,11176	3654,1	7,4069	0,10860	3653,3	7,3932
600	0,12013	3678,5	7,4618	0,11654	3677,7	7,4473	0,11315	3676,9	7,4332	0,10996	3676,1	7,4195
610	0,12160	3701,3	7,4877	0,11797	3700,5	7,4732	0,11454	3699,7	7,4592	0,11131	3698,9	7,4455
620	0,12307	3724,1	7,5134	0,11939	3723,3	7,4989	0,11593	3722,6	7,4849	0,11266	3721,8	7,4712
630	0,12453	3747,0	7,5388	0,12082	3746,2	7,5244	0,11731	3745,5	7,5104	0,11401	3744,7	7,4968
640	0,12599	3769,9	7,5641	0,12224	3769,2	7,5497	0,11869	3768,4	7,5357	0,11535	3767,7	7,5221
650	0,12745	3792,9	7,5891	0,12365	3792,1	7,5747	0,12007	3791,4	7,5607	0,11669	3790,7	7,5471
660	0,12891	3815,9	7,6139	0,12507	3815,2	7,5995	0,12145	3814,5	7,5856	0,11803	3813,8	7,5720
670	0,13036	3838,9	7,6385	0,12648	3838,3	7,6241	0,12283	3837,6	7,6102	0,11937	3836,9	7,5966
680	0,13181	3862,0	7,6628	0,12789	3861,4	7,6485	0,12420	3860,7	7,6346	0,12071	3860,1	7,6211
690	0,13327	3885,2	7,6870	0,12930	3884,6	7,6727	0,12557	3883,9	7,6588	0,12204	3883,3	7,6453
700	0,13471	3908,4	7,7110	0,13071	3907,8	7,6967	0,12694	3907,2	7,6828	0,12337	3906,5	7,6693
710	0,13616	3931,7	7,7348	0,13212	3931,1	7,7205	0,12831	3930,5	7,7066	0,12470	3929,9	7,6932
720	0,13761	3955,0	7,7583	0,13352	3954,4	7,7441	0,12967	3953,8	7,7303	0,12603	3953,2	7,7168
730	0,13905	3978,3	7,7818	0,13492	3977,8	7,7675	0,13103	3977,2	7,7537	0,12736	3976,6	7,7403
740	0,14049	4001,8	7,8050	0,13633	4001,2	7,7908	0,13240	4000,6	7,7770	0,12869	4000,1	7,7635
750	0,14193	4025,2	7,8280	0,13773	4024,7	7,8138	0,13376	4024,1	7,8000	0,13001	4023,6	7,7866
760	0,14337	4048,7	7,8509	0,13912	4048,2	7,8367	0,13512	4047,7	7,8229	0,13133	4047,2	7,8095
770	0,14481	4072,3	7,8736	0,14052	4071,8	7,8594	0,13647	4071,3	7,8457	0,13265	4070,8	7,8323
780	0,14625	4095,9	7,8961	0,14192	4095,4	7,8820	0,13783	4094,9	7,8682	0,13397	4094,4	7,8549
790	0,14768	4119,6	7,9185	0,14331	4119,1	7,9044	0,13919	4118,6	7,8906	0,13529	4118,1	7,8773
800	0,14912	4143,3	7,9407	0,14470	4142,8	7,9266	0,14054	4142,4	7,9128	0,13661	4141,9	7,8995

Table 3. Water and Superheated Steam (Continuation) Wasser und überhitzter Dampf (Fortsetzung)

t	37 bar $t_s = 245{,}75\,°C$			38 bar $t_s = 247{,}31\,°C$			39 bar $t_s = 248{,}84\,°C$			40 bar $t_s = 250{,}33\,°C$		
	v'' 0,05389	h'' 2801,4	s'' 6,1004	v'' 0,05244	h'' 2801,1	s'' 6,0896	v'' 0,05106	h'' 2800,8	s'' 6,0789	v'' 0,04975	h'' 2800,3	s'' 6,0685
°C	v	h	s	v	h	s	v	h	s	v	h	s
0	0,0009984	3,7	0,0001	0,0009983	3,8	0,0001	0,0009983	3,9	0,0002	0,0009982	4,0	0,0002
10	0,0009985	45,6	0,1507	0,0009985	45,7	0,1507	0,0009984	45,8	0,1507	0,0009984	45,9	0,1506
20	0,0010000	87,3	0,2955	0,0010000	87,4	0,2955	0,0010000	87,5	0,2955	0,0009999	87,6	0,2955
30	0,0010027	129,0	0,4354	0,0010026	129,1	0,4353	0,0010026	129,2	0,4353	0,0010025	129,3	0,4353
40	0,0010062	170,7	0,5707	0,0010061	170,8	0,5706	0,0010061	170,9	0,5706	0,0010060	171,0	0,5706
50	0,0010105	212,4	0,7018	0,0010104	212,5	0,7017	0,0010104	212,6	0,7017	0,0010103	212,7	0,7016
60	0,0010155	254,2	0,8290	0,0010154	254,3	0,8290	0,0010154	254,3	0,8289	0,0010153	254,4	0,8289
70	0,0010211	296,0	0,9526	0,0010211	296,0	0,9526	0,0010210	296,1	0,9525	0,0010210	296,2	0,9524
80	0,0010274	337,8	1,0728	0,0010274	337,9	1,0728	0,0010273	338,0	1,0727	0,0010273	338,1	1,0726
90	0,0010343	379,7	1,1899	0,0010343	379,8	1,1898	0,0010342	379,9	1,1898	0,0010342	380,0	1,1897
100	0,0010418	421,8	1,3041	0,0010418	421,8	1,3040	0,0010417	421,9	1,3039	0,0010417	422,0	1,3038
110	0,0010499	463,9	1,4155	0,0010499	464,0	1,4154	0,0010498	464,0	1,4153	0,0010498	464,1	1,4152
120	0,0010586	506,2	1,5244	0,0010586	506,3	1,5243	0,0010585	506,3	1,5242	0,0010584	506,4	1,5242
130	0,0010679	548,6	1,6310	0,0010679	548,7	1,6309	0,0010678	548,8	1,6308	0,0010677	548,8	1,6308
140	0,0010779	591,3	1,7355	0,0010778	591,3	1,7354	0,0010778	591,4	1,7353	0,0010777	591,5	1,7352
150	0,0010886	634,2	1,8380	0,0010885	634,2	1,8379	0,0010884	634,3	1,8378	0,0010883	634,3	1,8377
160	0,0010999	677,3	1,9388	0,0010999	677,3	1,9387	0,0010998	677,4	1,9386	0,0010997	677,5	1,9385
170	0,0011121	720,7	2,0379	0,0011121	720,8	2,0378	0,0011120	720,8	2,0377	0,0011119	720,9	2,0376
180	0,0011252	764,5	2,1356	0,0011251	764,5	2,1355	0,0011250	764,6	2,1354	0,0011249	764,6	2,1352
190	0,0011392	808,6	2,2320	0,0011391	808,7	2,2319	0,0011390	808,7	2,2317	0,0011389	808,8	2,2316
200	0,0011543	853,2	2,3273	0,0011542	853,3	2,3271	0,0011541	853,3	2,3270	0,0011540	853,4	2,3268
210	0,0011706	898,4	2,4216	0,0011704	898,4	2,4215	0,0011703	898,4	2,4213	0,0011702	898,5	2,4211
220	0,0011882	944,1	2,5152	0,0011881	944,1	2,5151	0,0011880	944,1	2,5149	0,0011878	944,1	2,5147
230	0,0012074	990,4	2,6083	0,0012073	990,5	2,6081	0,0012072	990,5	2,6079	0,0012070	990,5	2,6077
240	0,0012285	1037,6	2,7012	0,0012284	1037,6	2,7010	0,0012282	1037,7	2,7008	0,0012280	1037,7	2,7006
250	0,05481	2816,8	6,1299	0,05302	2811,0	6,1085	0,05131	2805,1	6,0872	0,0012512	1085,8	2,7934
260	0,05689	2851,3	6,1951	0,05508	2846,1	6,1750	0,05336	2840,9	6,1551	0,05172	2835,6	6,1353
270	0,05886	2883,7	6,2555	0,05703	2879,1	6,2364	0,05529	2874,5	6,2175	0,05363	2869,8	6,1988
280	0,06074	2914,5	6,3117	0,05888	2910,4	6,2935	0,05712	2906,3	6,2754	0,05544	2902,0	6,2576
290	0,06254	2944,0	6,3645	0,06066	2940,3	6,3469	0,05887	2936,5	6,3296	0,05717	2932,7	6,3126
300	0,06428	2972,3	6,4143	0,06237	2968,9	6,3973	0,06056	2965,5	6,3806	0,05883	2962,0	6,3642
310	0,06597	2999,6	6,4615	0,06403	2996,5	6,4451	0,06219	2993,4	6,4289	0,06044	2990,2	6,4130
320	0,06761	3026,1	6,5066	0,06564	3023,3	6,4906	0,06377	3020,4	6,4748	0,06200	3017,5	6,4593
330	0,06921	3052,0	6,5498	0,06721	3049,3	6,5341	0,06531	3046,7	6,5187	0,06351	3044,0	6,5036
340	0,07077	3077,2	6,5913	0,06875	3074,8	6,5760	0,06682	3072,3	6,5609	0,06499	3069,8	6,5461
350	0,07231	3102,0	6,6314	0,07025	3099,7	6,6163	0,06830	3097,4	6,6015	0,06645	3095,1	6,5870
360	0,07382	3126,3	6,6702	0,07174	3124,2	6,6553	0,06975	3122,0	6,6408	0,06787	3119,9	6,6265
370	0,07532	3150,3	6,7078	0,07320	3148,3	6,6931	0,07119	3146,3	6,6788	0,06927	3144,3	6,6647
380	0,07679	3174,1	6,7444	0,07464	3172,2	6,7299	0,07260	3170,3	6,7157	0,07066	3168,4	6,7019
390	0,07825	3197,5	6,7801	0,07606	3195,7	6,7657	0,07399	3194,0	6,7517	0,07202	3192,1	6,7380
400	0,07969	3220,8	6,8149	0,07747	3219,1	6,8007	0,07537	3217,4	6,7868	0,07338	3215,7	6,7733
410	0,08112	3243,9	6,8490	0,07887	3242,3	6,8349	0,07674	3240,7	6,8212	0,07471	3239,1	6,8077
420	0,08253	3266,9	6,8824	0,08025	3265,4	6,8684	0,07809	3263,8	6,8548	0,07604	3262,3	6,8414
430	0,08394	3289,8	6,9152	0,08163	3288,3	6,9013	0,07944	3286,8	6,8877	0,07735	3285,4	6,8745
440	0,08534	3312,6	6,9473	0,08300	3311,2	6,9336	0,08077	3309,7	6,9201	0,07866	3308,3	6,9069
450	0,08673	3335,3	6,9790	0,08435	3333,9	6,9653	0,08210	3332,6	6,9519	0,07996	3331,2	6,9388
460	0,08811	3357,9	7,0101	0,08570	3356,6	6,9965	0,08342	3355,3	6,9832	0,08125	3354,0	6,9702
470	0,08949	3380,6	7,0408	0,08705	3379,3	7,0272	0,08473	3378,1	7,0140	0,08253	3376,8	7,0010
480	0,09085	3403,2	7,0710	0,08838	3402,0	7,0575	0,08604	3400,8	7,0443	0,08381	3399,6	7,0314
490	0,09222	3425,8	7,1008	0,08971	3424,6	7,0873	0,08733	3423,4	7,0742	0,08508	3422,3	7,0614
500	0,09358	3448,4	7,1302	0,09104	3447,2	7,1168	0,08863	3446,1	7,1037	0,08634	3445,0	7,0909
510	0,09493	3470,9	7,1592	0,09236	3469,9	7,1459	0,08992	3468,8	7,1328	0,08760	3467,7	7,1201
520	0,09628	3493,5	7,1879	0,09367	3492,5	7,1746	0,09120	3491,4	7,1616	0,08886	3490,4	7,1489
530	0,09762	3516,2	7,2162	0,09499	3515,1	7,2030	0,09248	3514,1	7,1900	0,09011	3513,1	7,1774
540	0,09896	3538,8	7,2442	0,09629	3537,8	7,2310	0,09376	3536,8	7,2181	0,09135	3535,8	7,2055
550	0,10030	3561,5	7,2719	0,09760	3560,5	7,2587	0,09503	3559,5	7,2459	0,09260	3558,6	7,2333

Table 3. **Water and Superheated Steam** (Continuation) **Wasser und überhitzter Dampf** (Fortsetzung)

t	37 bar $t_s = 245{,}75\,°C$			38 bar $t_s = 247{,}31\,°C$			39 bar $t_s = 248{,}84\,°C$			40 bar $t_s = 250{,}33\,°C$		
	v'' 0,05389	h'' 2801,4	s'' 6,1004	v'' 0,05244	h'' 2801,1	s'' 6,0896	v'' 0,05106	h'' 2800,8	s'' 6,0789	v'' 0,04975	h'' 2800,3	s'' 6,0685
°C	v	h	s	v	h	s	v	h	s	v	h	s
550	0,10030	3561,5	7,2719	0,09760	3560,5	7,2587	0,09503	3559,5	7,2459	0,09260	3558,6	7,2333
560	0,10163	3584,1	7,2993	0,09890	3583,2	7,2862	0,09630	3582,3	7,2733	0,09384	3581,4	7,2608
570	0,10296	3606,9	7,3264	0,10019	3606,0	7,3133	0,09757	3605,1	7,3005	0,09507	3604,2	7,2880
580	0,10429	3629,6	7,3532	0,10149	3628,7	7,3402	0,09883	3627,9	7,3274	0,09631	3627,0	7,3149
590	0,10561	3652,4	7,3798	0,10278	3651,6	7,3667	0,10009	3650,7	7,3540	0,09754	3649,9	7,3416
600	0,10693	3675,2	7,4061	0,10406	3674,4	7,3931	0,10135	3673,6	7,3804	0,09876	3672,8	7,3680
610	0,10825	3698,1	7,4321	0,10535	3697,3	7,4191	0,10260	3696,5	7,4065	0,09999	3695,7	7,3941
620	0,10956	3721,0	7,4579	0,10663	3720,2	7,4450	0,10385	3719,5	7,4323	0,10121	3718,7	7,4200
630	0,11088	3744,0	7,4835	0,10791	3743,2	7,4705	0,10510	3742,5	7,4579	0,10243	3741,7	7,4456
640	0,11219	3767,0	7,5088	0,10919	3766,2	7,4959	0,10635	3765,5	7,4833	0,10364	3764,8	7,4710
650	0,11349	3790,0	7,5339	0,11046	3789,3	7,5210	0,10759	3788,6	7,5084	0,10486	3787,9	7,4961
660	0,11480	3813,1	7,5588	0,11174	3812,4	7,5459	0,10883	3811,7	7,5333	0,10607	3811,0	7,5211
670	0,11610	3836,2	7,5834	0,11301	3835,6	7,5706	0,11007	3834,9	7,5580	0,10728	3834,2	7,5458
680	0,11740	3859,4	7,6079	0,11428	3858,8	7,5951	0,11131	3858,1	7,5825	0,10849	3857,4	7,5703
690	0,11870	3882,6	7,6321	0,11554	3882,0	7,6193	0,11254	3881,4	7,6068	0,10970	3880,7	7,5946
700	0,12000	3905,9	7,6562	0,11681	3905,3	7,6434	0,11378	3904,7	7,6309	0,11090	3904,1	7,6187
710	0,12130	3929,3	7,6800	0,11807	3928,7	7,6673	0,11501	3928,0	7,6548	0,11210	3927,4	7,6426
720	0,12259	3952,6	7,7037	0,11933	3952,0	7,6909	0,11624	3951,5	7,6785	0,11330	3950,9	7,6663
730	0,12389	3976,1	7,7272	0,12059	3975,5	7,7144	0,11747	3974,9	7,7020	0,11450	3974,4	7,6899
740	0,12518	3999,5	7,7505	0,12185	3999,0	7,7377	0,11870	3998,4	7,7253	0,11570	3997,9	7,7132
750	0,12647	4023,1	7,7736	0,12311	4022,5	7,7608	0,11992	4022,0	7,7484	0,11689	4021,4	7,7363
760	0,12775	4046,6	7,7965	0,12436	4046,1	7,7838	0,12115	4045,6	7,7714	0,11809	4045,1	7,7593
770	0,12904	4070,3	7,8193	0,12562	4069,8	7,8066	0,12237	4069,2	7,7942	0,11928	4068,7	7,7821
780	0,13033	4093,9	7,8418	0,12687	4093,4	7,8292	0,12359	4092,9	7,8168	0,12047	4092,4	7,8047
790	0,13161	4117,7	7,8643	0,12812	4117,2	7,8516	0,12481	4116,7	7,8392	0,12166	4116,2	7,8272
800	0,13289	4141,4	7,8865	0,12937	4141,0	7,8739	0,12603	4140,5	7,8615	0,12285	4140,0	7,8495

Table 3. Water and Superheated Steam (Continuation) Wasser und überhitzter Dampf (Fortsetzung)

t	41 bar $t_s = 251,80$ °C			42 bar $t_s = 253,24$ °C			43 bar $t_s = 254,66$ °C			44 bar $t_s = 256,05$ °C		
	v'' 0,04850	h'' 2799,9	s'' 6,0583	v'' 0,04731	h'' 2799,4	s'' 6,0482	v'' 0,04617	h'' 2798,9	s'' 6,0383	v'' 0,04508	h'' 2798,3	s'' 6,0286
°C	v	h	s	v	h	s	v	h	s	v	h	s
0	0,0009982	4,1	0,0002	0,0009981	4,2	0,0002	0,0009981	4,3	0,0002	0,0009980	4,4	0,0002
10	0,0009983	46,0	0,1506	0,0009983	46,1	0,1506	0,0009982	46,2	0,1506	0,0009982	46,3	0,1506
20	0,0009999	87,7	0,2954	0,0009998	87,8	0,2954	0,0009998	87,9	0,2954	0,0009997	88,0	0,2954
30	0,0010025	129,4	0,4352	0,0010024	129,5	0,4352	0,0010024	129,6	0,4352	0,0010024	129,7	0,4352
40	0,0010060	171,1	0,5705	0,0010060	171,2	0,5705	0,0010059	171,2	0,5704	0,0010059	171,3	0,5704
50	0,0010103	212,8	0,7016	0,0010102	212,9	0,7016	0,0010102	212,9	0,7015	0,0010102	213,0	0,7015
60	0,0010153	254,5	0,8288	0,0010152	254,6	0,8287	0,0010152	254,7	0,8287	0,0010152	254,8	0,8286
70	0,0010210	296,3	0,9524	0,0010209	296,4	0,9523	0,0010209	296,5	0,9523	0,0010208	296,5	0,9522
80	0,0010272	338,1	1,0726	0,0010272	338,2	1,0725	0 0010271	338,3	1,0724	0,0010271	338,4	1,0724
90	0,0010341	380,1	1,1896	0,0010341	380,1	1,1895	0,0010340	380,2	1,1895	0,0010340	380,3	1,1894
100	0,0010416	422,1	1,3037	0,0010416	422,1	1,3037	0,0010415	422,2	1,3036	0,0010415	422,3	1,3035
110	0,0010497	464,2	1,4152	0,0010496	464,3	1,4151	0,0010496	464,3	1,4150	0,0010495	464,4	1,4149
120	0,0010584	506,5	1,5241	0,0010583	506,5	1,5240	0,0010583	506,6	1,5239	0,0010582	506,7	1,5238
130	0,0010677	548,9	1,6307	0,0010676	549,0	1,6306	0,0010676	549,0	1,6305	0,0010675	549,1	1,6304
140	0,0010776	591,5	1,7351	0,0010776	591,6	1,7350	0,0010775	591,7	1,7349	0,0010775	591,7	1,7348
150	0,0010883	634,4	1,8376	0,0010882	634,5	1,8375	0,0010881	634,5	1,8374	0,0010881	634,6	1,8373
160	0,0010997	677,5	1,9384	0,0010996	677,6	1,9382	0,0010995	677,6	1,9381	0,0010994	677,7	1,9380
170	0,0011118	720,9	2,0374	0,0011117	721,0	2,0373	0,0011117	721,1	2,0372	0,0011116	721,1	2,0371
180	0,0011248	764,7	2,1351	0,0011248	764,7	2,1349	0,0011247	764,8	2,1348	0,0011246	764,9	2,1347
190	0,0011388	808,8	2,2314	0,0011387	808,9	2,2313	0,0011386	808,9	2,2311	0,0011386	809,0	2,2310
200	0,0011539	853,4	2,3267	0,0011538	853,5	2,3265	0,0011537	853,5	2,3264	0,0011536	853,5	2,3262
210	0,0011701	898,5	2,4210	0,0011700	898,5	2,4208	0,0011699	898,6	2,4206	0,0011698	898,6	2,4204
220	0,0011877	944,2	2,5145	0,0011876	944,2	2,5143	0,0011875	944,2	2,5141	0,0011873	944,3	2,5140
230	0,0012069	990,5	2,6075	0,0012067	990,5	2,6073	0,0012066	990,6	2,6071	0,0012064	990,6	2,6069
240	0,0012279	1037,7	2,7003	0,0012277	1037,7	2,7001	0,0012275	1037,7	2,6999	0,0012274	1037,7	2,6997
250	0,0012511	1085,8	2,7932	0,0012509	1085,8	2,7930	0,0012507	1085,8	2,7927	0,0012505	1085,8	2,7925
260	0,05015	2830,3	6,1157	0,04865	2824,8	6,0962	0,04722	2819,2	6,0768	0,04585	2813,6	6,0575
270	0,05205	2865,0	6,1803	0,05054	2860,2	6,1620	0,04909	2855,3	6,1438	0,04771	2850,3	6,1257
280	0,05384	2897,8	6,2401	0,05231	2893,5	6,2227	0,05086	2889,1	6,2054	0,04946	2884,7	6,1884
290	0,05555	2928,8	6,2957	0,05401	2925,0	6,2791	0,05253	2921,0	6,2627	0,05112	2917,1	6,2464
300	0,05719	2958,5	6,3480	0,05562	2955,0	6,3320	0,05413	2951,4	6,3162	0,05270	2947,8	6,3006
310	0,05877	2987,0	6,3973	0,05719	2983,8	6,3818	0,05567	2980,6	6,3666	0,05422	2977,3	6,3516
320	0,06031	3014,6	6,4441	0,05870	3011,6	6,4291	0,05716	3008,7	6,4143	0,05569	3005,7	6,3998
330	0,06180	3041,3	6,4888	0,06016	3038,6	6,4742	0,05861	3035,8	6,4598	0,05712	3033,1	6,4456
340	0,06325	3067,3	6,5316	0,06160	3064,8	6,5173	0,06002	3062,3	6,5033	0,05850	3059,7	6,4894
350	0,06468	3092,8	6,5727	0,06300	3090,4	6,5587	0,06139	3088,1	6,5450	0,05986	3085,7	6,5315
360	0,06608	3117,7	6,6125	0,06437	3115,5	6,5987	0,06274	3113,3	6,5852	0,06119	3111,1	6,5719
370	0,06746	3142,3	6,6509	0,06572	3140,2	6,6374	0,06407	3138,1	6,6241	0,06249	3136,1	6,6110
380	0,06881	3166,4	6,6882	0,06706	3164,5	6,6749	0,06538	3162,6	6,6618	0,06378	3160,6	6,6489
390	0,07015	3190,3	6,7246	0,06837	3188,5	6,7114	0,06667	3186,7	6,6984	0,06504	3184,9	6,6857
400	0,07148	3214,0	6,7600	0,06967	3212,3	6,7469	0,06794	3210,5	6,7341	0,06630	3208,8	6,7216
410	0,07279	3237,4	6,7945	0,07095	3235,8	6,7816	0,06920	3234,2	6,7690	0,06753	3232,5	6,7566
420	0,07409	3260,7	6,8284	0,07222	3259,2	6,8156	0,07045	3257,6	6,8030	0,06876	3256,0	6,7907
430	0,07537	3283,9	6,8615	0,07349	3282,4	6,8488	0,07169	3280,9	6,8364	0,06997	3279,4	6,8242
440	0,07665	3306,9	6,8941	0,07474	3305,5	6,8815	0,07291	3304,1	6,8691	0,07117	3302,6	6,8570
450	0,07792	3329,9	6,9260	0,07598	3328,5	6,9135	0,07413	3327,1	6,9012	0,07236	3325,8	6,8892
460	0,07918	3352,7	6,9574	0,07722	3351,4	6,9450	0,07534	3350,1	6,9328	0,07355	3348,8	6,9209
470	0,08044	3375,6	6,9884	0,07844	3374,3	6,9760	0,07654	3373,0	6,9639	0,07473	3371,8	6,9520
480	0,08169	3398,4	7,0188	0,07967	3397,1	7,0065	0,07774	3395,9	6,9944	0,07590	3394,7	6,9826
490	0,08293	3421,1	7,0488	0,08088	3419,9	7,0366	0,07893	3418,8	7,0246	0,07707	3417,6	7,0128
500	0,08417	3443,9	7,0785	0,08209	3442,7	7,0662	0,08012	3441,6	7,0543	0,07823	3440,5	7,0426
510	0,08540	3466,6	7,1077	0,08330	3465,5	7,0955	0,08130	3464,4	7,0836	0,07938	3463,3	7,0720
520	0,08662	3489,3	7,1365	0,08450	3488,3	7,1244	0,08247	3487,2	7,1126	0,08054	3486,2	7,1010
530	0,08785	3512,1	7,1650	0,08569	3511,1	7,1530	0,08364	3510,0	7,1411	0,08168	3509,0	7,1296
540	0,08907	3534,8	7,1932	0,08689	3533,9	7,1812	0,08481	3532,9	7,1694	0,08282	3531,9	7,1579
550	0,09028	3557,6	7,2210	0,08807	3556,7	7,2090	0,08597	3555,7	7,1973	0,08396	3554,7	7,1858

Table 3. Water and Superheated Steam (Continuation) **Wasser und überhitzter Dampf** (Fortsetzung)

t	41 bar $t_s = 251,80\,°C$			42 bar $t_s = 253,24\,°C$			43 bar $t_s = 254,66\,°C$			44 bar $t_s = 256,05\,°C$		
	v'' 0,04850	h'' 2799,9	s'' 6,0583	v'' 0,04731	h'' 2799,4	s'' 6,0482	v'' 0,04617	h'' 2798,9	s'' 6,0383	v'' 0,04508	h'' 2798,3	s'' 6,0286
°C	v	h	s	v	h	s	v	h	s	v	h	s
550	0,09028	3557,6	7,2210	0,08807	3556,7	7,2090	0,08597	3555,7	7,1973	0,08396	3554,7	7,1858
560	0,09149	3580,4	7,2486	0,08926	3579,5	7,2366	0,08713	3578,6	7,2249	0,08510	3577,6	7,2135
570	0,09270	3603,3	7,2758	0,09044	3602,4	7,2639	0,08829	3601,5	7,2522	0,08623	3600,6	7,2408
580	0,09391	3626,1	7,3028	0,09162	3625,2	7,2909	0,08944	3624,4	7,2792	0,08736	3623,5	7,2678
590	0,09511	3649,0	7,3294	0,09279	3648,2	7,3176	0,09059	3647,3	7,3060	0,08848	3646,5	7,2946
600	0,09631	3671,9	7,3558	0,09397	3671,1	7,3440	0,09173	3670,3	7,3324	0,08960	3669,5	7,3211
610	0,09750	3694,9	7,3820	0,09514	3694,1	7,3702	0,09288	3693,3	7,3586	0,09072	3692,5	7,3473
620	0,09870	3717,9	7,4079	0,09630	3717,1	7,3961	0,09402	3716,3	7,3846	0,09184	3715,6	7,3733
630	0,09989	3740,9	7,4335	0,09747	3740,2	7,4218	0,09516	3739,4	7,4103	0,09296	3738,7	7,3990
640	0,10107	3764,0	7,4590	0,09863	3763,3	7,4472	0,09629	3762,5	7,4358	0,09407	3761,8	7,4245
650	0,10226	3787,1	7,4842	0,09979	3786,4	7,4724	0,09743	3785,7	7,4610	0,09518	3785,0	7,4498
660	0,10345	3810,3	7,5091	0,10094	3809,6	7,4974	0,09856	3808,9	7,4860	0,09628	3808,2	7,4748
670	0,10463	3833,5	7,5339	0,10210	3832,9	7,5222	0,09969	3832,2	7,5108	0,09739	3831,5	7,4996
680	0,10581	3856,8	7,5584	0,10325	3856,1	7,5467	0,10082	3855,5	7,5353	0,09849	3854,8	7,5242
690	0,10699	3880,1	7,5827	0,10440	3879,5	7,5711	0,10194	3878,8	7,5597	0,09959	3878,2	7,5486
700	0,10816	3903,4	7,6068	0,10555	3902,8	7,5952	0,10307	3902,2	7,5839	0,10069	3901,6	7,5728
710	0,10934	3926,8	7,6308	0,10670	3926,2	7,6192	0,10419	3925,6	7,6078	0,10179	3925,0	7,5967
720	0,11051	3950,3	7,6545	0,10785	3949,7	7,6429	0,10531	3949,1	7,6316	0,10289	3948,5	7,6205
730	0,11168	3973,8	7,6780	0,10899	3973,2	7,6665	0,10643	3972,6	7,6551	0,10398	3972,1	7,6441
740	0,11285	3997,3	7,7014	0,11013	3996,8	7,6898	0,10754	3996,2	7,6785	0,10507	3995,7	7,6675
750	0,11402	4020,9	7,7245	0,11127	4020,4	7,7130	0,10866	4019,8	7,7017	0,10616	4019,3	7,6907
760	0,11518	4044,5	7,7475	0,11241	4044,0	7,7360	0,10977	4043,5	7,7247	0,10725	4043,0	7,7137
770	0,11635	4068,2	7,7703	0,11355	4067,7	7,7588	0,11089	4067,2	7,7476	0,10834	4066,7	7,7366
780	0,11751	4092,0	7,7930	0,11469	4091,5	7,7815	0,11200	4091,0	7,7703	0,10943	4090,5	7,7593
790	0,11867	4115,7	7,8154	0,11582	4115,3	7,8040	0,11311	4114,8	7,7927	0,11051	4114,3	7,7818
800	0,11983	4139,6	7,8378	0,11696	4139,1	7,8263	0,11422	4138,6	7,8151	0,11160	4138,2	7,8041

Table 3. Water and Superheated Steam (Continuation) **Wasser und überhitzter Dampf** (Fortsetzung)

t	45 bar $t_s = 257{,}41$ °C			46 bar $t_s = 258{,}75$ °C			47 bar $t_s = 260{,}07$ °C			48 bar $t_s = 261{,}37$ °C		
	v'' 0,04404	h'' 2797,7	s'' 6,0191	v'' 0,04304	h'' 2797,0	s'' 6,0097	v'' 0,04208	h'' 2796,4	s'' 6,0004	v'' 0,04116	h'' 2795,7	s'' 5,9913
°C	v	h	s	v	h	s	v	h	s	v	h	s
0	0,0009980	4,5	0,0002	0,0009979	4,7	0,0002	0,0009979	4,8	0,0002	0,0009978	4,9	0,0002
10	0,0009981	46,4	0,1506	0,0009981	46,5	0,1506	0,0009980	46,6	0,1506	0,0009980	46,7	0,1506
20	0,0009997	88,1	0,2954	0,0009996	88,2	0,2953	0,0009996	88,3	0,2953	0,0009996	88,4	0,2953
30	0,0010023	129,8	0,4351	0,0010023	129,8	0,4351	0,0010022	129,9	0,4351	0,0010022	130,0	0,4350
40	0,0010058	171,4	0,5704	0,0010058	171,5	0,5703	0,0010057	171,6	0,5703	0,0010057	171,7	0,5702
50	0,0010101	213,1	0,7014	0,0010101	213,2	0,7014	0,0010100	213,3	0,7013	0,0010100	213,4	0,7013
60	0,0010151	254,8	0,8286	0,0010151	254,9	0,8285	0,0010150	255,0	0,8285	0,0010150	255,1	0,8284
70	0,0010208	296,6	0,9521	0,0010207	296,7	0,9521	0,0010207	296,8	0,9520	0,0010206	296,9	0,9520
80	0,0010271	338,5	1,0723	0,0010270	338,5	1,0722	0,0010270	338,6	1,0722	0,0010269	338,7	1,0721
90	0,0010339	380,4	1,1893	0,0010339	380,4	1,1893	0,0010338	380,5	1,1892	0,0010338	380,6	1,1891
100	0,0010414	422,4	1,3034	0,0010414	422,4	1,3034	0,0010413	422,5	1,3033	0,0010413	422,6	1,3032
110	0,0010495	464,5	1,4148	0,0010494	464,6	1,4147	0,0010494	464,6	1,4147	0,0010493	464,7	1,4146
120	0,0010582	506,7	1,5237	0,0010581	506,8	1,5236	0,0010580	506,9	1,5235	0,0010580	507,0	1,5234
130	0,0010674	549,2	1,6303	0,0010674	549,2	1,6302	0,0010673	549,3	1,6301	0,0010673	549,4	1,6300
140	0,0010774	591,8	1,7347	0,0010773	591,9	1,7346	0,0010773	591,9	1,7345	0,0010772	592,0	1,7344
150	0,0010880	634,6	1,8372	0,0010879	634,7	1,8371	0,0010879	634,8	1,8370	0,0010878	634,8	1,8368
160	0,0010994	677,8	1,9379	0,0010993	677,8	1,9378	0,0010992	677,9	1,9376	0,0010991	677,9	1,9375
170	0,0011115	721,2	2,0369	0,0011114	721,2	2,0368	0,0011113	721,3	2,0367	0,0011113	721,3	2,0366
180	0,0011245	764,9	2,1345	0,0011244	765,0	2,1344	0,0011243	765,0	2,1343	0,0011243	765,1	2,1341
190	0,0011385	809,0	2,2308	0,0011384	809,1	2,2307	0,0011383	809,1	2,2306	0,0011382	809,2	2,2304
200	0,0011535	853,6	2,3261	0,0011534	853,6	2,3259	0,0011533	853,7	2,3257	0,0011532	853,7	2,3256
210	0,0011697	898,6	2,4203	0,0011695	898,7	2,4201	0,0011694	898,7	2,4199	0,0011693	898,8	2,4198
220	0,0011872	944,3	2,5138	0,0011871	944,3	2,5136	0,0011870	944,3	2,5134	0,0011868	944,4	2,5132
230	0,0012063	990,6	2,6067	0,0012062	990,6	2,6065	0,0012060	990,6	2,6063	0,0012059	990,7	2,6061
240	0,0012272	1037,7	2,6995	0,0012271	1037,7	2,6992	0,0012269	1037,7	2,6990	0,0012268	1037,7	2,6988
250	0,0012503	1085,8	2,7922	0,0012501	1085,8	2,7920	0,0012500	1085,8	2,7917	0,0012498	1085,8	2,7915
260	0,04454	2807,9	6,0382	0,04328	2802,0	6,0190	0,0012756	1134,9	2,8848	0,0012754	1134,9	2,8846
270	0,04639	2845,3	6,1078	0,04512	2840,2	6,0899	0,04391	2835,0	6,0722	0,04274	2829,7	6,0545
280	0,04813	2880,2	6,1714	0,04685	2875,6	6,1547	0,04562	2871,0	6,1380	0,04444	2866,4	6,1214
290	0,04977	2913,0	6,2303	0,04848	2909,0	6,2144	0,04724	2904,9	6,1986	0,04605	2900,7	6,1829
300	0,05134	2944,2	6,2852	0,05003	2940,5	6,2700	0,04877	2936,8	6,2549	0,04757	2933,1	6,2399
310	0,05284	2974,0	6,3367	0,05151	2970,7	6,3221	0,05024	2967,3	6,3076	0,04902	2963,9	6,2933
320	0,05429	3002,6	6,3854	0,05294	2999,6	6,3712	0,05166	2996,5	6,3573	0,05042	2993,4	6,3434
330	0,05569	3030,3	6,4317	0,05433	3027,5	6,4179	0,05303	3024,7	6,4043	0,05177	3021,8	6,3909
340	0,05706	3057,2	6,4758	0,05568	3054,6	6,4624	0,05435	3052,0	6,4492	0,05309	3049,4	6,4362
350	0,05840	3083,3	6,5182	0,05699	3080,9	6,5050	0,05565	3078,5	6,4921	0,05436	3076,1	6,4794
360	0,05970	3108,9	6,5589	0,05828	3106,7	6,5460	0,05692	3104,4	6,5334	0,05561	3102,2	6,5209
370	0,06099	3134,0	6,5982	0,05954	3131,9	6,5856	0,05816	3129,8	6,5732	0,05684	3127,7	6,5610
380	0,06225	3158,7	6,6363	0,06079	3156,7	6,6239	0,05938	3154,8	6,6117	0,05804	3152,8	6,5996
390	0,06349	3183,0	6,6733	0,06201	3181,2	6,6610	0,06059	3179,3	6,6490	0,05922	3177,5	6,6371
400	0,06472	3207,1	6,7093	0,06321	3205,3	6,6972	0,06177	3203,6	6,6853	0,06039	3201,8	6,6736
410	0,06593	3230,9	6,7444	0,06441	3229,2	6,7324	0,06294	3227,6	6,7207	0,06154	3225,9	6,7091
420	0,06714	3254,5	6,7787	0,06559	3252,9	6,7668	0,06410	3251,3	6,7552	0,06268	3249,7	6,7438
430	0,06833	3277,9	6,8122	0,06676	3276,4	6,8005	0,06525	3274,9	6,7890	0,06381	3273,4	6,7777
440	0,06951	3301,2	6,8451	0,06791	3299,8	6,8335	0,06639	3298,3	6,8221	0,06493	3296,9	6,8108
450	0,07068	3324,4	6,8774	0,06906	3323,0	6,8659	0,06752	3321,6	6,8545	0,06604	3320,3	6,8434
460	0,07184	3347,5	6,9092	0,07020	3346,2	6,8977	0,06864	3344,9	6,8864	0,06714	3343,5	6,8753
470	0,07300	3370,5	6,9404	0,07134	3369,3	6,9289	0,06975	3368,0	6,9177	0,06823	3366,7	6,9067
480	0,07415	3393,5	6,9711	0,07247	3392,3	6,9597	0,07086	3391,1	6,9486	0,06931	3389,8	6,9376
490	0,07529	3416,4	7,0013	0,07359	3415,3	6,9900	0,07196	3414,1	6,9789	0,07039	3412,9	6,9681
500	0,07643	3439,3	7,0311	0,07470	3438,2	7,0199	0,07305	3437,1	7,0089	0,07147	3435,9	6,9981
510	0,07756	3462,2	7,0606	0,07581	3461,1	7,0494	0,07414	3460,0	7,0384	0,07254	3458,9	7,0276
520	0,07869	3485,1	7,0896	0,07692	3484,1	7,0784	0,07522	3483,0	7,0675	0,07360	3481,9	7,0568
530	0,07981	3508,0	7,1183	0,07802	3507,0	7,1072	0,07630	3505,9	7,0963	0,07466	3504,9	7,0856
540	0,08093	3530,9	7,1466	0,07912	3529,9	7,1355	0,07738	3528,9	7,1247	0,07572	3527,9	7,1140
550	0,08204	3553,8	7,1746	0,08021	3552,8	7,1636	0,07845	3551,9	7,1527	0,07677	3550,9	7,1422

Table 3. Water and Superheated Steam (Continuation) **Wasser und überhitzter Dampf** (Fortsetzung)

	45 bar $t_s = 257{,}41\,°C$			46 bar $t_s = 258{,}75\,°C$			47 bar $t_s = 260{,}07\,°C$			48 bar $t_s = 261{,}37\,°C$		
t	v'' 0,04404	h'' 2797,7	s'' 6,0191	v'' 0,04304	h'' 2797,0	s'' 6,0097	v'' 0,04208	h'' 2796,4	s'' 6,0004	v'' 0,04116	h'' 2795,7	s'' 5,9913
°C	v	h	s	v	h	s	v	h	s	v	h	s
550	0,08204	3553,8	7,1746	0,08021	3552,8	7,1636	0,07845	3551,9	7,1527	0,07677	3550,9	7,1422
560	0,08316	3576,7	7,2022	0,08130	3575,8	7,1913	0,07952	3574,8	7,1805	0,07782	3573,9	7,1699
570	0,08426	3599,6	7,2296	0,08238	3598,7	7,2187	0,08058	3597,8	7,2079	0,07886	3596,9	7,1974
580	0,08537	3622,6	7,2567	0,08347	3621,7	7,2458	0,08165	3620,9	7,2351	0,07990	3620,0	7,2246
590	0,08647	3645,6	7,2835	0,08455	3644,8	7,2726	0,08270	3643,9	7,2619	0,08094	3643,1	7,2515
600	0,08757	3668,6	7,3100	0,08562	3667,8	7,2991	0,08376	3667,0	7,2885	0,08197	3666,2	7,2781
610	0,08867	3691,7	7,3363	0,08670	3690,9	7,3254	0,08481	3690,1	7,3148	0,08301	3689,3	7,3044
620	0,08976	3714,8	7,3623	0,08777	3714,0	7,3515	0,08586	3713,2	7,3409	0,08404	3712,4	7,3305
630	0,09085	3737,9	7,3880	0,08884	3737,2	7,3772	0,08691	3736,4	7,3667	0,08506	3735,6	7,3563
640	0,09194	3761,1	7,4135	0,08990	3760,3	7,4028	0,08795	3759,6	7,3922	0,08609	3758,9	7,3819
650	0,09303	3784,3	7,4388	0,09097	3783,6	7,4281	0,08900	3782,9	7,4176	0,08711	3782,1	7,4072
660	0,09411	3807,5	7,4639	0,09203	3806,8	7,4531	0,09004	3806,1	7,4426	0,08813	3805,5	7,4324
670	0,09519	3830,8	7,4887	0,09309	3830,2	7,4780	0,09108	3829,5	7,4675	0,08915	3828,8	7,4572
680	0,09627	3854,2	7,5133	0,09415	3853,5	7,5026	0,09211	3852,9	7,4922	0,09016	3852,2	7,4819
690	0,09735	3877,5	7,5377	0,09520	3876,9	7,5270	0,09315	3876,3	7,5166	0,09118	3875,6	7,5064
700	0,09843	3901,0	7,5619	0,09626	3900,3	7,5513	0,09418	3899,7	7,5408	0,09219	3899,1	7,5306
710	0,09950	3924,4	7,5859	0,09731	3923,8	7,5753	0,09521	3923,2	7,5649	0,09320	3922,6	7,5547
720	0,10057	3947,9	7,6097	0,09836	3947,4	7,5991	0,09624	3946,8	7,5887	0,09421	3946,2	7,5785
730	0,10164	3971,5	7,6333	0,09941	3970,9	7,6227	0,09727	3970,4	7,6123	0,09521	3969,8	7,6022
740	0,10271	3995,1	7,6567	0,10045	3994,5	7,6461	0,09829	3994,0	7,6358	0,09622	3993,4	7,6256
750	0,10378	4018,8	7,6799	0,10150	4018,2	7,6694	0,09932	4017,7	7,6590	0,09722	4017,1	7,6489
760	0,10485	4042,4	7,7030	0,10254	4041,9	7,6924	0,10034	4041,4	7,6821	0,09822	4040,9	7,6720
770	0,10591	4066,2	7,7258	0,10359	4065,7	7,7153	0,10136	4065,2	7,7050	0,09923	4064,7	7,6949
780	0,10697	4090,0	7,7485	0,10463	4089,5	7,7380	0,10238	4089,0	7,7277	0,10022	4088,5	7,7176
790	0,10804	4113,8	7,7711	0,10567	4113,3	7,7605	0,10340	4112,8	7,7503	0,10122	4112,4	7,7402
800	0,10910	4137,7	7,7934	0,10670	4137,2	7,7829	0,10441	4136,8	7,7726	0,10222	4136,3	7,7626

Table 3. Water and Superheated Steam (Continuation) Wasser und überhitzter Dampf (Fortsetzung)

	49 bar $t_s = 262,65$ °C			50 bar $t_s = 263,91$ °C			52 bar $t_s = 266,37$ °C			54 bar $t_s = 268,76$ °C		
t	v'' 0,04028	h'' 2794,9	s'' 5,9824	v'' 0,03943	h'' 2794,2	s'' 5,9735	v'' 0,03782	h'' 2792,6	s'' 5,9561	v'' 0,03633	h'' 2790,8	s'' 5,9392
°C	v	h	s	v	h	s	v	h	s	v	h	s
0	0,0009978	5,0	0,0002	0,0009977	5,1	0,0002	0,0009976	5,3	0,0002	0,0009975	5,5	0,0003
10	0,0009980	46,8	0,1506	0,0009979	46,9	0,1505	0,0009978	47,1	0,1505	0,0009977	47,3	0,1505
20	0,0009995	88,5	0,2953	0,0009995	88,6	0,2952	0,0009994	88,7	0,2952	0,0009993	88,9	0,2952
30	0,0010021	130,1	0,4350	0,0010021	130,2	0,4350	0,0010020	130,4	0,4349	0,0010019	130,6	0,4348
40	0,0010056	171,8	0,5702	0,0010056	171,9	0,5702	0,0010055	172,0	0,5701	0,0010054	172,2	0,5700
50	0,0010099	213,5	0,7012	0,0010099	213,5	0,7012	0,0010098	213,7	0,7011	0,0010097	213,9	0,7010
60	0,0010149	255,2	0,8284	0,0010149	255,3	0,8283	0,0010148	255,4	0,8282	0,0010147	255,6	0,8281
70	0,0010206	296,9	0,9519	0,0010205	297,0	0,9518	0,0010204	297,2	0,9517	0,0010204	297,4	0,9516
80	0,0010269	338,8	1,0720	0,0010268	338,8	1,0720	0,0010267	339,0	1,0718	0,0010266	339,2	1,0717
90	0,0010337	380,7	1,1890	0,0010337	380,7	1,1890	0,0010336	380,9	1,1888	0,0010335	381,1	1,1887
100	0,0010412	422,7	1,3031	0,0010412	422,7	1,3030	0,0010411	422,9	1,3029	0,0010410	423,0	1,3027
110	0,0010493	464,8	1,4145	0,0010492	464,9	1,4144	0,0010491	465,0	1,4142	0,0010490	465,1	1,4141
120	0,0010579	507,0	1,5233	0,0010579	507,1	1,5233	0,0010578	507,2	1,5231	0,0010576	507,4	1,5229
130	0,0010672	549,4	1,6299	0,0010671	549,5	1,6298	0,0010670	549,7	1,6296	0,0010669	549,8	1,6294
140	0,0010771	592,1	1,7343	0,0010771	592,1	1,7342	0,0010769	592,3	1,7340	0,0010768	592,4	1,7338
150	0,0010877	634,9	1,8367	0,0010877	635,0	1,8366	0,0010875	635,1	1,8364	0,0010874	635,2	1,8362
160	0,0010991	678,0	1,9374	0,0010990	678,1	1,9373	0,0010989	678,2	1,9371	0,0010987	678,3	1,9368
170	0,0011112	721,4	2,0364	0,0011111	721,4	2,0363	0,0011110	721,6	2,0361	0,0011108	721,7	2,0358
180	0,0011242	765,1	2,1340	0,0011241	765,2	2,1339	0,0011239	765,3	2,1336	0,0011237	765,4	2,1333
190	0,0011381	809,2	2,2303	0,0011380	809,3	2,2301	0,0011378	809,4	2,2298	0,0011376	809,4	2,2295
200	0,0011531	853,7	2,3254	0,0011530	853,8	2,3253	0,0011528	853,9	2,3250	0,0011526	854,0	2,3247
210	0,0011692	898,8	2,4196	0,0011691	898,8	2,4194	0,0011689	898,9	2,4191	0,0011687	899,0	2,4188
220	0,0011867	944,4	2,5130	0,0011866	944,4	2,5129	0,0011863	944,5	2,5125	0,0011861	944,5	2,5121
230	0,0012057	990,7	2,6059	0,0012056	990,7	2,6057	0,0012053	990,7	2,6053	0,0012050	990,8	2,6050
240	0,0012266	1037,8	2,6986	0,0012264	1037,8	2,6984	0,0012261	1037,8	2,6979	0,0012258	1037,8	2,6975
250	0,0012496	1085,8	2,7913	0,0012494	1085,8	2,7910	0,0012490	1085,8	2,7905	0,0012487	1085,8	2,7900
260	0,0012752	1134,9	2,8843	0,0012750	1134,9	2,8840	0,0012746	1134,8	2,8835	0,0012741	1134,8	2,8829
270	0,04161	2824,3	6,0368	0,04053	2818,9	6,0192	0,03847	2807,7	5,9841	0,03655	2796,1	5,9490
280	0,04331	2861,7	6,1049	0,04222	2856,9	6,0886	0,04016	2847,1	6,0560	0,03823	2837,0	6,0236
290	0,04490	2896,5	6,1674	0,04380	2892,2	6,1519	0,04172	2883,6	6,1214	0,03979	2874,7	6,0912
300	0,04641	2929,3	6,2252	0,04530	2925,5	6,2105	0,04320	2917,8	6,1815	0,04125	2909,8	6,1530
310	0,04785	2960,5	6,2791	0,04673	2957,0	6,2651	0,04460	2950,0	6,2374	0,04263	2942,9	6,2102
320	0,04924	2990,3	6,3298	0,04810	2987,2	6,3163	0,04595	2980,8	6,2897	0,04395	2974,3	6,2636
330	0,05057	3019,0	6,3777	0,04942	3016,1	6,3647	0,04724	3010,3	6,3390	0,04522	3004,4	6,3138
340	0,05187	3046,7	6,4233	0,05070	3044,1	6,4106	0,04849	3038,7	6,3857	0,04644	3033,3	6,3614
350	0,05313	3073,6	6,4669	0,05194	3071,2	6,4545	0,04970	3066,2	6,4302	0,04763	3061,2	6,4066
360	0,05436	3099,9	6,5087	0,05316	3097,6	6,4966	0,05089	3093,0	6,4729	0,04878	3088,3	6,4498
370	0,05557	3125,6	6,5489	0,05435	3123,4	6,5371	0,05204	3119,1	6,5138	0,04991	3114,8	6,4912
380	0,05675	3150,8	6,5878	0,05551	3148,8	6,5762	0,05318	3144,8	6,5533	0,05102	3140,7	6,5312
390	0,05792	3175,6	6,6255	0,05666	3173,7	6,6140	0,05429	3169,9	6,5916	0,05210	3166,1	6,5698
400	0,05906	3200,0	6,6621	0,05779	3198,3	6,6508	0,05539	3194,7	6,6287	0,05317	3191,1	6,6072
410	0,06020	3224,2	6,6977	0,05891	3222,5	6,6866	0,05647	3219,2	6,6647	0,05422	3215,7	6,6436
420	0,06132	3248,1	6,7325	0,06001	3246,6	6,7215	0,05754	3243,3	6,6999	0,05525	3240,1	6,6790
430	0,06243	3271,9	6,7665	0,06110	3270,4	6,7556	0,05860	3267,3	6,7342	0,05628	3264,2	6,7135
440	0,06353	3295,4	6,7998	0,06218	3294,0	6,7890	0,05964	3291,1	6,7678	0,05729	3288,2	6,7473
450	0,06461	3318,9	6,8324	0,06325	3317,5	6,8217	0,06068	3314,7	6,8007	0,05830	3311,9	6,7804
460	0,06570	3342,2	6,8645	0,06431	3340,9	6,8538	0,06171	3338,2	6,8330	0,05929	3335,5	6,8128
470	0,06677	3365,4	6,8960	0,06537	3364,2	6,8853	0,06273	3361,6	6,8647	0,06028	3359,0	6,8447
480	0,06784	3388,6	6,9269	0,06642	3387,4	6,9164	0,06374	3384,9	6,8958	0,06126	3382,5	6,8760
490	0,06890	3411,7	6,9574	0,06746	3410,5	6,9469	0,06475	3408,2	6,9265	0,06224	3405,8	6,9068
500	0,06995	3434,8	6,9874	0,06849	3433,7	6,9770	0,06575	3431,4	6,9567	0,06320	3429,1	6,9371
510	0,07100	3457,8	7,0171	0,06952	3456,7	7,0067	0,06674	3454,5	6,9865	0,06417	3452,3	6,9669
520	0,07204	3480,9	7,0463	0,07055	3479,8	7,0360	0,06773	3477,7	7,0158	0,06513	3475,6	6,9964
530	0,07308	3503,9	7,0751	0,07157	3502,9	7,0648	0,06872	3500,8	7,0448	0,06608	3498,8	7,0255
540	0,07412	3526,9	7,1036	0,07259	3525,9	7,0934	0,06970	3523,9	7,0734	0,06703	3521,9	7,0542
550	0,07515	3549,9	7,1318	0,07360	3549,0	7,1215	0,07068	3547,1	7,1017	0,06797	3545,1	7,0825

Table 3. Water and Superheated Steam (Continuation) Wasser und überhitzter Dampf (Fortsetzung)

t	49 bar $t_s = 262,65$ °C			50 bar $t_s = 263,91$ °C			52 bar $t_s = 266,37$ °C			54 bar $t_s = 268,76$ °C		
	v'' 0,04028	h'' 2794,9	s'' 5,9824	v'' 0,03943	h'' 2794,2	s'' 5,9735	v'' 0,03782	h'' 2792,6	s'' 5,9561	v'' 0,03633	h'' 2790,8	s'' 5,9392
°C	v	h	s	v	h	s	v	h	s	v	h	s
550	0,07515	3549,9	7,1318	0,07360	3549,0	7,1215	0,07068	3547,1	7,1017	0,06797	3545,1	7,0825
560	0,07618	3573,0	7,1596	0,07461	3572,0	7,1494	0,07165	3570,2	7,1296	0,06891	3568,3	7,1105
570	0,07721	3596,0	7,1871	0,07562	3595,1	7,1769	0,07262	3593,3	7,1572	0,06985	3591,5	7,1382
580	0,07823	3619,1	7,2143	0,07662	3618,2	7,2042	0,07359	3616,5	7,1845	0,07079	3614,7	7,1655
590	0,07924	3642,2	7,2412	0,07762	3641,3	7,2311	0,07455	3639,6	7,2115	0,07172	3637,9	7,1926
600	0,08026	3665,3	7,2678	0,07862	3664,5	7,2578	0,07552	3662,8	7,2382	0,07265	3661,2	7,2194
610	0,08127	3688,5	7,2942	0,07961	3687,7	7,2842	0,07647	3686,1	7,2647	0,07357	3684,5	7,2459
620	0,08228	3711,7	7,3203	0,08060	3710,9	7,3103	0,07743	3709,3	7,2909	0,07449	3707,8	7,2721
630	0,08329	3734,9	7,3462	0,08159	3734,1	7,3362	0,07838	3732,6	7,3168	0,07541	3731,1	7,2981
640	0,08429	3758,1	7,3718	0,08258	3757,4	7,3618	0,07933	3755,9	7,3425	0,07633	3754,5	7,3238
650	0,08530	3781,4	7,3971	0,08356	3780,7	7,3872	0,08028	3779,3	7,3679	0,07725	3777,8	7,3493
660	0,08630	3804,8	7,4223	0,08454	3804,1	7,4124	0,08123	3802,7	7,3931	0,07816	3801,3	7,3745
670	0,08730	3828,1	7,4472	0,08552	3827,5	7,4373	0,08217	3826,1	7,4181	0,07907	3824,7	7,3996
680	0,08829	3851,5	7,4719	0,08650	3850,9	7,4620	0,08311	3849,6	7,4428	0,07998	3848,3	7,4244
690	0,08929	3875,0	7,4963	0,08747	3874,3	7,4865	0,08405	3873,1	7,4674	0,08089	3871,8	7,4489
700	0,09028	3898,5	7,5206	0,08845	3897,9	7,5108	0,08499	3896,6	7,4917	0,08179	3895,4	7,4733
710	0,09127	3922,0	7,5447	0,08942	3921,4	7,5349	0,08593	3920,2	7,5158	0,08270	3919,0	7,4974
720	0,09226	3945,6	7,5685	0,09039	3945,0	7,5587	0,08686	3943,8	7,5397	0,08360	3942,7	7,5214
730	0,09325	3969,2	7,5922	0,09136	3968,7	7,5824	0,08780	3967,5	7,5634	0,08450	3966,4	7,5451
740	0,09423	3992,9	7,6157	0,09232	3992,3	7,6059	0,08873	3991,2	7,5870	0,08540	3990,1	7,5687
750	0,09522	4016,6	7,6390	0,09329	4016,1	7,6292	0,08966	4015,0	7,6103	0,08629	4013,9	7,5920
760	0,09620	4040,4	7,6621	0,09425	4039,8	7,6523	0,09058	4038,8	7,6334	0,08719	4037,7	7,6152
770	0,09718	4064,2	7,6850	0,09521	4063,6	7,6753	0,09151	4062,6	7,6564	0,08808	4061,6	7,6382
780	0,09816	4088,0	7,7077	0,09617	4087,5	7,6980	0,09244	4086,5	7,6792	0,08898	4085,5	7,6610
790	0,09914	4111,9	7,7303	0,09713	4111,4	7,7206	0,09336	4110,4	7,7018	0,08987	4109,5	7,6837
800	0,10011	4135,8	7,7527	0,09809	4135,3	7,7431	0,09428	4134,4	7,7243	0,09076	4133,5	7,7061

Table 3. Water and Superheated Steam (Continuation) Wasser und überhitzter Dampf (Fortsetzung)

t	56 bar $t_s = 271,09$ °C			58 bar $t_s = 273,35$ °C			60 bar $t_s = 275,55$ °C			62 bar $t_s = 277,70$ °C		
	v'' 0,03495	h'' 2789,0	s'' 5,9227	v'' 0,03365	h'' 2787,0	s'' 5,9066	v'' 0,03244	h'' 2785,0	s'' 5,8908	v'' 0,03130	h'' 2782,9	s'' 5,8753
°C	v	h	s	v	h	s	v	h	s	v	h	s
0	0,0009974	5,7	0,0003	0,0009973	5,9	0,0003	0,0009972	6,1	0,0003	0,0009971	6,3	0,0003
10	0,0009976	47,4	0,1505	0,0009975	47,6	0,1505	0,0009974	47,8	0,1505	0,0009973	48,0	0,1504
20	0,0009992	89,1	0,2951	0,0009991	89,3	0,2951	0,0009990	89,5	0,2950	0,0009989	89,7	0,2950
30	0,0010018	130,8	0,4348	0,0010017	130,9	0,4347	0,0010016	131,1	0,4347	0,0010016	131,3	0,4346
40	0,0010053	172,4	0,5699	0,0010053	172,6	0,5699	0,0010052	172,7	0,5698	0,0010051	172,9	0,5697
50	0,0010096	214,1	0,7009	0,0010095	214,2	0,7008	0,0010094	214,4	0,7007	0,0010094	214,6	0,7006
60	0,0010146	255,8	0,8280	0,0010145	255,9	0,8279	0,0010144	256,1	0,8278	0,0010143	256,3	0,8277
70	0,0010203	297,5	0,9515	0,0010202	297,7	0,9514	0,0010201	297,8	0,9512	0,0010200	298,0	0,9511
80	0,0010265	339,3	1,0716	0,0010264	339,5	1,0715	0,0010263	339,6	1,0713	0,0010262	339,8	1,0712
90	0,0010334	381,2	1,1885	0,0010333	381,4	1,1884	0,0010332	381,5	1,1883	0,0010331	381,7	1,1881
100	0,0010408	423,2	1,3026	0,0010407	423,3	1,3024	0,0010406	423,5	1,3023	0,0010405	423,6	1,3021
110	0,0010489	465,3	1,4139	0,0010488	465,4	1,4137	0,0010487	465,6	1,4136	0,0010486	465,7	1,4134
120	0,0010575	507,5	1,5227	0,0010574	507,7	1,5225	0,0010573	507,8	1,5224	0,0010572	507,9	1,5222
130	0,0010668	549,9	1,6292	0,0010667	550,1	1,6290	0,0010665	550,2	1,6288	0,0010664	550,3	1,6286
140	0,0010767	592,5	1,7336	0,0010766	592,6	1,7334	0,0010764	592,8	1,7332	0,0010763	592,9	1,7330
150	0,0010873	635,3	1,8360	0,0010871	635,5	1,8357	0,0010870	635,6	1,8355	0,0010869	635,7	1,8353
160	0,0010986	678,4	1,9366	0,0010984	678,5	1,9364	0,0010983	678,6	1,9361	0,0010981	678,8	1,9359
170	0,0011106	721,8	2,0356	0,0011105	721,9	2,0353	0,0011103	722,0	2,0351	0,0011102	722,1	2,0348
180	0,0011236	765,5	2,1331	0,0011234	765,6	2,1328	0,0011232	765,7	2,1325	0,0011231	765,8	2,1323
190	0,0011374	809,5	2,2293	0,0011373	809,6	2,2290	0,0011371	809,7	2,2287	0,0011369	809,8	2,2284
200	0,0011524	854,0	2,3243	0,0011522	854,1	2,3240	0,0011519	854,2	2,3237	0,0011517	854,3	2,3234
210	0,0011684	899,0	2,4184	0,0011682	899,1	2,4181	0,0011680	899,2	2,4178	0,0011678	899,3	2,4174
220	0,0011858	944,6	2,5118	0,0011856	944,7	2,5114	0,0011853	944,7	2,5110	0,0011851	944,8	2,5107
230	0,0012048	990,8	2,6046	0,0012045	990,9	2,6042	0,0012042	990,9	2,6038	0,0012039	990,9	2,6034
240	0,0012255	1037,8	2,6971	0,0012252	1037,9	2,6966	0,0012249	1037,9	2,6962	0,0012245	1037,9	2,6958
250	0,0012483	1085,8	2,7896	0,0012480	1085,8	2,7891	0,0012476	1085,8	2,7886	0,0012472	1085,8	2,7881
260	0,0012737	1134,8	2,8824	0,0012733	1134,8	2,8819	0,0012729	1134,7	2,8813	0,0012724	1134,7	2,8808
270	0,0013023	1185,2	2,9761	0,0013018	1185,1	2,9755	0,0013013	1185,1	2,9748	0,0013008	1185,0	2,9742
280	0,03644	2826,7	5,9914	0,03476	2816,0	5,9592	0,03317	2804,9	5,9270	0,03168	2793,5	5,8946
290	0,03799	2865,6	6,0612	0,03630	2856,3	6,0314	0,03472	2846,7	6,0017	0,03323	2836,8	5,9721
300	0,03943	2901,7	6,1248	0,03774	2893,5	6,0969	0,03614	2885,0	6,0692	0,03465	2876,3	6,0418
310	0,04080	2935,6	6,1834	0,03908	2928,2	6,1570	0,03748	2920,7	6,1310	0,03598	2913,0	6,1051
320	0,04210	2967,7	6,2380	0,04036	2961,0	6,2128	0,03874	2954,2	6,1880	0,03722	2947,3	6,1635
330	0,04334	2998,4	6,2892	0,04159	2992,3	6,2650	0,03995	2986,1	6,2412	0,03841	2979,8	6,2178
340	0,04454	3027,7	6,3375	0,04276	3022,2	6,3142	0,04111	3016,5	6,2913	0,03955	3010,8	6,2688
350	0,04570	3056,1	6,3834	0,04390	3051,0	6,3608	0,04222	3045,8	6,3386	0,04065	3040,5	6,3168
360	0,04683	3083,6	6,4272	0,04501	3078,9	6,4052	0,04330	3074,0	6,3836	0,04171	3069,2	6,3625
370	0,04793	3110,4	6,4692	0,04608	3106,0	6,4477	0,04436	3101,5	6,4267	0,04274	3097,0	6,4061
380	0,04901	3136,6	6,5096	0,04713	3132,4	6,4885	0,04539	3128,3	6,4680	0,04375	3124,0	6,4478
390	0,05006	3162,2	6,5486	0,04816	3158,4	6,5279	0,04639	3154,4	6,5077	0,04473	3150,5	6,4880
400	0,05110	3187,5	6,5863	0,04918	3183,8	6,5660	0,04738	3180,1	6,5462	0,04570	3176,4	6,5268
410	0,05212	3212,3	6,6230	0,05017	3208,9	6,6030	0,04835	3205,4	6,5834	0,04664	3201,9	6,5644
420	0,05313	3236,9	6,6586	0,05115	3233,6	6,6389	0,04931	3230,3	6,6196	0,04758	3227,0	6,6009
430	0,05413	3261,2	6,6934	0,05212	3258,0	6,6739	0,05025	3254,9	6,6549	0,04850	3251,8	6,6364
440	0,05511	3285,2	6,7274	0,05308	3282,3	6,7081	0,05118	3279,3	6,6893	0,04941	3276,3	6,6710
450	0,05609	3309,1	6,7607	0,05403	3306,3	6,7416	0,05210	3303,5	6,7230	0,05030	3300,6	6,7049
460	0,05705	3332,9	6,7933	0,05496	3330,2	6,7743	0,05302	3327,4	6,7559	0,05119	3324,7	6,7380
470	0,05801	3356,5	6,8253	0,05589	3353,9	6,8065	0,05392	3351,3	6,7882	0,05207	3348,7	6,7704
480	0,05896	3380,0	6,8567	0,05682	3377,5	6,8381	0,05482	3375,0	6,8199	0,05295	3372,5	6,8023
490	0,05990	3403,4	6,8876	0,05773	3401,0	6,8691	0,05571	3398,6	6,8511	0,05381	3396,2	6,8336
500	0,06084	3426,8	6,9181	0,05864	3424,5	6,8996	0,05659	3422,2	6,8818	0,05467	3419,9	6,8644
510	0,06178	3450,1	6,9480	0,05955	3447,9	6,9297	0,05747	3445,7	6,9119	0,05553	3443,5	6,8947
520	0,06270	3473,4	6,9776	0,06045	3471,3	6,9594	0,05834	3469,1	6,9417	0,05637	3467,0	6,9245
530	0,06363	3496,7	7,0068	0,06134	3494,6	6,9886	0,05921	3492,5	6,9710	0,05722	3490,5	6,9540
540	0,06455	3519,9	7,0355	0,06223	3517,9	7,0175	0,06008	3515,9	7,0000	0,05806	3513,9	6,9830
550	0,06546	3543,2	7,0639	0,06312	3541,2	7,0460	0,06094	3539,3	7,0285	0,05889	3537,4	7,0116

Table 3. Water and Superheated Steam (Continuation) Wasser und überhitzter Dampf (Fortsetzung)

t	56 bar $t_s = 271{,}09$ °C			58 bar $t_s = 273{,}35$ °C			60 bar $t_s = 275{,}55$ °C			62 bar $t_s = 277{,}70$ °C		
	v'' 0,03495	h'' 2789,0	s'' 5,9227	v'' 0,03365	h'' 2787,0	s'' 5,9066	v'' 0,03244	h'' 2785,0	s'' 5,8908	v'' 0,03130	h'' 2782,9	s'' 5,8753
°C	v	h	s	v	h	s	v	h	s	v	h	s
550	0,06546	3543,2	7,0639	0,06312	3541,2	7,0460	0,06094	3539,3	7,0285	0,05889	3537,4	7,0116
560	0,06637	3566,4	7,0920	0,06400	3564,6	7,0741	0,06179	3562,7	7,0568	0,05973	3560,8	7,0399
570	0,06728	3589,7	7,1198	0,06488	3587,9	7,1019	0,06265	3586,0	7,0846	0,06055	3584,2	7,0679
580	0,06818	3612,9	7,1472	0,06576	3611,2	7,1294	0,06349	3609,4	7,1122	0,06138	3607,7	7,0955
590	0,06908	3636,2	7,1743	0,06663	3634,5	7,1566	0,06434	3632,8	7,1395	0,06220	3631,1	7,1228
600	0,06998	3659,5	7,2011	0,06750	3657,9	7,1835	0,06518	3656,2	7,1664	0,06302	3654,5	7,1498
610	0,07088	3682,8	7,2277	0,06837	3681,2	7,2101	0,06602	3679,6	7,1931	0,06383	3678,0	7,1766
620	0,07177	3706,2	7,2540	0,06923	3704,6	7,2365	0,06686	3703,1	7,2195	0,06465	3701,5	7,2030
630	0,07266	3729,6	7,2800	0,07009	3728,1	7,2625	0,06770	3726,5	7,2456	0,06546	3725,0	7,2292
640	0,07355	3753,0	7,3058	0,07095	3751,5	7,2884	0,06853	3750,0	7,2715	0,06626	3748,6	7,2551
650	0,07443	3776,4	7,3313	0,07181	3775,0	7,3139	0,06936	3773,5	7,2971	0,06707	3772,1	7,2808
660	0,07531	3799,9	7,3566	0,07266	3798,5	7,3393	0,07019	3797,1	7,3225	0,06787	3795,7	7,3062
670	0,07619	3823,4	7,3817	0,07351	3822,0	7,3644	0,07101	3820,7	7,3476	0,06867	3819,3	7,3314
680	0,07707	3846,9	7,4065	0,07436	3845,6	7,3892	0,07184	3844,3	7,3725	0,06947	3843,0	7,3563
690	0,07795	3870,5	7,4311	0,07521	3869,2	7,4139	0,07266	3868,0	7,3972	0,07027	3866,7	7,3810
700	0,07882	3894,1	7,4555	0,07606	3892,9	7,4383	0,07348	3891,7	7,4217	0,07106	3890,4	7,4056
710	0,07970	3917,8	7,4797	0,07690	3916,6	7,4625	0,07430	3915,4	7,4459	0,07186	3914,2	7,4298
720	0,08057	3941,5	7,5037	0,07775	3940,3	7,4866	0,07511	3939,1	7,4700	0,07265	3938,0	7,4539
730	0,08144	3965,2	7,5275	0,07859	3964,1	7,5104	0,07593	3962,9	7,4938	0,07344	3961,8	7,4778
740	0,08231	3989,0	7,5510	0,07943	3987,9	7,5340	0,07674	3986,8	7,5175	0,07423	3985,7	7,5015
750	0,08317	4012,8	7,5744	0,08026	4011,7	7,5574	0,07755	4010,7	7,5409	0,07501	4009,6	7,5250
760	0,08404	4036,7	7,5976	0,08110	4035,6	7,5807	0,07836	4034,6	7,5642	0,07580	4033,5	7,5483
770	0,08490	4060,6	7,6207	0,08194	4059,6	7,6037	0,07917	4058,6	7,5873	0,07658	4057,5	7,5714
780	0,08576	4084,5	7,6435	0,08277	4083,5	7,6266	0,07998	4082,6	7,6102	0,07736	4081,6	7,5943
790	0 08662	4108,5	7,6662	0,08360	4107,6	7,6493	0,08078	4106,6	7,6329	0,07814	4105,6	7,6171
800	0,08748	4132,5	7,6887	0,08443	4131,6	7,6718	0,08159	4130,7	7,6554	0,07892	4129,7	7,6396

Table 3. Water and Superheated Steam (Continuation) Wasser und überhitzter Dampf (Fortsetzung)

t	64 bar $t_s = 279{,}79$ °C			66 bar $t_s = 281{,}84$ °C			68 bar $t_s = 283{,}84$ °C			70 bar $t_s = 285{,}79$ °C		
	v'' 0,03023	h'' 2780,6	s'' 5,8601	v'' 0,02922	h'' 2778,3	s'' 5,8452	v'' 0,02827	h'' 2775,9	s'' 5,8306	v'' 0,02737	h'' 2773,5	s'' 5,8162
°C	v	h	s	v	h	s	v	h	s	v	h	s
0	0,0009970	6,5	0,0003	0,0009969	6,7	0,0003	0,0009968	6,9	0,0003	0,0009967	7,1	0,0004
10	0,0009973	48,2	0,1504	0,0009972	48,4	0,1504	0,0009971	48,6	0,1504	0,0009970	48,8	0,1504
20	0,0009988	89,9	0,2950	0,0009987	90,1	0,2949	0,0009987	90,2	0,2949	0,0009986	90,4	0,2948
30	0,0010015	131,5	0,4345	0,0010014	131,7	0,4345	0,0010013	131,8	0,4344	0,0010012	132,0	0,4343
40	0,0010050	173,1	0,5696	0,0010049	173,3	0,5695	0,0010048	173,5	0,5695	0,0010047	173,6	0,5694
50	0,0010093	214,7	0,7005	0,0010092	214,9	0,7004	0,0010091	215,1	0,7003	0,0010090	215,3	0,7003
60	0,0010143	256,4	0,8276	0,0010142	256,6	0,8275	0,0010141	256,8	0,8274	0,0010140	256,9	0,8273
70	0,0010199	298,4	0,9510	0,0010198	298,3	0,9509	0,0010197	298,5	0,9508	0,0010196	298,7	0,9506
80	0,0010261	340,0	1,0711	0,0010261	340,1	1,0709	0,0010260	340,3	1,0708	0,0010259	340,4	1,0707
90	0,0010330	381,8	1,1880	0,0010329	382,0	1,1878	0,0010328	382,1	1,1877	0,0010327	382,3	1,1875
100	0,0010404	423,8	1,3020	0,0010403	423,9	1,3018	0,0010402	424,1	1,3016	0,0010401	424,2	1,3015
110	0,0010485	465,9	1,4132	0,0010484	466,0	1,4131	0,0010482	466,2	1,4129	0,0010481	466,3	1,4127
120	0,0010571	508,1	1,5220	0,0010570	508,2	1,5218	0,0010569	508,4	1,5216	0,0010567	508,5	1,5215
130	0,0010663	550,5	1,6285	0,0010662	550,6	1,6283	0,0010661	550,7	1,6281	0,0010660	550,9	1,6279
140	0,0010762	593,0	1,7328	0,0010761	593,2	1,7326	0,0010759	593,3	1,7324	0,0010758	593,4	1,7322
150	0,0010867	635,8	1,8351	0,0010866	636,0	1,8349	0,0010865	636,1	1,8347	0,0010863	636,2	1,8344
160	0,0010980	678,9	1,9357	0,0010978	679,0	1,9354	0,0010977	679,1	1,9352	0,0010976	679,2	1,9350
170	0,0011100	722,2	2,0346	0,0011099	722,3	2,0343	0,0011097	722,4	2,0341	0,0011096	722,6	2,0338
180	0,0011229	765,9	2,1320	0,0011227	766,0	2,1317	0,0011226	766,1	2,1315	0,0011224	766,2	2,1312
190	0,0011367	809,9	2,2281	0,0011365	810,0	2,2278	0,0011363	810,1	2,2275	0,0011362	810,2	2,2273
200	0,0011515	854,4	2,3231	0,0011513	854,5	2,3228	0,0011512	854,5	2,3225	0,0011510	854,6	2,3222
210	0,0011675	899,3	2,4171	0,0011673	899,4	2,4168	0,0011671	899,5	2,4164	0,0011669	899,5	2,4161
220	0,0011849	944,8	2,5103	0,0011846	944,9	2,5100	0,0011844	945,0	2,5096	0,0011841	945,0	2,5092
230	0,0012037	991,0	2,6030	0,0012034	991,0	2,6026	0,0012031	991,1	2,6022	0,0012028	991,1	2,6018
240	0,0012242	1037,9	2,6953	0,0012239	1037,9	2,6949	0,0012236	1038,0	2,6945	0,0012233	1038,0	2,6940
250	0,0012469	1085,8	2,7877	0,0012465	1085,8	2,7872	0,0012462	1085,8	2,7867	0,0012458	1085,8	2,7862
260	0,0012720	1134,7	2,8803	0,0012716	1134,6	2,8797	0,0012712	1134,6	2,8792	0,0012708	1134,6	2,8787
270	0,0013003	1184,9	2,9736	0,0012998	1184,9	2,9730	0,0012993	1184,8	2,9724	0,0012988	1184,7	2,9718
280	0,03026	2781,6	5,8619	0,0013319	1236,7	3,0677	0,0013313	1236,6	3,0670	0,0013307	1236,5	3,0663
290	0,03182	2826,6	5,9425	0,03049	2816,1	5,9129	0,02923	2805,3	5,8830	0,02802	2794,1	5,8530
300	0,03324	2867,5	6,0144	0,03191	2858,4	5,9872	0,03065	2849,0	5,9599	0,02946	2839,4	5,9327
310	0,03456	2905,1	6,0796	0,03322	2897,1	6,0542	0,03196	2888,9	6,0289	0,03076	2880,5	6,0037
320	0,03580	2940,3	6,1393	0,03445	2933,1	6,1154	0,03318	2925,8	6,0917	0,03198	2918,3	6,0681
330	0,03697	2973,4	6,1948	0,03561	2966,9	6,1720	0,03433	2960,3	6,1495	0,03312	2953,6	6,1272
340	0,03809	3004,9	6,2466	0,03672	2999,0	6,2248	0,03542	2993,1	6,2033	0,03420	2987,0	6,1820
350	0,03917	3035,1	6,2955	0,03778	3029,7	6,2744	0,03647	3024,2	6,2537	0,03523	3018,7	6,2333
360	0,04021	3064,2	6,3418	0,03881	3059,2	6,3214	0,03748	3054,2	6,3014	0,03623	3049,1	6,2817
370	0,04123	3092,4	6,3859	0,03980	3087,8	6,3661	0,03846	3083,1	6,3467	0,03719	3078,4	6,3276
380	0,04221	3119,8	6,4282	0,04077	3115,5	6,4089	0,03941	3111,1	6,3899	0,03812	3106,7	6,3714
390	0,04318	3146,5	6,4688	0,04171	3142,5	6,4499	0,04033	3138,4	6,4314	0,03903	3134,3	6,4133
400	0,04412	3172,7	6,5079	0,04264	3168,9	6,4894	0,04124	3165,1	6,4713	0,03992	3161,2	6,4536
410	0,04505	3198,4	6,5458	0,04354	3194,8	6,5277	0,04213	3191,2	6,5099	0,04079	3187,6	6,4925
420	0,04596	3223,7	6,5826	0,04443	3220,3	6,5647	0,04300	3216,9	6,5472	0,04165	3213,5	6,5301
430	0,04685	3248,6	6,6183	0,04531	3245,4	6,6007	0,04386	3242,2	6,5835	0,04249	3239,0	6,5666
440	0,04774	3273,3	6,6532	0,04618	3270,3	6,6358	0,04470	3267,2	6,6188	0,04331	3264,2	6,6022
450	0,04862	3297,7	6,6872	0,04703	3294,9	6,6700	0,04554	3292,0	6,6532	0,04413	3289,1	6,6368
460	0,04948	3322,0	6,7205	0,04788	3319,2	6,7035	0,04636	3316,5	6,6869	0,04494	3313,7	6,6707
470	0,05034	3346,1	6,7531	0,04871	3343,4	6,7363	0,04718	3340,8	6,7168	0,04574	3338,1	6,7038
480	0,05119	3370,0	6,7851	0,04954	3367,5	6,7684	0,04799	3365,0	6,7521	0,04653	3362,4	6,7362
490	0,05203	3393,8	6,8166	0,05036	3391,4	6,8000	0,04879	3389,0	6,7838	0,04731	3386,5	6,7680
500	0,05287	3417,6	6,8475	0,05118	3415,2	6,8310	0,04959	3412,9	6,8150	0,04809	3410,6	6,7993
510	0,05370	3441,2	6,8779	0,05199	3439,0	6,8615	0,05038	3436,7	6,8456	0,04886	3434,5	6,8300
520	0,05453	3464,8	6,9078	0,05279	3462,7	6,8916	0,05116	3460,5	6,8757	0,04962	3458,3	6,8603
530	0,05535	3488,4	6,9373	0,05359	3486,3	6,9212	0,05194	3484,2	6,9055	0,05038	3482,1	6,8901
540	0,05617	3511,9	6,9665	0,05439	3509,9	6,9504	0,05272	3507,9	6,9347	0,05114	3505,9	6,9195
550	0,05698	3535,4	6,9952	0,05518	3533,5	6,9792	0,05349	3531,5	6,9636	0,05189	3529,6	6,9485

Table 3. Water and Superheated Steam (Continuation) Wasser und überhitzter Dampf (Fortsetzung)

t	64 bar $t_s = 279,79\,°C$			66 bar $t_s = 281,84\,°C$			68 bar $t_s = 283,84\,°C$			70 bar $t_s = 285,79\,°C$		
	v'' 0,03023	h'' 2780,6	s'' 5,8601	v'' 0,02922	h'' 2778,3	s'' 5,8452	v'' 0,02827	h'' 2775,9	s'' 5,8306	v'' 0,02737	h'' 2773,5	s'' 5,8162
°C	v	h	s	v	h	s	v	h	s	v	h	s
550	0,05698	3535,4	6,9952	0,05518	3533,5	6,9792	0,05349	3531,5	6,9636	0,05189	3529,6	6,9485
560	0,05779	3558,9	7,0236	0,05597	3557,0	7,0076	0,05425	3555,1	6,9922	0,05264	3553,3	6,9771
570	0,05859	3582,4	7,0516	0,05675	3580,6	7,0357	0,05501	3578,7	7,0203	0,05338	3576,9	7,0053
580	0,05939	3605,9	7,0793	0,05753	3604,1	7,0635	0,05577	3602,3	7,0481	0,05412	3600,6	7,0332
590	0,06019	3629,4	7,1067	0,05831	3627,7	7,0909	0,05653	3625,9	7,0756	0,05486	3624,2	7,0608
600	0,06099	3652,9	7,1337	0,05908	3651,2	7,1181	0,05728	3649,6	7,1028	0,05559	3647,9	7,0880
610	0,06178	3676,4	7,1605	0,05985	3674,8	7,1449	0,05803	3673,2	7,1297	0,05632	3671,6	7,1150
620	0,06257	3699,9	7,1870	0,06062	3698,4	7,1715	0,05878	3696,8	7,1563	0,05705	3695,2	7,1416
630	0,06336	3723,5	7,2132	0,06138	3722,0	7,1977	0,05953	3720,5	7,1827	0,05777	3718,9	7,1680
640	0,06414	3747,1	7,2392	0,06214	3745,6	7,2238	0,06027	3744,1	7,2087	0,05850	3742,6	7,1941
650	0,06492	3770,7	7,2649	0,06291	3769,2	7,2495	0,06101	3767,8	7,2345	0,05922	3766,4	7,2200
660	0,06570	3794,3	7,2904	0,06366	3792,9	7,2750	0,06174	3791,5	7,2601	0,05994	3790,1	7,2456
670	0,06648	3818,0	7,3156	0,06442	3816,6	7,3003	0,06248	3815,3	7,2854	0,06065	3813,9	7,2709
680	0,06726	3841,7	7,3406	0,06517	3840,4	7,3253	0,06321	3839,0	7,3105	0,06137	3837,7	7,2960
690	0,06803	3865,4	7,3654	0,06592	3864,1	7,3501	0,06394	3862,8	7,3353	0,06208	3861,6	7,3209
700	0,06880	3889,2	7,3899	0,06667	3887,9	7,3747	0,06467	3886,7	7,3599	0,06279	3885,4	7,3456
710	0,06957	3913,0	7,4142	0,06742	3911,8	7,3991	0,06540	3910,6	7,3843	0,06350	3909,3	7,3700
720	0,07034	3936,8	7,4384	0,06817	3935,6	7,4232	0,06613	3934,5	7,4085	0,06420	3933,3	7,3942
730	0,07111	3960,7	7,4623	0,06891	3959,5	7,4472	0,06685	3958,4	7,4325	0,06491	3957,2	7,4182
740	0,07187	3984,6	7,4860	0,06966	3983,5	7,4709	0,06757	3982,4	7,4563	0,06561	3981,2	7,4420
750	0,07263	4008,5	7,5095	0,07040	4007,4	7,4945	0,06829	4006,4	7,4799	0,06631	4005,3	7,4657
760	0,07340	4032,5	7,5328	0,07114	4031,5	7,5178	0,06901	4030,4	7,5032	0,06701	4029,4	7,4891
770	0,07416	4056,5	7,5560	0,07188	4055,5	7,5410	0,06973	4054,5	7,5264	0,06771	4053,5	7,5123
780	0,07491	4080,6	7,5789	0,07261	4079,6	7,5640	0,07045	4078,6	7,5494	0,06841	4077,6	7,5353
790	0,07567	4104,7	7,6017	0,07335	4103,7	7,5868	0,07116	4102,7	7,5723	0,06910	4101,8	7,5582
800	0,07643	4128,8	7,6243	0,07408	4127,9	7,6094	0,07188	4126,9	7,5949	0,06980	4126,0	7,5808

Table 3. Water and Superheated Steam (Continuation) Wasser und überhitzter Dampf (Fortsetzung)

t	72 bar t_s = 287,70 °C			74 bar t_s = 289,57 °C			76 bar t_s = 291,41 °C			78 bar t_s = 293,21 °C		
	v'' 0,02652	h'' 2770,9	s'' 5,8020	v'' 0,02572	h'' 2768,3	s'' 5,7880	v'' 0,02495	h'' 2765,5	s'' 5,7742	v'' 0,02422	h'' 2762,8	s'' 5,7606
°C	v	h	s	v	h	s	v	h	s	v	h	s
0	0,0009966	7,3	0,0004	0,0009965	7,5	0,0004	0,0009964	7,7	0,0004	0,0009963	7,9	0,0004
10	0,0009969	49,0	0,1503	0,0009968	49,2	0,1503	0,0009967	49,4	0,1503	0,0009966	49,6	0,1503
20	0,0009985	90,6	0,2948	0,0009984	90,8	0,2947	0,0009983	91,0	0,2947	0,0009982	91,2	0,2946
30	0,0010011	132,2	0,4343	0,0010010	132,4	0,4342	0,0010009	132,6	0,4342	0,0010009	132,7	0,4341
40	0,0010046	173,8	0,5693	0,0010046	174,0	0,5692	0,0010045	174,2	0,5691	0,0010044	174,3	0,5691
50	0,0010089	215,4	0,7002	0,0010088	215,6	0,7001	0,0010087	215,8	0,7000	0,0010087	216,0	0,6999
60	0,0010139	257,1	0,8271	0,0010138	257,3	0,8270	0,0010137	257,4	0,8269	0,0010136	257,6	0,8268
70	0,0010195	298,8	0,9505	0,0010194	299,0	0,9504	0,0010193	299,1	0,9503	0,0010193	299,3	0,9502
80	0,0010258	340,6	1,0705	0,0010257	340,8	1,0704	0,0010256	340,9	1,0703	0,0010255	341,1	1,0701
90	0,0010326	382,5	1,1874	0,0010325	382,6	1,1873	0,0010324	382,8	1,1871	0,0010323	382,9	1,1870
100	0,0010400	424,4	1,3013	0,0010399	424,5	1,3012	0,0010398	424,7	1,3010	0,0010397	424,8	1,3009
110	0,0010480	466,5	1,4126	0,0010479	466,6	1,4124	0,0010478	466,7	1,4122	0,0010477	466,9	1,4121
120	0,0010566	508,7	1,5213	0,0010565	508,8	1,5211	0,0010564	508,9	1,5209	0,0010563	509,1	1,5208
130	0,0010658	551,0	1,6277	0,0010657	551,2	1,6275	0,0010656	551,3	1,6273	0,0010655	551,4	1,6271
140	0,0010757	593,6	1,7319	0,0010755	593,7	1,7317	0,0010754	593,8	1,7315	0,0010753	594,0	1,7313
150	0,0010862	636,3	1,8342	0,0010860	636,5	1,8340	0,0010859	636,6	1,8338	0,0010858	636,7	1,8336
160	0,0010974	679,4	1,9347	0,0010973	679,5	1,9345	0,0010971	679,6	1,9343	0,0010970	679,7	1,9340
170	0,0011094	722,7	2,0336	0,0011092	722,8	2,0333	0,0011091	722,9	2,0331	0,0011089	723,0	2,0328
180	0,0011222	766,3	2,1309	0,0011221	766,4	2,1307	0,0011219	766,5	2,1304	0,0011217	766,6	2,1301
190	0,0011360	810,3	2,2270	0,0011358	810,4	2,2267	0,0011356	810,5	2,2264	0,0011354	810,6	2,2261
200	0,0011508	854,7	2,3219	0,0011506	854,8	2,3216	0,0011504	854,9	2,3213	0,0011502	855,0	2,3210
210	0,0011667	899,6	2,4158	0,0011664	899,7	2,4154	0,0011662	899,8	2,4151	0,0011660	899,8	2,4148
220	0,0011839	945,1	2,5089	0,0011836	945,1	2,5085	0,0011834	945,2	2,5082	0,0011831	945,3	2,5078
230	0,0012026	991,2	2,6014	0,0012023	991,2	2,6010	0,0012020	991,3	2,6006	0,0012018	991,3	2,6002
240	0,0012230	1038,0	2,6936	0,0012227	1038,0	2,6932	0,0012224	1038,1	2,6928	0,0012221	1038,1	2,6923
250	0,0012455	1085,8	2,7858	0,0012451	1085,8	2,7853	0,0012448	1085,8	2,7848	0,0012444	1085,8	2,7843
260	0,0012704	1134,6	2,8781	0,0012700	1134,5	2,8776	0,0012696	1134,5	2,8771	0,0012692	1134,5	2,8766
270	0,0012983	1184,7	2,9713	0,0012978	1184,6	2,9707	0,0012973	1184,5	2,9701	0,0012969	1184,5	2,9695
280	0,0013301	1236,4	3,0656	0,0013295	1236,3	3,0650	0,0013289	1236,2	3,0643	0,0013283	1236,1	3,0636
290	0,02688	2782,5	5,8226	0,02578	2770,5	5,7919	0,0013654	1289,9	3,1605	0,0013647	1289,7	3,1597
300	0,02832	2829,5	5,9054	0,02724	2819,3	5,8779	0,02620	2808,8	5,8503	0,02521	2798,0	5,8224
310	0,02963	2871,9	5,9786	0,02855	2863,1	5,9536	0,02752	2854,0	5,9285	0,02654	2844,8	5,9033
320	0,03084	2910,7	6,0447	0,02976	2903,0	6,0214	0,02873	2895,0	5,9982	0,02775	2887,0	5,9751
330	0,03197	2946,8	6,1051	0,03088	2939,9	6,0832	0,02985	2932,9	6,0615	0,02886	2925,7	6,0399
340	0,03304	2980,8	6,1610	0,03194	2974,6	6,1402	0,03090	2968,2	6,1196	0,02991	2961,8	6,0992
350	0,03406	3013,1	6,2132	0,03295	3007,4	6,1933	0,03190	3001,6	6,1737	0,03090	2995,8	6,1542
360	0,03504	3043,9	6,2623	0,03392	3038,7	6,2432	0,03286	3033,4	6,2243	0,03185	3028,1	6,2056
370	0,03599	3073,6	6,3088	0,03486	3068,8	6,2903	0,03378	3063,9	6,2721	0,03276	3059,0	6,2541
380	0,03691	3102,3	6,3531	0,03576	3097,8	6,3351	0,03467	3093,3	6,3174	0,03364	3088,8	6,3000
390	0,03780	3130,2	6,3955	0,03664	3126,0	6,3780	0,03554	3121,8	6,3607	0,03449	3117,6	6,3438
400	0,03868	3157,4	6,4362	0,03750	3153,5	6,4191	0,03638	3149,6	6,4022	0,03532	3145,6	6,3857
410	0,03953	3184,0	6,4754	0,03834	3180,3	6,4586	0,03720	3176,6	6,4422	0,03613	3172,9	6,4260
420	0,04037	3210,1	6,5133	0,03916	3206,6	6,4969	0,03801	3203,2	6,4807	0,03692	3199,7	6,4649
430	0,04119	3235,8	6,5501	0,03997	3232,5	6,5340	0,03880	3229,2	6,5181	0,03770	3225,9	6,5025
440	0,04200	3261,1	6,5859	0,04076	3258,0	6,5700	0,03958	3254,9	6,5543	0,03847	3251,8	6,5390
450	0,04280	3286,1	6,6208	0,04154	3283,2	6,6050	0,04035	3280,3	6,5896	0,03922	3277,3	6,5745
460	0,04359	3310,9	6,6548	0,04232	3308,1	6,6393	0,04111	3305,3	6,6240	0,03996	3302,5	6,6091
470	0,04437	3335,5	6,6881	0,04308	3332,8	6,6727	0,04186	3330,1	6,6577	0,04070	3327,4	6,6429
480	0,04514	3359,9	6,7207	0,04384	3357,3	6,7054	0,04260	3354,7	6,6906	0,04142	3352,2	6,6760
490	0,04591	3384,1	6,7526	0,04459	3381,7	6,7376	0,04333	3379,2	6,7228	0,04214	3376,7	6,7084
500	0,04667	3408,2	6,7840	0,04533	3405,9	6,7691	0,04406	3403,5	6,7545	0,04285	3401,1	6,7402
510	0,04742	3432,2	6,8149	0,04606	3430,0	6,8000	0,04478	3427,7	6,7856	0,04355	3425,4	6,7714
520	0,04817	3456,2	6,8452	0,04679	3454,0	6,8305	0,04549	3451,8	6,8161	0,04425	3449,6	6,8021
530	0,04891	3480,0	6,8751	0,04752	3477,9	6,8605	0,04620	3475,8	6,8462	0,04495	3473,7	6,8322
540	0,04965	3503,8	6,9046	0,04824	3501,8	6,8901	0,04690	3499,8	6,8759	0,04564	3497,7	6,8620
550	0,05038	3527,6	6,9337	0,04895	3525,7	6,9192	0,04760	3523,7	6,9051	0,04632	3521,7	6,8913

Table 3. Water and Superheated Steam (Continuation) **Wasser und überhitzter Dampf** (Fortsetzung)

t	72 bar $t_s = 287,70$ °C			74 bar $t_s = 289,57$ °C			76 bar $t_s = 291,41$ °C			78 bar $t_s = 293,21$ °C		
	v'' 0,02652	h'' 2770,9	s'' 5,8020	v'' 0,02572	h'' 2768,3	s'' 5,7880	v'' 0,02495	h'' 2765,5	s'' 5,7742	v'' 0,02422	h'' 2762,8	s'' 5,7606
°C	v	h	s	v	h	s	v	h	s	v	h	s
550	0,05038	3527,6	6,9337	0,04895	3525,7	6,9192	0,04760	3523,7	6,9051	0,04632	3521,7	6,8913
560	0,05111	3551,4	6,9623	0,04967	3549,5	6,9480	0,04830	3547,6	6,9339	0,04700	3545,7	6,9202
570	0,05184	3575,1	6,9906	0,05037	3573,2	6,9763	0,04899	3571,4	6,9624	0,04768	3569,6	6,9487
580	0,05256	3598,8	7,0186	0,05108	3597,0	7,0044	0,04968	3595,2	6,9905	0,04835	3593,5	6,9769
590	0,05328	3622,5	7,0462	0,05178	3620,8	7,0321	0,05036	3619,1	7,0182	0,04902	3617,3	7,0047
600	0,05399	3646,2	7,0735	0,05248	3644,5	7,0594	0,05105	3642,9	7,0457	0,04969	3641,2	7,0322
610	0,05470	3669,9	7,1006	0,05317	3668,3	7,0865	0,05172	3666,7	7,0728	0,05035	3665,1	7,0594
620	0,05541	3693,7	7,1273	0,05387	3692,1	7,1133	0,05240	3690,5	7,0996	0,05101	3688,9	7,0863
630	0,05612	3717,4	7,1537	0,05456	3715,9	7,1398	0,05307	3714,4	7,1262	0,05167	3712,8	7,1129
640	0,05682	3741,2	7,1799	0,05524	3739,7	7,1660	0,05374	3738,2	7,1524	0,05232	3736,7	7,1392
650	0,05753	3764,9	7,2058	0,05593	3763,5	7,1919	0,05441	3762,1	7,1784	0,05298	3760,6	7,1652
660	0,05823	3788,7	7,2314	0,05661	3787,3	7,2176	0,05508	3785,9	7,2041	0,05363	3784,5	7,1910
670	0,05892	3812,6	7,2568	0,05729	3811,2	7,2430	0,05574	3809,8	7,2296	0,05428	3808,5	7,2165
680	0,05962	3836,4	7,2820	0,05797	3835,1	7,2682	0,05641	3833,8	7,2549	0,05492	3832,5	7,2418
690	0,06031	3860,3	7,3069	0,05865	3859,0	7,2932	0,05707	3857,7	7,2799	0,05557	3856,4	7,2668
700	0,06101	3884,2	7,3316	0,05932	3883,0	7,3179	0,05772	3881,7	7,3046	0,05621	3880,5	7,2916
710	0,06170	3908,1	7,3560	0,05999	3906,9	7,3424	0,05838	3905,7	7,3292	0,05685	3904,5	7,3162
720	0,06238	3932,1	7,3803	0,06066	3930,9	7,3667	0,05903	3929,8	7,3535	0,05749	3928,6	7,3406
730	0,06307	3956,1	7,4043	0,06133	3955,0	7,3908	0,05969	3953,8	7,3776	0,05813	3952,7	7,3647
740	0,06375	3980,1	7,4282	0,06200	3979,0	7,4147	0,06034	3977,9	7,4015	0,05876	3976,8	7,3887
750	0,06444	4004,2	7,4518	0,06267	4003,1	7,4383	0,06099	4002,1	7,4252	0,05939	4001,0	7,4124
760	0,06512	4028,3	7,4753	0,06333	4027,3	7,4618	0,06164	4026,2	7,4487	0,06003	4025,2	7,4359
770	0,06580	4052,4	7,4985	0,06399	4051,4	7,4851	0,06228	4050,4	7,4720	0,06066	4049,4	7,4593
780	0,06648	4076,6	7,5216	0,06466	4075,6	7,5082	0,06293	4074,6	7,4951	0,06129	4073,6	7,4824
790	0,06716	4100,8	7,5445	0,06532	4099,9	7,5311	0,06357	4098,9	7,5181	0,06192	4097,9	7,5054
800	0,06783	4125,1	7,5671	0,06597	4124,1	7,5538	0,06421	4123,2	7,5408	0,06254	4122,3	7,5281

Table 3. Water and Superheated Steam (Continuation) Wasser und überhitzter Dampf (Fortsetzung)

t	80 bar $t_s = 294{,}97\ °C$			82 bar $t_s = 296{,}70\ °C$			84 bar $t_s = 298{,}39\ °C$			86 bar $t_s = 300{,}06\ °C$		
	v'' 0,02353	h'' 2759,9	s'' 5,7471	v'' 0,02286	h'' 2757,0	s'' 5,7338	v'' 0,02223	h'' 2754,0	s'' 5,7207	v'' 0,02163	h'' 2750,9	s'' 5,7076
°C	v	h	s	v	h	s	v	h	s	v	h	s
0	0,0009962	8,1	0,0004	0,0009961	8,3	0,0004	0,0009961	8,5	0,0004	0,0009960	8,7	0,0004
10	0,0009965	49,8	0,1503	0,0009964	50,0	0,1502	0,0009963	50,2	0,1502	0,0009962	50,4	0,1502
20	0,0009981	91,4	0,2946	0,0009980	91,5	0,2946	0,0009979	91,7	0,2945	0,0009979	91,9	0,2945
30	0,0010008	132,9	0,4340	0,0010007	133,1	0,4340	0,0010006	133,3	0,4339	0,0010005	133,5	0,4339
40	0,0010043	174,5	0,5690	0,0010042	174,7	0,5689	0,0010041	174,9	0,5688	0,0010040	175,0	0,5688
50	0,0010086	216,1	0,6998	0,0010085	216,3	0,6997	0,0010084	216,5	0,6996	0,0010083	216,6	0,6995
60	0,0010135	257,8	0,8267	0,0010135	257,9	0,8266	0,0010134	258,1	0,8265	0,0010133	258,3	0,8264
70	0,0010192	299,5	0,9500	0,0010191	299,6	0,9499	0,0010190	299,8	0,9498	0,0010189	300,0	0,9497
80	0,0010254	341,2	1,0700	0,0010253	341,4	1,0699	0,0010252	341,6	1,0697	0,0010251	341,7	1,0696
90	0,0010322	383,1	1,1868	0,0010321	383,2	1,1867	0,0010320	383,4	1,1865	0,0010319	383,5	1,1864
100	0,0010396	425,0	1,3007	0,0010395	425,1	1,3006	0,0010394	425,3	1,3004	0,0010393	425,4	1,3003
110	0,0010476	467,0	1,4119	0,0010475	467,2	1,4117	0,0010474	467,3	1,4116	0,0010473	467,5	1,4114
120	0,0010562	509,2	1,5206	0,0010561	509,4	1,5204	0,0010560	509,5	1,5202	0,0010558	509,6	1,5200
130	0,0010654	551,6	1,6269	0,0010652	551,7	1,6267	0,0010651	551,8	1,6265	0,0010650	552,0	1,6264
140	0,0010752	594,1	1,7311	0,0010750	594,2	1,7309	0,0010749	594,4	1,7307	0,0010748	594,5	1,7305
150	0,0010856	636,8	1,8334	0,0010855	637,0	1,8331	0,0010854	637,1	1,8329	0,0010852	637,2	1,8327
160	0,0010968	679,8	1,9338	0,0010967	680,0	1,9336	0,0010965	680,1	1,9333	0,0010964	680,2	1,9331
170	0,0011088	723,1	2,0326	0,0011086	723,2	2,0323	0,0011085	723,3	2,0321	0,0011083	723,5	2,0318
180	0,0011216	766,7	2,1299	0,0011214	766,8	2,1296	0,0011212	766,9	2,1293	0,0011211	767,0	2,1291
190	0,0011353	810,7	2,2258	0,0011351	810,8	2,2256	0,0011349	810,9	2,2253	0,0011347	811,0	2,2250
200	0,0011500	855,1	2,3207	0,0011498	855,1	2,3204	0,0011496	855,2	2,3201	0,0011494	855,3	2,3197
210	0,0011658	899,9	2,4145	0,0011656	900,0	2,4141	0,0011654	900,1	2,4138	0,0011651	900,1	2,4135
220	0,0011829	945,3	2,5075	0,0011827	945,4	2,5071	0,0011824	945,4	2,5067	0,0011822	945,5	2,5064
230	0,0012015	991,3	2,5999	0,0012012	991,4	2,5995	0,0012009	991,4	2,5991	0,0012007	991,5	2,5987
240	0,0012218	1038,1	2,6919	0,0012215	1038,1	2,6915	0,0012212	1038,2	2,6911	0,0012209	1038,2	2,6906
250	0,0012441	1085,8	2,7839	0,0012437	1085,8	2,7834	0,0012434	1085,8	2,7829	0,0012430	1085,8	2,7825
260	0,0012687	1134,5	2,8761	0,0012683	1134,4	2,8755	0,0012679	1134,4	2,8750	0,0012675	1134,4	2,8745
270	0,0012964	1184,4	2,9689	0,0012959	1184,4	2,9683	0,0012954	1184,3	2,9677	0,0012950	1184,2	2,9671
280	0,0013277	1236,0	3,0629	0,0013272	1235,9	3,0623	0,0013266	1235,8	3,0616	0,0013260	1235,7	3,0609
290	0,0013639	1289,5	3,1589	0,0013632	1289,4	3,1581	0,0013625	1289,2	3,1573	0,0013618	1289,0	3,1566
300	0,02426	2786,8	5,7942	0,02335	2775,2	5,7656	0,02247	2763,1	5,7366	0,0014040	1345,0	3,2551
310	0,02560	2835,2	5,8780	0,02470	2825,4	5,8526	0,02384	2815,4	5,8270	0,02301	2805,0	5,8012
320	0,02681	2878,7	5,9519	0,02592	2870,2	5,9288	0,02506	2861,6	5,9056	0,02424	2852,7	5,8823
330	0,02792	2918,4	6,0183	0,02703	2911,0	5,9969	0,02617	2903,4	5,9755	0,02535	2895,6	5,9541
340	0,02896	2955,3	6,0790	0,02806	2948,6	6,0588	0,02720	2941,9	6,0388	0,02638	2935,0	6,0189
350	0,02995	2989,9	6,1349	0,02904	2983,9	6,1159	0,02817	2977,8	6,0969	0,02734	2971,6	6,0781
360	0,03088	3022,7	6,1872	0,02997	3017,2	6,1689	0,02909	3011,7	6,1509	0,02826	3006,1	6,1330
370	0,03178	3054,0	6,2363	0,03086	3049,0	6,2188	0,02997	3043,9	6,2014	0,02913	3038,8	6,1843
380	0,03265	3084,2	6,2828	0,03171	3079,5	6,2659	0,03082	3074,8	6,2491	0,02997	3070,1	6,2326
390	0,03349	3113,3	6,3271	0,03254	3109,0	6,3106	0,03164	3104,7	6,2944	0,03078	3100,3	6,2784
400	0,03431	3141,6	6,3694	0,03335	3137,6	6,3534	0,03243	3133,5	6,3376	0,03156	3129,4	6,3220
410	0,03511	3169,2	6,4101	0,03414	3165,4	6,3945	0,03321	3161,6	6,3791	0,03232	3157,8	6,3639
420	0,03589	3196,2	6,4493	0,03490	3192,6	6,4340	0,03397	3189,1	6,4189	0,03307	3185,5	6,4041
430	0,03665	3222,6	6,4872	0,03566	3219,3	6,4722	0,03471	3215,9	6,4574	0,03380	3212,5	6,4429
440	0,03740	3248,7	6,5240	0,03639	3245,5	6,5092	0,03543	3242,3	6,4947	0,03451	3239,1	6,4804
450	0,03814	3274,3	6,5597	0,03712	3271,3	6,5452	0,03615	3268,3	6,5309	0,03522	3265,3	6,5168
460	0,03887	3299,7	6,5945	0,03784	3296,8	6,5802	0,03685	3293,9	6,5661	0,03591	3291,1	6,5522
470	0,03959	3324,7	6,6285	0,03854	3322,0	6,6143	0,03754	3319,3	6,6004	0,03659	3316,5	6,5868
480	0,04030	3349,6	6,6617	0,03924	3347,0	6,6477	0,03823	3344,4	6,6340	0,03726	3341,8	6,6205
490	0,04101	3374,3	6,6942	0,03993	3371,8	6,6804	0,03891	3369,3	6,6668	0,03793	3366,8	6,6534
500	0,04170	3398,8	6,7262	0,04061	3396,4	6,7124	0,03958	3394,0	6,6990	0,03859	3391,6	6,6858
510	0,04239	3423,1	6,7575	0,04129	3420,9	6,7439	0,04024	3418,6	6,7305	0,03924	3416,3	6,7175
520	0,04308	3447,4	6,7883	0,04196	3445,2	6,7748	0,04090	3443,0	6,7616	0,03988	3440,8	6,7486
530	0,04376	3471,6	6,8186	0,04263	3469,5	6,8052	0,04155	3467,3	6,7921	0,04052	3465,2	6,7792
540	0,04443	3495,7	6,8484	0,04329	3493,7	6,8351	0,04220	3491,6	6,8221	0,04116	3489,5	6,8093
550	0,04510	3519,7	6,8778	0,04394	3517,8	6,8646	0,04284	3515,8	6,8516	0,04179	3513,8	6,8390

Table 3. Water and Superheated Steam (Continuation) Wasser und überhitzter Dampf (Fortsetzung)

t	80 bar $t_s = 294{,}97\ °C$			82 bar $t_s = 296{,}70\ °C$			84 bar $t_s = 298{,}39\ °C$			86 bar $t_s = 300{,}06\ °C$		
	v'' 0,02353	h'' 2759,9	s'' 5,7471	v'' 0,02286	h'' 2757,0	s'' 5,7338	v'' 0,02223	h'' 2754,0	s'' 5,7207	v'' 0,02163	h'' 2750,9	s'' 5,7076
°C	v	h	s	v	h	s	v	h	s	v	h	s
550	0,04510	3519,7	6,8778	0,04394	3517,8	6,8646	0,04284	3515,8	6,8516	0,04179	3513,8	6,8390
560	0,04577	3543,8	6,9068	0,04459	3541,8	6,8937	0,04348	3539,9	6,8808	0,04241	3538,0	6,8682
570	0,04643	3567,7	6,9354	0,04524	3565,9	6,9223	0,04411	3564,0	6,9095	0,04304	3562,2	6,8970
580	0,04709	3591,7	6,9636	0,04589	3589,9	6,9506	0,04474	3588,1	6,9379	0,04365	3586,3	6,9255
590	0,04774	3615,6	6,9915	0,04653	3613,9	6,9786	0,04537	3612,1	6,9659	0,04427	3610,4	6,9535
600	0,04839	3639,5	7,0191	0,04717	3637,9	7,0062	0,04600	3636,2	6,9936	0,04488	3634,5	6,9813
610	0,04904	3663,4	7,0463	0,04780	3661,8	7,0335	0,04662	3660,2	7,0210	0,04549	3658,6	7,0087
620	0,04969	3687,4	7,0732	0,04843	3685,8	7,0605	0,04724	3684,2	7,0480	0,04609	3682,6	7,0358
630	0,05033	3711,3	7,0999	0,04906	3709,8	7,0872	0,04785	3708,2	7,0748	0,04670	3706,7	7,0626
640	0,05097	3735,2	7,1262	0,04969	3733,7	7,1136	0,04846	3732,3	7,1012	0,04730	3730,8	7,0891
650	0,05161	3759,2	7,1523	0,05031	3757,7	7,1397	0,04908	3756,3	7,1274	0,04790	3754,9	7,1153
660	0,05225	3783,1	7,1781	0,05093	3781,8	7,1656	0,04968	3780,4	7,1533	0,04849	3779,0	7,1413
670	0,05288	3807,1	7,2037	0,05155	3805,8	7,1912	0,05029	3804,4	7,1790	0,04909	3803,1	7,1670
680	0,05351	3831,1	7,2290	0,05217	3829,8	7,2166	0,05089	3828,5	7,2044	0,04968	3827,2	7,1924
690	0,05414	3855,2	7,2541	0,05279	3853,9	7,2417	0,05150	3852,6	7,2295	0,05027	3851,3	7,2176
700	0,05477	3879,2	7,2790	0,05340	3878,0	7,2666	0,05210	3876,7	7,2544	0,05085	3875,5	7,2426
710	0,05540	3903,3	7,3036	0,05401	3902,1	7,2912	0,05270	3900,9	7,2791	0,05144	3899,7	7,2673
720	0,05602	3927,4	7,3280	0,05462	3926,2	7,3156	0,05329	3925,1	7,3036	0,05202	3923,9	7,2918
730	0,05664	3951,5	7,3521	0,05523	3950,4	7,3399	0,05389	3949,3	7,3278	0,05261	3948,1	7,3161
740	0,05726	3975,7	7,3761	0,05584	3974,6	7,3639	0,05448	3973,5	7,3519	0,05319	3972,4	7,3402
750	0,05788	3999,9	7,3999	0,05644	3998,8	7,3876	0,05507	3997,7	7,3757	0,05377	3996,7	7,3640
760	0,05850	4024,1	7,4234	0,05705	4023,1	7,4112	0,05566	4022,0	7,3993	0,05434	4021,0	7,3877
770	0,05912	4048,4	7,4468	0,05765	4047,4	7,4346	0,05625	4046,3	7,4227	0,05492	4045,3	7,4111
780	0,05973	4072,7	7,4700	0,05825	4071,7	7,4578	0,05684	4070,7	7,4460	0,05549	4069,7	7,4344
790	0,06034	4097,0	7,4930	0,05885	4096,0	7,4808	0,05742	4095,1	7,4690	0,05607	4094,1	7,4574
800	0,06096	4121,3	7,5158	0,05943	4120,4	7,5037	0,05801	4119,5	7,4918	0,05664	4118,5	7,4803

Table 3. Water and Superheated Steam (Continuation) Wasser und überhitzter Dampf (Fortsetzung)

	88 bar $t_s = 301{,}70\ °C$			90 bar $t_s = 303{,}31\ °C$			92 bar $t_s = 304{,}89\ °C$			94 bar $t_s = 306{,}44\ °C$		
t	v'' 0,02105	h'' 2747,8	s'' 5,6948	v'' 0,02050	h'' 2744,6	s'' 5,6820	v'' 0,01996	h'' 2741,4	s'' 5,6694	v'' 0,01945	h'' 2738,0	s'' 5,6568
°C	v	h	s	v	h	s	v	h	s	v	h	s
0	0,0009959	8,9	0,0005	0,0009958	9,1	0,0005	0,0009957	9,3	0,0005	0,0009956	9,5	0,0005
10	0,0009961	50,5	0,1502	0,0009960	50,7	0,1502	0,0009960	50,9	0,1501	0,0009959	51,1	0,1501
20	0,0009978	92,1	0,2944	0,0009977	92,3	0,2944	0,0009976	92,5	0,2943	0,0009975	92,7	0,2943
30	0,0010004	133,7	0,4338	0,0010003	133,8	0,4337	0,0010002	134,0	0,4337	0,0010002	134,2	0,4336
40	0,0010039	175,2	0,5687	0,0010039	175,4	0,5686	0,0010038	175,6	0,5685	0,0010037	175,7	0,5684
50	0,0010082	216,8	0,6994	0,0010081	217,0	0,6993	0,0010080	217,2	0,6992	0,0010080	217,3	0,6991
60	0,0010132	258,4	0,8263	0,0010131	258,6	0,8262	0,0010130	258,8	0,8261	0,0010129	258,9	0,8260
70	0,0010188	300,1	0,9496	0,0010187	300,3	0,9495	0,0010186	300,5	0,9493	0,0010185	300,6	0,9492
80	0,0010250	341,9	1,0695	0,0010249	342,0	1,0694	0,0010248	342,2	1,0692	0,0010247	342,3	1,0691
90	0,0010318	383,7	1,1863	0,0010317	383,8	1,1861	0,0010316	384,0	1,1860	0,0010315	384,2	1,1858
100	0,0010392	425,6	1,3001	0,0010391	425,7	1,3000	0,0010390	425,9	1,2998	0,0010389	426,1	1,2996
110	0,0010472	467,6	1,4112	0,0010471	467,8	1,4111	0,0010470	467,9	1,4109	0,0010469	468,1	1,4108
120	0,0010557	509,8	1,5199	0,0010556	509,9	1,5197	0,0010555	510,1	1,5195	0,0010554	510,2	1,5193
130	0,0010649	552,1	1,6262	0,0010648	552,2	1,6260	0,0010647	552,4	1,6258	0,0010645	552,5	1,6256
140	0,0010747	594,6	1,7303	0,0010745	594,7	1,7301	0,0010744	594,9	1,7299	0,0010743	595,0	1,7297
150	0,0010851	637,3	1,8325	0,0010850	637,5	1,8323	0,0010849	637,6	1,8321	0,0010847	637,7	1,8318
160	0,0010963	680,3	1,9329	0,0010961	680,4	1,9327	0,0010960	680,6	1,9324	0,0010958	680,7	1,9322
170	0,0011082	723,6	2,0316	0,0011080	723,7	2,0313	0,0011079	723,8	2,0311	0,0011077	723,9	2,0308
180	0,0011209	767,1	2,1288	0,0011207	767,2	2,1286	0,0011206	767,3	2,1283	0,0011204	767,5	2,1280
190	0,0011345	811,1	2,2247	0,0011344	811,2	2,2244	0,0011342	811,3	2,2241	0,0011340	811,4	2,2239
200	0,0011492	855,4	2,3194	0,0011490	855,5	2,3191	0,0011488	855,6	2,3188	0,0011486	855,7	2,3185
210	0,0011649	900,2	2,4131	0,0011647	900,3	2,4128	0,0011645	900,4	2,4125	0,0011643	900,4	2,4122
220	0,0011819	945,6	2,5060	0,0011817	945,6	2,5057	0,0011815	945,7	2,5053	0,0011812	945,7	2,5050
230	0,0012004	991,5	2,5983	0,0012001	991,6	2,5979	0,0011999	991,6	2,5975	0,0011996	991,7	2,5972
240	0,0012206	1038,2	2,6902	0,0012203	1038,3	2,6898	0,0012200	1038,3	2,6894	0,0012197	1038,3	2,6890
250	0,0012427	1085,8	2,7820	0,0012423	1085,8	2,7815	0,0012420	1085,8	2,7811	0,0012416	1085,8	2,7806
260	0,0012671	1134,4	2,8740	0,0012667	1134,3	2,8735	0,0012663	1134,3	2,8729	0,0012659	1134,3	2,8724
270	0,0012945	1184,2	2,9666	0,0012940	1184,1	2,9660	0,0012935	1184,1	2,9654	0,0012931	1184,0	2,9648
280	0,0013255	1235,6	3,0603	0,0013249	1235,5	3,0596	0,0013243	1235,4	3,0590	0,0013238	1235,3	3,0583
290	0,0013611	1288,9	3,1558	0,0013604	1288,7	3,1550	0,0013597	1288,6	3,1543	0,0013590	1288,4	3,1535
300	0,0014031	1344,8	3,2542	0,0014022	1344,5	3,2533	0,0014013	1344,3	3,2524	0,0014005	1344,1	3,2515
310	0,02221	2794,3	5,7751	0,02143	2783,2	5,7486	0,02069	2771,7	5,7217	0,01996	2759,9	5,6944
320	0,02345	2843,6	5,8590	0,02269	2834,3	5,8355	0,02195	2824,7	5,8118	0,02124	2814,8	5,7879
330	0,02456	2887,8	5,9328	0,02381	2879,7	5,9114	0,02308	2871,4	5,8899	0,02238	2863,0	5,8684
340	0,02559	2928,0	5,9990	0,02484	2920,9	5,9792	0,02411	2913,7	5,9594	0,02341	2906,3	5,9397
350	0,02655	2965,4	6,0594	0,02579	2959,0	6,0408	0,02506	2952,6	6,0223	0,02436	2946,0	6,0039
360	0,02746	3000,4	6,1152	0,02669	2994,7	6,0976	0,02596	2988,9	6,0801	0,02526	2983,0	6,0627
370	0,02832	3033,6	6,1673	0,02755	3028,4	6,1504	0,02681	3023,1	6,1337	0,02610	3017,7	6,1172
380	0,02915	3065,3	6,2162	0,02837	3060,5	6,2000	0,02763	3055,7	6,1840	0,02691	3050,7	6,1681
390	0,02995	3095,8	6,2626	0,02916	3091,4	6,2469	0,02841	3086,9	6,2314	0,02768	3082,3	6,2161
400	0,03073	3125,3	6,3067	0,02993	3121,2	6,2915	0,02917	3117,0	6,2765	0,02843	3112,8	6,2617
410	0,03148	3154,0	6,3489	0,03067	3150,1	6,3341	0,02990	3146,2	6,3195	0,02916	3142,2	6,3051
420	0,03222	3181,9	6,3894	0,03140	3178,2	6,3750	0,03062	3174,6	6,3608	0,02987	3170,9	6,3467
430	0,03293	3209,1	6,4285	0,03211	3205,7	6,4144	0,03131	3202,3	6,4005	0,03055	3198,8	6,3868
440	0,03364	3235,9	6,4663	0,03280	3232,7	6,4525	0,03200	3229,4	6,4388	0,03123	3226,2	6,4254
450	0,03433	3262,2	6,5030	0,03348	3259,2	6,4894	0,03267	3256,1	6,4760	0,03189	3253,0	6,4628
460	0,03501	3288,5	6,5386	0,03415	3285,3	6,5252	0,03333	3282,4	6,5120	0,03254	3279,4	6,4991
470	0,03568	3313,8	6,5733	0,03481	3311,0	6,5601	0,03398	3308,3	6,5471	0,03318	3305,5	6,5343
480	0,03634	3339,1	6,6072	0,03546	3336,5	6,5942	0,03462	3333,9	6,5814	0,03381	3331,2	6,5688
490	0,03699	3364,3	6,6403	0,03610	3361,7	6,6275	0,03525	3359,2	6,6148	0,03443	3356,7	6,6023
500	0,03764	3389,2	6,6728	0,03674	3386,8	6,6600	0,03587	3384,4	6,6475	0,03504	3381,9	6,6352
510	0,03828	3413,9	6,7046	0,03736	3411,6	6,6920	0,03649	3409,3	6,6796	0,03565	3407,0	6,6674
520	0,03891	3438,6	6,7358	0,03799	3436,3	6,7234	0,03710	3434,1	6,7111	0,03625	3431,9	6,6990
530	0,03954	3463,1	6,7666	0,03860	3460,9	6,7542	0,03771	3458,8	6,7420	0,03685	3456,6	6,7300
540	0,04016	3487,5	6,7968	0,03922	3485,4	6,7845	0,03831	3483,4	6,7724	0,03744	3481,3	6,7605
550	0,04078	3511,8	6,8265	0,03982	3509,8	6,8143	0,03890	3507,8	6,8023	0,03802	3505,9	6,7906

Table 3. Water and Superheated Steam (Continuation) **Wasser und überhitzter Dampf** (Fortsetzung)

t	88 bar $t_s = 301{,}70\ °C$			90 bar $t_s = 303{,}31\ °C$			92 bar $t_s = 304{,}89\ °C$			94 bar $t_s = 306{,}44\ °C$		
	v'' 0,02105	h'' 2747,8	s'' 5,6948	v'' 0,02050	h'' 2744,6	s'' 5,6820	v'' 0,01996	h'' 2741,4	s'' 5,6694	v'' 0,01945	h'' 2738,0	s'' 5,6568
°C	v	h	s	v	h	s	v	h	s	v	h	s
550	0,04078	3511,8	6,8265	0,03982	3509,8	6,8143	0,03890	3507,8	6,8023	0,03802	3505,9	6,7906
560	0,04140	3536,1	6,8558	0,04042	3534,2	6,8437	0,03950	3532,3	6,8318	0,03861	3530,3	6,8201
570	0,04201	3560,3	6,8847	0,04102	3558,5	6,8727	0,04008	3556,6	6,8609	0,03918	3554,8	6,8493
580	0,04261	3584,5	6,9133	0,04162	3582,7	6,9013	0,04067	3580,9	6,8895	0,03976	3579,1	6,8780
590	0,04322	3608,7	6,9414	0,04221	3606,9	6,9295	0,04125	3605,2	6,9178	0,04033	3603,5	6,9063
600	0,04382	3632,8	6,9692	0,04280	3631,1	6,9574	0,04182	3629,5	6,9457	0,04089	3627,8	6,9343
610	0,04441	3656,9	6,9967	0,04338	3655,3	6,9849	0,04240	3653,7	6,9733	0,04146	3652,0	6,9620
620	0,04501	3681,1	7,0238	0,04396	3679,5	7,0121	0,04297	3677,9	7,0006	0,04202	3676,3	6,9893
630	0,04560	3705,2	7,0507	0,04454	3703,6	7,0390	0,04354	3702,1	7,0276	0,04257	3700,6	7,0163
640	0,04619	3729,3	7,0772	0,04512	3727,8	7,0656	0,04410	3726,3	7,0542	0,04313	3724,8	7,0430
650	0,04677	3753,4	7,1035	0,04570	3752,0	7,0919	0,04467	3750,5	7,0806	0,04368	3749,1	7,0695
660	0,04735	3777,6	7,1295	0,04627	3776,1	7,1180	0,04523	3774,7	7,1067	0,04423	3773,3	7,0956
670	0,04794	3801,7	7,1553	0,04684	3800,3	7,1438	0,04579	3799,0	7,1325	0,04478	3797,6	7,1215
680	0,04852	3825,9	7,1807	0,04741	3824,5	7,1693	0,04634	3823,2	7,1581	0,04533	3821,9	7,1471
690	0,04909	3850,0	7,2060	0,04797	3848,8	7,1946	0,04690	3847,5	7,1834	0,04587	3846,2	7,1724
700	0,04967	3874,2	7,2310	0,04853	3873,0	7,2196	0,04745	3871,7	7,2085	0,04641	3870,5	7,1975
710	0,05024	3898,5	7,2557	0,04910	3897,2	7,2444	0,04800	3896,0	7,2333	0,04695	3894,8	7,2224
720	0,05081	3922,7	7,2803	0,04966	3921,5	7,2690	0,04855	3920,3	7,2579	0,04749	3919,2	7,2470
730	0,05138	3947,0	7,3046	0,05021	3945,8	7,2933	0,04910	3944,7	7,2823	0,04803	3943,5	7,2715
740	0,05195	3971,3	7,3287	0,05077	3970,2	7,3174	0,04964	3969,0	7,3064	0,04856	3967,9	7,2957
750	0,05252	3995,6	7,3526	0,05133	3994,5	7,3414	0,05019	3993,4	7,3304	0,04909	3992,3	7,3196
760	0,05308	4019,9	7,3762	0,05188	4018,9	7,3651	0,05073	4017,8	7,3541	0,04963	4016,8	7,3434
770	0,05365	4044,3	7,3997	0,05243	4043,3	7,3886	0,05127	4042,3	7,3777	0,05016	4041,2	7,3670
780	0,05421	4068,7	7,4230	0,05298	4067,7	7,4119	0,05181	4066,7	7,4010	0,05069	4065,7	7,3903
790	0,05477	4093,1	7,4461	0,05353	4092,2	7,4350	0,05235	4091,2	7,4241	0,05121	4090,3	7,4135
800	0,05533	4117,6	7,4690	0,05408	4116,7	7,4579	0,05288	4115,7	7,4471	0,05174	4114,8	7,4365

Table 3. Water and Superheated Steam (Continuation) Wasser und überhitzter Dampf (Fortsetzung)

t	96 bar t_s = 307,97 °C			98 bar t_s = 309,48 °C			100 bar t_s = 310,96 °C			105 bar t_s = 314,57 °C		
	v'' 0,01897	h'' 2734,7	s'' 5,6444	v'' 0,01849	h'' 2731,2	s'' 5,6321	v'' 0,01804	h'' 2727,7	s'' 5,6198	v'' 0,01698	h'' 2718,7	s'' 5,5895
°C	v	h	s	v	h	s	v	h	s	v	h	s
0	0,0009955	9,7	0,0005	0,0009954	9,9	0,0005	0,0009953	10,1	0,0005	0,0009950	10,6	0,0005
10	0,0009958	51,3	0,1501	0,0009957	51,5	0,1501	0,0009956	51,7	0,1501	0,0009954	52,2	0,1500
20	0,0009974	92,9	0,2943	0,0009973	93,0	0,2942	0,0009972	93,2	0,2942	0,0009970	93,7	0,2941
30	0,0010001	134,4	0,4335	0,0010000	134,6	0,4335	0,0009999	134,7	0,4334	0,0009997	135,2	0,4333
40	0,0010036	175,9	0,5684	0,0010035	176,1	0,5683	0,0010034	176,3	0,5682	0,0010032	176,7	0,5680
50	0,0010079	217,5	0,6990	0,0010078	217,7	0,6990	0,0010077	217,8	0,6989	0,0010075	218,3	0,6986
60	0,0010128	259,1	0,8259	0,0010127	259,3	0,8258	0,0010127	259,4	0,8257	0,0010124	259,9	0,8254
70	0,0010184	300,8	0,9491	0,0010183	300,9	0,9490	0,0010183	301,1	0,9489	0,0010180	301,5	0,9486
80	0,0010246	342,5	1,0690	0,0010245	342,7	1,0688	0,0010245	342,8	1,0687	0,0010242	343,2	1,0684
90	0,0010314	384,3	1,1857	0,0010313	384,5	1,1856	0,0010312	384,6	1,1854	0,0010310	385,0	1,1851
100	0,0010388	426,2	1,2995	0,0010387	426,4	1,2993	0,0010386	426,5	1,2992	0,0010384	426,9	1,2988
110	0,0010468	468,2	1,4106	0,0010467	468,4	1,4104	0,0010465	468,5	1,4103	0,0010463	468,9	1,4098
120	0,0010553	510,4	1,5192	0,0010552	510,5	1,5190	0,0010551	510,6	1,5188	0,0010548	511,0	1,5184
130	0,0010644	552,7	1,6254	0,0010643	552,8	1,6252	0,0010642	552,9	1,6250	0,0010639	553,3	1,6246
140	0,0010742	595,1	1,7295	0,0010740	595,3	1,7293	0,0010739	595,4	1,7291	0,0010736	595,7	1,7286
150	0,0010846	637,8	1,8316	0,0010845	638,0	1,8314	0,0010843	638,1	1,8312	0,0010840	638,4	1,8307
160	0,0010957	680,8	1,9320	0,0010956	680,9	1,9317	0,0010954	681,0	1,9315	0,0010951	681,3	1,9309
170	0,0011076	724,0	2,0306	0,0011074	724,1	2,0304	0,0011073	724,2	2,0301	0,0011069	724,5	2,0295
180	0,0011202	767,6	2,1278	0,0011201	767,7	2,1275	0,0011199	767,8	2,1272	0,0011195	768,0	2,1266
190	0,0011338	811,4	2,2236	0,0011336	811,5	2,2233	0,0011335	811,6	2,2230	0,0011330	811,9	2,2223
200	0,0011484	855,8	2,3182	0,0011482	855,8	2,3179	0,0011480	855,9	2,3176	0,0011475	856,1	2,3169
210	0,0011641	900,5	2,4118	0,0011639	900,6	2,4115	0,0011636	900,7	2,4112	0,0011631	900,9	2,4104
220	0,0011810	945,8	2,5046	0,0011808	945,9	2,5043	0,0011805	945,9	2,5039	0,0011799	946,1	2,5030
230	0,0011993	991,7	2,5968	0,0011991	991,8	2,5964	0,0011988	991,8	2,5960	0,0011982	991,9	2,5951
240	0,0012194	1038,3	2,6885	0,0012191	1038,4	2,6881	0,0012188	1038,4	2,6877	0,0012180	1038,5	2,6867
250	0,0012413	1085,8	2,7802	0,0012410	1085,8	2,7797	0,0012406	1085,8	2,7792	0,0012398	1085,9	2,7781
260	0,0012655	1134,3	2,8719	0,0012652	1134,3	2,8714	0,0012648	1134,2	2,8709	0,0012638	1134,2	2,8696
270	0,0012926	1184,0	2,9643	0,0012922	1183,9	2,9637	0,0012917	1183,9	2,9631	0,0012905	1183,7	2,9617
280	0,0013232	1235,2	3,0576	0,0013227	1235,1	3,0570	0,0013221	1235,0	3,0563	0,0013207	1234,7	3,0547
290	0,0013583	1288,2	3,1527	0,0013577	1288,1	3,1520	0,0013570	1287,9	3,1512	0,0013553	1287,6	3,1494
300	0,0013996	1343,8	3,2506	0,0013987	1343,6	3,2497	0,0013979	1343,4	3,2488	0,0013957	1342,8	3,2466
310	0,01926	2747,5	5,6665	0,01857	2734,7	5,6379	0,0014472	1402,2	3,3505	0,0014444	1401,3	3,3478
320	0,02056	2804,7	5,7637	0,01990	2794,3	5,7393	0,01926	2783,5	5,7145	0,01773	2754,9	5,6508
330	0,02171	2854,4	5,8468	0,02105	2845,6	5,8251	0,02042	2836,5	5,8032	0,01894	2812,8	5,7476
340	0,02274	2898,8	5,9199	0,02209	2891,2	5,9001	0,02147	2883,4	5,8803	0,02000	2863,1	5,8303
350	0,02369	2939,4	5,9855	0,02304	2932,6	5,9672	0,02242	2925,8	5,9489	0,02096	2908,1	5,9031
360	0,02458	2977,0	6,0454	0,02393	2971,0	6,0282	0,02331	2964,8	6,0110	0,02184	2949,1	5,9684
370	0,02542	3012,3	6,1007	0,02477	3006,8	6,0844	0,02414	3001,3	6,0682	0,02266	2987,1	6,0280
380	0,02622	3045,8	6,1524	0,02556	3040,8	6,1368	0,02493	3035,7	6,1213	0,02344	3022,8	6,0831
390	0,02699	3077,8	6,2010	0,02632	3073,1	6,1860	0,02568	3068,5	6,1711	0,02418	3056,6	6,1345
400	0,02773	3108,5	6,2470	0,02706	3104,2	6,2325	0,02641	3099,9	6,2182	0,02489	3089,0	6,1829
410	0,02845	3138,3	6,2909	0,02777	3134,3	6,2768	0,02711	3130,3	6,2629	0,02558	3120,1	6,2287
420	0,02915	3167,2	6,3329	0,02846	3163,4	6,3192	0,02779	3159,7	6,3057	0,02624	3150,2	6,2725
430	0,02983	3195,3	6,3732	0,02913	3191,8	6,3599	0,02846	3188,3	6,3467	0,02689	3179,4	6,3143
440	0,03049	3222,9	6,4121	0,02979	3219,6	6,3990	0,02911	3216,2	6,3861	0,02752	3207,9	6,3545
450	0,03114	3249,9	6,4498	0,03043	3246,8	6,4369	0,02974	3243,6	6,4243	0,02814	3235,7	6,3933
460	0,03179	3276,5	6,4863	0,03106	3273,5	6,4737	0,03036	3270,5	6,4612	0,02874	3263,1	6,4309
470	0,03241	3302,7	6,5218	0,03168	3299,9	6,5093	0,03098	3297,0	6,4973	0,02933	3289,9	6,4673
480	0,03304	3328,5	6,5563	0,03229	3325,9	6,5441	0,03158	3323,2	6,5321	0,02992	3316,4	6,5027
490	0,03365	3354,1	6,5901	0,03290	3351,6	6,5780	0,03217	3349,0	6,5661	0,03049	3342,6	6,5372
500	0,03425	3379,5	6,6231	0,03349	3377,0	6,6112	0,03276	3374,6	6,5994	0,03105	3368,4	6,5708
510	0,03485	3404,7	6,6554	0,03408	3402,3	6,6436	0,03334	3400,0	6,6320	0,03161	3394,1	6,6038
520	0,03544	3429,6	6,6871	0,03466	3427,4	6,6755	0,03391	3425,1	6,6640	0,03216	3419,5	6,6360
530	0,03603	3454,5	6,7183	0,03524	3452,3	6,7067	0,03448	3450,2	6,6953	0,03271	3444,7	6,6677
540	0,03661	3479,2	6,7489	0,03581	3477,1	6,7374	0,03504	3475,1	6,7261	0,03325	3469,8	6,6987
550	0,03718	3503,9	6,7790	0,03637	3501,9	6,7676	0,03560	3499,8	6,7564	0,03379	3494,8	6,7293

Table 3. Water and Superheated Steam (Continuation) Wasser und überhitzter Dampf (Fortsetzung)

t	96 bar $t_s = 307{,}97\,°C$			98 bar $t_s = 309{,}48\,°C$			100 bar $t_s = 310{,}96\,°C$			105 bar $t_s = 314{,}57\,°C$		
	v'' 0,01897	h'' 2734,7	s'' 5,6444	v'' 0,01849	h'' 2731,2	s'' 5,6321	v'' 0,01804	h'' 2727,7	s'' 5,6198	v'' 0·01698	h'' 2718,7	s'' 5,5895
°C	v	h	s	v	h	s	v	h	s	v	h	s
550	0,03718	3503,9	6,7790	0,03637	3501,9	6,7676	0,03560	3499,8	6,7564	0,03379	3494,8	6,7293
560	0,03775	3528,4	6,8086	0,03693	3526,5	6,7974	0,03615	3524,5	6,7863	0,03432	3519,7	6,7593
570	0,03832	3552,9	6,8379	0,03749	3551,0	6,8266	0,03670	3549,2	6,8156	0,03484	3544,5	6,7889
580	0,03888	3577,3	6,8667	0,03805	3575,5	6,8555	0,03724	3573,7	6,8446	0,03537	3569,2	6,8180
590	0,03944	3601,7	6,8951	0,03860	3600,0	6,8840	0,03778	3598,2	6,8731	0,03589	3593,8	6,8467
600	0,04000	3626,1	6,9231	0,03914	3624,4	6,9121	0,03832	3622,7	6,9013	0,03640	3618,5	6,8751
610	0,04055	3650,4	6,9509	0,03969	3648,8	6,9399	0,03885	3647,1	6,9292	0,03691	3643,0	6,9031
620	0,04110	3674,7	6,9782	0,04023	3673,1	6,9674	0,03939	3671,6	6,9567	0,03742	3667,6	6,9307
630	0,04165	3699,0	7,0053	0,04076	3697,5	6,9945	0,03991	3696,0	6,9838	0,03793	3692,1	6,9580
640	0,04220	3723,3	7,0321	0,04130	3721,8	7,0213	0,04044	3720,4	7,0107	0,03843	3716,6	6,9850
650	0,04274	3747,6	7,0585	0,04183	3746,2	7,0478	0,04096	3744,7	7,0373	0,03894	3741,1	7,0117
660	0,04328	3771,9	7,0847	0,04236	3770,5	7,0740	0,04148	3769,1	7,0635	0,03943	3765,6	7,0381
670	0,04382	3796,3	7,1106	0,04289	3794,9	7,1000	0,04200	3793,5	7,0895	0,03993	3790,1	7,0642
680	0,04435	3820,6	7,1363	0,04342	3819,2	7,1257	0,04252	3817,9	7,1153	0,04043	3814,6	7,0901
690	0,04489	3844,9	7,1617	0,04394	3843,6	7,1511	0,04303	3842,3	7,1408	0,04092	3839,1	7,1156
700	0,04542	3869,3	7,1868	0,04446	3868,0	7,1763	0,04355	3866,8	7,1660	0,04141	3863,6	7,1410
710	0,04595	3893,6	7,2117	0,04498	3892,4	7,2012	0,04406	3891,2	7,1910	0,04190	3888,2	7,1660
720	0,04647	3918,0	7,2364	0,04550	3916,8	7,2260	0,04457	3915,6	7,2157	0,04238	3912,7	7,1909
730	0,04700	3942,4	7,2608	0,04602	3941,3	7,2504	0,04507	3940,1	7,2402	0,04287	3937,3	7,2155
740	0,04753	3966,8	7,2851	0,04653	3965,7	7,2747	0,04558	3964,6	7,2645	0,04335	3961,8	7,2398
750	0,04805	3991,3	7,3091	0,04705	3990,2	7,2987	0,04608	3989,1	7,2886	0,04384	3986,4	7,2640
760	0,04857	4015,7	7,3329	0,04756	4014,7	7,3226	0,04658	4013,6	7,3124	0,04432	4011,0	7,2879
770	0,04909	4040,2	7,3565	0,04807	4039,2	7,3462	0,04709	4038,2	7,3361	0,04479	4035,6	7,3116
780	0,04961	4064,7	7,3799	0,04858	4063,8	7,3696	0,04759	4062,8	7,3595	0,04527	4060,3	7,3352
790	0,05013	4089,3	7,4031	0,04908	4088,3	7,3928	0,04808	4087,4	7,3828	0,04575	4085,0	7,3585
800	0,05064	4113,9	7,4261	0,04959	4112,9	7,4159	0,04858	4112,0	7,4058	0,04622	4109,7	7,3816

Table 3. Water and Superheated Steam (Continuation) Wasser und überhitzter Dampf (Fortsetzung)

	110 bar $t_s = 318{,}05$ °C			115 bar $t_s = 321{,}40$ °C			120 bar $t_s = 324{,}65$ °C			125 bar $t_s = 327{,}79$ °C		
t	v'' 0,01601	h'' 2709,3	s'' 5,5595	v'' 0,01511	h'' 2699,5	s'' 5,5298	v'' 0,01428	h'' 2689,2	s'' 5,5002	v'' 0,01351	h'' 2678,4	s'' 5,4706
°C	v	h	s	v	h	s	v	h	s	v	h	s
0	0,0009948	11,1	0,0006	0,0009945	11,6	0,0006	0,0009943	12,1	0,0006	0,0009940	12,6	0,0006
10	0,0009951	52,7	0,1499	0,0009949	53,2	0,1499	0,0009947	53,6	0,1498	0,0009944	54,1	0,1498
20	0,0009968	94,2	0,2939	0,0009966	94,6	0,2938	0,0009963	95,1	0,2937	0,0009961	95,6	0,2936
30	0,0009995	135,6	0,4331	0,0009993	136,1	0,4329	0,0009990	136,6	0,4328	0,0009988	137,0	0,4326
40	0,0010030	177,2	0,5678	0,0010028	177,6	0,5676	0,0010026	178,0	0,5674	0,0010023	178,5	0,5672
50	0,0010073	218,7	0,6984	0,0010070	219,1	0,6982	0,0010068	219,6	0,6979	0,0010066	220,0	0,6977
60	0,0010122	260,3	0,8251	0,0010120	260,7	0,8249	0,0010118	261,1	0,8246	0,0010116	261,5	0,8243
70	0,0010178	301,9	0,9483	0,0010176	302,3	0,9480	0,0010174	302,7	0,9477	0,0010171	303,1	0,9474
80	0,0010240	343,6	1,0681	0,0010238	344,0	1,0677	0,0010235	344,4	1,0674	0,0010233	344,8	1,0671
90	0,0010308	385,4	1,1847	0,0010305	385,8	1,1843	0,0010303	386,2	1,1840	0,0010300	386,6	1,1836
100	0,0010381	427,3	1,2984	0,0010379	427,6	1,2980	0,0010376	428,0	1,2977	0,0010374	428,4	1,2973
110	0,0010460	469,2	1,4094	0,0010458	469,6	1,4090	0,0010455	470,0	1,4086	0,0010452	470,3	1,4082
120	0,0010545	511,3	1,5179	0,0010542	511,7	1,5175	0,0010540	512,1	1,5171	0,0010537	512,4	1,5166
130	0,0010636	553,6	1,6241	0,0010633	554,0	1,6236	0,0010630	554,3	1,6232	0,0010627	554,6	1,6227
140	0,0010733	596,1	1,7281	0,0010730	596,4	1,7276	0,0010727	596,7	1,7271	0,0010724	597,1	1,7266
150	0,0010837	638,7	1,8301	0,0010833	639,0	1,8296	0,0010830	639,4	1,8291	0,0010827	639,7	1,8285
160	0,0010947	681,6	1,9304	0,0010944	681,9	1,9298	0,0010940	682,2	1,9292	0,0010937	682,5	1,9287
170	0,0011065	724,8	2,0289	0,0011061	725,1	2,0283	0,0011058	725,4	2,0277	0,0011054	725,7	2,0271
180	0,0011191	768,3	2,1259	0,0011187	768,6	2,1253	0,0011183	768,8	2,1246	0,0011179	769,1	2,1240
190	0,0011326	812,1	2,2216	0,0011321	812,4	2,2209	0,0011317	812,6	2,2202	0,0011313	812,9	2,2195
200	0,0011470	856,4	2,3161	0,0011466	856,6	2,3154	0,0011461	856,8	2,3146	0,0011456	857,0	2,3139
210	0,0011626	901,0	2,4096	0,0011621	901,2	2,4088	0,0011615	901,4	2,4080	0,0011610	901,6	2,4072
220	0,0011793	946,3	2,5022	0,0011788	946,4	2,5013	0,0011782	946,6	2,5004	0,0011776	946,7	2,4996
230	0,0011975	992,0	2,5941	0,0011969	992,2	2,5932	0,0011962	992,3	2,5922	0,0011956	992,4	2,5913
240	0,0012173	1038,6	2,6856	0,0012165	1038,6	2,6846	0,0012158	1038,7	2,6836	0,0012151	1038,8	2,6825
250	0,0012389	1085,9	2,7770	0,0012381	1085,9	2,7758	0,0012373	1085,9	2,7747	0,0012364	1086,0	2,7736
260	0,0012628	1134,2	2,8684	0,0012619	1134,1	2,8671	0,0012609	1134,1	2,8659	0,0012600	1134,1	2,8646
270	0,0012894	1183,6	2,9603	0,0012883	1183,5	2,9589	0,0012872	1183,4	2,9575	0,0012861	1183,3	2,9561
280	0,0013194	1234,5	3,0531	0,0013180	1234,3	3,0515	0,0013167	1234,1	3,0500	0,0013154	1233,9	3,0484
290	0,0013536	1287,2	3,1475	0,0013520	1286,8	3,1457	0,0013504	1286,5	3,1439	0,0013488	1286,2	3,1421
300	0,0013936	1342,2	3,2444	0,0013915	1341,7	3,2422	0,0013895	1341,2	3,2401	0,0013875	1340,6	3,2380
310	0,0014416	1400,5	3,3451	0,0014389	1399,6	3,3424	0,0014362	1398,8	3,3398	0,0014336	1398,1	3,3373
320	0,01628	2723,5	5,5835	0,0014979	1462,1	3,4486	0,0014941	1460,8	3,4453	0,0014905	1459,7	3,4420
330	0,01755	2787,4	5,6904	0,01625	2760,0	5,6310	0,01502	2730,2	5,5686	0,01383	2697,2	5,5018
340	0,01864	2841,7	5,7797	0,01738	2819,0	5,7279	0,01619	2794,7	5,6747	0,01508	2768,7	5,6195
350	0,01961	2889,6	5,8571	0,01836	2870,1	5,8107	0,01721	2849,7	5,7636	0,01612	2828,0	5,7155
360	0,02049	2932,8	5,9259	0,01926	2915,8	5,8835	0,01811	2898,1	5,8408	0,01704	2879,6	5,7976
370	0,02131	2972,5	5,9882	0,02007	2957,4	5,9487	0,01893	2941,8	5,9093	0,01787	2925,7	5,8698
380	0,02208	3009,6	6,0454	0,02084	2996,0	6,0082	0,01969	2982,0	5,9712	0,01863	2967,6	5,9345
390	0,02281	3044,5	6,0985	0,02156	3032,1	6,0631	0,02041	3019,4	6,0281	0,01934	3006,4	5,9935
400	0,02351	3077,8	6,1483	0,02225	3066,4	6,1144	0,02108	3054,8	6,0810	0,02001	3042,9	6,0481
410	0,02418	3109,7	6,1954	0,02291	3099,2	6,1627	0,02173	3088,4	6,1306	0,02065	3077,5	6,0991
420	0,02483	3140,5	6,2401	0,02354	3130,7	6,2085	0,02236	3120,7	6,1775	0,02126	3110,5	6,1471
430	0,02546	3170,3	6,2828	0,02416	3161,1	6,2521	0,02296	3151,8	6,2221	0,02186	3142,3	6,1927
440	0,02608	3199,4	6,3238	0,02476	3190,7	6,2939	0,02355	3182,0	6,2647	0,02243	3173,1	6,2362
450	0,02668	3227,7	6,3633	0,02534	3219,6	6,3341	0,02412	3211,4	6,3056	0,02299	3203,0	6,2778
460	0,02726	3255,5	6,4014	0,02591	3247,8	6,3728	0,02467	3240,0	6,3450	0,02353	3232,2	6,3179
470	0,02784	3282,7	6,4384	0,02647	3275,5	6,4103	0,02522	3268,1	6,3831	0,02406	3260,7	6,3565
480	0,02840	3309,6	6,4742	0,02702	3302,7	6,4467	0,02575	3295,7	6,4199	0,02458	3288,7	6,3939
490	0,02896	3336,1	6,5092	0,02756	3329,5	6,4821	0,02627	3322,9	6,4557	0,02509	3316,2	6,4302
500	0,02950	3362,2	6,5432	0,02809	3356,0	6,5165	0,02679	3349,6	6,4906	0,02559	3343,3	6,4654
510	0,03004	3388,1	6,5765	0,02861	3382,1	6,5502	0,02729	3376,1	6,5246	0,02608	3370,0	6,4998
520	0,03058	3413,8	6,6091	0,02912	3408,1	6,5830	0,02779	3402,3	6,5578	0,02657	3396,5	6,5334
530	0,03110	3439,3	6,6410	0,02963	3433,8	6,6153	0,02829	3428,2	6,5903	0,02705	3422,7	6,5662
540	0,03162	3464,6	6,6723	0,03014	3459,3	6,6468	0,02877	3454,0	6,6222	0,02752	3448,6	6,5983
550	0,03214	3489,7	6,7031	0,03064	3484,7	6,6779	0,02926	3479,6	6,6535	0,02799	3474,4	6,6298

Table 3. Water and Superheated Steam (Continuation) Wasser und überhitzter Dampf (Fortsetzung)

t	110 bar $t_s = 318,05$ °C			115 bar $t_s = 321,40$ °C			120 bar $t_s = 324,65$ °C			125 bar $t_s = 327,79$ °C		
	v'' 0,01601	h'' 2709,3	s'' 5,5595	v'' 0,01511	h'' 2699,5	s'' 5,5298	v'' 0,01428	h'' 2689,2	s'' 5,5002	v'' 0,01351	h'' 2678,4	s'' 5,4706
°C	v	h	s	v	h	s	v	h	s	v	h	s
550	0,03214	3489,7	6,7031	0,03064	3484,7	6,6779	0,02926	3479,6	6,6535	0,02799	3474,4	6,6298
560	0,03265	3514,8	6,7334	0,03113	3509,9	6,7083	0,02973	3505,0	6,6842	0,02845	3500,0	6,6608
570	0,03316	3539,8	6,7631	0,03162	3535,0	6,7383	0,03021	3530,3	6,7144	0,02891	3525,5	6,6912
580	0,03366	3564,6	6,7925	0,03210	3560,1	6,7679	0,03068	3555,5	6,7441	0,02936	3550,9	6,7211
590	0,03416	3589,5	6,8214	0,03258	3585,0	6,7970	0,03114	3580,6	6,7734	0,02981	3576,2	6,7506
600	0,03466	3614,2	6,8499	0,03306	3610,0	6,8256	0,03160	3605,7	6,8022	0,03026	3601,4	6,7796
610	0,03515	3638,9	6,8780	0,03354	3634,8	6,8539	0,03206	3630,7	6,8307	0,03070	3626,5	6,8082
620	0,03564	3663,6	6,9058	0,03401	3659,6	6,8819	0,03252	3655,6	6,8588	0,03114	3651,6	6,8365
630	0,03613	3688,3	6,9333	0,03448	3684,4	6,9095	0,03297	3680,5	6,8865	0,03158	3676,6	6,8643
640	0,03661	3712,9	6,9604	0,03494	3709,1	6,9367	0,03342	3705,4	6,9139	0,03201	3701,6	6,8919
650	0,03709	3737,5	6,9872	0,03541	3733,9	6,9636	0,03387	3730,2	6,9409	0,03245	3726,6	6,9190
660	0,03757	3762,1	7,0137	0,03587	3758,6	6,9903	0,03431	3755,0	6,9677	0,03287	3751,5	6,9459
670	0,03805	3786,7	7,0399	0,03633	3783,3	7,0166	0,03475	3779,9	6,9941	0,03330	3776,4	6,9725
680	0,03852	3811,3	7,0659	0,03678	3808,0	7,0427	0,03519	3804,7	7,0203	0,03373	3801,3	6,9987
690	0,03900	3835,9	7,0916	0,03724	3832,7	7,0684	0,03563	3829,5	7,0462	0,03415	3826,2	7,0247
700	0,03947	3860,5	7,1170	0,03769	3857,4	7,0939	0,03607	3854,3	7,0718	0,03457	3851,1	7,0504
710	0,03993	3885,1	7,1421	0,03814	3882,1	7,1192	0,03650	3879,1	7,0971	0,03499	3876,0	7,0759
720	0,04040	3909,8	7,1671	0,03859	3906,8	7,1442	0,03693	3903,9	7,1222	0,03541	3900,9	7,1011
730	0,04087	3934,4	7,1917	0,03904	3931,5	7,1690	0,03736	3928,7	7,1471	0,03582	3925,8	7,1260
740	0,04133	3959,0	7,2162	0,03948	3956,3	7,1935	0,03779	3953,5	7,1717	0,03623	3950,7	7,1507
750	0,04179	3983,7	7,2404	0,03993	3981,0	7,2178	0,03822	3978,3	7,1961	0,03665	3975,6	7,1752
760	0,04225	4008,4	7,2644	0,04037	4005,8	7,2419	0,03864	4003,1	7,2203	0,03706	4000,5	7,1994
770	0,04271	4033,1	7,2882	0,04081	4030,5	7,2658	0,03907	4028,0	7,2442	0,03746	4025,5	7,2234
780	0,04317	4057,8	7,3118	0,04125	4055,3	7,2894	0,03949	4052,9	7,2679	0,03787	4050,4	7,2472
790	0,04363	4082,6	7,3352	0,04169	4080,2	7,3129	0,03991	4077,8	7,2914	0,03828	4075,3	7,2708
800	0,04408	4107,3	7,3584	0,04212	4105,0	7,3361	0,04033	4102,7	7,3147	0,03868	4100,3	7,2942

Table 3. Water and Superheated Steam (Continuation) Wasser und überhitzter Dampf (Fortsetzung)

t	130 bar $t_s = 330{,}83$ °C			135 bar $t_s = 333{,}78$ °C			140 bar $t_s = 336{,}64$ °C			145 bar $t_s = 339{,}42$ °C		
	v'' 0,01280	h'' 2667,0	s'' 5,4408	v'' 0,01213	h'' 2655,0	s'' 5,4107	v'' 0,01150	h'' 2642,4	s'' 5,3803	v'' 0,01090	h'' 2629,1	s'' 5,3493
°C	v	h	s	v	h	s	v	h	s	v	h	s
0	0,0009938	13,1	0,0007	0,0009936	13,6	0,0007	0,0009933	14,1	0,0007	0,0009931	14,6	0,0007
10	0,0009942	54,6	0,1497	0,0009940	55,1	0,1497	0,0009938	55,6	0,1496	0,0009935	56,0	0,1496
20	0,0009959	96,0	0,2935	0,0009957	96,5	0,2934	0,0009955	97,0	0,2933	0,0009952	97,4	0,2932
30	0,0009986	137,5	0,4325	0,0009984	137,9	0,4323	0,0009982	138,4	0,4322	0,0009980	138,8	0,4320
40	0,0010021	178,9	0,5670	0,0010019	179,4	0,5668	0,0010017	179,8	0,5666	0,0010015	180,2	0,5665
50	0,0010064	220,4	0,6975	0,0010062	220,8	0,6972	0,0010060	221,3	0,6970	0,0010058	221,7	0,6968
60	0,0010113	262,0	0,8241	0,0010111	262,4	0,8238	0,0010109	262,8	0,8236	0,0010107	263,2	0,8233
70	0,0010169	303,6	0,9471	0,0010167	304,0	0,9468	0,0010165	304,4	0,9465	0,0010162	304,8	0,9462
80	0,0010231	345,2	1,0668	0,0010228	345,6	1,0664	0,0010226	346,0	1,0661	0,0010224	346,4	1,0658
90	0,0010298	386,9	1,1833	0,0010296	387,3	1,1829	0,0010293	387,7	1,1826	0,0010291	388,1	1,1822
100	0,0010371	428,8	1,2969	0,0010369	429,1	1,2965	0,0010366	429,5	1,2961	0,0010364	429,9	1,2958
110	0,0010450	470,7	1,4078	0,0010447	471,1	1,4074	0,0010445	471,4	1,4070	0,0010442	471,8	1,4066
120	0,0010534	512,8	1,5162	0,0010531	513,1	1,5157	0,0010529	513,5	1,5153	0,0010526	513,8	1,5149
130	0,0010624	555,0	1,6222	0,0010622	555,3	1,6218	0,0010619	555,7	1,6213	0,0010616	556,0	1,6208
140	0,0010721	597,4	1,7261	0,0010718	597,7	1,7256	0,0010715	598,0	1,7251	0,0010712	598,4	1,7246
150	0,0010824	640,0	1,8280	0,0010820	640,3	1,8275	0,0010817	640,6	1,8269	0,0010814	640,9	1,8264
160	0,0010933	682,8	1,9281	0,0010930	683,1	1,9275	0,0010926	683,4	1,9270	0,0010923	683,7	1,9264
170	0,0011050	725,9	2,0265	0,0011046	726,2	2,0259	0,0011043	726,5	2,0253	0,0011039	726,8	2,0247
180	0,0011175	769,4	2,1233	0,0011171	769,6	2,1227	0,0011167	769,9	2,1221	0,0011163	770,2	2,1214
190	0,0011308	813,1	2,2188	0,0011304	813,4	2,2182	0,0011300	813,6	2,2175	0,0011295	813,8	2,2168
200	0,0011451	857,2	2,3131	0,0011447	857,5	2,3124	0,0011442	857,7	2,3117	0,0011437	857,9	2,3109
210	0,0011605	901,8	2,4064	0,0011600	902,0	2,4056	0,0011595	902,2	2,4048	0,0011589	902,4	2,4040
220	0,0011770	946,9	2,4987	0,0011765	947,1	2,4979	0,0011759	947,2	2,4970	0,0011753	947,4	2,4961
230	0,0011949	992,6	2,5904	0,0011943	992,7	2,5894	0,0011937	992,8	2,5885	0,0011930	992,9	2,5876
240	0,0012144	1038,9	2,6815	0,0012137	1039,0	2,6805	0,0012130	1039,1	2,6795	0,0012122	1039,2	2,6785
250	0,0012356	1086,0	2,7725	0,0012348	1086,0	2,7714	0,0012340	1086,1	2,7703	0,0012332	1086,1	2,7692
260	0,0012590	1134,0	2,8634	0,0012581	1134,0	2,8622	0,0012572	1134,0	2,8610	0,0012562	1134,0	2,8597
270	0,0012850	1183,2	2,9547	0,0012839	1183,1	2,9533	0,0012828	1183,0	2,9520	0,0012817	1182,9	2,9506
280	0,0013141	1233,7	3,0468	0,0013128	1233,5	3,0453	0,0013115	1233,3	3,0438	0,0013103	1233,1	3,0423
290	0,0013472	1285,8	3,1403	0,0013456	1285,5	3,1385	0,0013441	1285,2	3,1368	0,0013426	1284,9	3,1350
300	0,0013855	1340,1	3,2359	0,0013836	1339,6	3,2338	0,0013817	1339,2	3,2318	0,0013798	1338,7	3,2297
310	0,0014310	1397,3	3,3348	0,0014285	1396,6	3,3323	0,0014260	1395,9	3,3298	0,0014236	1395,2	3,3274
320	0,0014870	1458,5	3,4389	0,0014835	1457,4	3,4357	0,0014801	1456,3	3,4327	0,0014769	1455,3	3,4297
330	0,0015600	1526,0	3,5517	0,0015548	1524,3	3,5474	0,0015497	1522,6	3,5433	0,0015449	1520,9	3,5394
340	0,01401	2740,6	5,5618	0,01299	2709,9	5,5007	0,01200	2675,7	5,4348	0,01100	2635,8	5,3603
350	0,01510	2805,0	5,6661	0,01413	2780,5	5,6150	0,01321	2754,2	5,5618	0,01232	2725,8	5,5060
360	0,01604	2860,2	5,7539	0,01510	2839,7	5,7093	0,01421	2818,1	5,6636	0,01337	2795,2	5,6165
370	0,01688	2908,8	5,8301	0,01595	2891,3	5,7901	0,01508	2873,0	5,7496	0,01426	2853,8	5,7083
380	0,01764	2952,7	5,8979	0,01672	2937,3	5,8612	0,01586	2921,4	5,8243	0,01505	2904,9	5,7872
390	0,01835	2993,1	5,9592	0,01743	2979,3	5,9250	0,01657	2965,2	5,8909	0,01576	2950,7	5,8568
400	0,01902	3030,7	6,0155	0,01809	3018,3	5,9833	0,01723	3005,6	5,9513	0,01642	2992,5	5,9194
410	0,01965	3066,3	6,0680	0,01872	3054,9	6,0373	0,01785	3043,3	6,0069	0,01704	3031,4	5,9768
420	0,02025	3100,2	6,1173	0,01931	3089,7	6,0878	0,01844	3079,0	6,0588	0,01762	3068,1	6,0301
430	0,02083	3132,7	6,1639	0,01989	3123,0	6,1355	0,01900	3113,0	6,1076	0,01818	3103,0	6,0800
440	0,02140	3164,1	6,2082	0,02044	3155,0	6,1808	0,01955	3145,8	6,1538	0,01872	3136,4	6,1272
450	0,02194	3194,6	6,2506	0,02098	3186,0	6,2240	0,02008	3177,4	6,1978	0,01924	3168,6	6,1721
460	0,02247	3224,2	6,2913	0,02150	3216,2	6,2654	0,02059	3208,1	6,2399	0,01974	3199,8	6,2150
470	0,02299	3253,2	6,3306	0,02200	3245,6	6,3052	0,02108	3237,9	6,2804	0,02023	3230,2	6,2561
480	0,02350	3281,6	6,3685	0,02250	3274,4	6,3437	0,02157	3267,1	6,3194	0,02070	3259,8	6,2957
490	0,02400	3309,4	6,4052	0,02298	3302,6	6,3809	0,02204	3295,7	6,3572	0,02117	3288,8	6,3339
500	0,02440	3336,8	6,4409	0,02346	3330,4	6,4171	0,02251	3323,8	6,3937	0,02162	3317,2	6,3710
510	0,02496	3363,9	6,4757	0,02393	3357,7	6,4522	0,02297	3351,5	6,4293	0,02207	3345,2	6,4069
520	0,02544	3390,6	6,5096	0,02439	3384,7	6,4865	0,02342	3378,8	6,4639	0,02251	3372,8	6,4419
530	0,02590	3417,0	6,5427	0,02484	3411,4	6,5199	0,02386	3405,7	6,4977	0,02294	3400,0	6,4760
540	0,02636	3443,3	6,5752	0,02529	3437,8	6,5526	0,02429	3432,4	6,5307	0,02337	3426,9	6,5093
550	0,02682	3469,3	6,6069	0,02573	3464,1	6,5847	0,02472	3458,8	6,5630	0,02378	3453,6	6,5419

Table 3. Water and Superheated Steam (Continuation) **Wasser und überhitzter Dampf** (Fortsetzung)

	130 bar $t_s = 330{,}83\,°C$			135 bar $t_s = 333{,}78\,°C$			140 bar $t_s = 336{,}64\,°C$			145 bar $t_s = 339{,}42\,°C$		
t	v'' 0,01280	h'' 2667,0	s'' 5,4408	v'' 0,01213	h'' 2655,0	s'' 5,4107	v'' 0,01150	h'' 2642,4	s'' 5,3803	v'' 0,01090	h'' 2629,1	s'' 5,3493
°C	v	h	s	v	h	s	v	h	s	v	h	s
550	0,02682	3469,3	6,6069	0,02573	3464,1	6,5847	0,02472	3458,8	6,5630	0,02378	3453,6	6,5419
560	0,02727	3495,1	6,6381	0,02617	3490,1	6,6161	0,02515	3485,1	6,5947	0,02420	3480,0	6,5738
570	0,02771	3520,8	6,6688	0,02660	3516,0	6,6470	0,02557	3511,1	6,6258	0,02461	3506,3	6,6052
580	0,02815	3546,3	6,6989	0,02703	3541,7	6,6773	0,02598	3537,0	6,6563	0,02501	3532,4	6,6359
590	0,02859	3571,7	6,7285	0,02745	3567,3	6,7071	0,02640	3562,8	6,6863	0,02541	3558,3	6,6661
600	0,02902	3597,1	6,7577	0,02787	3592,8	6,7365	0,02680	3588,5	6,7159	0,02581	3584,1	6,6959
610	0,02945	3622,4	6,7865	0,02829	3618,2	6,7655	0,02721	3614,0	6,7450	0,02620	3609,8	6,7252
620	0,02987	3647,6	6,8149	0,02870	3643,6	6,7940	0,02761	3639,5	6,7737	0,02659	3635,5	6,7540
630	0,03030	3672,7	6,8429	0,02911	3668,8	6,8222	0,02801	3664,9	6,8020	0,02698	3661,0	6,7825
640	0,03072	3697,9	6,8706	0,02952	3694,1	6,8499	0,02840	3690,3	6,8299	0,02737	3686,5	6,8105
650	0,03114	3722,9	6,8979	0,02992	3719,3	6,8774	0,02880	3715,6	6,8575	0,02775	3711,9	6,8382
660	0,03155	3748,0	6,9249	0,03032	3744,4	6,9045	0,02919	3740,9	6,8847	0,02813	3737,3	6,8656
670	0,03196	3773,0	6,9515	0,03072	3769,6	6,9313	0,02957	3766,1	6,9117	0,02850	3762,7	6,8926
680	0,03237	3798,0	6,9779	0,03112	3794,7	6,9578	0,02996	3791,3	6,9383	0,02888	3788,0	6,9193
690	0,03278	3823,0	7,0040	0,03152	3819,8	6,9840	0,03034	3816,5	6,9645	0,02925	3813,3	6,9457
700	0,03319	3848,0	7,0298	0,03191	3844,9	7,0099	0,03072	3841,7	6,9906	0,02962	3838,6	6,9718
710	0,03359	3873,0	7,0553	0,03230	3869,9	7,0355	0,03110	3866,9	7,0163	0,02999	3863,8	6,9977
720	0,03400	3898,0	7,0806	0,03269	3895,0	7,0609	0,03148	3892,0	7,0417	0,03035	3889,1	7,0232
730	0,03440	3922,9	7,1057	0,03308	3920,1	7,0860	0,03186	3917,2	7,0669	0,03072	3914,3	7,0485
740	0,03480	3947,9	7,1304	0,03347	3945,1	7,1108	0,03223	3942,3	7,0919	0,03108	3939,6	7,0735
750	0,03519	3972,9	7,1550	0,03385	3970,2	7,1355	0,03260	3967,5	7,1166	0,03144	3964,8	7,0983
760	0,03559	3997,9	7,1793	0,03423	3995,3	7,1599	0,03297	3992,6	7,1411	0,03180	3990,0	7,1228
770	0,03598	4022,9	7,2034	0,03461	4020,4	7,1840	0,03334	4017,8	7,1653	0,03216	4015,3	7,1472
780	0,03638	4047,9	7,2272	0,03499	4045,4	7,2080	0,03371	4043,0	7,1893	0,03251	4040,5	7,1712
790	0,03677	4072,9	7,2509	0,03537	4070,5	7,2317	0,03408	4068,1	7,2131	0,03287	4065,7	7,1951
800	0,03716	4098,0	7,2743	0,03575	4095,6	7,2552	0,03444	4093,3	7,2367	0,03322	4091,0	7,2187

Table 3. Water and Superheated Steam (Continuation) Wasser und überhitzter Dampf (Fortsetzung)

	150 bar $t_s = 342{,}13$ °C			155 bar $t_s = 344{,}76$ °C			160 bar $t_s = 347{,}33$ °C			165 bar $t_s = 349{,}83$ °C		
t	v'' 0,01034	h'' 2615,0	s'' 5,3178	v'' 0,009810	h'' 2600,3	s'' 5,2858	v'' 0,009308	h'' 2584,9	s'' 5,2531	v'' 0,008832	h'' 2568,8	s'' 5,2200
°C	v	h	s	v	h	s	v	h	s	v	h	s
0	0,0009928	15,1	0,0007	0,0009926	15,6	0,0007	0,0009923	16,1	0,0008	0,0009921	16,6	0,0008
10	0,0009933	56,5	0,1495	0,0009931	57,0	0,1494	0,0009928	57,5	0,1494	0,0009926	58,0	0,1493
20	0,0009950	97,9	0,2931	0,0009948	98,3	0,2929	0,0009946	98,8	0,2928	0,0009944	99,3	0,2927
30	0,0009977	139,3	0,4318	0,0009975	139,7	0,4317	0,0009973	140,2	0,4315	0,0009971	140,6	0,4314
40	0,0010013	180,7	0,5663	0,0010011	181,1	0,5661	0,0010009	181,6	0,5659	0,0010006	182,0	0,5657
50	0,0010055	222,1	0,6966	0,0010053	222,6	0,6963	0,0010051	223,0	0,6961	0,0010049	223,4	0,6959
60	0,0010105	263,6	0,8230	0,0010102	264,0	0,8228	0,0010100	264,5	0,8225	0,0010098	264,9	0,8223
70	0,0010160	305,2	0,9459	0,0010158	305,6	0,9456	0,0010156	306,0	0,9453	0,0010153	306,4	0,9451
80	0,0010221	346,8	1,0655	0,0010219	347,2	1,0652	0,0010217	347,6	1,0648	0,0010215	348,0	1,0645
90	0,0010289	388,5	1,1819	0,0010286	388,9	1,1815	0,0010284	389,3	1,1812	0,0010281	389,7	1,1808
100	0,0010361	430,3	1,2954	0,0010359	430,6	1,2950	0,0010356	431,0	1,2946	0,0010354	431,4	1,2943
110	0,0010439	472,2	1,4062	0,0010437	472,5	1,4058	0,0010434	472,9	1,4054	0,0010432	473,3	1,4050
120	0,0010523	514,2	1,5144	0,0010521	514,5	1,5140	0,0010518	514,9	1,5136	0,0010515	515,3	1,5131
130	0,0010613	556,4	1,6204	0,0010610	556,7	1,6199	0,0010607	557,0	1,6194	0,0010604	557,4	1,6190
140	0,0010709	598,7	1,7241	0,0010706	599,0	1,7236	0,0010703	599,4	1,7231	0,0010700	599,7	1,7227
150	0,0010811	641,3	1,8259	0,0010807	641,6	1,8254	0,0010804	641,9	1,8248	0,0010801	642,2	1,8243
160	0,0010919	684,0	1,9258	0,0010916	684,3	1,9253	0,0010913	684,6	1,9247	0,0010909	685,0	1,9242
170	0,0011035	727,1	2,0241	0,0011032	727,4	2,0235	0,0011028	727,7	2,0229	0,0011024	727,9	2,0223
180	0,0011159	770,4	2,1208	0,0011155	770,7	2,1201	0,0011151	771,0	2,1195	0,0011147	771,2	2,1189
190	0,0011291	814,1	2,2161	0,0011287	814,3	2,2154	0,0011283	814,6	2,2147	0,0011278	814,8	2,2141
200	0,0011433	858,1	2,3102	0,0011428	858,4	2,3095	0,0011423	858,6	2,3087	0,0011419	858,8	2,3080
210	0,0011584	902,6	2,4032	0,0011579	902,8	2,4024	0,0011574	903,0	2,4016	0,0011569	903,2	2,4008
220	0,0011748	947,6	2,4953	0,0011742	947,7	2,4945	0,0011736	947,9	2,4936	0,0011731	948,1	2,4928
230	0,0011924	993,1	2,5867	0,0011918	993,2	2,5857	0,0011912	993,4	2,5848	0,0011905	993,5	2,5839
240	0,0012115	1039,2	2,6775	0,0012108	1039,3	2,6765	0,0012102	1039,4	2,6755	0,0012095	1039,5	2,6745
250	0,0012324	1086,2	2,7681	0,0012316	1086,2	2,7670	0,0012308	1086,3	2,7659	0,0012301	1086,3	2,7648
260	0,0012553	1133,9	2,8585	0,0012544	1133,9	2,8573	0,0012535	1133,9	2,8561	0,0012526	1133,9	2,8550
270	0,0012807	1182,8	2,9493	0,0012796	1182,7	2,9480	0,0012786	1182,6	2,9466	0,0012776	1182,6	2,9453
280	0,0013090	1232,9	3,0407	0,0013078	1232,7	3,0392	0,0013065	1232,6	3,0378	0,0013053	1232,4	3,0363
290	0,0013411	1284,6	3,1333	0,0013396	1284,3	3,1316	0,0013381	1284,0	3,1299	0,0013366	1283,7	3,1282
300	0,0013779	1338,2	3,2277	0,0013761	1337,8	3,2258	0,0013743	1337,4	3,2238	0,0013725	1336,9	3,2219
310	0,0014212	1394,5	3,3250	0,0014189	1393,8	3,3227	0,0014166	1393,2	3,3204	0,0014143	1392,6	3,3181
320	0,0014736	1454,3	3,4267	0,0014705	1453,3	3,4239	0,0014674	1452,4	3,4210	0,0014644	1451,4	3,4182
330	0,0015402	1519,4	3,5355	0,0015357	1517,9	3,5317	0,0015313	1516,4	3,5280	0,0015270	1515,0	3,5244
340	0,0016324	1593,3	3,6571	0,0016248	1590,8	3,6515	0,0016176	1588,3	3,6462	0,0016108	1586,0	3,6411
350	0,01146	2694,8	5,4467	0,01062	2660,4	5,3826	0,009764	2620,8	5,3109	0,008866	2571,7	5,2247
360	0,01256	2770,8	5,5677	0,01179	2744,6	5,5168	0,01104	2716,5	5,4634	0,01030	2686,0	5,4067
370	0,01348	2833,6	5,6662	0,01274	2812,4	5,6230	0,01203	2789,9	5,5784	0,01135	2766,1	5,5323
380	0,01428	2887,7	5,7497	0,01356	2869,8	5,7116	0,01287	2851,1	5,6729	0,01222	2831,6	5,6333
390	0,01500	2935,7	5,8225	0,01429	2920,1	5,7881	0,01361	2904,1	5,7534	0,01297	2887,4	5,7183
400	0,01566	2979,1	5,8876	0,01495	2965,4	5,8558	0,01427	2951,3	5,8240	0,01364	2936,7	5,7920
410	0,01628	3019,3	5,9469	0,01556	3006,9	5,9171	0,01489	2994,3	5,8874	0,01425	2981,3	5,8578
420	0,01686	3057,0	6,0016	0,01614	3045,7	5,9735	0,01546	3034,2	5,9455	0,01483	3022,5	5,9176
430	0,01741	3092,7	6,0528	0,01669	3082,3	6,0259	0,01601	3071,8	5,9993	0,01537	3061,0	5,9728
440	0,01794	3126,9	6,1010	0,01721	3117,2	6,0752	0,01653	3107,5	6,0497	0,01588	3097,5	6,0244
450	0,01845	3159,7	6,1468	0,01772	3150,7	6,1219	0,01703	3141,6	6,0972	0,01638	3132,4	6,0729
460	0,01895	3191,5	6,1904	0,01820	3183,1	6,1663	0,01751	3174,5	6,1425	0,01685	3165,9	6,1190
470	0,01943	3222,3	6,2322	0,01868	3214,4	6,2087	0,01797	3206,4	6,1856	0,01731	3198,3	6,1629
480	0,01989	3252,4	6,2724	0,01913	3244,9	6,2495	0,01842	3237,4	6,2270	0,01775	3229,8	6,2049
490	0,02035	3281,8	6,3112	0,01958	3274,7	6,2888	0,01886	3267,6	6,2669	0,01819	3260,4	6,2453
500	0,02080	3310,6	6,3487	0,02002	3303,9	6,3268	0,01929	3297,1	6,3054	0,01861	3290,3	6,2843
510	0,02123	3338,9	6,3850	0,02045	3332,5	6,3636	0,01971	3326,1	6,3426	0,01902	3319,7	6,3220
520	0,02166	3366,8	6,4204	0,02087	3360,7	6,3993	0,02013	3354,6	6,3787	0,01943	3348,5	6,3585
530	0,02208	3394,3	6,4548	0,02128	3388,5	6,4341	0,02053	3382,7	6,4139	0,01982	3376,8	6,3940
540	0,02250	3421,4	6,4885	0,02169	3415,9	6,4681	0,02093	3410,3	6,4481	0,02021	3404,7	6,4286
550	0,02291	3448,3	6,5213	0,02209	3443,0	6,5012	0,02132	3437,7	6,4816	0,02060	3432,4	6,4624

Table 3. Water and Superheated Steam (Continuation) Wasser und überhitzter Dampf (Fortsetzung)

t	150 bar $t_s = 342,13$ °C			155 bar $t_s = 344,76$ °C			160 bar $t_s = 347,33$ °C			165 bar $t_s = 349,83$ °C		
	v'' 0,01034	h'' 2615,0	s'' 5,3178	v'' 0,009810	h'' 2600,3	s'' 5,2858	v'' 0,009308	h'' 2584,9	s'' 5,2531	v'' 0,008832	h'' 2568,8	s'' 5,2200
°C	v	h	s	v	h	s	v	h	s	v	h	s
550	0,02291	3448,3	6,5213	0,02209	3443,0	6,5012	0,02132	3437,7	6,4816	0,02060	3432,4	6,4624
560	0,02331	3475,0	6,5535	0,02248	3469,9	6,5337	0,02171	3464,8	6,5143	0,02098	3459,7	6,4953
570	0,02371	3501,4	6,5851	0,02287	3496,5	6,5655	0,02209	3491,6	6,5463	0,02135	3486,7	6,5276
580	0,02411	3527,7	6,6160	0,02326	3523,0	6,5967	0,02246	3518,3	6,5777	0,02172	3513,5	6,5592
590	0,02450	3553,8	6,6465	0,02364	3549,3	6,6273	0,02284	3544,7	6,6086	0,02208	3540,2	6,5903
600	0,02488	3579,8	6,6764	0,02402	3575,4	6,6574	0,02320	3571,0	6,6389	0,02244	3566,6	6,6208
610	0,02527	3605,6	6,7058	0,02439	3601,4	6,6870	0,02357	3597,2	6,6687	0,02280	3593,0	6,6507
620	0,02565	3631,4	6,7349	0,02476	3627,3	6,7162	0,02393	3623,3	6,6980	0,02315	3619,2	6,6802
630	0,02602	3657,1	6,7635	0,02513	3653,2	6,7449	0,02429	3649,2	6,7269	0,02350	3645,3	6,7093
640	0,02640	3682,7	6,7917	0,02549	3678,9	6,7733	0,02464	3675,1	6,7554	0,02384	3671,3	6,7379
650	0,02677	3708,3	6,8195	0,02585	3704,6	6,8013	0,02499	3700,9	6,7835	0,02419	3697,2	6,7662
660	0,02714	3733,8	6,8470	0,02621	3730,2	6,8289	0,02534	3726,6	6,8112	0,02453	3723,1	6,7940
670	0,02750	3759,2	6,8741	0,02657	3755,8	6,8561	0,02569	3752,3	6,8386	0,02487	3748,9	6,8215
680	0,02787	3784,7	6,9009	0,02692	3781,3	6,8831	0,02603	3778,0	6,8656	0,02520	3774,6	6,8487
690	0,02823	3810,1	6,9274	0,02727	3806,8	6,9097	0,02638	3803,6	6,8924	0,02554	3800,3	6,8755
700	0,02859	3835,4	6,9536	0,02762	3832,3	6,9360	0,02672	3829,1	6,9188	0,02587	3826,0	6,9020
710	0,02894	3860,8	6,9796	0,02797	3857,7	6,9620	0,02706	3854,7	6,9449	0,02620	3851,6	6,9282
720	0,02930	3886,1	7,0052	0,02832	3883,2	6,9877	0,02739	3880,2	6,9707	0,02653	3877,2	6,9542
730	0,02965	3911,5	7,0306	0,02866	3908,6	7,0132	0,02773	3905,7	6,9963	0,02685	3902,8	6,9798
740	0,03001	3936,8	7,0557	0,02900	3934,0	7,0384	0,02806	3931,2	7,0216	0,02718	3928,4	7,0052
750	0,03036	3962,1	7,0806	0,02934	3959,4	7,0633	0,02839	3956,7	7,0466	0,02750	3954,0	7,0303
760	0,03070	3987,4	7,1052	0,02968	3984,8	7,0880	0,02872	3982,1	7,0714	0,02782	3979,5	7,0551
770	0,03105	4012,7	7,1296	0,03002	4010,2	7,1125	0,02905	4007,6	7,0959	0,02814	4005,0	7,0797
780	0,03140	4038,0	7,1537	0,03035	4035,5	7,1367	0,02938	4033,1	7,1202	0,02846	4030,6	7,1041
790	0,03174	4063,3	7,1776	0,03069	4060,9	7,1607	0,02970	4058,5	7,1442	0,02877	4056,1	7,1282
800	0,03209	4088,6	7,2013	0,03102	4086,3	7,1845	0,03002	4084,0	7,1681	0,02909	4081,6	7,1521

Table 3. Water and Superheated Steam (Continuation) Wasser und überhitzter Dampf (Fortsetzung)

t	170 bar $t_s = 352,26\,°C$			175 bar $t_s = 354,64\,°C$			180 bar $t_s = 356,96\,°C$			185 bar $t_s = 359,22\,°C$		
	v'' 0,008371	h'' 2551,6	s'' 5,1855	v'' 0,007926	h'' 2533,3	s'' 5,1498	v'' 0,007498	h'' 2513,9	s'' 5,1128	v'' 0,007083	h'' 2493,1	s'' 5,0741
°C	v	h	s	v	h	s	v	h	s	v	h	s
0	0,0009919	17,1	0,0008	0,0009916	17,6	0,0008	0,0009914	18,1	0,0008	0,0009911	18,6	0,0008
10	0,0009924	58,4	0,1493	0,0009922	58,9	0,1492	0,0009919	59,4	0,1491	0,0009917	59,9	0,1491
20	0,0009942	99,7	0,2926	0,0009939	100,2	0,2925	0,0009937	100,7	0,2924	0,0009935	101,1	0,2923
30	0,0009969	141,1	0,4312	0,0009967	141,5	0,4311	0,0009965	142,0	0,4309	0,0009963	142,4	0,4307
40	0,0010004	182,4	0,5655	0,0010002	182,9	0,5653	0,0010000	183,3	0,5651	0,0009998	183,7	0,5649
50	0,0010047	223,8	0,6956	0,0010045	224,3	0,6954	0,0010043	224,7	0,6952	0,0010041	225,1	0,6950
60	0,0010096	265,3	0,8220	0,0010094	265,7	0,8217	0,0010092	266,1	0,8215	0,0010090	266,6	0,8212
70	0,0010151	306,8	0,9448	0,0010149	307,2	0,9445	0,0010147	307,6	0,9442	0,0010145	308,0	0,9439
80	0,0010212	348,4	1,0642	0,0010210	348,8	1,0639	0,0010208	349,2	1,0636	0,0010206	349,6	1,0632
90	0,0010279	390,0	1,1805	0,0010277	390,4	1,1802	0,0010274	390,8	1,1798	0,0010272	391,2	1,1795
100	0,0010351	431,8	1,2939	0,0010349	432,2	1,2935	0,0010346	432,5	1,2931	0,0010344	432,9	1,2928
110	0,0010429	473,6	1,4046	0,0010427	474,0	1,4042	0,0010424	474,4	1,4038	0,0010422	474,7	1,4034
120	0,0010513	515,6	1,5127	0,0010510	516,0	1,5123	0,0010507	516,3	1,5118	0,0010505	516,7	1,5114
130	0,0010602	557,7	1,6185	0,0010599	558,1	1,6181	0,0010596	558,4	1,6176	0,0010593	558,8	1,6171
140	0,0010697	600,0	1,7222	0,0010694	600,4	1,7217	0,0010691	600,7	1,7212	0,0010688	601,0	1,7207
150	0,0010798	642,5	1,8238	0,0010795	642,9	1,8233	0,0010792	643,2	1,8227	0,0010788	643,5	1,8222
160	0,0010906	685,3	1,9236	0,0010902	685,6	1,9230	0,0010899	685,9	1,9225	0,0010896	686,2	1,9219
170	0,0011021	728,2	2,0217	0,0011017	728,5	2,0211	0,0011014	728,8	2,0205	0,0011010	729,1	2,0199
180	0,0011143	771,5	2,1182	0,0011139	771,8	2,1176	0,0011136	772,0	2,1170	0,0011132	772,3	2,1163
190	0,0011274	815,1	2,2134	0,0011270	815,3	2,2127	0,0011266	815,6	2,2120	0,0011262	815,8	2,2114
200	0,0011414	859,0	2,3073	0,0011410	859,3	2,3066	0,0011405	859,5	2,3058	0,0011400	859,7	2,3051
210	0,0011564	903,4	2,4001	0,0011559	903,6	2,3993	0,0011554	903,8	2,3985	0,0011549	904,0	2,3977
220	0,0011725	948,3	2,4919	0,0011720	948,4	2,4911	0,0011714	948,6	2,4903	0,0011709	948,8	2,4894
230	0,0011899	993,6	2,5830	0,0011893	993,8	2,5821	0,0011887	993,9	2,5812	0,0011881	994,1	2,5803
240	0,0012088	1039,6	2,6735	0,0012081	1039,7	2,6726	0,0012074	1039,8	2,6716	0,0012067	1039,9	2,6706
250	0,0012293	1086,4	2,7637	0,0012285	1086,4	2,7627	0,0012278	1086,5	2,7616	0,0012270	1086,5	2,7605
260	0,0012517	1133,9	2,8538	0,0012509	1133,9	2,8526	0,0012500	1133,9	2,8514	0,0012491	1133,9	2,8503
270	0,0012765	1182,5	2,9440	0,0012755	1182,4	2,9427	0,0012745	1182,4	2,9414	0,0012735	1182,3	2,9401
280	0,0013041	1232,3	3,0348	0,0013029	1232,1	3,0334	0,0013018	1232,0	3,0319	0,0013006	1231,8	3,0305
290	0,0013352	1283,5	3,1266	0,0013338	1283,2	3,1249	0,0013324	1283,0	3,1233	0,0013310	1282,7	3,1217
300	0,0013707	1336,5	3,2200	0,0013690	1336,1	3,2181	0,0013673	1335,7	3,2162	0,0013656	1335,4	3,2143
310	0,0014121	1392,0	3,3158	0,0014099	1391,4	3,3136	0,0014077	1390,8	3,3114	0,0014056	1390,2	3,3092
320	0,0014615	1450,5	3,4155	0,0014586	1449,7	3,4128	0,0014558	1448,8	3,4101	0,0014531	1448,0	3,4075
330	0,0015229	1513,7	3,5209	0,0015189	1512,3	3,5175	0,0015150	1511,1	3,5141	0,0015112	1509,8	3,5108
340	0,0016042	1583,8	3,6362	0,0015980	1581,7	3,6315	0,0015920	1579,7	3,6269	0,0015863	1577,8	3,6225
350	0,0017283	1667,7	3,7720	0,0017158	1663,6	3,7640	0,0017043	1659,8	3,7566	0,0016938	1656,3	3,7496
360	0,009584	2652,4	5,3458	0,008864	2614,8	5,2791	0,008104	2569,1	5,2002	0,007267	251c,8	5,1020
370	0,01069	2740,7	5,4842	0,01005	2713,5	5,4339	0,009430	2684,2	5,3808	0,008817	2652,5	5,3244
380	0,01159	2811,0	5,5928	0,01099	2789,4	5,5510	0,01040	2766,6	5,5079	0,009844	2742,4	5,4632
390	0,01235	2870,2	5,6827	0,01177	2852,2	5,6464	0,01121	2833,4	5,6095	0,01068	2813,9	5,5717
400	0,01303	2921,7	5,7599	0,01246	2906,3	5,7274	0,01191	2890,3	5,6947	0,01139	2873,8	5,6615
410	0,01365	2968,0	5,8281	0,01308	2954,4	5,7984	0,01254	2940,4	5,7686	0,01202	2926,1	5,7386
420	0,01423	3010,5	5,8899	0,01366	2998,3	5,8622	0,01311	2985,8	5,8345	0,01260	2973,0	5,8068
430	0,01476	3050,1	5,9466	0,01419	3039,0	5,9205	0,01365	3027,6	5,8945	0,01314	3016,1	5,8685
440	0,01527	3087,5	5,9993	0,01470	3077,2	5,9745	0,01416	3066,9	5,9498	0,01364	3056,3	5,9253
450	0,01576	3123,1	6,0489	0,01518	3113,6	6,0251	0,01464	3104,0	6,0015	0,01412	3094,3	5,9781
460	0,01623	3157,2	6,0958	0,01565	3148,4	6,0729	0,01510	3139,4	6,0502	0,01457	3130,4	6,0278
470	0,01668	3190,2	6,1404	0,01610	3181,9	6,1183	0,01554	3173,5	6,0964	0,01501	3165,1	6,0748
480	0,01712	3222,1	6,1831	0,01653	3214,3	6,1617	0,01597	3206,5	6,1405	0,01543	3198,6	6,1195
490	0,01755	3253,1	6,2241	0,01695	3245,8	6,2032	0,01638	3238,4	6,1826	0,01584	3231,0	6,1623
500	0,01797	3283,5	6,2636	0,01736	3276,5	6,2432	0,01678	3269,6	6,2232	0,01624	3262,5	6,2034
510	0,01837	3313,1	6,3017	0,01776	3306,6	6,2818	0,01718	3300,0	6,2622	0,01663	3293,3	6,2429
520	0,01877	3342,3	6,3387	0,01851	3336,0	6,3192	0,01756	3329,8	6,3000	0,01701	3323,4	6,2812
530	0,01916	3370,9	6,3746	0,01853	3365,0	6,3554	0,01794	3359,0	6,3366	0,01738	3353,0	6,3182
540	0,01954	3399,1	6,4095	0,01891	3393,5	6,3907	0,01831	3387,8	6,3722	0,01774	3382,0	6,3541
550	0,01992	3427,0	6,4435	0,01928	3421,6	6,4250	0,01867	3416,1	6,4069	0,01810	3410,6	6,3891

Table 3. Water and Superheated Steam (Continuation) Wasser und überhitzter Dampf (Fortsetzung)

t	170 bar $t_s = 352,26\,°C$			175 bar $t_s = 354,64\,°C$			180 bar $t_s = 356,96\,°C$			185 bar $t_s = 359,22\,°C$		
	v'' 0,008371	h'' 2551,6	s'' 5,1855	v'' 0,007926	h'' 2533,3	s'' 5,1498	v'' 0,007498	h'' 2513,9	s'' 5,1128	v'' 0,007083	h'' 2493,1	s'' 5,0741
°C	v	h	s	v	h	s	v	h	s	v	h	s
550	0,01992	3427,0	6,4435	0,01928	3421,6	6,4250	0,01867	3416,1	6,4069	0,01810	3410,6	6,3891
560	0,02029	3454,5	6,4768	0,01964	3449,3	6,4586	0,01903	3444,1	6,4407	0,01845	3438,9	6,4232
570	0,02065	3481,8	6,5093	0,02000	3476,8	6,4913	0,01938	3471,8	6,4737	0,01879	3466,8	6,4565
580	0,02101	3508,8	6,5411	0,02035	3504,0	6,5234	0,01972	3499,2	6,5061	0,01913	3494,4	6,4891
590	0,02137	3535,6	6,5724	0,02070	3531,0	6,5549	0,02007	3526,4	6,5378	0,01947	3521,8	6,5210
600	0,02172	3562,2	6,6031	0,02104	3557,8	6,5858	0,02040	3553,4	6,5688	0,01980	3548,9	6,5522
610	0,02207	3588,7	6,6332	0,02138	3584,5	6,6161	0,02074	3580,2	6,5994	0,02012	3575,9	6,5829
620	0,02241	3615,1	6,6629	0,02172	3611,0	6,6460	0,02107	3606,8	6,6294	0,02045	3602,7	6,6131
630	0,02275	3641,3	6,6921	0,02205	3637,3	6,6753	0,02139	3633,4	6,6589	0,02077	3629,4	6,6428
640	0,02309	3667,4	6,7209	0,02238	3663,6	6,7042	0,02172	3659,8	6,6880	0,02108	3655,9	6,6720
650	0,02343	3693,5	6,7493	0,02271	3689,8	6,7328	0,02204	3686,1	6,7166	0,02140	3682,4	6,7008
660	0,02376	3719,5	6,7773	0,02304	3715,9	6,7609	0,02235	3712,3	6,7449	0,02171	3708,7	6,7292
670	0,02409	2745,4	6,8049	0,02336	3741,6	6,7886	0,02267	3738,4	6,7727	0,02202	3734,9	6,7572
680	0,02442	3771,2	6,8322	0,02368	3767,9	6,8160	0,02298	3764,5	6,8002	0,02232	3761,1	6,7848
690	0,02474	3797,1	6,8591	0,02400	3793,8	6,8431	0,02329	3790,5	6,8274	0,02263	3787,3	6,8121
700	0,02507	3822,8	6,8857	0,02431	3819,7	6,8698	0,02360	3816,5	6,8542	0,02293	3813,3	6,8390
710	0,02539	3848,6	6,9120	0,02463	3845,5	6,8962	0,02391	3842,4	6,8807	0,02323	3839,4	6,8656
720	0,02571	3874,3	6,9380	0,02494	3871,3	6,9223	0,02421	3868,3	6,9069	0,02353	3865,3	6,8919
730	0,02603	3899,9	6,9638	0,02525	3897,1	6,9481	0,02452	3894,2	6,9328	0,02382	3891,3	6,9179
740	0,02634	3925,6	6,9892	0,02556	3922,8	6,9736	0,02482	3920,0	6,9585	0,02412	3917,2	6,9436
750	0,02666	3951,2	7,0144	0,02587	3948,5	6,9989	0,02512	3945,8	6,9838	0,02441	3943,1	6,9691
760	0,02697	3976,9	7,0393	0,02617	3974,2	7,0239	0,02541	3971,6	7,0089	0,02470	3969,0	6,9942
770	0,02728	4002,5	7,0640	0,02647	3999,9	7,0487	0,02571	3997,4	7,0337	0,02499	3994,8	7,0191
780	0,02759	4028,1	7,0884	0,02678	4025,6	7,0732	0,02601	4023,1	7,0583	0,02528	4020,7	7,0438
790	0,02790	4053,7	7,1126	0,02708	4051,3	7,0974	0,02630	4048,9	7,0826	0,02556	4046,5	7,0682
800	0,02821	4079,3	7,1366	0,02738	4077,0	7,1215	0,02659	4074,6	7,1067	0,02585	4072,3	7,0923

Table 3. Water and Superheated Steam (Continuation) Wasser und überhitzter Dampf (Fortsetzung)

	190 bar $t_s = 361{,}43$ °C			195 bar $t_s = 363{,}59$ °C			200 bar $t_s = 365{,}70$ °C			210 bar $t_s = 369{,}78$ °C		
t	v'' 0,006677	h'' 2470,6	s'' 5,0332	v'' 0,006278	h'' 2446,0	s'' 4,9893	v'' 0,005877	h'' 2418,4	s'' 4,9412	v'' 0,005023	h'' 2347,6	s'' 4,8223
°C	v	h	s	v	h	s	v	h	s	v	h	s
0	0,0009909	19,1	0,0008	0,0009907	19,6	0,0008	0,0009904	20,1	0,0008	0,0009900	21,1	0,0009
10	0,0009915	60,4	0,1490	0,0009913	60,8	0,1490	0,0009910	61,3	0,1489	0,0009906	62,3	0,1488
20	0,0009933	101,6	0,2921	0,0009931	102,1	0,2920	0,0009929	102,5	0,2919	0,0009924	103,4	0,2917
30	0,0009960	142,9	0,4306	0,0009958	143,3	0,4304	0,0009956	143,8	0,4303	0,0009952	144,7	0,4299
40	0,0009996	184,2	0,5647	0,0009994	184,6	0,5645	0,0009992	185,1	0,5643	0,0009988	185,9	0,5639
50	0,0010038	225,6	0,6947	0,0010036	226,0	0,6945	0,0010034	226,4	0,6943	0,0010030	227,3	0,6938
60	0,0010087	267,0	0,8209	0,0010085	267,4	0,8207	0,0010083	267,8	0,8204	0,0010079	268,6	0,8199
70	0,0010142	308,4	0,9436	0,0010140	308,9	0,9433	0,0010138	309,3	0,9430	0,0010134	310,1	0,9424
80	0,0010203	350,0	1,0629	0,0010201	350,4	1,0626	0,0010199	350,8	1,0623	0,0010194	351,6	1,0616
90	0,0010270	391,6	1,1791	0,0010267	392,0	1,1788	0,0010265	392,4	1,1784	0,0010260	393,1	1,1777
100	0,0010342	433,3	1,2924	0,0010339	433,7	1,2920	0,0010337	434,0	1,2916	0,0010332	434,8	1,2909
110	0,0010419	475,1	1,4030	0,0010416	475,5	1,4026	0,0010414	475,8	1,4022	0,0010409	476,6	1,4014
120	0,0010502	517,0	1,5110	0,0010499	517,4	1,5106	0,0010497	517,7	1,5101	0,0010491	518,5	1,5093
130	0,0010590	559,1	1,6167	0,0010588	559,5	1,6162	0,0010585	559,8	1,6158	0,0010579	560,5	1,6149
140	0,0010685	601,4	1,7202	0,0010682	601,7	1,7197	0,0010679	602,0	1,7192	0,0010673	602,7	1,7183
150	0,0010785	643,8	1,8217	0,0010782	644,1	1,8212	0,0010779	644,5	1,8207	0,0010773	645,1	1,8196
160	0,0010892	686,5	1,9214	0,0010889	686,8	1,9208	0,0010886	687,1	1,9203	0,0010879	687,7	1,9192
170	0,0011006	729,4	2,0193	0,0011003	729,7	2,0187	0,0010999	730,0	2,0181	0,0010992	730,6	2,0170
180	0,0011128	772,6	2,1157	0,0011124	772,9	2,1151	0,0011120	773,1	2,1145	0,0011113	773,7	2,1132
190	0,0011257	816,1	2,2107	0,0011253	816,4	2,2100	0,0011249	816,6	2,2093	0,0011241	817,1	2,2080
200	0,0011396	860,0	2,3044	0,0011391	860,2	2,3037	0,0011387	860,4	2,3030	0,0011378	860,9	2,3016
210	0,0011544	904,2	2,3970	0,0011539	904,4	2,3962	0,0011534	904,6	2,3954	0,0011525	905,1	2,3939
220	0,0011704	949,0	2,4886	0,0011698	949,1	2,4878	0,0011693	949,3	2,4870	0,0011682	949,7	2,4853
230	0,0011875	994,2	2,5794	0,0011869	994,3	2,5785	0,0011863	994,5	2,5776	0,0011851	994,8	2,5759
240	0,0012061	1040,1	2,6696	0,0012054	1040,2	2,6687	0,0012047	1040,3	2,6677	0,0012034	1040,5	2,6658
250	0,0012262	1086,6	2,7595	0,0012255	1086,7	2,7584	0,0012247	1086,7	2,7574	0,0012233	1086,9	2,7553
260	0,0012483	1133,9	2,8491	0,0012474	1133,9	2,8480	0,0012466	1134,0	2,8468	0,0012449	1134,0	2,8445
270	0,0012725	1182,2	2,9389	0,0012715	1182,2	2,9376	0,0012706	1182,1	2,9363	0,0012686	1182,0	2,9338
280	0,0012994	1231,7	3,0291	0,0012983	1231,5	3,0276	0,0012971	1231,4	3,0262	0,0012949	1231,2	3,0234
290	0,0013296	1282,5	3,1201	0,0013282	1282,2	3,1185	0,0013269	1282,0	3,1169	0,0013242	1281,6	3,1137
300	0,0013639	1335,0	3,2125	0,0013622	1334,6	3,2107	0,0013606	1334,3	3,2088	0,0013574	1333,6	3,2053
310	0,0014035	1389,7	3,3071	0,0014015	1389,1	3,3049	0,0013994	1388,6	3,3028	0,0013955	1387,6	3,2987
320	0,0014504	1447,2	3,4049	0,0014477	1446,4	3,4023	0,0014451	1445,6	3,3998	0,0014400	1444,1	3,3949
330	0,0015075	1508,7	3,5076	0,0015039	1507,5	3,5044	0,0015004	1506,4	3,5013	0,0014936	1504,2	3,4953
340	0,0015808	1575,9	3,6182	0,0015755	1574,2	3,6140	0,0015704	1572,5	3,6100	0,0015607	1569,2	3,6022
350	0,0016839	1653,1	3,7430	0,0016748	1650,0	3,7368	0,0016662	1647,2	3,7308	0,0016503	1641,9	3,7197
360	0,001875	1758,0	3,9102	0,001849	1749,8	3,8959	0,001827	1742,9	3,8835	0,001791	1731,4	3,8625
370	0,008210	2617,7	5,2637	0,007575	2576,6	5,1935	0,006908	2527,6	5,1117	0,005142	2362,0	4,8447
380	0,009298	2716,8	5,4166	0,008767	2689,4	5,3678	0,008246	2660,2	5,3165	0,007208	2592,8	5,2015
390	0,01016	2793,4	5,5330	0,009663	2771,9	5,4931	0,009181	2749,3	5,4520	0,008257	2700,5	5,3652
400	0,01089	2856,7	5,6278	0,01041	2838,9	5,5935	0,009947	2820,5	5,5585	0,009071	2781,3	5,4863
410	0,01153	2911,3	5,7083	0,01106	2896,1	5,6778	0,01061	2880,4	5,6470	0,009758	2847,6	5,5840
420	0,01211	2960,0	5,7791	0,01164	2946,6	5,7512	0,01120	2932,9	5,7232	0,01036	2904,5	5,6667
430	0,01265	3004,4	5,8427	0,01218	2992,4	5,8168	0,01174	2980,2	5,7910	0,01090	2955,0	5,7391
440	0,01315	3045,6	5,9009	0,01268	3034,7	5,8766	0,01224	3023,7	5,8523	0,01141	3001,0	5,8040
450	0,01362	3084,4	5,9549	0,01315	3074,4	5,9319	0,01271	3064,3	5,9089	0,01187	3043,6	5,8633
460	0,01408	3121,3	6,0056	0,01360	3112,0	5,9835	0,01315	3102,7	5,9616	0,01232	3083,6	5,9182
470	0,01451	3156,6	6,0534	0,01403	3148,0	6,0322	0,01358	3139,2	6,0112	0,01274	3121,5	5,9697
480	0,01493	3190,6	6,0988	0,01445	3182,5	6,0784	0,01399	3174,4	6,0581	0,01314	3157,8	6,0182
490	0,01533	3223,5	6,1423	0,01485	3215,9	6,1224	0,01439	3208,3	6,1028	0,01353	3192,8	6,0643
500	0,01573	3255,4	6,1839	0,01524	3248,3	6,1646	0,01477	3241,1	6,1456	0,01391	3226,5	6,1082
510	0,01611	3286,6	6,2239	0,01561	3279,9	6,2052	0,01514	3273,1	6,1867	0,01427	3259,3	6,1503
520	0,01648	3317,1	6,2626	0,01598	3310,7	6,2443	0,01551	3304,2	6,2262	0,01463	3291,2	6,1908
530	0,01685	3346,9	6,3000	0,01634	3340,9	6,2821	0,01586	3334,7	6,2644	0,01497	3322,4	6,2299
540	0,01720	3376,3	6,3363	0,01669	3370,5	6,3188	0,01621	3364,7	6,3015	0,01531	3352,9	6,2677
550	0,01755	3405,2	6,3716	0,01704	3399,6	6,3544	0,01655	3394,1	6,3374	0,01564	3382,9	6,3043

Table 3. Water and Superheated Steam (Continuation) Wasser und überhitzter Dampf (Fortsetzung)

	190 bar $t_s = 361{,}43\,°C$			195 bar $t_s = 363{,}59\,°C$			200 bar $t_s = 365{,}70\,°C$			210 bar $t_s = 369{,}78\,°C$		
t	v'' 0,006677	h'' 2470,6	s'' 5,0332	v'' 0,006278	h'' 2446,0	s'' 4,9893	v'' 0,005877	h'' 2418,4	s'' 4,9412	v'' 0,005023	h'' 2347,6	s'' 4,8223
°C	v	h	s	v	h	s	v	h	s	v	h	s
550	0,01755	3405,2	6,3716	0,01704	3399,6	6,3544	0,01655	3394,1	6,3374	0,01564	3382,9	6,3043
560	0,01790	3433,6	6,4060	0,01738	3428,3	6,3891	0,01688	3423,0	6,3724	0,01596	3412,4	6,3399
570	0,01824	3461,8	6,4395	0,01771	3456,7	6,4229	0,01721	3451,6	6,4065	0,01628	3441,4	6,3745
580	0,01857	3489,6	6,4723	0,01804	3484,7	6,4559	0,01753	3479,9	6,4398	0,01659	3470,1	6,4084
590	0,01890	3517,1	6,5045	0,01836	3512,5	6,4883	0,01785	3507,8	6,4724	0,01690	3498 4	6,4414
600	0,01922	3544,5	6,5360	0,01868	3540,0	6,5200	0,01816	3535,5	6,5043	0,01720	3526,5	6,4737
610	0,01954	3571,6	6,5669	0,01899	3567,3	6,5511	0,01847	3563,0	6,5356	0,01750	3554,3	6,5054
620	0,01986	3598,6	6,5972	0,01930	3594,4	6,5816	0,01878	3590,3	6,5663	0,01779	3581,9	6,5365
630	0,02017	3625,4	6,6271	0,01961	3621,4	6,6116	0,01908	3617,4	6,5965	0,01809	3609,3	6,5670
640	0,02048	3652,1	6,6564	0,01992	3648,2	6,6411	0,01938	3644,3	6,6261	0,01837	3636,6	6,5970
650	0,02079	3678,6	6,6854	0,02022	3674,9	6,6702	0,01967	3671,1	6,6554	0,01866	3663,6	6,6265
660	0,02110	3705,1	6,7139	0,02052	3701,5	6,6989	0,01996	3697,9	6,6841	0,01894	3690,6	6,6555
670	0,02140	3731,5	6,7420	0,02081	3728,0	6,7271	0,02025	3724,5	6,7125	0,01922	3717,5	6,6841
680	0,02170	3757,8	6,7697	0,02111	3754,4	6,7550	0,02054	3751,0	6,7405	0,01950	3744,2	6,7124
690	0,02200	3784,0	6,7971	0,02140	3780,7	6,7824	0,02083	3777,4	6,7681	0,01977	3770,9	6,7402
700	0,02229	3810,2	6,8241	0,02169	3807,0	6,8096	0,02111	3803,8	6,7953	0,02004	3797,5	6,7677
710	0,02258	3836,3	6,8508	0,02197	3833,2	6,8364	0,02139	3830,1	6,8222	0,02031	3824,0	6,7948
720	0,02288	3862,4	6,8772	0,02226	3859,4	6,8629	0,02167	3856,4	6,8488	0,02058	3850,4	6,8215
730	0,02317	3888,4	6,9033	0,02254	3885,5	6,8891	0,02195	3882,6	6,8751	0,02085	3876,8	6,8480
740	0,02345	3914,4	6,9291	0,02282	3911,6	6,9149	0,02222	3908,8	6,9011	0,02111	3903,2	6,8741
750	0,02374	3940,4	6,9546	0,02310	3937,7	6,9405	0,02250	3935,0	6,9267	0,02138	3929,5	6,9000
760	0,02402	3966,3	6,9799	0,02338	3963,7	6,9659	0,02277	3961,1	6,9521	0,02164	3955,8	6,9255
770	0,02431	3992,3	7,0049	0,02366	3989,7	6,9909	0,02304	3987,2	6,9773	0,02190	3982,0	6,9508
780	0,02459	4018,2	7,0296	0,02393	4015,7	7,0157	0,02331	4013,2	7,0021	0,02216	4008,2	6,9758
790	0,02487	4044,1	7,0540	0,02421	4041,7	7,0402	0,02358	4039,3	7,0267	0,02241	4034,4	7,0006
800	0,02515	4070,0	7,0783	0,02448	4067,6	7,0645	0,02385	4065,3	7,0511	0,02267	4060,6	7,0251

Table 3. Water and Superheated Steam (Continuation) Wasser und überhitzter Dampf (Fortsetzung)

t	220 bar $t_s = 373{,}69$ °C			230 bar			240 bar			250 bar		
	v'' 0,003728	h'' 2195,6	s'' 4,5799									
°C	v	h	s	v	h	s	v	h	s	v	h	s
0	0,0009895	22,1	0,0009	0,0009890	23,1	0,0009	0,0009885	24,1	0,0009	0,0009881	25,1	0,0009
10	0,0009901	63,2	0,1486	0,0009897	64,2	0,1485	0,0009893	65,1	0,1484	0,0009888	66,1	0,1482
20	0,0009920	104,4	0,2914	0,0009916	105,3	0,2912	0,0009911	106,2	0,2910	0,0009907	107,1	0,2907
30	0,0009948	145,6	0,4296	0,0009944	146,5	0,4293	0,0009939	147,4	0,4290	0,0009935	148,3	0,4287
40	0,0009983	186,8	0,5635	0,0009979	187,7	0,5631	0,0009975	188,6	0,5627	0,0009971	189,4	0,5623
50	0,0010026	228,1	0,6933	0,0010022	229,0	0,6929	0,0010018	229,8	0,6924	0,0010013	230,7	0,6920
60	0,0010075	269,5	0,8194	0,0010070	270,3	0,8189	0,0010066	271,1	0,8184	0,0010062	272,0	0,8178
70	0,0010129	310,9	0,9419	0,0010125	311,7	0,9413	0,0010121	312,5	0,9407	0,0010116	313,3	0,9401
80	0,0010190	352,4	1,0610	0,0010185	353,2	1,0604	0,0010181	354,0	1,0598	0,0010177	354,8	1,0591
90	0,0010256	393,9	1,1770	0,0010251	394,7	1,1764	0,0010247	395,5	1,1757	0,0010242	396,2	1,1750
100	0,0010327	435,6	1,2902	0,0010322	436,3	1,2894	0,0010318	437,1	1,2887	0,0010313	437,8	1,2879
110	0,0010404	477,3	1,4006	0,0010399	478,0	1,3998	0,0010394	478,8	1,3990	0,0010389	479,5	1,3982
120	0,0010486	519,2	1,5084	0,0010481	519,9	1,5076	0,0010476	520,6	1,5067	0,0010470	521,3	1,5059
130	0,0010574	561,2	1,6140	0,0010568	561,9	1,6131	0,0010563	562,6	1,6121	0,0010557	563,3	1,6113
140	0,0010667	603,4	1,7173	0,0010661	604,0	1,7163	0,0010656	604,7	1,7154	0,0010650	605,4	1,7144
150	0,0010767	645,7	1,8186	0,0010760	646,4	1,8176	0,0010754	647,0	1,8166	0,0010748	647,7	1,8155
160	0,0010872	688,2	1,9181	0,0010866	688,9	1,9170	0,0010859	689,5	1,9159	0,0010853	690,2	1,9148
170	0,0010985	731,1	2,0158	0,0010978	731,7	2,0147	0,0010971	732,3	2,0135	0,0010964	732,9	2,0123
180	0,0011105	774,2	2,1120	0,0011097	774,8	2,1107	0,0011090	775,3	2,1095	0,0011083	775,9	2,1083
190	0,0011233	817,6	2,2067	0,0011225	818,1	2,2054	0,0011217	818,7	2,2041	0,0011209	819,2	2,2028
200	0,0011369	861,4	2,3001	0,0011360	861,8	2,2987	0,0011352	862,3	2,2973	0,0011343	862,8	2,2960
210	0,0011515	905,5	2,3924	0,0011505	905,9	2,3909	0,0011496	906,3	2,3894	0,0011487	906,8	2,3879
220	0,0011671	950,0	2,4837	0,0011661	950,4	2,4821	0,0011651	950,8	2,4805	0,0011640	951,2	2,4789
230	0,0011840	995,1	2,5741	0,0011828	995,4	2,5724	0,0011817	995,7	2,5707	0,0011805	996,0	2,5689
240	0,0012021	1040,7	2,6639	0,0012008	1041,0	2,6620	0,0011996	1041,2	2,6602	0,0011983	1041,5	2,6583
250	0,0012218	1087,0	2,7532	0,0012204	1087,2	2,7512	0,0012189	1087,3	2,7492	0,0012175	1087,5	2,7472
260	0,0012432	1134,0	2,8423	0,0012416	1134,1	2,8401	0,0012400	1134,1	2,8379	0,0012384	1134,2	2,8357
270	0,0012667	1182,0	2,9313	0,0012649	1181,9	2,9289	0,0012630	1181,8	2,9265	0,0012612	1181,8	2,9241
280	0,0012927	1230,9	3,0207	0,0012905	1230,7	3,0180	0,0012884	1230,5	3,0153	0,0012863	1230,3	3,0126
290	0,0013216	1281,2	3,1107	0,0013191	1280,8	3,1076	0,0013166	1280,4	3,1046	0,0013141	1280,0	3,1017
300	0,0013543	1332,9	3,2018	0,0013512	1332,3	3,1983	0,0013482	1331,7	3,1949	0,0013453	1331,1	3,1916
310	0,0013916	1386,6	3,2946	0,0013879	1385,7	3,2907	0,0013842	1384,8	3,2868	0,0013807	1383,9	3,2829
320	0,0014351	1442,7	3,3901	0,0014304	1441,4	3,3854	0,0014259	1440,1	3,3809	0,0014214	1438,9	3,3764
330	0,0014872	1502,2	3,4895	0,0014810	1500,3	3,4838	0,0014751	1498,5	3,4783	0,0014694	1496,7	3,4730
340	0,0015516	1566,2	3,5947	0,0015431	1563,4	3,5876	0,0015350	1560,8	3,5808	0,0015273	1558,3	3,5743
350	0,0016361	1637,2	3,7095	0,0016231	1632,8	3,6999	0,0016111	1628,8	3,6909	0,0016000	1625,1	3,6824
360	0,001762	1722,0	3,8449	0,001738	1714,1	3,8296	0,001717	1707,2	3,8159	0,001698	1701,1	3,8036
370	0,002033	1842,3	4,0333	0,001949	1818,1	3,9926	0,001894	1801,6	3,9638	0,001852	1788,8	3,9411
380	0,006111	2504,4	5,0559	0,004747	2362,5	4,8302	0,002692	2044,0	4,3370	0,002240	1941,0	4,1757
390	0,007377	2645,9	5,2711	0,006506	2581,3	5,1632	0,005603	2500,5	5,0322	0,004609	2391,3	4,8599
400	0,008251	2738,8	5,4102	0,007476	2692,3	5,3294	0,006739	2641,2	5,2430	0,006014	2582,0	5,1455
410	0,008969	2812,6	5,5190	0,008232	2775,0	5,4515	0,007540	2734,7	5,3809	0,006887	2691,3	5,3069
420	0,009588	2874,6	5,6091	0,008872	2843,0	5,5502	0,008205	2809,5	5,4896	0,007580	2774,1	5,4271
430	0,01014	2928,8	5,6868	0,009435	2901,3	5,6338	0,008781	2872,6	5,5801	0,008172	2842,5	5,5252
440	0,01064	2977,5	5,7556	0,009944	2953,2	5,7070	0,009297	2927,9	5,6581	0,008696	2901,7	5,6087
450	0,01111	3022,3	5,8179	0,01041	3000,3	5,7727	0,009768	2977,7	5,7275	0,009171	2954,3	5,6821
460	0,01155	3064,0	5,8753	0,01085	3044,0	5,8327	0,01021	3023,4	5,7903	0,009609	3002,3	5,7479
470	0,01197	3103,4	5,9287	0,01126	3085,0	5,8882	0,01062	3066,1	5,8481	0,01002	3046,7	5,8082
480	0,01237	3141,0	5,9789	0,01166	3123,8	5,9402	0,01101	3106,3	5,9019	0,01041	3088,5	5,8640
490	0,01275	3177,0	6,0264	0,01204	3161,0	5,9892	0,01138	3144,7	5,9525	0,01078	3128,1	5,9162
500	0,01312	3211,7	6,0716	0,01240	3196,7	6,0357	0,01174	3181,4	6,0003	0,01113	3165,9	5,9655
510	0,01348	3245,3	6,1148	0,01275	3231,2	6,0800	0,01208	3216,9	6,0458	0,01147	3202,3	6,0122
520	0,01382	3278,0	6,1563	0,01309	3264,7	6,1225	0,01242	3251,1	6,0893	0,01180	3237,5	6,0568
530	0,01416	3309,9	6,1962	0,01342	3297,2	6,1633	0,01274	3284,5	6,1311	0,01211	3271,5	6,0995
540	0,01449	3341,0	6,2347	0,01374	3329,0	6,2026	0,01305	3316,9	6,1713	0,01242	3304,7	6,1405
550	0,01481	3371,6	6,2721	0,01405	3360,2	6,2407	0,01336	3348,7	6,2101	0,01272	3337,0	6,1801

Table 3. Water and Superheated Steam (Continuation) Wasser und überhitzter Dampf (Fortsetzung)

t	220 bar $t_s = 373{,}69\ °C$			230 bar			240 bar			250 bar		
	v'' 0,003728	h'' 2195,6	s'' 4,5799									
°C	v	h	s	v	h	s	v	h	s	v	h	s
550	0,01481	3371,6	6,2721	0,01405	3360,2	6,2407	0,01336	3348,7	6,2101	0,01272	3337,0	6,1801
560	0,01512	3401,6	6,3083	0,01436	3390,7	6,2776	0,01366	3379,7	6,2476	0,01301	3368,7	6,2183
570	0,01543	3431,1	6,3435	0,01466	3420,7	6,3134	0,01395	3410,3	6,2840	0,01330	3399,7	6,2553
580	0,01574	3460,2	6,3779	0,01496	3450,3	6,3482	0,01424	3440,3	6,3194	0,01358	3430,2	6,2913
590	0,01603	3489,0	6,4114	0,01524	3479,5	6,3822	0,01452	3469,9	6,3539	0,01386	3460,3	6,3263
600	0,01633	3517,4	6,4441	0,01553	3508,3	6,4154	0,01480	3499,1	6,3876	0,01413	3489,9	6,3604
610	0,01662	3545,6	6,4762	0,01581	3536,8	6,4479	0,01507	3528,0	6,4205	0,01439	3519,1	6,3938
620	0,01690	3573,5	6,5076	0,01609	3565,1	6,4797	0,01534	3556,6	6,4527	0,01465	3548,1	6,4263
630	0,01718	3601,2	6,5385	0,01636	3593,1	6,5109	0,01561	3584,9	6,4842	0,01491	3576,7	6,4583
640	0,01746	3628,7	6,5688	0,01663	3620,9	6,5416	0,01587	3613,1	6,5152	0,01517	3605,2	6,4896
650	0,01774	3656,1	6,5986	0,01690	3648,6	6,5717	0,01613	3641,0	6,5456	0,01542	3633,4	6,5203
660	0,01801	3683,3	6,6279	0,01716	3676,0	6,6013	0,01638	3668,7	6,5755	0,01566	3661,4	6,5504
670	0,01828	3710,4	6,6568	0,01742	3703,4	6,6304	0,01663	3696,3	6,6049	0,01591	3689,2	6,5801
680	0,01855	3737,4	6,6853	0,01768	3730,6	6,6591	0,01688	3723,8	6,6338	0,01615	3716,9	6,6093
690	0,01881	3764,3	6,7133	0,01793	3757,7	6,6874	0,01713	3751,1	6,6624	0,01639	3744,5	6,6381
700	0,01907	3791,1	6,7410	0,01819	3784,7	6,7153	0,01738	3778,3	6,6905	0,01663	3771,9	6,6664
710	0,01934	3817,8	6,7683	0,01844	3811,6	6,7428	0,01762	3805,4	6,7182	0,01687	3799,2	6,6944
720	0,01959	3844,5	6,7953	0,01869	3838,5	6,7700	0,01786	3832,5	6,7456	0,01710	3826,5	6,7219
730	0,01985	3871,0	6,8219	0,01894	3865,2	6,7968	0,01810	3859,4	6,7726	0,01733	3853,6	6,7491
740	0,02010	3897,6	6,8483	0,01918	3892,0	6,8233	0,01834	3886,3	6,7993	0,01756	3880,7	6,7760
750	0,02036	3924,1	6,8743	0,01943	3918,6	6,8495	0,01857	3913,2	6,8256	0,01779	3907,7	6,8025
760	0,02061	3950,5	6,9000	0,01967	3945,2	6,8754	0,01881	3939,9	6,8517	0,01802	3934,6	6,8287
770	0,02086	3976,9	6,9254	0,01991	3971,8	6,9010	0,01904	3966,7	6,8774	0,01824	3961,5	6,8546
780	0,02111	4003,3	6,9506	0,02015	3998,3	6,9263	0,01927	3993,3	6,9029	0,01846	3988,4	6,8802
790	0,02135	4029,6	6,9755	0,02039	4024,8	6,9513	0,01950	4020,0	6,9280	0,01869	4015,1	6,9055
800	0,02160	4055,9	7,0001	0,02062	4051,2	6,9761	0,01973	4046,6	6,9529	0,01891	4041,9	6,9306

Table 3. Water and Superheated Steam (Continuation) Wasser und überhitzter Dampf (Fortsetzung)

t	260 bar			270 bar			280 bar			290 bar		
°C	v	h	s	v	h	s	v	h	s	v	h	s
0	0,0009876	26,1	0,0009	0,0009871	27,1	0,0009	0,0009867	28,0	0,0008	0,0009862	29,0	0,0008
10	0,0009884	67,0	0,1481	0,0009879	68,0	0,1479	0,0009875	68,9	0,1478	0,0009870	69,9	0,1477
20	0,0009903	108,1	0,2905	0,0009899	109,0	0,2902	0 0009895	109,9	0,2900	0,0009890	110,8	0,2898
30	0,0009931	149,2	0,4284	0,0009927	150,0	0,4280	0,0009923	150,9	0,4277	0,0009919	151,8	0,4274
40	0,0009967	190,3	0,5620	0,0009963	191,2	0,5616	0,0009959	192,1	0,5612	0,0009955	192,9	0,5608
50	0,0010009	231,5	0,6915	0,0010005	232,4	0,6911	0,0010001	233,2	0,6906	0,0009997	234,1	0,6902
60	0,0010058	272,8	0,8173	0,0010054	273,6	0,8168	0,0010049	274,5	0,8163	0,0010045	275,3	0,8158
70	0,0010112	314,1	0,9396	0,0010108	315,0	0,9390	0,0010104	315,8	0,9384	0,0010099	316,6	0,9379
80	0,0010172	355,6	1,0585	0,0010168	356,3	1,0579	0,0010163	357,1	1,0572	0,0010159	357,9	1,0566
90	0,0010237	397,0	1,1743	0,0010233	397,8	1,1736	0,0010228	398,6	1,1730	0,0010224	399,4	1,1723
100	0,0010308	438,6	1,2872	0,0010303	439,3	1,2865	0,0010299	440,1	1,2858	0,0010294	440,9	1,2850
110	0,0010384	480,3	1,3974	0,0010379	481,0	1,3966	0,0010374	481,7	1,3958	0,0010369	482,5	1,3951
120	0,0010465	522,0	1,5051	0,0010460	522,8	1,5042	0,0010455	523,5	1,5034	0,0010450	524,2	1,5025
130	0,0010552	564,0	1,6104	0,0010546	564,7	1,6095	0,0010541	565,4	1,6086	0,0010536	566,1	1,6077
140	0,0010644	606,0	1,7135	0,0010638	606,7	1,7125	0,0010633	607,4	1,7116	0,0010627	608,1	1,7106
150	0,0010742	648,3	1,8145	0,0010736	649,0	1,8135	0,0010730	649,6	1,8125	0,0010724	650,3	1,8115
160	0,0010846	690,8	1,9138	0,0010840	691,4	1,9127	0,0010834	692,0	1,9116	0,0010827	692,6	1,9105
170	0,0010957	733,5	2,0112	0,0010950	734,1	2,0101	0,0010944	734,7	2,0089	0,0010937	735,3	2,0078
180	0,0011075	776,4	2,1071	0,0011068	777,0	2,1059	0,0011060	777,6	2,1046	0,0011053	778,1	2,1034
190	0,0011201	819,7	2,2015	0,0011193	820,2	2,2002	0,0011185	820,7	2,1989	0,0011177	821,3	2,1976
200	0,0011334	863,3	2,2946	0,0011326	863,7	2,2932	0,0011317	864,2	2,2918	0,0011309	864,7	2,2905
210	0,0011477	907,2	2,3864	0,0011468	907,6	2,3850	0,0011459	908,1	2,3835	0,0011450	908,5	2,3820
220	0,0011630	951,5	2,4773	0,0011620	951,9	2,4757	0,0011610	952,3	2,4741	0,0011600	952,7	2,4726
230	0,0011794	996,4	2,5672	0,0011783	996,7	2,5655	0,0011772	997,0	2,5639	0,0011761	997,3	2,5622
240	0,0011970	1041,7	2,6565	0,0011958	1042,0	2,6547	0,0011946	1042,2	2,6528	0,0011934	1042,5	2,6510
250	0,0012161	1087,7	2,7452	0,0012147	1087,8	2,7432	0,0012134	1088,0	2,7412	0,0012120	1088,2	2,7393
260	0,0012368	1134,3	2,8335	0,0012353	1134,4	2,8313	0,0012337	1134,5	2,8292	0,0012322	1134,6	2,8271
270	0,0012594	1181,8	2,9217	0,0012576	1181,7	2,9193	0,0012559	1181,7	2,9170	0,0012542	1181,7	2,9147
280	0,0012842	1230,2	3,0100	0,0012822	1230,0	3,0074	0,0012802	1229,9	3,0049	0,0012782	1229,8	3,0023
290	0,0013117	1279,7	3,0987	0,0013093	1279,4	3,0959	0,0013070	1279,1	3,0930	0,0013047	1278,8	3,0902
300	0,0013424	1330,6	3,1883	0,0013396	1330,1	3,1851	0,0013369	1329,6	3,1819	0,0013342	1329,1	3,1787
310	0,0013772	1383,1	3,2792	0,0013738	1382,4	3,2755	0,0013705	1381,6	3,2719	0,0013673	1380,9	3,2683
320	0,0014171	1437,8	3,3721	0,0014130	1436,6	3,3678	0,0014089	1435,6	3,3637	0,0014050	1434,6	3,3596
330	0,0014639	1495,1	3,4679	0,0014586	1493,5	3,4628	0,0014535	1492,0	3,4580	0,0014486	1490,6	3,4532
340	0,0015200	1555,9	3,5680	0,0015130	1553,7	3,5619	0,0015064	1551,6	3,5560	0,0015000	1549,6	3,5503
350	0,0015896	1621,7	3,6744	0,0015799	1618,5	3,6667	0,0015708	1615,5	3,6593	0,0015622	1612,7	3,6523
360	0,001682	1695,6	3,7923	0,001667	1690,6	3,7818	0,001653	1686,1	3,7720	0,001640	1681,9	3,7628
370	0,001819	1778,4	3,9220	0,001791	1769,6	3,9055	0,001767	1761,9	3,8908	0,001746	1755,1	3,8774
380	0,002097	1903,0	4,1141	0,002013	1879,5	4,0750	0,001955	1862,4	4,0458	0,001910	1848,9	4,0221
390	0,003534	2241,2	4,6274	0,002792	2109,4	4,4239	0,002436	2032,6	4,3041	0,002254	1988,3	4,2338
400	0,005281	2511,7	5,0326	0,004539	2427,5	4,9003	0,003822	2330,7	4,7503	0,003236	2237,0	4,6059
410	0,006267	2644,6	5,2289	0,005665	2592,2	5,1434	0,005069	2532,2	5,0477	0,004492	2465,6	4,9431
420	0,006992	2736,5	5,3624	0,006437	2696,6	5,2952	0,005913	2654,3	5,2253	0,005413	2609,0	5,1517
430	0,007602	2811,0	5,4692	0,007067	2777,9	5,4117	0,006564	2743,2	5,3526	0,006090	2706,9	5,2919
440	0,008136	2874,4	5,5587	0,007612	2846,0	5,5079	0,007122	2816,5	5,4562	0,006661	2785,9	5,4035
450	0,008616	2930,2	5,6364	0,008098	2905,3	5,5904	0,007615	2879,6	5,5440	0,007161	2853,0	5,4970
460	0,009056	2980,6	5,7056	0,008541	2958,3	5,6632	0,008060	2935,4	5,6207	0,007611	2911,9	5,5779
470	0,009465	3027,0	5,7685	0,008950	3006,8	5,7289	0,008471	2986,1	5,6893	0,008022	2964,9	5,6498
480	0,009850	3070,3	5,8264	0,009334	3051,8	5,7890	0,008853	3032,9	5,7519	0,008405	3013,6	5,7148
490	0,01022	3111,2	5,8804	0,009697	3094,1	5,8449	0,009214	3076,6	5,8096	0,008764	3058,9	5,7746
500	0,01056	3150,2	5,9311	0,01004	3134,2	5,8972	0,009557	3118,1	5,8636	0,009104	3101,6	5,8302
510	0,01090	3187,6	5,9792	0,01037	3172,7	5,9465	0,009884	3157,5	5,9143	0,009428	3142,2	5,8824
520	0,01122	3223,6	6,0249	0,01069	3209,6	5,9934	0,01020	3195,4	5,9623	0,009739	3181,1	5,9317
530	0,01153	3258,5	6,0685	0,01100	3245,2	6,0381	0,01050	3231,9	6,0081	0,01004	3218,4	5,9785
540	0,01184	3292,3	6,1104	0,01130	3279,8	6,0809	0,01079	3267,2	6,0518	0,01033	3254,5	6,0231
550	0,01213	3325,3	6,1508	0,01159	3313,5	6,1220	0,01108	3301,5	6,0937	0,01061	3289,5	6,0659

Table 3. Water and Superheated Steam (Continuation) Wasser und überhitzter Dampf (Fortsetzung)

t	260 bar			270 bar			280 bar			290 bar		
°C	v	h	s	v	h	s	v	h	s	v	h	s
550	0,01213	3325,3	6,1508	0,01159	3313,5	6,1220	0,01108	3301,5	6,0937	0,01061	3289,5	6,0659
560	0,01242	3357,5	6,1897	0,01187	3346,3	6,1616	0,01135	3335,0	6,1341	0,01088	3323,6	6,1071
570	0,01270	3389,1	6,2273	0,01214	3378,4	6,1999	0,01162	3367,6	6,1731	0,01114	3356,8	6,1467
580	0,01297	3420,1	6,2639	0,01241	3409,9	6,2370	0,01189	3399,6	6,2108	0,01140	3389,3	6,1850
590	0,01324	3450,6	6,2994	0,01267	3440,8	6,2731	0,01214	3431,0	6,2474	0,01165	3421,1	6,2221
600	0,01350	3480,6	6,3340	0,01293	3471,3	6,3082	0,01240	3461,9	6,2829	0,01190	3452,5	6,2582
610	0,01376	3510,2	6,3677	0,01318	3501,3	6,3424	0,01264	3492,3	6,3176	0,01214	3483,3	6,2933
620	0,01402	3539,5	6,4007	0,01343	3530,9	6,3757	0,01289	3522,3	6,3514	0,01238	3513,7	6,3275
630	0,01427	3568,5	6,4330	0,01368	3560,3	6,4084	0,01313	3552,0	6,3844	0,01261	3543,7	6,3609
640	0,01452	3597,2	6,4646	0,01392	3589,3	6,4404	0,01336	3581,3	6,4167	0,01284	3573,3	6,3936
650	0,01476	3625,7	6,4957	0,01416	3618,1	6,4717	0,01359	3610,4	6,4484	0,01307	3602,7	6,4256
660	0,01500	3654,0	6,5261	0,01439	3646,6	6,5025	0,01382	3639,2	6,4794	0,01329	3631,8	6,4569
670	0,01524	3682,1	6,5561	0,01462	3675,0	6,5327	0,01405	3667,8	6,5099	0,01351	3660,7	6,4877
680	0,01548	3710,0	6,5855	0,01485	3703,2	6,5624	0,01427	3696,3	6,5399	0,01373	3689,4	6,5180
690	0,01571	3737,8	6,6145	0,01508	3731,2	6,5916	0,01449	3724,5	6,5694	0,01395	3717,8	6,5477
700	0,01594	3765,5	6,6431	0,01530	3759,0	6,6204	0,01471	3752,6	6,5984	0,01416	3746,2	6,5769
710	0,01617	3793,0	6,6713	0,01553	3786,8	6,6488	0,01493	3780,6	6,6270	0,01437	3774,3	6,6057
720	0,01640	3820,5	6,6990	0,01575	3814,4	6,6768	0,01514	3808,4	6,6552	0,01458	3802,4	6,6341
730	0,01662	3847,8	6,7264	0,01596	3842,0	6,7044	0,01535	3836,1	6,6829	0,01479	3830,3	6,6621
740	0,01684	3875,1	6,7534	0,01618	3869,4	6,7316	0,01556	3863,8	6,7103	0,01499	3858,1	6,6897
750	0,01707	3902,2	6,7801	0,01640	3896,8	6,7585	0,01577	3891,3	6,7374	0,01519	3885,8	6,7169
760	0,01728	3929,3	6,8065	0,01661	3924,0	6,7850	0,01598	3918,7	6,7641	0,01540	3913,4	6,7438
770	0,01750	3956,4	6,8326	0,01682	3951,3	6,8112	0,01619	3946,1	6,7905	0,01560	3941,0	6,7703
780	0,01772	3983,4	6,8583	0,01703	3978,4	6,8371	0,01639	3973,4	6,8165	0,01579	3968,4	6,7965
790	0,01793	4010,3	6,8838	0,01724	4005,5	6,8627	0,01659	4000,7	6,8423	0,01599	3995,8	6,8224
800	0,01815	4037,2	6,9090	0,01744	4032,5	6,8880	0,01679	4027,8	6,8677	0,01619	4023,2	6,8480

Table 3. Water and Superheated Steam (Continuation) Wasser und überhitzter Dampf (Fortsetzung)

t	300 bar			310 bar			320 bar			330 bar		
°C	v	h	s	v	h	s	v	h	s	v	h	s
0	0,0009857	30,0	0,0008	0,0009853	31,0	0,0008	0,0009848	32,0	0,0008	0,0009843	32,9	0,0007
10	0,0009866	70,8	0,1475	0,0009862	71,8	0,1474	0 0009857	72,7	0,1472	0,0009853	73,6	0,1470
20	0,0009886	111,7	0,2895	0,0009882	112,6	0,2893	0,0009878	113,6	0,2890	0,0009874	114,5	0,2888
30	0,0009915	152,7	0,4271	0,0009911	153,6	0,4267	0,0009907	154,5	0,4264	0,0009902	155,4	0,4261
40	0,0009951	193,8	0,5604	0,0009946	194,7	0,5600	0,0009942	195,6	0,5596	0,0009938	196,4	0,5592
50	0,0009993	235,0	0,6897	0,0009989	235,8	0,6893	0,0009985	236,7	0,6888	0,0009981	237,5	0,6883
60	0,0010041	276,1	0,8153	0,0010037	277,0	0,8148	0,0010033	277,8	0,8142	0,0010029	278,6	0,8137
70	0,0010095	317,4	0,9373	0,0010091	318,2	0,9367	0,0010087	319,0	0,9362	0,0010083	319,9	0,9356
80	0,0010155	358,7	1,0560	0,0010150	359,5	1,0554	0,0010146	360,3	1,0548	0,0010142	361,1	1,0541
90	0,0010219	400,1	1,1716	0,0010215	400,9	1,1709	0,0010211	401,7	1,1703	0,0010206	402,5	1,1696
100	0,0010289	441,6	1,2843	0,0010285	442,4	1,2836	0,0010280	443,1	1,2829	0,0010276	443,9	1,2821
110	0,0010364	483,2	1,3943	0,0010360	483,9	1,3935	0,0010355	484,7	1,3927	0,0010350	485,4	1,3920
120	0,0010445	524,9	1,5017	0,0010440	525,6	1,5009	0,0010435	526,3	1,5001	0,0010430	527,1	1,4992
130	0,0010530	566,7	1,6068	0,0010525	567,4	1,6059	0,0010520	568,1	1,6050	0,0010514	568,8	1,6042
140	0,0010621	608,7	1,7097	0,0010616	609,4	1,7088	0,0010610	610,1	1,7078	0,0010605	610,8	1,7069
150	0,0010718	650,9	1,8105	0,0010712	651,6	1,8095	0,0010706	652,2	1,8085	0,0010701	652,9	1,8075
160	0,0010821	693,3	1,9095	0,0010815	693,9	1,9084	0,0010808	694,5	1,9074	0,0010802	695,1	1,9063
170	0,0010930	735,9	2,0067	0,0010923	736,4	2,0055	0,0010917	737,0	2,0044	0,0010910	737,6	2,0033
180	0,0011046	778,7	2,1022	0,0011039	779,2	2,1011	0,0011032	779,8	2,0999	0,0011025	780,4	2,0987
190	0,0011169	821,8	2,1963	0,0011162	822,3	2,1951	0,0011154	822,8	2,1938	0,0011146	823,4	2,1925
200	0,0011301	865,2	2,2891	0,0011292	865,7	2,2877	0,0011284	866,2	2,2864	0,0011276	866,7	2,2851
210	0,0011440	909,0	2,3806	0,0011431	909,4	2,3792	0,0011423	909,9	2,3777	0,0011414	910,3	2,3763
220	0,0011590	953,1	2,4710	0,0011580	953,5	2,4695	0,0011570	953,9	2,4680	0,0011561	954,3	2,4664
230	0,0011750	997,7	2,5605	0,0011739	998,0	2,5589	0,0011728	998,4	2,5572	0,0011718	998,7	2,5556
240	0,0011922	1042,8	2,6492	0,0011910	1043,1	2,6475	0,0011898	1043,3	2,6457	0,0011886	1043,6	2,6440
250	0,0012107	1088,4	2,7374	0,0012094	1088,6	2,7354	0,0012081	1088,8	2,7335	0,0012068	1089,1	2,7317
260	0,0012307	1134,7	2,8250	0,0012292	1134,8	2,8229	0,0012278	1135,0	2,8209	0,0012263	1135,1	2,8188
270	0,0012525	1181,8	2,9124	0,0012508	1181,8	2,9102	0,0012492	1181,8	2,9079	0,0012475	1181,9	2,9057
280	0,0012763	1229,7	2,9998	0,0012744	1229,6	2,9973	0,0012725	1229,5	2,9949	0,0012706	1229,4	2,9925
290	0,0013025	1278,6	3,0874	0,0013003	1278,3	3,0847	0,0012981	1278,1	3,0820	0,0012960	1277,9	3,0793
300	0,0013316	1328,7	3,1756	0,0013290	1328,3	3,1726	0,0013264	1327,9	3,1696	0,0013240	1327,5	3,1666
310	0,0013642	1380,3	3,2648	0,0013611	1379,6	3,2614	0,0013581	1379,0	3,2580	0,0013552	1378,4	3,2547
320	0,0014012	1433,6	3,3556	0,0013975	1432,7	3,3517	0,0013938	1431,8	3,3478	0,0013903	1430,9	3,3440
330	0,0014438	1489,2	3,4485	0,0014392	1487,9	3,4440	0,0014347	1486,7	3,4395	0,0014304	1485,5	3,4352
340	0,0014939	1547,7	3,5447	0,0014880	1545,9	3,5394	0,0014824	1544,2	3,5341	0,0014769	1542,6	3,5290
350	0,0015540	1610,0	3,6455	0,0015463	1607,5	3,6390	0,0015389	1605,1	3,6327	0,0015318	1602,9	3,6266
360	0,001628	1678,0	3,7541	0,001618	1674,5	3,7459	0,001607	1671,1	3,7381	0,001598	1668,0	3,7306
370	0,001728	1749,0	3,8653	0,001711	1743,5	3,8540	0,001695	1738,4	3,8435	0,001681	1733,8	3,8337
380	0,001874	1837,7	4,0021	0,001843	1828,2	3,9847	0,001817	1819,9	3,9692	0,001794	1812,5	3,9551
390	0,002144	1959,1	4,1865	0,002067	1937,9	4,1513	0,002009	1921,3	4,1232	0,001963	1907,7	4,0997
400	0,002831	2161,8	4,4896	0,002566	2106,1	4,4029	0,002390	2065,5	4,3389	0,002269	2035,2	4,2905
410	0,003956	2394,5	4,8329	0,003495	2324,8	4,7255	0,003131	2262,7	4,6297	0,002855	2210,5	4,5490
420	0,004921	2558,0	5,0706	0,004457	2503,7	4,9855	0,004031	2447,9	4,8989	0,003656	2393,1	4,8144
430	0,005643	2668,8	5,2295	0,005221	2629,2	5,1654	0,004813	2586,3	5,0972	0,004429	2541,0	5,0263
440	0,006227	2754,0	5,3499	0,005819	2721,0	5,2952	0,005434	2686,9	5,2394	0,005072	2651,7	5,1827
450	0,006735	2825,6	5,4495	0,006335	2797,3	5,4014	0,005958	2768,2	5,3526	0,005603	2738,3	5,3033
460	0,007189	2887,7	5,5349	0,006793	2863,0	5,4916	0,006421	2837,6	5,4479	0,006071	2811,6	5,4040
470	0,007602	2943,3	5,6102	0,007209	2921,2	5,5704	0,006839	2898,6	5,5306	0,006491	2875,6	5,4906
480	0,007985	2993,9	5,6779	0,007591	2973,9	5,6410	0,007222	2953,6	5,6041	0,006874	2932,8	5,5672
490	0,008343	3040,9	5,7398	0,007948	3022,6	5,7052	0,007578	3004,0	5,6706	0,007230	2985,1	5,6362
500	0,008681	3085,0	5,7972	0,008285	3068,1	5,7644	0,007913	3051,0	5,7318	0,007564	3033,6	5,6993
510	0,009002	3126,7	5,8508	0,008604	3111,0	5,8196	0,008230	3095,1	5,7885	0,007879	3079,1	5,7578
520	0,009310	3166,6	5,9014	0,008908	3151,9	5,8714	0,008532	3137,1	5,8418	0,008178	3122,1	5,8124
530	0,009605	3204,8	5,9493	0,009200	3191,1	5,9205	0,008821	3177,2	5,8920	0,008465	3163,2	5,8638
540	0,009890	3241,7	5,9949	0,009482	3228,7	5,9671	0,009099	3215,7	5,9397	0,008740	3202,5	5,9125
550	0,01017	3277,4	6,0386	0,009754	3265,2	6,0116	0,009367	3252,9	5,9851	0,009005	3240,5	5,9589

Table 3. Water and Superheated Steam (Continuation) **Wasser und überhitzter Dampf** (Fortsetzung)

t	300 bar			310 bar			320 bar			330 bar		
°C	v	h	s	v	h	s	v	h	s	v	h	s
550	0,01017	3277,4	6,0386	0,009754	3265,2	6,0116	0,009367	3252,9	5,9851	0,009005	3240,5	5,9589
560	0,01043	3312,1	6,0805	0,01002	3300,5	6,0543	0,009627	3288,9	6,0286	0,009261	3277,1	6,0032
570	0,01069	3345,9	6,1208	0,01027	3334,9	6,0954	0,009879	3323,8	6,0703	0,009509	3312,7	6,0456
580	0,01095	3378,9	6,1597	0,01052	3368,4	6,1349	0,01012	3357,9	6,1105	0,009751	3347,4	6,0865
590	0,01119	3411,2	6,1974	0,01077	3401,3	6,1732	0,01036	3391,3	6,1494	0,009986	3381,2	6,1259
600	0,01144	3443,0	6,2340	0,01100	3433,5	6,2103	0,01060	3423,9	6,1870	0,01022	3414,3	6,1641
610	0,01167	3474,2	6,2696	0,01124	3465,1	6,2463	0,01083	3456,0	6,2235	0,01044	3446,8	6,2010
620	0,01191	3505,0	6,3042	0,01146	3496,2	6,2813	0,01105	3487,5	6,2589	0,01066	3478,7	6,2370
630	0,01214	3535,3	6,3380	0,01169	3526,9	6,3155	0,01127	3518,6	6,2935	0,01088	3510,1	6,2720
640	0,01236	3565,3	6,3710	0,01191	3557,3	6,3489	0,01149	3549,2	6,3273	0,01109	3541,1	6,3061
650	0,01258	3595,0	6,4033	0,01213	3587,2	6,3816	0,01170	3579,5	6,3603	0,01130	3571,7	6,3394
660	0,01280	3624,4	6,4350	0,01234	3616,9	6,4135	0,01191	3609,4	6,3926	0,01150	3602,0	6,3720
670	0,01302	3653,5	6,4661	0,01255	3646,3	6,4449	0,01211	3639,1	6,4242	0,01170	3631,9	6,4039
680	0,01323	3682,4	6,4966	0,01276	3675,5	6,4757	0,01232	3668,6	6,4552	0,01190	3661,6	6,4352
690	0,01344	3711,2	6,5265	0,01296	3704,5	6,5059	0,01252	3697,8	6,4857	0,01210	3691,1	6,4660
700	0,01365	3739,7	6,5560	0,01317	3733,2	6,5356	0,01272	3726,8	6,5157	0,01229	3720,3	6,4962
710	0,01385	3768,1	6,5850	0,01337	3761,8	6,5648	0,01291	3755,6	6,5451	0,01248	3749,3	6,5258
720	0,01406	3796,3	6,6136	0,01357	3790,3	6,5936	0,01311	3784,2	6,5741	0,01267	3778,2	6,5550
730	0,01426	3824,4	6,6418	0,01376	3818,6	6,6220	0,01330	3812,7	6,6027	0,01286	3806,9	6,5838
740	0,01446	3852,4	6,6696	0,01396	3846,8	6,6500	0,01349	3841,1	6,6308	0,01305	3835,4	6,6121
750	0,01465	3880,3	6,6970	0,01415	3874,9	6,6775	0,01368	3869,4	6,6586	0,01323	3863,9	6,6401
760	0,01485	3908,1	6,7240	0,01434	3902,8	6,7047	0,01386	3897,5	6,6859	0,01341	3892,2	6,6676
770	0,01504	3935,8	6,7507	0,01453	3930,7	6,7316	0,01405	3925,5	6,7130	0,01359	3920,4	6,6948
780	0,01524	3963,5	6,7770	0,01472	3958,5	6,7581	0,01423	3953,5	6,7396	0,01377	3948,5	6,7216
790	0,01543	3991,0	6,8031	0,01490	3986,2	6,7843	0,01441	3981,3	6,7659	0,01395	3976,5	6,7481
800	0,01562	4018,5	6,8288	0,01509	4013,8	6,8101	0,01459	4009,1	6,7919	0,01413	4004,4	6,7742

Table 3. Water and Superheated Steam (Continuation) Wasser und überhitzter Dampf (Fortsetzung)

t	340 bar			350 bar			360 bar			370 bar		
°C	v	h	s	v	h	s	v	h	s	v	h	s
0	0,0009839	33,9	0,0007	0,0009834	34,9	0,0007	0,0009830	35,9	0,0006	0,0009825	36,8	0,0006
10	0,0009849	74,6	0,1469	0,0009844	75,5	0,1467	0,0009840	76,5	0,1466	0,0009836	77,4	0,1464
20	0,0009869	115,4	0,2885	0,0009865	116,3	0,2883	0,0009861	117,2	0,2880	0,0009857	118,1	0,2878
30	0,0009898	156,3	0,4258	0,0009894	157,2	0,4254	0,0009890	158,1	0,4251	0,0009886	159,0	0,4248
40	0,0009934	197,3	0,5588	0,0009930	198,2	0,5584	0,0009926	199,0	0,5580	0,0009922	199,9	0,5576
50	0,0009977	238,4	0,6879	0,0009973	239,2	0,6874	0,0009969	240,1	0,6870	0,0009965	240,9	0,6865
60	0,0010025	279,5	0,8132	0,0010021	280,3	0,8127	0,0010017	281,1	0,8122	0,0010013	282,0	0,8117
70	0,0010078	320,7	0,9350	0,0010074	321,5	0,9345	0,0010070	322,3	0,9339	0,0010066	323,1	0,9333
80	0,0010137	361,9	1,0535	0,0010133	362,7	1,0529	0,0010129	363,5	1,0523	0,0010125	364,3	1,0517
90	0,0010202	403,2	1,1689	0,0010197	404,0	1,1683	0,0010193	404,8	1,1676	0,0010189	405,6	1,1669
100	0,0010271	444,7	1,2814	0,0010266	445,4	1,2807	0,0010262	446,2	1,2800	0,0010257	446,9	1,2793
110	0,0010345	486,2	1,3912	0,0010341	486,9	1,3904	0,0010336	487,6	1,3897	0,0010331	488,4	1,3889
120	0,0010425	527,8	1,4984	0,0010420	528,5	1,4976	0,0010415	529,2	1,4968	0,0010410	529,9	1,4960
130	0,0010509	569,5	1,6033	0,0010504	570,2	1,6024	0,0010499	570,9	1,6016	0,0010494	571,6	1,6007
140	0,0010599	611,4	1,7060	0,0010594	612,1	1,7050	0,0010588	612,8	1,7041	0,0010583	613,5	1,7032
150	0,0010695	653,5	1,8066	0,0010689	654,2	1,8056	0,0010683	654,8	1,8046	0,0010677	655,5	1,8036
160	0,0010796	695,8	1,9053	0,0010790	696,4	1,9042	0,0010784	697,0	1,9032	0,0010778	697,7	1,9022
170	0,0010904	738,2	2,0022	0,0010897	738,8	2,0011	0,0010890	739,4	2,0000	0,0010884	740,0	1,9989
180	0,0011018	780,9	2,0975	0,0011011	781,5	2,0963	0,0011004	782,1	2,0952	0,0010997	782,7	2,0940
190	0,0011139	823,9	2,1913	0,0011131	824,4	2,1900	0,0011124	825,0	2,1888	0,0011116	825,5	2,1876
200	0,0011268	867,2	2,2837	0,0011260	867,7	2,2824	0,0011252	868,2	2,2811	0,0011244	868,7	2,2798
210	0,0011405	910,8	2,3749	0,0011396	911,2	2,3735	0,0011387	911,7	2,3721	0,0011379	912,1	2,3707
220	0,0011551	954,7	2,4649	0,0011542	955,1	2,4634	0,0011532	955,5	2,4619	0,0011523	956,0	2,4604
230	0,0011707	999,1	2,5540	0,0011697	999,4	2,5524	0,0011687	999,8	2,5508	0,0011676	1000,2	2,5492
240	0,0011875	1043,9	2,6422	0,0011863	1044,2	2,6405	0,0011852	1044,5	2,6388	0,0011841	1044,8	2,6371
250	0,0012055	1089,3	2,7298	0,0012042	1089,5	2,7279	0,0012030	1089,8	2,7261	0,0012017	1090,0	2,7242
260	0,0012249	1135,2	2,8168	0,0012235	1135,4	2,8148	0,0012221	1135,6	2,8128	0,0012207	1135,7	2,8108
270	0,0012459	1181,9	2,9035	0,0012443	1182,0	2,9013	0,0012428	1182,0	2,8991	0,0012412	1182,1	2,8970
280	0,0012688	1229,3	2,9901	0,0012670	1229,3	2,9877	0,0012652	1229,3	2,9853	0,0012635	1229,2	2,9830
290	0,0012939	1277,7	3,0767	0,0012918	1277,5	3,0741	0,0012898	1277,4	3,0715	0,0012878	1277,2	3,0689
300	0,0013215	1327,1	3,1637	0,0013191	1326,8	3,1608	0,0013168	1326,5	3,1579	0,0013144	1326,2	3,1551
310	0,0013523	1377,9	3,2514	0,0013494	1377,3	3,2482	0,0013467	1376,8	3,2450	0,0013440	1376,3	3,2419
320	0,0013868	1430,1	3,3403	0,0013835	1429,4	3,3367	0,0013802	1428,6	3,3331	0,0013770	1427,9	3,3296
330	0,0014262	1484,4	3,4309	0,0014221	1483,3	3,4268	0,0014181	1482,2	3,4227	0,0014142	1481,2	3,4187
340	0,0014716	1541,0	3,5241	0,0014666	1539,5	3,5192	0,0014616	1538,0	3,5145	0,0014569	1536,7	3,5098
350	0,0015251	1600,7	3,6207	0,0015186	1598,7	3,6149	0,0015123	1596,7	3,6094	0,0015063	1594,8	3,6039
360	0,001589	1665,0	3,7234	0,001580	1662,3	3,7166	0,001572	1659,6	3,7099	0,001564	1657,2	3,7035
370	0,001668	1729,5	3,8244	0,001656	1725,5	3,8156	0,001644	1721,8	3,8073	0,001634	1718,3	3,7993
380	0,001773	1805,9	3,9423	0,001754	1799,9	3,9304	0,001737	1794,5	3,9194	0,001722	1789,4	3,9091
390	0,001925	1896,2	4,0795	0,001892	1886,3	4,0617	0,001864	1877,6	4,0457	0,001839	1869,9	4,0312
400	0,002179	2011,8	4,2525	0,002111	1993,1	4,2214	0,002056	1977,5	4,1953	0,002010	1964,4	4,1727
410	0,002649	2167,8	4,4824	0,002494	2133,1	4,4278	0,002375	2104,8	4,3829	0,002282	2081,5	4,3453
420	0,003340	2342,2	4,7359	0,003082	2296,7	4,6656	0,002874	2257,0	4,6040	0,002707	2222,6	4,5504
430	0,004076	2495,4	4,9553	0,003761	2450,6	4,8861	0,003487	2408,1	4,8205	0,003253	2368,9	4,7599
440	0,004731	2615,6	5,1252	0,004404	2577,2	5,0649	0,004105	2538,8	5,0050	0,003835	2501,1	4,9467
450	0,005270	2707,7	5,2535	0,004956	2676,4	5,2031	0,004661	2644,5	5,1524	0,004384	2612,1	5,1013
460	0,005741	2785,0	5,3597	0,005430	2758,0	5,3151	0,005138	2730,4	5,2704	0,004863	2702,5	5,2255
470	0,006163	2852,1	5,4506	0,005854	2828,2	5,4103	0,005563	2804,0	5,3700	0,005289	2779,4	5,3297
480	0,006547	2911,8	5,5303	0,006239	2890,4	5,4934	0,005948	2868,7	5,4566	0,005674	2846,7	5,4197
490	0,006903	2966,0	5,6018	0,006594	2946,6	5,5676	0,006303	2926,9	5,5334	0,006028	2907,1	5,4993
500	0,007235	3016,0	5,6670	0,006925	2998,3	5,6349	0,006633	2980,3	5,6028	0,006357	2962,1	5,5710
510	0,007548	3062,8	5,7272	0,007237	3046,4	5,6968	0,006943	3029,9	5,6666	0,006665	3013,1	5,6365
520	0,007846	3107,0	5,7832	0,007532	3091,8	5,7543	0,007236	3076,4	5,7256	0,006957	3060,9	5,6971
530	0,008130	3149,1	5,8359	0,007814	3134,8	5,8083	0,007516	3120,5	5,7809	0,007234	3106,0	5,7537
540	0,008402	3189,3	5,8857	0,008083	3176,0	5,8592	0,007783	3162,5	5,8329	0,007499	3149,0	5,8069
550	0,008664	3228,0	5,9330	0,008342	3215,4	5,9074	0,008039	3202,8	5,8821	0,007753	3190,1	5,8571

Table 3. Water and Superheated Steam (Continuation) Wasser und überhitzter Dampf (Fortsetzung)

t	340 bar			350 bar			360 bar			870 bar		
°C	v	h	s	v	h	s	v	h	s	v	h	s
550	0,008664	3228,0	5,9330	0,008342	3215,4	5,9074	0,008039	3202,8	5,8821	0,007753	3190,1	5,8571
560	0,008917	3265,4	5,9781	0,008592	3253,5	5,9534	0,008286	3241,6	5,9290	0,007997	3229,6	5,9048
570	0,009162	3301,6	6,0213	0,008834	3290,4	5,9974	0,008525	3279,1	5,9737	0,008233	3267,8	5,9503
580	0,009400	3336,8	6,0629	0,009069	3326,2	6,0396	0,008757	3315,5	6,0166	0,008462	3304,7	5,9939
590	0,009631	3371,1	6,1029	0,009297	3361,0	6,0802	0,008982	3350,9	6,0578	0,008683	3340,7	6,0358
600	0,009857	3404,7	6,1416	0,009519	3395,1	6,1194	0,009201	3385,4	6,0976	0,008899	3375,7	6,0761
610	0,01008	3437,6	6,1790	0,009737	3428,4	6,1574	0,009415	3419,1	6,1361	0,009110	3409,9	6,1151
620	0,01029	3469,9	6,2154	0,009949	3461,1	6,1942	0,009624	3452,3	6,1733	0,009316	3443,4	6,1528
630	0,01051	3501,7	6,2508	0,01016	3493,2	6,2300	0,009828	3484,8	6,2095	0,009517	3476,3	6,1894
640	0,01071	3533,0	6,2853	0,01036	3524,9	6,2648	0,01003	3516,8	6,2448	0,009715	3508,6	6,2251
650	0,01092	3563,9	6,3189	0,01056	3556,1	6,2988	0,01023	3548,3	6,2791	0,009909	3540,5	6,2598
660	0,01112	3594,5	6,3518	0,01076	3587,0	6,3321	0,01042	3579,5	6,3127	0,01010	3571,9	6,2936
670	0,01132	3624,7	6,3841	0,01095	3617,5	6,3646	0,01061	3610,2	6,3455	0,01029	3603,0	6,3267
680	0,01151	3654,6	6,4156	0,01115	3647,7	6,3965	0,01080	3640,7	6,3776	0,01047	3633,7	6,3591
690	0,01171	3684,3	6,4466	0,01133	3677,6	6,4277	0,01098	3670,9	6,4091	0,01065	3664,2	6,3909
700	0,01190	3713,8	6,4771	0,01152	3707,3	6,4584	0,01117	3700,8	6,4400	0,01083	3694,3	6,4221
710	0,01208	3743,0	6,5070	0,01170	3736,8	6,4885	0,01135	3730,5	6,4704	0,01101	3724,2	6,4526
720	0,01227	3772,1	6,5364	0,01189	3766,1	6,5181	0,01153	3760,0	6,5002	0,01118	3753,9	6,4827
730	0,01245	3801,0	6,5653	0,01207	3795,2	6,5473	0,01170	3789,3	6,5296	0,01136	3783,4	6,5123
740	0,01263	3829,8	6,5939	0,01224	3824,1	6,5760	0,01188	3818,4	6,5585	0,01153	3812,8	6,5413
750	0,01281	3858,4	6,6220	0,01242	3852,9	6,6043	0,01205	3847,4	6,5870	0,01170	3841,9	6,5700
760	0,01299	3886,9	6,6497	0,01259	3881,6	6,6322	0,01222	3876,3	6,6150	0,01186	3870,9	6,5982
770	0,01317	3915,3	6,6770	0,01277	3910,1	6,6597	0,01239	3905,0	6,6427	0,01203	3899,8	6,6260
780	0,01334	3943,5	6,7040	0,01294	3938,5	6,6868	0,01255	3933,6	6,6699	0,01219	3928,6	6,6535
790	0,01352	3971,7	6,7306	0,01311	3966,9	6,7135	0,01272	3962,0	6,6969	0,01235	3957,2	6,6805
800	0,01369	3999,8	6,7569	0,01327	3995,1	6,7400	0,01288	3990,4	6,7234	0,01252	3985,7	6,7072

Table 3. Water and Superheated Steam (Continuation) Wasser und überhitzter Dampf (Fortsetzung)

t	380 bar			390 bar			400 bar			410 bar		
°C	v	h	s	v	h	s	v	h	s	v	h	s
0	0,0009820	37,8	0,0005	0,0009816	38,8	0,0005	0,0009811	39,7	0,0004	,0009807	40,7	0,0004
10	0,0009831	78,3	0,1462	0,0009827	79,3	0,1461	0,0009823	80,2	0,1459	0,0009819	81,1	0,1457
20	0,0009853	119,0	0,2875	0,0009849	119,9	0,2872	0,0009845	120,8	0,2870	0,0009841	121,8	0,2867
30	0,0009882	159,9	0,4245	0,0009878	160,7	0,4241	0,0009874	161,6	0,4238	0,0009870	162,5	0,4235
40	0,0009918	200,8	0,5573	0,0009914	201,6	0,5569	0,0009910	202,5	0,5565	0,0009907	203,4	0,5561
50	0,0009961	241,8	0,6861	0,0009957	242,6	0,6856	0,0009953	243,5	0,6852	0,0009949	244,3	0,6847
60	0,0010009	282,8	0,8112	0,0010005	283,6	0,8107	0,0010001	284,5	0,8102	0,0009997	285,3	0,8097
70	0,0010062	323,9	0,9328	0,0010058	324,7	0,9322	0,0010054	325,6	0,9317	0,0010050	326,4	0,9311
80	0,0010121	365,1	1,0511	0,0010116	365,9	1,0505	0,0010112	366,7	1,0498	0,0010108	367,5	1,0492
90	0,0010184	406,4	1,1663	0,0010180	407,1	1,1656	0,0010175	407,9	1,1649	0,0010171	408,7	1,1643
100	0,0010253	447,7	1,2786	0,0010248	448,5	1,2778	0,0010244	449,2	1,2771	0,0010239	450,0	1,2764
110	0,0010326	489,1	1,3881	0,0010322	489,9	1,3874	0,0010317	490,6	1,3866	0,0010312	491,4	1,3859
120	0,0010405	530,7	1,4952	0,0010400	531,4	1,4944	0,0010395	532,1	1,4935	0,0010390	532,8	1,4927
130	0,0010489	572,3	1,5998	0,0010483	573,0	1,5990	0,0010478	573,7	1,5981	0,0010473	574,4	1,5972
140	0,0010577	614,2	1,7023	0,0010572	614,8	1,7014	0,0010567	615,5	1,7005	0,0010561	616,2	1,6995
150	0,0010672	656,1	1,8026	0,0010666	656,8	1,8017	0,0010660	657,4	1,8007	0,0010655	658,1	1,7997
160	0,0010772	698,3	1,9011	0,0010766	698,9	1,9001	0,0010760	699,6	1,8991	0,0010754	700,2	1,8981
170	0,0010878	740,6	1,9978	0,0010871	741,3	1,9967	0,0010865	741,9	1,9956	0,0010858	742,5	1,9946
180	0,0010990	783,2	2,0928	0,0010983	783,8	2,0917	0,0010976	784,4	2,0905	0,0010969	785,0	2,0894
190	0,0011109	826,1	2,1863	0,0011102	826,6	2,1851	0,0011094	827,2	2,1839	0,0011087	827,7	2,1827
200	0,0011236	869,2	2,2785	0,0011228	869,7	2,2772	0,0011220	870,2	2,2759	0,0011212	870,7	2,2746
210	0,0011370	912,6	2,3693	0,0011362	913,1	2,3679	0,0011353	913,5	2,3665	0,0011345	914,0	2,3651
220	0,0011513	956,4	2,4589	0,0011504	956,8	2,4575	0,0011495	957,2	2,4560	0,0011486	957,7	2,4545
230	0,0011666	1000,5	2,5476	0,0011656	1000,9	2,5460	0,0011646	1001,3	2,5444	0,0011636	1001,7	2,5429
240	0,0011830	1045,1	2,6354	0,0011819	1045,5	2,6337	0,0011808	1045,8	2,6320	0,0011797	1046,1	2,6303
250	0,0012005	1090,2	2,7224	0,0011993	1090,5	2,7206	0,0011981	1090,8	2,7188	0,0011969	1091,0	2,7170
260	0,0012193	1135,9	2,8089	0,0012180	1136,1	2,8069	0,0012166	1136,3	2,8050	0,0012153	1136,5	2,8031
270	0,0012397	1182,2	2,8949	0,0012382	1182,3	2,8928	0,0012367	1182,4	2,8907	0,0012352	1182,5	2,8886
280	0,0012617	1229,2	2,9807	0,0012600	1229,2	2,9784	0,0012583	1229,2	2,9761	0,0012567	1229,2	2,9739
290	0,0012858	1277,1	3,0664	0,0012838	1276,9	3,0639	0,0012819	1276,8	3,0614	0,0012800	1276,7	3,0590
300	0,0013122	1325,9	3,1523	0,0013099	1325,6	3,1496	0,0013077	1325,4	3,1469	0,0013055	1325,2	3,1442
310	0,0013413	1375,9	3,2388	0,0013387	1375,4	3,2357	0,0013361	1375,0	3,2327	0,0013336	1374,6	3,2298
320	0,0013738	1427,2	3,3261	0,0013707	1426,6	3,3227	0,0013677	1425,9	3,3193	0,0013648	1425,3	3,3160
330	0,0014104	1480,2	3,4147	0,0014068	1479,3	3,4109	0,0014032	1478,4	3,4071	0,0013997	1477,6	3,4033
340	0,0014522	1535,3	3,5053	0,0014477	1534,1	3,5009	0,0014434	1532,9	3,4965	0,0014391	1531,7	3,4923
350	0,0015006	1593,0	3,5987	0,0014950	1591,3	3,5935	0,0014896	1589,7	3,5885	0,0014844	1588,1	3,5836
360	0,001556	1654,8	3,6974	0,001549	1652,6	3,6914	0,001542	1650,5	3,6856	0,001536	1648,5	3,6800
370	0,001624	1715,0	3,7917	0,001614	1711,9	3,7844	0,001605	1709,0	3,7774	0,001597	1706,3	3,7706
380	0,001708	1784,8	3,8993	0,001694	1780,5	3,8901	0,001682	1776,4	3,8814	0,001670	1772,7	3,8730
390	0,001817	1862,9	4,0179	0,001797	1856,5	4,0056	0,001779	1850,7	3,9942	0,001762	1845,3	3,9834
400	0,001972	1953,0	4,1528	0,001938	1943,0	4,1351	0,001909	1934,1	4,1190	0,001883	1926,1	4,1043
410	0,002208	2062,0	4,3135	0,002147	2045,5	4,2861	0,002095	2031,2	4,2621	0,002052	2018,7	4,2409
420	0,002572	2193,1	4,5040	0,002462	2167,7	4,4637	0,002371	2145,7	4,4285	0,002295	2126,7	4,3977
430	0,003055	2333,3	4,7048	0,002889	2301,3	4,6551	0,002749	2272,8	4,6105	0,002631	2247,3	4,5705
440	0,003595	2465,0	4,8908	0,003383	2431,0	4,8382	0,003200	2399,4	4,7893	0,003040	2370,3	4,7441
450	0,004124	2579,1	5,0497	0,003888	2546,8	4,9995	0,003675	2515,6	4,9511	0,003484	2485,8	4,9050
460	0,004605	2674,2	5,1805	0,004363	2645,8	5,1355	0,004137	2617,1	5,0906	0,003927	2588,9	5,0466
470	0,005031	2754,6	5,2894	0,004788	2729,6	5,2491	0,004560	2704,4	5,2089	0,004346	2679,2	5,1690
480	0,005415	2824,6	5,3829	0,005171	2802,2	5,3462	0,004941	2779,8	5,3097	0,004725	2757,2	5,2734
490	0,005768	2887,0	5,4653	0,005523	2866,8	5,4314	0,005291	2846,5	5,3977	0,005072	2826,1	5,3642
500	0,006096	2943,8	5,5392	0,005849	2925,3	5,5076	0,005616	2906,8	5,4762	0,005395	2888,1	5,4450
510	0,006403	2996,3	5,6066	0,006154	2979,3	5,5769	0,005919	2962,2	5,5474	0,005696	2945,0	5,5181
520	0,006693	3045,3	5,6688	0,006442	3029,5	5,6407	0,006205	3013,7	5,6128	0,005980	2997,8	5,5851
530	0,006968	3091,5	5,7268	0,006716	3076,9	5,7000	0,006476	3062,1	5,6735	0,006250	3047,4	5,6471
540	0,007230	3135,4	5,7811	0,006976	3121,7	5,7555	0,006735	3108,0	5,7302	0,006506	3094,2	5,7051
550	0,007482	3177,3	5,8323	0,007225	3164,5	5,8078	0,006982	3151,6	5,7835	0,006750	3138,7	5,7595

Table 3. Water and Superheated Steam (Continuation) Wasser und überhitzter Dampf (Fortsetzung)

t	380 bar			390 bar			400 bar			410 bar		
°C	v	h	s	v	h	s	v	h	s	v	h	s
550	0,007482	3177,3	5,8323	0,007225	3164,5	5,8078	0,006982	3151,6	5,7835	0,006750	3138,7	5,7595
560	0,007723	3217,6	5,8810	0,007464	3205,5	5,8574	0,007219	3193,4	5,8340	0,006985	3181,2	5,8109
570	0,007957	3256,4	5,9273	0,007695	3245,0	5,9045	0,007447	3233,6	5,8819	0,007211	3222,1	5,8596
580	0,008182	3294,0	5,9716	0,007918	3283,2	5,9495	0,007667	3272,4	5,9276	0,007429	3261,5	5,9061
590	0,008401	3330,4	6,0140	0,008134	3320,2	5,9926	0,007881	3309,9	5,9714	0,007640	3299,6	5,9505
600	0,008614	3365,9	6,0549	0,008344	3356,2	6,0341	0,008088	3346,4	6,0135	0,007845	3336,6	5,9931
610	0,008822	3400,6	6,0944	0,008549	3391,3	6,0741	0,008290	3382,0	6,0540	0,008044	3372,7	6,0342
620	0,009025	3434,5	6,1326	0,008749	3425,6	6,1127	0,008487	3416,7	6,0931	0,008239	3407,8	6,0738
630	0,009223	3467,8	6,1697	0,008944	3459,3	6,1502	0,008680	3450,8	6,1310	0,008429	3442,3	6,1121
640	0,009418	3500,5	6,2057	0,009136	3492,3	6,1866	0,008868	3484,2	6,1678	0,008614	3476,0	6,1493
650	0,009608	3532,7	6,2407	0,009323	3524,8	6,2220	0,009053	3517,0	6,2035	0,008796	3509,2	6,1854
660	0,009795	3564,4	6,2749	0,009507	3556,9	6,2565	0,009234	3549,4	6,2384	0,008975	3541,8	6,2206
670	0,009980	3595,8	6,3083	0,009689	3588,5	6,2902	0,009412	3581,3	6,2724	0,009150	3574,0	6,2549
680	0,01016	3626,7	6,3410	0,009867	3619,8	6,3232	0,009588	3612,8	6,3056	0,009322	3605,8	6,2884
690	0,01034	3657,4	6,3730	0,01004	3650,7	6,3554	0,009760	3644,0	6,3382	0,009492	3637,2	6,3212
700	0,01052	3687,8	6,4044	0,01022	3681,3	6,3871	0,009930	3674,8	6,3701	0,009659	3668,3	6,3533
710	0,01069	3718,0	6,4352	0,01039	3711,7	6,4181	0,01010	3705,4	6,4013	0,009824	3699,1	6,3848
720	0,01086	3747,9	6,4655	0,01055	3741,8	6,4486	0,01026	3735,7	6,4320	0,009987	3729,7	6,4157
730	0,01103	3777,6	6,4953	0,01072	3771,7	6,4786	0,01043	3765,8	6,4622	0,01015	3760,0	6,4461
740	0,01120	3807,1	6,5245	0,01088	3801,4	6,5080	0,01059	3795,7	6,4918	0,01031	3790,1	6,4759
750	0,01136	3836,4	6,5534	0,01105	3830,9	6,5370	0,01075	3825,5	6,5210	0,01046	3820,0	6,5053
760	0,01153	3865,6	6,5818	0,01121	3860,3	6,5656	0,01091	3855,0	6,5498	0,01062	3849,7	6,5342
770	0,01169	3894,7	6,6097	0,01137	3889,5	6,5938	0,01106	3884,4	6,5781	0,01077	3879,3	6,5627
780	0,01185	3923,6	6,6373	0,01152	3918,6	6,6215	0,01122	3913,6	6,6060	0,01092	3908,7	6,5907
790	0,01201	3952,4	6,6645	0,01168	3947,6	6,6489	0,01137	3942,7	6,6335	0,01107	3937,9	6,6184
800	0,01217	3981,1	6,6914	0,01184	3976,4	6,6758	0,01152	3971,7	6,6606	0,01122	3967,1	6,6457

Table 3. Water and Superheated Steam (Continuation) Wasser und überhitzter Dampf (Fortsetzung)

t	420 bar			430 bar			440 bar			450 bar		
°C	v	h	s	v	h	s	v	h	s	v	h	s
0	0,0009802	41,7	0,0003	0,0009798	42,6	0,0003	0,0009793	43,6	0,0002	0,0009789	44,6	0,0001
10	0,0009814	82,1	0,1455	0,0009810	83,0	0,1454	0,0009806	83,9	0,1452	0,0009802	84,8	0,1450
20	0,0009837	122,7	0,2865	0,0009832	123,6	0,2862	0,0009828	124,5	0,2859	0,0009824	125,4	0,2857
30	0,0009866	163,4	0,4231	0,0009862	164,3	0,4228	0,0009858	165,2	0,4225	0,0009854	166,1	0,4221
40	0,0009903	204,2	0,5557	0,0009899	205,1	0,5553	0,0009895	206,0	0,5549	0,0009891	206,8	0,5545
50	0,0009945	245,2	0,6843	0,0009941	246,0	0,6838	0,0009937	246,9	0,6834	0,0009933	247,7	0,6829
60	0,0009993	286,1	0,8092	0,0009989	287,0	0,8087	0,0009985	287,8	0,8082	0,0009981	288,6	0,8077
70	0,0010046	327,2	0,9305	0,0010042	328,0	0,9300	0,0010038	328,8	0,9294	0,0010033	329,6	0,9289
80	0,0010104	368,3	1,0486	0,0010100	369,1	1,0480	0,0010095	369,9	1,0474	0,0010091	370,7	1,0468
90	0,0010167	409,5	1,1636	0,0010163	410,2	1,1630	0,0010158	411,0	1,1623	0,0010154	411,8	1,1617
100	0,0010235	450,7	1,2757	0,0010230	451,5	1,2750	0,0010226	452,3	1,2743	0,0010222	453,0	1,2736
110	0,0010308	492,1	1,3851	0,0010303	492,8	1,3844	0,0010298	493,6	1,3836	0,0010294	494,3	1,3829
120	0,0010385	533,6	1,4919	0,0010381	534,3	1,4911	0,0010376	535,0	1,4903	0,0010371	535,7	1,4895
130	0,0010468	575,2	1,5964	0,0010463	575,9	1,5955	0,0010458	576,6	1,5947	0,0010453	577,3	1,5939
140	0,0010556	616,9	1,6986	0,0010551	617,6	1,6977	0,0010545	618,2	1,6968	0,0010540	618,9	1,6959
150	0,0010649	658,8	1,7988	0,0010643	659,4	1,7978	0,0010638	660,1	1,7969	0,0010632	660,7	1,7959
160	0,0010748	700,8	1,8970	0,0010742	701,5	1,8960	0,0010736	702,1	1,8950	0,0010730	702,7	1,8940
170	0,0010852	743,1	1,9935	0,0010846	743,7	1,9924	0,0010840	744,3	1,9913	0,0010833	744,9	1,9903
180	0,0010963	785,5	2,0883	0,0010956	786,1	2,0871	0,0010949	786,7	2,0860	0,0010943	787,3	2,0849
190	0,0011080	828,3	2,1815	0,0011073	828,8	2,1803	0,0011066	829,4	2,1791	0,0011059	829,9	2,1779
200	0,0011204	871,2	2,2733	0,0011197	871,7	2,2720	0,0011189	872,3	2,2707	0,0011182	872,8	2,2695
210	0,0011337	914,5	2,3638	0,0011328	915,0	2,3624	0,0011320	915,5	2,3611	0,0011312	915,9	2,3597
220	0,0011477	958,1	2,4531	0,0011468	958,5	2,4516	0,0011459	959,0	2,4502	0,0011450	959,4	2,4488
230	0,0011627	1002,1	2,5413	0,0011617	1002,4	2,5398	0,0011607	1002,8	2,5383	0,0011598	1003,2	2,5367
240	0,0011786	1046,4	2,6287	0,0011776	1046,8	2,6270	0,0011765	1047,1	2,6254	0,0011754	1047,5	2,6238
250	0,0011957	1091,3	2,7153	0,0011945	1091,6	2,7135	0,0011934	1091,8	2,7117	0,0011922	1092,1	2,7100
260	0,0012140	1136,7	2,8011	0,0012127	1136,9	2,7993	0,0012114	1137,1	2,7974	0,0012102	1137,3	2,7955
270	0,0012337	1182,6	2,8866	0,0012323	1182,7	2,8845	0,0012308	1182,9	2,8825	0,0012294	1183,0	2,8805
280	0,0012550	1229,3	2,9716	0,0012534	1229,3	2,9694	0,0012518	1229,3	2,9673	0,0012502	1229,4	2,9651
290	0,0012782	1276,7	3,0565	0,0012763	1276,6	3,0541	0,0012745	1276,5	3,0518	0,0012727	1276,5	3,0494
300	0,0013034	1324,9	3,1415	0,0013013	1324,7	3,1389	0,0012992	1324,5	3,1363	0,0012972	1324,4	3,1337
310	0,0013312	1374,2	3,2268	0,0013287	1373,9	3,2239	0,0013263	1373,5	3,2211	0,0013240	1373,2	3,2182
320	0,0013619	1424,8	3,3127	0,0013590	1424,2	3,3095	0,0013563	1423,7	3,3064	0,0013535	1423,2	3,3032
330	0,0013962	1476,8	3,3997	0,0013929	1476,0	3,3960	0,0013896	1475,2	3,3925	0,0013864	1474,5	3,3890
340	0,0014350	1530,6	3,4881	0,0014310	1529,5	3,4840	0,0014271	1528,5	3,4800	0,0014233	1527,5	3,4760
350	0,0014793	1586,6	3,5788	0,0014744	1585,1	3,5741	0,0014697	1583,7	3,5694	0,0014651	1582,4	3,5649
360	0,001530	1646,5	3,6745	0,001524	1644,7	3,6692	0,001518	1642,9	3,6640	0,001512	1641,3	3,6590
370	0,001588	1703,7	3,7641	0,001581	1701,2	3,7578	0,001573	1698,8	3,7516	0,001566	1696,6	3,7457
380	0,001659	1769,1	3,8651	0,001649	1765,8	3,8574	0,001639	1762,6	3,8501	0,001630	1759,7	3,8430
390	0,001747	1840,4	3,9733	0,001732	1835,8	3,9637	0,001719	1831,5	3,9546	0,001706	1827,4	3,9459
400	0,001860	1918,9	4,0908	0,001839	1912,3	4,0782	0,001819	1906,2	4,0664	0,001801	1900,6	4,0554
410	0,002014	2007,7	4,2218	0,001980	1997,9	4,2044	0,001951	1989,0	4,1886	0,001924	1981,0	4,1739
420	0,002231	2110,0	4,3704	0,002177	2095,3	4,3460	0,002129	2082,3	4,3241	0,002088	2070,6	4,3042
430	0,002531	2224,7	4,5346	0,002445	2204,5	4,5024	0,002371	2186,5	4,4734	0,002307	2170,4	4,4471
440	0,002902	2343,6	4,7026	0,002782	2319,3	4,6645	0,002678	2297,2	4,6296	0,002587	2277,0	4,5977
450	0,003315	2457,7	4,8615	0,003164	2431,4	4,8206	0,003031	2406,9	4,7825	0,002913	2384,2	4,7469
460	0,003736	2561,5	5,0041	0,003563	2535,3	4,9633	0,003407	2510,2	4,9243	0,003266	2486,4	4,8874
470	0,004145	2653,9	5,1293	0,003957	2628,7	5,0900	0,003784	2604,2	5,0517	0,003626	2580,8	5,0152
480	0,004521	2734,7	5,2373	0,004330	2712,2	5,2015	0,004150	2689,8	5,1661	0,003982	2667,5	5,1312
490	0,004866	2805,7	5,3309	0,004671	2785,3	5,2979	0,004488	2764,9	5,2653	0,004315	2744,7	5,2330
500	0,005186	2869,5	5,4140	0,004988	2850,8	5,3832	0,004801	2832,1	5,3527	0,004625	2813,5	5,3226
510	0,005485	2927,8	5,4889	0,005285	2910,5	5,4600	0,005095	2893,3	5,4314	0,004915	2876,1	5,4030
520	0,005767	2981,9	5,5575	0,005565	2965,9	5,5302	0,005372	2949,8	5,5031	0,005190	2933,8	5,4763
530	0,006034	3032,5	5,6210	0,005830	3017,6	5,5951	0,005635	3002,7	5,5693	0,005450	2987,7	5,5439
540	0,006288	3080,3	5,6801	0,006082	3066,4	5,6554	0,005885	3052,4	5,6309	0,005698	3038,5	5,6066
550	0,006531	3125,7	5,7356	0,006322	3112,7	5,7120	0,006124	3099,6	5,6886	0,005934	3086,5	5,6654

Table 3. Water and Superheated Steam (Continuation) Wasser und überhitzter Dampf (Fortsetzung)

t	420 bar			430 bar			440 bar			450 bar		
°C	v	h	s	v	h	s	v	h	s	v	h	s
550	0,006531	3125,7	5,7356	0,006322	3112,7	5,7120	0,006124	3099,6	5,6886	0,005934	3086,5	5,6654
560	0,006763	3169,0	5,7880	0,006553	3156,8	5,7653	0,006352	3144,5	5,7428	0,006161	3132,2	5,7206
570	0,006987	3210,6	5,8376	0,006774	3199,1	5,8157	0,006571	3187,5	5,7941	0,006378	3175,9	5,7727
580	0,007203	3250,6	5,8848	0,006987	3239,7	5,8637	0,006782	3228,8	5,8428	0,006587	3217,9	5,8222
590	0,007411	3289,3	5,9298	0,007194	3279,0	5,9094	0,006986	3268,7	5,8892	0,006789	3258,3	5,8693
600	0,007614	3326,8	5,9731	0,007394	3317,0	5,9532	0,007184	3307,2	5,9336	0,006984	3297,4	5,9143
610	0,007811	3363,3	6,0146	0,007588	3354,0	5,9953	0,007376	3344,6	5,9763	0,007174	3335,3	5,9575
620	0,008002	3398,9	6,0547	0,007777	3390,0	6,0359	0,007563	3381,1	6,0173	0,007359	3372,2	5,9990
630	0,008190	3433,7	6,0935	0,007962	3425,2	6,0751	0,007745	3416,7	6,0569	0,007538	3408,2	6,0391
640	0,008373	3467,8	6,1310	0,008143	3459,7	6,1130	0,007923	3451,5	6,0953	0,007714	3443,4	6,0778
650	0,008552	3501,3	6,1675	0,008319	3493,5	6,1499	0,008098	3485,7	6,1325	0,007886	3477,8	6,1154
660	0,008728	3534,3	6,2030	0,008493	3526,8	6,1857	0,008268	3519,2	6,1687	0,008055	3511,7	6,1519
670	0,008900	3566,8	6,2376	0,008663	3559,5	6,2206	0,008436	3552,3	6,2039	0,008220	3545,1	6,1874
680	0,009070	3598,8	6,2714	0,008830	3591,8	6,2547	0,008601	3584,9	6,2383	0,008382	3577,9	6,2221
690	0,009237	3630,5	6,3045	0,008994	3623,8	6,2880	0,008763	3617,0	6,2718	0,008542	3610,3	6,2559
700	0,009402	3661,8	6,3369	0,009156	3655,3	6,3207	0,008922	3648,9	6,3047	0,008699	3642,4	6,2890
710	0,009564	3692,9	6,3686	0,009316	3686,6	6,3526	0,009079	3680,3	6,3369	0,008854	3674,1	6,3214
720	0,009724	3723,6	6,3997	0,009473	3717,6	6,3839	0,009234	3711,5	6,3684	0,009006	3705,5	6,3532
730	0,009882	3754,1	6,4303	0,009629	3748,3	6,4147	0,009387	3742,4	6,3994	0,009157	3736,6	6,3843
740	0,01004	3784,4	6,4603	0,009782	3778,8	6,4450	0,009538	3773,1	6,4298	0,009305	3767,5	6,4150
750	0,01019	3814,5	6,4899	0,009933	3809,0	6,4747	0,009687	3803,6	6,4597	0,009452	3798,1	6,4451
760	0,01034	3844,4	6,5189	0,01008	3839,1	6,5039	0,009834	3833,8	6,4892	0,009597	3828,5	6,4746
770	0,01049	3874,1	6,5476	0,01023	3869,0	6,5327	0,009980	3863,9	6,5181	0,009740	3858,7	6,5038
780	0,01064	3903,7	6,5758	0,01038	3898,7	6,5611	0,01012	3893,8	6,5466	0,009881	3888,8	6,5324
790	0,01079	3933,1	6,6036	0,01052	3928,3	6,5890	0,01027	3923,5	6,5747	0,01002	3918,7	6,5607
800	0,01094	3962,4	6,6310	0,01067	3957,7	6,6166	0,01041	3953,1	6,6024	0,01016	3948,4	6,5885

Table 3. Water and Superheated Steam (Continuation) Wasser und überhitzter Dampf (Fortsetzung)

t	460 bar			470 bar			480 bar			490 bar		
°C	v	h	s	v	h	s	v	h	s	v	h	s
0	0,0009784	45,5	0,0001	0,0009780	46,5	—0,0000	0,0009776	47,4	—0,0001	0,0009771	48,4	—0,0002
10	0,0009798	85,8	0,1448	0,0009793	86,7	0,1446	0,0009789	87,6	0,1444	0,0009785	88,6	0,1443
20	0,0009820	126,3	0,2854	0,0009816	127,2	0,2851	0,0009812	128,1	0,2849	0,0009808	129,0	0,2846
30	0,0009851	166,9	0,4218	0,0009847	167,8	0,4215	0,0009843	168,7	0,4211	0,0009839	169,6	0,4208
40	0,0009887	207,7	0,5541	0,0009883	208,6	0,5537	0,0009879	209,4	0,5533	0,0009875	210,3	0,5529
50	0,0009929	248,6	0,6825	0,0009925	249,4	0,6820	0,0009921	250,2	0,6816	0,0009918	251,1	0,6811
60	0,0009977	289,5	0,8072	0,0009973	290,3	0,8067	0,0009969	291,1	0,8062	0,0009965	292,0	0,8057
70	0,0010029	330,4	0,9283	0,0010025	331,2	0,9278	0,0010021	332,1	0,9272	0,0010017	332,9	0,9267
80	0,0010087	371,5	1,0462	0,0010083	372,3	1,0456	0,0010079	373,1	1,0450	0,0010075	373,9	1,0444
90	0,0010150	412,6	1,1610	0,0010146	413,4	1,1604	0,0010141	414,1	1,1597	0,0010137	414,9	1,1591
100	0,0010217	453,8	1,2729	0,0010213	454,5	1,2722	0,0010208	455,3	1,2715	0,0010204	456,1	1,2708
110	0,0010289	495,1	1,3821	0,0010285	495,8	1,3814	0,0010280	496,6	1,3806	0,0010276	497,3	1,3799
120	0,0010366	536,5	1,4887	0,0010362	537,2	1,4880	0,0010357	537,9	1,4872	0,0010352	538,6	1,4864
130	0,0010448	578,0	1,5930	0,0010443	578,7	1,5922	0,0010438	579,4	1,5913	0,0010433	580,1	1,5905
140	0,0010535	619,6	1,6950	0,0010530	620,3	1,6942	0,0010524	621,0	1,6933	0,0010519	621,7	1,6924
150	0,0010627	661,4	1,7950	0,0010621	662,1	1,7940	0,0010616	662,7	1,7931	0,0010610	663,4	1,7922
160	0,0010724	703,4	1,8930	0,0010718	704,0	1,8920	0,0010713	704,6	1,8910	0,0010707	705,3	1,8900
170	0,0010827	745,5	1,9892	0,0010821	746,1	1,9881	0,0010815	746,7	1,9871	0,0010809	747,4	1,9860
180	0,0010936	787,9	2,0837	0,0010930	788,5	2,0826	0,0010923	789,0	2,0815	0,0010917	789,6	2,0804
190	0,0011052	830,5	2,1767	0,0011045	831,0	2,1755	0,0011038	831,6	2,1743	0,0011031	832,1	2,1731
200	0,0011174	873,3	2,2682	0,0011167	873,8	2,2670	0,0011159	874,3	2,2657	0,0011152	874,9	2,2645
210	0,0011304	916,4	2,3584	0,0011296	916,9	2,3570	0,0011288	917,4	2,3557	0,0011280	917,9	2,3544
220	0,0011442	959,8	2,4474	0,0011433	960,3	2,4459	0,0011424	960,7	2,4445	0,0011416	961,2	2,4431
230	0,0011588	1003,6	2,5352	0,0011579	1004,0	2,5337	0,0011569	1004,4	2,5322	0,0011560	1004,8	2,5307
240	0,0011744	1047,8	2,6222	0,0011734	1048,2	2,6206	0,0011724	1048,5	2,6190	0,0011713	1048,9	2,6174
250	0,0011911	1092,4	2,7083	0,0011899	1092,7	2,7066	0,0011888	1093,0	2,7049	0,0011877	1093,3	2,7032
260	0,0012089	1137,5	2,7937	0,0012076	1137,7	2,7918	0,0012064	1138,0	2,7900	0,0012052	1138,2	2,7882
270	0,0012280	1183,2	2,8785	0,0012267	1183,3	2,8765	0,0012253	1183,5	2,8745	0,0012239	1183,6	2,8726
280	0,0012487	1229,4	2,9629	0,0012471	1229,5	2,9608	0,0012456	1229,6	2,9587	0,0012441	1229,7	2,9566
290	0,0012709	1276,4	3,0471	0,0012692	1276,4	3,0448	0,0012675	1276,4	3,0425	0,0012658	1276,4	3,0402
300	0,0012952	1324,2	3,1312	0,0012932	1324,1	3,1287	0,0012913	1323,9	3,1262	0,0012893	1323,8	3,1237
310	0,0013217	1372,9	3,2154	0,0013194	1372,6	3,2127	0,0013172	1372,4	3,2100	0,0013150	1372,1	3,2073
320	0,0013509	1422,7	3,3001	0,0013482	1422,2	3,2971	0,0013457	1421,8	3,2941	0,0013431	1421,4	3,2911
330	0,0013833	1473,8	3,3855	0,0013802	1473,2	3,3822	0,0013772	1472,5	3,3788	0,0013742	1471,9	3,3755
340	0,0014195	1526,5	3,4722	0,0014159	1525,6	3,4683	0,0014123	1524,7	3,4646	0,0014089	1523,8	3,4609
350	0,0014606	1581,1	3,5605	0,0014562	1579,9	3,5562	0,0014520	1578,7	3,5519	0,0014479	1577,5	3,5477
360	0,001507	1639,7	3,6541	0,001501	1638,1	3,6493	0,001496	1636,6	3,6446	0,001491	1635,2	3,6400
370	0,001559	1694,4	3,7399	0,001552	1692,4	3,7343	0,001546	1690,4	3,7289	0,001540	1688,6	3,7236
380	0,001621	1756,8	3,8362	0,001612	1754,1	3,8296	0,001604	1751,6	3,8232	0,001596	1749,2	3,8170
390	0,001694	1823,6	3,9377	0,001683	1820,1	3,9297	0,001673	1816,7	3,9221	0,001663	1813,5	3,9148
400	0,001785	1895,3	4,0450	0,001769	1890,5	4,0351	0,001755	1885,9	4,0258	0,001742	1881,7	4,0168
410	0,001900	1973,6	4,1604	0,001878	1966,8	4,1477	0,001858	1960,6	4,1358	0,001839	1954,8	4,1246
420	0,002051	2060,1	4,2861	0,002018	2050,6	4,2694	0,001989	2041,9	4,2540	0,001962	2033,9	4,2396
430	0,002251	2155,8	4,4232	0,002202	2142,7	4,4014	0,002159	2130,8	4,3814	0,002120	2120,0	4,3629
440	0,002507	2258,6	4,5683	0,002436	2241,8	4,5413	0,002374	2226,5	4,5165	0,002319	2212,5	4,4935
450	0,002809	2363,1	4,7138	0,002716	2343,5	4,6829	0,002633	2325,5	4,6542	0,002559	2308,7	4,6275
460	0,003140	2464,0	4,8524	0,003026	2442,9	4,8195	0,002923	2423,1	4,7884	0,002831	2404,6	4,7592
470	0,003482	2558,3	4,9802	0,003351	2536,8	4,9466	0,003231	2516,3	4,9146	0,003122	2496,9	4,8842
480	0,003824	2645,4	5,0966	0,003676	2623,5	5,0626	0,003544	2603,2	5,0308	0,003421	2583,7	5,0003
490	0,004153	2724,6	5,2012	0,004000	2704,8	5,1698	0,003857	2685,1	5,1389	0,003722	2665,7	5,1085
500	0,004458	2795,0	5,2928	0,004301	2776,7	5,2635	0,004153	2758,6	5,2346	0,004013	2740,7	5,2061
510	0,004745	2858,9	5,3750	0,004584	2841,9	5,3473	0,004432	2825,0	5,3199	0,004288	2808,3	5,2930
520	0,005017	2917,9	5,4497	0,004852	2901,9	5,4234	0,004696	2886,1	5,3975	0,004548	2870,4	5,3719
530	0,005274	2972,8	5,5186	0,005107	2957,9	5,4936	0,004948	2943,1	5,4688	0,004797	2928,3	5,4444
540	0,005520	3024,5	5,5826	0,005350	3010,6	5,5587	0,005189	2996,6	5,5351	0,005035	2982,7	5,5117
550	0,005754	3073,4	5,6423	0,005582	3060,3	5,6195	0,005418	3047,2	5,5970	0,005262	3034,2	5,5746

Table 3. Water and Superheated Steam (Continuation) Wasser und überhitzter Dampf (Fortsetzung)

t	460 bar			470 bar			480 bar			490 bar		
°C	v	h	s	v	h	s	v	h	s	v	h	s
550	0,005754	3073,4	5,6423	0,005582	3060,3	5,6195	0,005418	3047,2	5,5970	0,005262	3034,2	5,5746
560	0,005978	3119,9	5,6985	0,005804	3107,6	5,6767	0,005638	3095,3	5,6550	0,005480	3083,0	5,6336
570	0,006193	3164,4	5,7515	0,006017	3152,8	5,7305	0,005849	3141,2	5,7097	0,005689	3129,6	5,6891
580	0,006400	3207,0	5,8017	0,006222	3196,0	5,7815	0,006052	3185,1	5,7615	0,005890	3174,1	5,7417
590	0,006600	3248,0	5,8495	0,006420	3237,6	5,8300	0,006248	3227,3	5,8107	0,006084	3216,9	5,7915
600	0,006793	3287,6	5,8951	0,006611	3277,7	5,8762	0,006437	3267,9	5,8575	0,006271	3258,1	5,8390
610	0,006981	3326,0	5,9389	0,006797	3316,6	5,9205	0,006621	3307,3	5,9023	0,006452	3298,0	5,8844
620	0,007163	3363,3	5,9809	0,006977	3354,4	5,9630	0,006799	3345,5	5,9453	0,006628	3336,6	5,9279
630	0,007341	3399,7	6,0214	0,007152	3391,1	6,0040	0,006972	3382,7	5,9867	0,006799	3374,2	5,9697
640	0,007514	3435,2	6,0605	0,007324	3427,1	6,0435	0,007141	3418,9	6,0267	0,006966	3410,8	6,0101
650	0,007684	3470,0	6,0985	0,007491	3462,2	6,0818	0,007306	3454,4	6,0653	0,007130	3446,7	6,0491
660	0,007850	3504,2	6,1353	0,007655	3496,7	6,1190	0,007468	3489,2	6,1028	0,007289	3481,8	6,0869
670	0,008013	3537,8	6,1711	0,007816	3530,6	6,1551	0,007627	3523,4	6,1393	0,007446	3516,2	6,1237
680	0,008173	3571,0	6,2061	0,007974	3564,0	6,1903	0,007783	3557,1	6,1748	0,007600	3550,2	6,1595
690	0,008331	3603,6	6,2402	0,008129	3596,9	6,2247	0,007936	3590,3	6,2094	0,007751	3583,6	6,1943
700	0,008485	3635,9	6,2735	0,008281	3629,5	6,2583	0,008086	3623,0	6,2432	0,007899	3616,6	6,2284
710	0,008638	3667,8	6,3062	0,008432	3661,6	6,2911	0,008234	3655,4	6,2763	0,008045	3649,2	6,2617
720	0,008788	3699,4	6,3381	0,008580	3693,4	6,3233	0,008380	3687,4	6,3087	0,008189	3681,4	6,2944
730	0,008936	3730,8	6,3695	0,008726	3724,9	6,3549	0,008524	3719,1	6,3405	0,008331	3713,3	6,3263
740	0,009083	3761,8	6,4003	0,008870	3756,2	6,3859	0,008666	3750,6	6,3717	0,008471	3744,9	6,3577
750	0,009227	3792,6	6,4306	0,009012	3787,2	6,4164	0,008806	3781,7	6,4023	0,008609	3776,3	6,3885
760	0,009370	3823,2	6,4604	0,009152	3818,0	6,4463	0,008944	3812,7	6,4324	0,008745	3807,4	6,4188
770	0,009511	3853,6	6,4896	0,009291	3848,5	6,4757	0,009081	3843,4	6,4620	0,008880	3838,3	6,4486
780	0,009650	3883,9	6,5185	0,009428	3878,9	6,5047	0,009216	3874,0	6,4912	0,009013	3869,0	6,4778
790	0,009788	3913,9	6,5469	0,009564	3909,1	6,5333	0,009350	3904,3	6,5199	0,009145	3899,6	6,5067
800	0,009924	3943,8	6,5749	0,009698	3939,2	6,5614	0,009482	3934,5	6,5481	0,009275	3929,9	6,5351

Table 3. Water and Superheated Steam (Continuation) Wasser und überhitzter Dampf (Fortsetzung)

t	500 bar			520 bar			540 bar			560 bar		
°C	v	h	s	v	h	s	v	h	s	v	h	s
0	0,0009767	49,3	—0,0002	0,0009758	51,3	—0,0004	0,0009749	53,2	—0,0006	0,0009741	55,1	—0,0008
10	0,0009781	89,5	0,1441	0,0009773	91,3	0,1437	0,0009764	93,2	0,1433	0,0009756	95,0	0,1429
20	0,0009804	129,9	0,2843	0,0009796	131,7	0,2838	0,0009788	133,5	0,2832	0,0009781	135,3	0,2827
30	0,0009835	170,5	0,4205	0,0009827	172,2	0,4198	0,0009819	174,0	0,4191	0,0009812	175,8	0,4184
40	0,0009872	211,2	0,5525	0,0009864	212,9	0,5517	0,0009856	214,6	0,5510	0,0009849	216,3	0,5502
50	0,0009914	251,9	0,6807	0,0009906	253,6	0,6798	0,0009898	255,3	0,6789	0,0009891	257,0	0,6780
60	0,0009961	292,8	0,8052	0,0009953	294,4	0,8042	0,0009946	296,1	0,8032	0,0009938	297,8	0,8022
70	0,0010014	333,7	0,9261	0,0010006	335,3	0,9250	0,0009998	336,9	0,9239	0,0009990	338,6	0,9228
80	0,0010071	374,7	1,0438	0,0010063	376,2	1,0426	0,0010055	377,8	1,0414	0,0010047	379,4	1,0403
90	0,0010133	415,7	1,1584	0,0010125	417,3	1,1571	0,0010116	418,8	1,1559	0,0010108	420,4	1,1546
100	0,0010200	456,8	1,2701	0,0010191	458,4	1,2688	0,0010183	459,9	1,2674	0,0010174	461,4	1,2660
110	0,0010271	498,0	1,3791	0,0010262	499,5	1,3777	0,0010253	501,0	1,3762	0,0010245	502,5	1,3747
120	0,0010347	539,4	1,4856	0,0010338	540,8	1,4840	0,0010329	542,3	1,4825	0,0010320	543,7	1,4809
130	0,0010428	580,8	1,5897	0,0010419	582,2	1,5880	0,0010409	583,6	1,5863	0,0010399	585,0	1,5847
140	0,0010514	622,4	1,6915	0,0010504	623,7	1,6897	0,0010494	625,1	1,6880	0,0010484	626,5	1,6862
150	0,0010605	664,1	1,7912	0,0010594	665,4	1,7894	0,0010583	666,7	1,7875	0,0010573	668,1	1,7857
160	0,0010701	705,9	1,8890	0,0010690	707,2	1,8871	0,0010678	708,5	1,8851	0,0010667	709,8	1,8832
170	0,0010803	748,0	1,9850	0,0010791	749,2	1,9829	0,0010779	750,5	1,9808	0,0010767	751,7	1,9788
180	0,0010910	790,2	2,0793	0,0010897	791,4	2,0771	0,0010885	792,6	2,0749	0,0010872	793,8	2,0727
190	0,0011024	832,7	2,1720	0,0011010	833,8	2,1696	0,0010997	834,9	2,1673	0,0010983	836,1	2,1650
200	0,0011144	875,4	2,2632	0,0011130	876,5	2,2607	0,0011115	877,5	2,2583	0,0011101	878,6	2,2559
210	0,0011272	918,4	2,3531	0,0011256	919,4	2,3505	0,0011241	920,4	2,3479	0,0011225	921,4	2,3453
220	0,0011407	961,6	2,4417	0,0011390	962,6	2,4390	0,0011373	963,5	2,4362	0,0011357	964,4	2,4335
230	0,0011551	1005,2	2,5292	0,0011532	1006,1	2,5263	0,0011514	1006,9	2,5234	0,0011496	1007,8	2,5205
240	0,0011703	1049,2	2,6158	0,0011683	1050,0	2,6127	0,0011664	1050,7	2,6096	0,0011645	1051,5	2,6065
250	0,0011866	1093,6	2,7015	0,0011844	1094,2	2,6981	0,0011823	1094,9	2,6948	0,0011802	1095,5	2,6916
260	0,0012040	1138,4	2,7864	0,0012016	1138,9	2,7828	0,0011992	1139,5	2,7793	0,0011969	1140,0	2,7758
270	0,0012226	1183,8	2,8707	0,0012199	1184,2	2,8668	0,0012173	1184,6	2,8631	0,0012148	1185,0	2,8593
280	0,0012426	1229,8	2,9545	0,0012396	1230,0	2,9504	0,0012367	1230,2	2,9463	0,0012339	1230,5	2,9423
290	0,0012641	1276,4	3,0380	0,0012608	1276,4	3,0335	0,0012576	1276,4	3,0292	0,0012545	1276,5	3,0249
300	0,0012874	1323,7	3,1213	0,0012837	1323,5	3,1164	0,0012801	1323,3	3,1117	0,0012766	1323,2	3,1071
310	0,0013128	1371,9	3,2046	0,0013086	1371,4	3,1993	0,0013045	1371,0	3,1942	0,0013005	1370,7	3,1892
320	0,0013406	1421,0	3,2882	0,0013358	1420,3	3,2824	0,0013311	1419,6	3,2768	0,0013265	1419,0	3,2713
330	0,0013713	1471,3	3,3723	0,0013657	1470,2	3,3659	0,0013603	1469,2	3,3597	0,0013551	1468,2	3,3536
340	0,0014055	1523,0	3,4573	0,0013989	1521,5	3,4501	0,0013926	1520,0	3,4432	0,0013865	1518,7	3,4365
350	0,0014438	1576,4	3,5436	0,0014361	1574,3	3,5356	0,0014287	1572,3	3,5278	0,0014216	1570,5	3,5203
360	0,001486	1633,9	3,6355	0,001477	1631,3	3,6267	0,001468	1628,9	3,6183	0,001460	1626,7	3,6102
370	0,001534	1686,8	3,7184	0,001523	1683,4	3,7084	0,001512	1680,3	3,6988	0,001502	1677,4	3,6897
380	0,001589	1746,8	3,8110	0,001575	1742,5	3,7996	0,001562	1738,5	3,7887	0,001550	1734,9	3,7783
390	0,001653	1810,5	3,9077	0,001635	1804,8	3,8943	0,001619	1799,7	3,8817	0,001604	1795,1	3,8698
400	0,001729	1877,7	4,0083	0,001706	1870,3	3,9922	0,001685	1863,7	3,9774	0,001666	1857,7	3,9635
410	0,001822	1949,4	4,1140	0,001791	1939,6	4,0944	0,001763	1930,9	4,0765	0,001739	1923,2	4,0601
420	0,001938	2026,6	4,2262	0,001895	2013,4	4,2017	0,001858	2002,0	4,1798	0,001825	1992,0	4,1600
430	0,002084	2110,1	4,3458	0,002024	2092,6	4,3150	0,001973	2077,5	4,2880	0,001929	2064,5	4,2639
440	0,002269	2199,7	4,4723	0,002184	2177,0	4,4343	0,002114	2157,6	4,4011	0,002055	2140,9	4,3718
450	0,002492	2293,2	4,6026	0,002378	2265,5	4,5576	0,002284	2241,7	4,5181	0,002206	2221,0	4,4833
460	0,002747	2387,2	4,7316	0,002602	2355,6	4,6813	0,002483	2327,9	4,6366	0,002382	2303,6	4,5967
470	0,003023	2478,4	4,8552	0,002849	2444,4	4,8016	0,002703	2414,0	4,7532	0,002581	2386,9	4,7096
480	0,003308	2564,9	4,9709	0,003109	2529,9	4,9158	0,002939	2497,9	4,8653	0,002795	2468,9	4,8192
490	0,003596	2646,6	5,0786	0,003374	2610,8	5,0226	0,003183	2578,3	4,9714	0,003019	2548,4	4,9241
500	0,003882	2723,0	5,1782	0,003643	2688,5	5,1239	0,003431	2655,3	5,0717	0,003248	2624,5	5,0232
510	0,004152	2791,8	5,2665	0,003902	2759,5	5,2150	0,003680	2728,3	5,1655	0,003482	2698,3	5,1180
520	0,004408	2854,9	5,3466	0,004149	2824,3	5,2973	0,003917	2794,7	5,2498	0,003710	2766,2	5,2042
530	0,004653	2913,7	5,4202	0,004386	2884,7	5,3730	0,004146	2856,5	5,3272	0,003929	2829,1	5,2831
540	0,004888	2968,9	5,4886	0,004615	2941,5	5,4432	0,004367	2914,6	5,3991	0,004142	2888,3	5,3564
550	0,005113	3021,1	5,5525	0,004834	2995,2	5,5089	0,004581	2969,7	5,4664	0,004349	2944,5	5,4251

Table 3. Water and Superheated Steam (Continuation) Wasser und überhitzter Dampf (Fortsetzung)

t	500 bar			520 bar			540 bar			560 bar		
°C	v	h	s	v	h	s	v	h	s	v	h	s
550	0,005113	3021,1	5,5525	0,004834	2995,2	5,5089	0,004581	2969,7	5,4664	0,004349	2944,5	5,4251
560	0,005328	3070,7	5,6124	0,005045	3046,3	5,5706	0,004786	3022,0	5,5297	0,004550	2998,1	5,4897
570	0,005535	3118,0	5,6688	0,005248	3094,9	5,6286	0,004985	3072,0	5,5892	0,004743	3049,2	5,5508
580	0,005734	3163,2	5,7221	0,005443	3141,4	5,6834	0,005176	3119,7	5,6455	0,004930	3098,1	5,6084
590	0,005926	3206,6	5,7726	0,005631	3185,9	5,7353	0,005360	3165,4	5,6988	0,005110	3145,0	5,6630
600	0,006111	3248,3	5,8207	0,005812	3228,8	5,7846	0,005538	3209,3	5,7493	0,005284	3189,9	5,7147
610	0,006291	3288,6	5,8666	0,005988	3270,1	5,8317	0,005710	3251,6	5,7975	0,005453	3233,1	5,7640
620	0,006465	3327,7	5,9106	0,006158	3310,0	5,8767	0,005876	3292,4	5,8435	0,005616	3274,9	5,8110
630	0,006634	3365,7	5,9529	0,006324	3348,8	5,9199	0,006038	3332,0	5,8876	0,005774	3315,3	5,8560
640	0,006799	3402,7	5,9937	0,006485	3386,6	5,9614	0,006196	3370,5	5,9299	0,005929	3354,5	5,8992
650	0,006960	3438,9	6,0331	0,006642	3423,4	6,0016	0,006349	3408,0	5,9708	0,006079	3392,7	5,9408
660	0,007118	3474,3	6,0712	0,006796	3459,4	6,0404	0,006500	3444,7	6,0103	0,006226	3430,0	5,9809
670	0,007273	3509,1	6,1083	0,006947	3494,8	6,0781	0,006647	3480,6	6,0486	0,006369	3466,5	6,0198
680	0,007424	3543,3	6,1443	0,007095	3529,5	6,1147	0,006791	3515,8	6,0858	0,006510	3502,2	6,0575
690	0,007573	3576,9	6,1795	0,007240	3563,7	6,1503	0,006932	3550,5	6,1219	0,006648	3537,4	6,0942
700	0,007720	3610,2	6,2138	0,007383	3597,4	6,1851	0,007072	3584,6	6,1572	0,006784	3572,0	6,1300
710	0,007864	3643,0	6,2473	0,007523	3630,6	6,2191	0,007208	3618,3	6,1916	0,006917	3606,1	6,1649
720	0,008006	3675,4	6,2802	0,007661	3663,5	6,2524	0,007343	3651,6	6,2253	0,007048	3639,8	6,1989
730	0,008146	3707,5	6,3123	0,007797	3696,0	6,2849	0,007475	3684,5	6,2583	0,007178	3673,1	6,2323
740	0,008284	3739,3	6,3439	0,007931	3728,2	6,3169	0,007606	3717,0	6,2906	0,007305	3706,0	6,2649
750	0,008420	3770,9	6,3749	0,008064	3760,1	6,3482	0,007735	3749,3	6,3223	0,007431	3738,6	6,2970
760	0,008554	3802,2	6,4053	0,008194	3791,7	6,3790	0,007862	3781,3	6,3534	0,007555	3770,9	6,3284
770	0,008687	3833,3	6,4353	0,008324	3823,1	6,4092	0,007988	3813,0	6,3839	0,007677	3803,0	6,3593
780	0,008818	3864,1	6,4647	0,008451	3854,3	6,4390	0,008112	3844,5	6,4140	0,007798	3834,8	6,3896
790	0,008948	3894,8	6,4937	0,008577	3885,3	6,4683	0,008235	3875,8	6,4435	0,007917	3866,4	6,4195
800	0,009076	3925,3	6,5222	0,008702	3916,1	6,4971	0,008356	3906,9	6,4726	0,008035	3897,7	6,4489

Table 3. Water and Superheated Steam (Continuation) Wasser und überhitzter Dampf (Fortsetzung)

t	580 bar			600 bar			620 bar			640 bar		
°C	v	h	s	v	h	s	v	h	s	v	h	s
0	0,0009732	56,9	−0,0010	0,0009723	58,8	−0,0012	0,0009715	60,7	−0,0014	0,0009706	62,6	−0,0016
10	0,0009748	96,8	0,1425	0,0009740	98,7	0,1420	0,0009732	100,5	0,1416	0,0009724	102,3	0,1412
20	0,0009773	137,1	0,2821	0,0009765	138,9	0,2815	0,0009757	140,6	0,2810	0,0009749	142,4	0,2804
30	0,0009804	177,5	0,4177	0,0009796	179,3	0,4170	0,0009789	181,0	0,4164	0,0009781	182,8	0,4157
40	0,0009841	218,1	0,5494	0,0009834	219,8	0,5486	0,0009826	221,5	0,5478	0,0009819	223,2	0,5470
50	0,0009883	258,7	0,6771	0,0009876	260,4	0,6762	0,0009868	262,1	0,6753	0,0009861	263,8	0,6744
60	0,0009930	299,4	0,8012	0,0009923	301,1	0,8002	0,0009915	302,7	0,7992	0,0009908	304,4	0,7982
70	0,0009982	340,2	0,9218	0,0009975	341,8	0,9207	0,0009967	343,4	0,9196	0,0009959	345,1	0,9185
80	0,0010039	381,0	1,0391	0,0010031	382,6	1,0379	0,0010023	384,2	1,0367	0,0010015	385,8	1,0356
90	0,0010100	421,9	1,1533	0,0010092	423,5	1,1520	0,0010084	425,1	1,1508	0,0010076	426,6	1,1495
100	0,0010166	462,9	1,2647	0,0010157	464,5	1,2633	0,0010149	466,0	1,2620	0,0010141	467,5	1,2606
110	0,0010236	504,0	1,3733	0,0010227	505,5	1,3719	0,0010218	507,0	1,3704	0,0010210	508,5	1,3690
120	0,0010310	545,2	1,4794	0,0010301	546,6	1,4778	0,0010292	548,1	1,4763	0,0010283	549,6	1,4748
130	0,0010390	586,5	1,5831	0,0010380	587,9	1,5814	0,0010371	589,3	1,5798	0,0010361	590,7	1,5782
140	0,0010474	627,9	1,6845	0,0010464	629,2	1,6828	0,0010454	630,6	1,6811	0,0010444	632,0	1,6794
150	0,0010562	669,4	1,7838	0,0010552	670,7	1,7820	0,0010542	672,1	1,7802	0,0010531	673,4	1,7784
160	0,0010656	711,1	1,8812	0,0010645	712,4	1,8793	0,0010634	713,7	1,8774	0,0010623	715,0	1,8755
170	0,0010755	752,9	1,9767	0,0010744	754,2	1,9747	0,0010732	755,4	1,9727	0,0010721	756,7	1,9707
180	0,0010860	795,0	2,0706	0,0010847	796,2	2,0684	0,0010835	797,4	2,0663	0,0010823	798,6	2,0642
190	0,0010970	837,2	2,1627	0,0010957	838,4	2,1605	0,0010944	839,5	2,1582	0,0010931	840,7	2,1560
200	0,0011087	879,7	2,2535	0,0011073	880,8	2,2511	0,0011059	881,9	2,2487	0,0011046	882,9	2,2464
210	0,0011210	922,4	2,3428	0,0011195	923,4	2,3402	0,0011181	924,4	2,3377	0,0011166	925,5	2,3352
220	0,0011341	965,3	2,4308	0,0011325	966,3	2,4281	0,0011309	967,3	2,4255	0,0011293	968,2	2,4228
230	0,0011479	1008,6	2,5176	0,0011462	1009,5	2,5148	0,0011445	1010,4	2,5120	0,0011428	1011,3	2,5092
240	0,0011626	1052,2	2,6035	0,0011607	1053,0	2,6005	0,0011588	1053,8	2,5975	0,0011570	1054,6	2,5945
250	0,0011781	1096,2	2,6883	0,0011761	1096,9	2,6851	0,0011741	1097,6	2,6820	0,0011721	1098,3	2,6789
260	0,0011947	1140,6	2,7724	0,0011924	1141,1	2,7690	0,0011903	1141,7	2,7656	0,0011881	1142,3	2,7623
270	0,0012123	1185,4	2,8557	0,0012099	1185,9	2,8521	0,0012075	1186,3	2,8485	0,0012051	1186,8	2,8449
280	0,0012312	1230,8	2,9384	0,0012285	1231,1	2,9345	0,0012258	1231,4	2,9307	0,0012232	1231,8	2,9270
290	0,0012514	1276,7	3,0206	0,0012484	1276,8	3,0165	0,0012455	1277,0	3,0124	0,0012426	1277,2	3,0084
300	0,0012731	1323,2	3,1025	0,0012698	1323,2	3,0980	0,0012665	1323,2	3,0936	0,0012633	1323,2	3,0893
310	0,0012966	1370,4	3,1842	0,0012929	1370,2	3,1794	0,0012892	1370,0	3,1746	0,0012856	1369,8	3,1700
320	0,0013221	1418,4	3,2659	0,0013179	1417,9	3,2606	0,0013137	1417,5	3,2555	0,0013097	1417,1	3,2504
330	0,0013500	1467,4	3,3477	0,0013451	1466,6	3,3419	0,0013404	1465,9	3,3363	0,0013358	1465,2	3,3308
340	0,0013807	1517,4	3,4300	0,0013751	1516,3	3,4236	0,0013697	1515,2	3,4174	0,0013645	1514,2	3,4113
350	0,0014148	1568,7	3,5130	0,0014083	1567,1	3,5059	0,0014021	1565,6	3,4990	0,0013960	1564,2	3,4923
360	0,001452	1624,7	3,6024	0,001444	1622,8	3,5948	0,001437	1621,0	3,5875	0,001430	1619,4	3,5804
370	0,001493	1674,8	3,6809	0,001484	1672,3	3,6725	0,001475	1670,0	3,6643	0,001467	1667,9	3,6564
380	0,001538	1731,5	3,7684	0,001528	1728,4	3,7589	0,001517	1725,5	3,7498	0,001508	1722,8	3,7411
390	0,001590	1790,8	3,8585	0,001577	1786,9	3,8478	0,001565	1783,3	3,8376	0,001553	1779,9	3,8279
400	0,001648	1852,3	3,9505	0,001632	1847,3	3,9383	0,001618	1842,7	3,9267	0,001604	1838,5	3,9156
410	0,001717	1916,2	4,0448	0,001696	1909,9	4,0306	0,001678	1904,2	4,0173	0,001661	1898,9	4,0047
420	0,001797	1983,0	4,1419	0,001771	1975,0	4,1252	0,001748	1967,7	4,1096	0,001727	1961,1	4,0951
430	0,001892	2053,0	4,2421	0,001858	2042,8	4,2222	0,001829	2033,6	4,2040	0,001802	2025,4	4,1871
440	0,002005	2126,3	4,3456	0,001962	2113,5	4,3221	0,001924	2102,0	4,3006	0,001890	2091,8	4,2809
450	0,002140	2202,9	4,4523	0,002084	2187,1	4,4246	0,002035	2173,0	4,3994	0,001992	2160,5	4,3766
460	0,002298	2282,2	4,5611	0,002226	2263,2	4,5291	0,002163	2246,4	4,5002	0,002109	2231,4	4,4739
470	0,002476	2362,6	4,6701	0,002387	2340,9	4,6344	0,002310	2321,5	4,6019	0,002243	2304,0	4,5723
480	0,002671	2442,6	4,7771	0,002565	2418,8	4,7385	0,002472	2397,3	4,7032	0,002391	2377,7	4,6708
490	0,002877	2520,9	4,8804	0,002754	2495,7	4,8400	0,002647	2472,6	4,8026	0,002552	2451,5	4,7681
500	0,003090	2596,5	4,9788	0,002952	2570,6	4,9374	0,002830	2546,5	4,8988	0,002722	2524,3	4,8629
510	0,003307	2669,5	5,0727	0,003153	2642,7	5,0301	0,003018	2618,2	4,9909	0,002898	2595,3	4,9542
520	0,003524	2738,8	5,1605	0,003359	2712,6	5,1189	0,003210	2687,6	5,0791	0,003077	2664,1	5,0414
530	0,003734	2802,7	5,2407	0,003559	2777,4	5,2001	0,003402	2753,3	5,1614	0,003260	2730,3	5,1245
540	0,003939	2862,9	5,3151	0,003755	2838,3	5,2755	0,003589	2814,8	5,2375	0,003439	2792,3	5,2012
550	0,004139	2920,0	5,3850	0,003947	2896,2	5,3463	0,003773	2873,3	5,3090	0,003615	2851,2	5,2732

Table 3. Water and Superheated Steam (Continuation) Wasser und überhitzter Dampf (Fortsetzung)

t	580 bar			600 bar			620 bar			640 bar		
°C	v	h	s	v	h	s	v	h	s	v	h	s
550	0,004139	2920,0	5,3850	0,003947	2896,2	5,3463	0,003773	2873,3	5,3090	0,003615	2851,2	5,2732
560	0,004333	2974,6	5,4509	0,004135	2951,7	5,4132	0,003954	2929,4	5,3767	0,003789	2907,8	5,3415
570	0,004522	3026,8	5,5132	0,004318	3004,8	5,4766	0,004132	2983,3	5,4410	0,003960	2962,3	5,4066
580	0,004704	3076,8	5,5721	0,004496	3055,8	5,5367	0,004304	3035,1	5,5021	0,004128	3014,9	5,4685
590	0,004880	3124,7	5,6279	0,004668	3104,7	5,5937	0,004472	3084,9	5,5602	0,004291	3065,5	5,5275
600	0,005051	3170,7	5,6809	0,004835	3151,6	5,6477	0,004635	3132,7	5,6153	0,004450	3114,1	5,5836
610	0,005216	3214,8	5,7312	0,004996	3196,7	5,6991	0,004793	3178,7	5,6677	0,004604	3161,0	5,6369
620	0,005375	3257,5	5,7792	0,005153	3240,2	5,7480	0,004946	3223,1	5,7176	0,004754	3206,1	5,6877
630	0,005531	3298,7	5,8251	0,005305	3282,2	5,7948	0,005095	3265,8	5,7652	0,004900	3249,6	5,7362
640	0,005681	3338,6	5,8691	0,005452	3322,9	5,8396	0,005239	3307,2	5,8108	0,005041	3291,8	5,7826
650	0,005828	3377,5	5,9114	0,005596	3362,4	5,8827	0,005380	3347,4	5,8546	0,005179	3332,6	5,8271
660	0,005972	3415,4	5,9522	0,005736	3400,9	5,9242	0,005517	3386,5	5,8967	0,005313	3372,3	5,8699
670	0,006112	3452,4	5,9917	0,005874	3438,5	5,9642	0,005652	3424,7	5,9374	0,005445	3411,0	5,9111
680	0,006250	3488,7	6,0300	0,006008	3475,3	6,0031	0,005783	3462,0	5,9768	0,005573	3448,9	5,9510
690	0,006385	3524,4	6,0672	0,006140	3511,5	6,0408	0,005912	3498,6	6,0150	0,005699	3485,9	5,9897
700	0,006517	3559,4	6,1034	0,006269	3547,0	6,0775	0,006038	3534,6	6,0521	0,005823	3522,3	6,0273
710	0,006647	3594,0	6,1387	0,006396	3581,9	6,1132	0,006162	3570,0	6,0883	0,005944	3558,1	6,0639
720	0,006775	3628,0	6,1732	0,006521	3616,4	6,1481	0,006285	3604,8	6,1236	0,006064	3593,4	6,0996
730	0,006901	3661,7	6,2069	0,006644	3650,4	6,1822	0,006405	3639,3	6,1580	0,006181	3628,2	6,1344
740	0,007026	3695,0	6,2400	0,006766	3684,1	6,2156	0,006523	3673,3	6,1918	0,006297	3662,5	6,1685
750	0,007148	3728,0	6,2723	0,006885	3717,4	6,2483	0,006640	3706,9	6,2248	0,006411	3696,5	6,2019
760	0,007269	3760,6	6,3041	0,007003	3750,4	6,2804	0,006755	3740,2	6,2572	0,006524	3730,1	6,2346
770	0,007388	3793,0	6,3353	0,007120	3783,1	6,3119	0,006869	3773,2	6,2890	0,006635	3763,4	6,2667
780	0,007506	3825,1	6,3659	0,007235	3815,5	6,3428	0,006981	3805,9	6,3202	0,006744	3796,4	6,2982
790	0,007623	3857,0	6,3961	0,007348	3847,7	6,3732	0,007092	3838,4	6,3509	0,006853	3829,2	6,3291
800	0,007738	3888,7	6,4257	0,007460	3879,6	6,4031	0,007202	3870,6	6,3811	0,006960	3861,7	6,3595

Table 3. Water and Superheated Steam (Continuation) Wasser und überhitzter Dampf (Fortsetzung)

t	660 bar			680 bar			700 bar			720 bar		
°C	v	h	s	v	h	s	v	h	s	v	h	s
0	0,0009698	64,5	−0,0019	0,0009690	66,3	−0,0021	0,0009682	68,2	−0,0024	0,0009673	70,1	−0,0026
10	0,0009716	104,1	0,1407	0,0009708	105,9	0,1403	0,0009700	107,8	0,1398	0,0009692	109,6	0,1393
20	0,0009742	144,2	0,2798	0,0009734	146,0	0,2792	0,0009726	147,7	0,2786	0,0009719	149,5	0,2780
30	0,0009774	184,5	0,4150	0,0009766	186,2	0,4143	0,0009759	188,0	0,4136	0,0009752	189,7	0,4129
40	0,0009811	224,9	0,5462	0,0009804	226,6	0,5454	0,0009797	228,4	0,5446	0,0009789	230,1	0,5438
50	0,0009853	265,4	0,6735	0,0009846	267,1	0,6727	0,0009839	268,8	0,6718	0,0009831	270,5	0,6709
60	0,0009900	306,0	0,7972	0,0009893	307,7	0,7962	0,0009885	309,3	0,7953	0,0009878	311,0	0,7943
70	0,0009952	346,7	0,9174	0,0009944	348,3	0,9164	0,0009937	349,9	0,9153	0,0009929	351,5	0,9142
80	0,0010008	387,4	1,0344	0,0010000	389,0	1,0332	0,0009992	390,6	1,0321	0,0009984	392,2	1,0309
90	0,0010068	428,2	1,1483	0,0010060	429,7	1,1470	0,0010052	431,3	1,1458	0,0010044	432,9	1,1446
100	0,0010132	469,0	1,2593	0,0010124	470,6	1,2579	0,0010116	472,1	1,2566	0,0010108	473,6	1,2553
110	0,0010201	510,0	1,3676	0,0010193	511,5	1,3661	0,0010184	513,0	1,3647	0,0010176	514,5	1,3633
120	0,0010275	551,0	1,4733	0,0010266	552,5	1,4718	0,0010257	553,9	1,4703	0,0010248	555,4	1,4688
130	0,0010352	592,2	1,5766	0,0010343	593,6	1,5750	0,0010334	595,0	1,5734	0,0010325	596,4	1,5718
140	0,0010434	633,4	1,6777	0,0010425	634,8	1,6760	0,0010415	636,2	1,6743	0,0010406	637,6	1,6726
150	0,0010521	674,8	1,7766	0,0010511	676,1	1,7748	0,0010501	677,5	1,7731	0,0010491	678,8	1,7713
160	0,0010613	716,3	1,8736	0,0010602	717,6	1,8717	0,0010592	718,9	1,8699	0,0010581	720,2	1,8680
170	0,0010709	758,0	1,9687	0,0010698	759,2	1,9667	0,0010687	760,5	1,9648	0,0010676	761,8	1,9628
180	0,0010811	799,8	2,0621	0,0010799	801,0	2,0600	0,0010788	802,2	2,0579	0,0010776	803,5	2,0559
190	0,0010919	841,8	2,1538	0,0010906	843,0	2,1516	0,0010894	844,2	2,1494	0,0010881	845,3	2,1472
200	0,0011032	884,1	2,2440	0,0011019	885,2	2,2417	0,0011005	886,3	2,2394	0,0010992	887,4	2,2371
210	0,0011152	926,5	2,3328	0,0011137	927,5	2,3303	0,0011123	928,6	2,3279	0,0011109	929,7	2,3255
220	0,0011278	969,2	2,4202	0,0011262	970,2	2,4177	0,0011247	971,2	2,4151	0,0011232	972,2	2,4125
230	0,0011411	1012,1	2,5065	0,0011395	1013,1	2,5037	0,0011378	1014,0	2,5010	0,0011362	1014,9	2,4984
240	0,0011552	1055,4	2,5916	0,0011534	1056,2	2,5887	0,0011517	1057,1	2,5859	0,0011500	1057,9	2,5830
250	0,0011701	1099,0	2,6758	0,0011682	1099,8	2,6727	0,0011663	1100,5	2,6697	0,0011645	1101,3	2,6667
260	0,0011860	1143,0	2,7590	0,0011839	1143,6	2,7558	0,0011818	1144,3	2,7526	0,0011798	1144,9	2,7494
270	0,0012028	1187,4	2,8415	0,0012005	1187,9	2,8380	0,0011983	1188,4	2,8346	0,0011961	1189,0	2,8312
280	0,0012207	1232,2	2,9232	0,0012182	1232,6	2,9196	0,0012157	1233,0	2,9159	0,0012133	1233,5	2,9124
290	0,0012398	1277,5	3,0044	0,0012370	1277,7	3,0005	0,0012343	1278,0	2,9966	0,0012317	1278,4	2,9928
300	0,0012602	1323,3	3,0851	0,0012571	1323,4	3,0809	0,0012541	1323,6	3,0767	0,0012512	1323,7	3,0727
310	0,0012821	1369,7	3,1654	0,0012787	1369,7	3,1609	0,0012754	1369,6	3,1564	0,0012721	1369,6	3,1521
320	0,0013057	1416,8	3,2454	0,0013019	1416,5	3,2406	0,0012982	1416,3	3,2358	0,0012946	1416,1	3,2311
330	0,0013314	1464,6	3,3254	0,0013271	1464,1	3,3201	0,0013229	1463,6	3,3149	0,0013188	1463,2	3,3098
340	0,0013594	1513,3	3,4054	0,0013545	1512,4	3,3996	0,0013498	1511,7	3,3939	0,0013452	1511,0	3,3883
350	0,0013903	1562,9	3,4857	0,0013847	1561,7	3,4793	0,0013793	1560,6	3,4730	0,0013740	1559,5	3,4669
360	0,001424	1617,8	3,5735	0,001417	1616,4	3,5667	0,001411	1615,1	3,5602	0,001405	1613,8	3,5537
370	0,001459	1665,9	3,6488	0,001452	1664,1	3,6414	0,001445	1662,4	3,6342	0,001438	1660,8	3,6273
380	0,001499	1720,3	3,7327	0,001490	1718,0	3,7245	0,001482	1715,8	3,7167	0,001474	1713,8	3,7091
390	0,001542	1776,8	3,8185	0,001532	1773,9	3,8095	0,001522	1771,2	3,8009	0,001513	1768,7	3,7925
400	0,001591	1834,7	3,9051	0,001579	1831,1	3,8951	0,001567	1827,8	3,8855	0,001556	1824,7	3,8763
410	0,001645	1894,1	3,9928	0,001631	1889,6	3,9814	0,001617	1885,5	3,9706	0,001604	1881,7	3,9603
420	0,001707	1955,1	4,0815	0,001689	1949,6	4,0686	0,001673	1944,5	4,0564	0,001657	1939,8	4,0448
430	0,001778	2017,9	4,1714	0,001756	2011,1	4,1567	0,001736	2004,8	4,1428	0,001717	1999,1	4,1297
440	0,001860	2082,6	4,2627	0,001833	2074,2	4,2458	0,001808	2066,6	4,2300	0,001785	2059,6	4,2152
450	0,001954	2149,3	4,3556	0,001920	2139,1	4,3362	0,001890	2129,9	4,3182	0,001862	2121,5	4,3014
460	0,002062	2217,9	4,4499	0,002020	2205,8	4,4278	0,001982	2194,8	4,4073	0,001948	2184,9	4,3884
470	0,002184	2288,4	4,5453	0,002132	2274,2	4,5204	0,002086	2261,4	4,4975	0,002045	2249,7	4,4762
480	0,002320	2360,0	4,6410	0,002258	2343,9	4,6136	0,002203	2329,3	4,5882	0,002153	2315,9	4,5647
490	0,002469	2432,1	4,7361	0,002395	2414,3	4,7065	0,002330	2398,1	4,6790	0,002271	2383,2	4,6534
500	0,002627	2503,7	4,8293	0,002542	2484,7	4,7980	0,002467	2467,1	4,7688	0,002399	2450,9	4,7416
510	0,002792	2574,0	4,9197	0,002697	2554,1	4,8873	0,002611	2535,6	4,8569	0,002534	2518,4	4,8283
520	0,002961	2642,4	5,0064	0,002855	2622,0	4,9734	0,002761	2602,9	4,9422	0,002675	2585,0	4,9128
530	0,003133	2708,4	5,0893	0,003018	2687,6	5,0557	0,002914	2668,3	5,0242	0,002820	2650,0	4,9943
540	0,003303	2770,8	5,1665	0,003180	2750,4	5,1335	0,003068	2731,1	5,1020	0,002967	2712,7	5,0719
550	0,003471	2830,1	5,2389	0,003341	2809,9	5,2061	0,003222	2790,7	5,1748	0,003113	2772,4	5,1449

Table 3. Water and Superheated Steam (Continuation) Wasser und überhitzter Dampf (Fortsetzung)

t	660 bar			680 bar			700 bar			720 bar		
°C	v	h	s	v	h	s	v	h	s	v	h	s
550	0,003471	2830,1	5,2389	0,003341	2809,9	5,2061	0,003222	2790,7	5,1748	0,003113	2772,4	5,1449
560	0,003638	2887,0	5,3077	0,003500	2867,1	5,2752	0,003375	2848,0	5,2440	0,003259	2829,7	5,2142
570	0,003803	2942,0	5,3733	0,003659	2922,4	5,3412	0,003527	2903,5	5,3103	0,003405	2885,4	5,2806
580	0,003965	2995,1	5,4359	0,003815	2976,0	5,4044	0,003677	2957,4	5,3739	0,003550	2939,6	5,3444
590	0,004124	3046,4	5,4957	0,003969	3027,8	5,4648	0,003826	3009,8	5,4348	0,003694	2992,2	5,4058
600	0,004279	3095,9	5,5526	0,004119	3077,9	5,5225	0,003972	3060,4	5,4931	0,003835	3043,3	5,4647
610	0,004429	3143,5	5,6069	0,004266	3126,2	5,5775	0,004115	3109,3	5,5489	0,003974	3092,8	5,5210
620	0,004576	3189,4	5,6585	0,004409	3172,8	5,6300	0,004254	3156,6	5,6021	0,004110	3140,6	5,5749
630	0,004718	3233,6	5,7078	0,004549	3217,8	5,6801	0,004391	3202,2	5,6529	0,004243	3186,9	5,6263
640	0,004856	3276,4	5,7550	0,004684	3261,3	5,7279	0,004523	3246,3	5,7015	0,004373	3231,6	5,6756
650	0,004991	3317,9	5,8001	0,004816	3303,4	5,7738	0,004652	3289,0	5,7480	0,004499	3274,8	5,7227
660	0,005123	3358,2	5,8436	0,004945	3344,3	5,8178	0,004779	3330,5	5,7926	0,004623	3316,8	5,7680
670	0,005252	3397,5	5,8854	0,005071	3384,1	5,8603	0,004902	3370,8	5,8356	0,004743	3357,7	5,8115
680	0,005377	3435,8	5,9259	0,005194	3422,9	5,9012	0,005023	3410,1	5,8771	0,004862	3397,5	5,8535
690	0,005501	3473,4	5,9651	0,005315	3460,9	5,9409	0,005141	3448,6	5,9172	0,004977	3436,4	5,8941
700	0,005622	3510,2	6,0031	0,005433	3498,2	5,9794	0,005257	3486,3	5,9562	0,005091	3474,5	5,9334
710	0,005740	3546,4	6,0401	0,005549	3534,8	6,0168	0,005370	3523,3	5,9940	0,005202	3511,9	5,9717
720	0,005857	3582,0	6,0762	0,005664	3570,8	6,0532	0,005482	3559,7	6,0308	0,005312	3548,6	6,0089
730	0,005972	3617,2	6,1114	0,005776	3606,3	6,0888	0,005592	3595,5	6,0667	0,005420	3584,8	6,0451
740	0,006085	3651,9	6,1458	0,005887	3641,3	6,1236	0,005701	3630,9	6,1018	0,005526	3620,5	6,0805
750	0,006197	3686,2	6,1795	0,005996	3675,9	6,1576	0,005808	3665,8	6,1361	0,005630	3655,8	6,1151
760	0,006307	3720,1	6,2125	0,006104	3710,2	6,1909	0,005913	3700,3	6,1697	0,005734	3690,6	6,1490
770	0,006416	3753,7	6,2448	0,006210	3744,1	6,2235	0,006017	3734,5	6,2026	0,005835	3725,1	6,1822
780	0,006523	3787,0	6,2766	0,006315	3777,7	6,2556	0,006119	3768,4	6,2350	0,005936	3759,2	6,2148
790	0,006629	3820,0	6,3078	0,006418	3811,0	6,2870	0,006221	3802,0	6,2667	0,006035	3793,0	6,2468
800	0,006733	3852,8	6,3385	0,006521	3844,0	6,3180	0,006321	3835,3	6,2979	0,006133	3826,6	6,2782

Table 3. Water and Superheated Steam (Continuation)   Wasser und überhitzter Dampf (Fortsetzung)

t °C	740 bar v	h	s	760 bar v	h	s	780 bar v	h	s	800 bar v	h	s
0	0,0009665	71,9	−0,0029	0,0009657	73,8	−0,0032	0,0009649	75,7	−0,0034	0,0009641	77,5	−0,0037
10	0,0009684	111,4	0,1389	0,0009676	113,2	0,1384	0,0009669	115,0	0,1379	0,0009661	116,8	0,1374
20	0,0009711	151,3	0,2774	0,0009704	153,0	0,2768	0,0009696	154,8	0,2762	0,0009689	156,6	0,2756
30	0,0009744	191,5	0,4122	0,0009737	193,2	0,4115	0,0009730	194,9	0,4107	0,0009722	196,6	0,4100
40	0,0009782	231,8	0,5430	0,0009775	233,5	0,5422	0,0009768	235,2	0,5414	0,0009760	236,9	0,5406
50	0,0009824	272,2	0,6700	0,0009817	273,8	0,6691	0,0009810	275,5	0,6682	0,0009803	277,2	0,6673
60	0,0009871	312,6	0,7933	0,0009863	314,3	0,7923	0,0009856	315,9	0,7914	0,0009849	317,6	0,7904
70	0,0009922	353,2	0,9132	0,0009914	354,8	0,9121	0,0009907	356,4	0,9110	0,0009900	358,0	0,9100
80	0,0009977	393,8	1,0298	0,0009969	395,3	1,0286	0,0009962	396,9	1,0275	0,0009954	398,5	1,0264
90	0,0010036	434,4	1,1433	0,0010029	436,0	1,1421	0,0010021	437,5	1,1409	0,0010013	439,1	1,1397
100	0,0010100	475,2	1,2540	0,0010092	476,7	1,2527	0,0010084	478,2	1,2514	0,0010076	479,7	1,2501
110	0,0010168	516,0	1,3619	0,0010159	517,5	1,3605	0,0010151	519,0	1,3591	0,0010143	520,5	1,3578
120	0,0010240	556,9	1,4673	0,0010231	558,3	1,4658	0,0010222	559,8	1,4643	0,0010214	561,3	1,4629
130	0,0010316	597,9	1,5703	0,0010307	599,3	1,5687	0,0010298	600,7	1,5672	0,0010289	602,2	1,5656
140	0,0010396	639,0	1,6710	0,0010387	640,4	1,6693	0,0010378	641,8	1,6677	0,0010368	643,2	1,6661
150	0,0010481	680,2	1,7696	0,0010471	681,6	1,7678	0,0010462	682,9	1,7661	0,0010452	684,3	1,7644
160	0,0010571	721,5	1,8662	0,0010561	722,9	1,8643	0,0010550	724,2	1,8625	0,0010540	725,5	1,8607
170	0,0010665	763,0	1,9609	0,0010654	764,3	1,9589	0,0010644	765,6	1,9570	0,0010633	766,9	1,9551
180	0,0010765	804,7	2,0538	0,0010753	805,9	2,0518	0,0010742	807,2	2,0498	0,0010731	808,4	2,0478
190	0,0010869	846,5	2,1451	0,0010857	847,7	2,1430	0,0010845	848,9	2,1408	0,0010833	850,1	2,1387
200	0,0010979	888,5	2,2348	0,0010966	889,6	2,2326	0,0010954	890,8	2,2304	0,0010941	891,9	2,2281
210	0,0011095	930,7	2,3231	0,0011082	931,8	2,3207	0,0011068	932,9	2,3184	0,0011055	934,0	2,3160
220	0,0011218	973,2	2,4100	0,0011203	974,2	2,4075	0,0011189	975,2	2,4050	0,0011174	976,2	2,4026
230	0,0011347	1015,8	2,4957	0,0011331	1016,8	2,4931	0,0011315	1017,7	2,4904	0,0011300	1018,7	2,4878
240	0,0011483	1058,8	2,5802	0,0011466	1059,7	2,5774	0,0011449	1060,5	2,5747	0,0011433	1061,4	2,5720
250	0,0011626	1102,0	2,6637	0,0011608	1102,8	2,6608	0,0011590	1103,6	2,6579	0,0011573	1104,4	2,6550
260	0,0011778	1145,6	2,7462	0,0011759	1146,3	2,7431	0,0011739	1147,0	2,7401	0,0011720	1147,7	2,7370
270	0,0011939	1189,6	2,8279	0,0011918	1190,2	2,8246	0,0011897	1190,8	2,8214	0,0011876	1191,4	2,8181
280	0,0012109	1233,9	2,9088	0,0012086	1234,4	2,9053	0,0012063	1234,9	2,9019	0,0012041	1235,5	2,8985
290	0,0012290	1278,7	2,9890	0,0012265	1279,1	2,9853	0,0012240	1279,5	2,9817	0,0012215	1279,9	2,9780
300	0,0012483	1323,9	3,0687	0,0012455	1324,2	3,0647	0,0012428	1324,4	3,0608	0,0012401	1324,7	3,0569
310	0,0012689	1369,7	3,1478	0,0012658	1369,7	3,1435	0,0012628	1369,8	3,1394	0,0012598	1370,0	3,1353
320	0,0012910	1415,9	3,2264	0,0012876	1415,8	3,2219	0,0012842	1415,7	3,2174	0,0012809	1415,7	3,2130
330	0,0013148	1462,8	3,3048	0,0013110	1462,5	3,2998	0,0013072	1462,2	3,2950	0,0013035	1461,9	3,2903
340	0,0013407	1510,3	3,3829	0,0013364	1509,7	3,3775	0,0013321	1509,1	3,3723	0,0013280	1508,6	3,3671
350	0,0013690	1558,5	3,4609	0,0013641	1557,6	3,4550	0,0013593	1556,7	3,4492	0,0013547	1555,9	3,4436
360	0,001399	1612,7	3,5475	0,001394	1611,6	3,5414	0,001388	1610,6	3,5354	0,001383	1609,7	3,5296
370	0,001431	1659,3	3,6205	0,001425	1657,9	3,6139	0,001419	1656,6	3,6074	0,001413	1655,4	3,6012
380	0,001466	1711,9	3,7017	0,001459	1710,1	3,6945	0,001452	1708,5	3,6875	0,001445	1707,0	3,6807
390	0,001504	1766,4	3,7844	0,001496	1764,2	3,7766	0,001488	1762,2	3,7691	0,001480	1760,3	3,7617
400	0,001546	1821,8	3,8674	0,001536	1819,1	3,8588	0,001527	1816,6	3,8505	0,001518	1814,2	3,8425
410	0,001592	1878,1	3,9505	0,001580	1874,8	3,9410	0,001570	1871,7	3,9318	0,001559	1868,8	3,9230
420	0,001643	1935,4	4,0337	0,001629	1931,4	4,0232	0,001617	1927,6	4,0131	0,001605	1924,1	4,0033
430	0,001700	1993,8	4,1173	0,001684	1988,8	4,1055	0,001669	1984,3	4,0942	0,001655	1980,0	4,0834
440	0,001764	2053,2	4,2012	0,001745	2047,2	4,1879	0,001727	2041,7	4,1753	0,001710	2036,6	4,1633
450	0,001837	2113,8	4,2856	0,001813	2106,7	4,2707	0,001792	2100,2	4,2567	0,001772	2094,1	4,2434
460	0,001918	2175,7	4,3707	0,001890	2167,4	4,3541	0,001864	2159,7	4,3384	0,001841	2152,5	4,3237
470	0,002008	2239,1	4,4565	0,001975	2229,3	4,4380	0,001944	2220,4	4,4207	0,001916	2212,1	4,4043
480	0,002109	2303,7	4,5429	0,002069	2292,5	4,5225	0,002032	2282,3	4,5034	0,001999	2272,8	4,4855
490	0,002219	2369,5	4,6296	0,002172	2356,9	4,6074	0,002129	2345,3	4,5866	0,002090	2334,7	4,5671
500	0,002338	2435,9	4,7161	0,002283	2422,0	4,6922	0,002233	2409,2	4,6698	0,002188	2397,4	4,6488
510	0,002465	2502,3	4,8015	0,002402	2487,4	4,7762	0,002345	2473,6	4,7525	0,002293	2460,7	4,7301
520	0,002598	2568,2	4,8850	0,002527	2552,4	4,8587	0,002463	2537,7	4,8339	0,002405	2524,0	4,8104
530	0,002735	2632,7	4,9658	0,002657	2616,4	4,9388	0,002586	2601,1	4,9132	0,002521	2586,7	4,8889
540	0,002874	2695,4	5,0434	0,002790	2678,8	5,0160	0,002712	2663,0	4,9899	0,002641	2648,2	4,9650
550	0,003014	2755,0	5,1164	0,002924	2738,6	5,0892	0,002841	2723,2	5,0634	0,002764	2708,0	5,0382

Table 3. Water and Superheated Steam (Continuation) **Wasser und überhitzter Dampf** (Fortsetzung)

t	740 bar			760 bar			780 bar			800 bar		
°C	v	h	s	v	h	s	v	h	s	v	h	s
550	0,003014	2755,0	5,1164	0,002924	2738,6	5,0892	0,002841	2723,2	5,0634	0,002764	2708,0	5,0382
560	0,003154	2812,4	5,1856	0,003057	2795,8	5,1583	0,002968	2780,1	5,1321	0,002886	2765,1	5,1072
570	0,003293	2868,1	5,2521	0,003191	2851,5	5,2247	0,003096	2835,6	5,1985	0,003009	2820,5	5,1733
580	0,003433	2922,4	5,3161	0,003325	2905,9	5,2889	0,003225	2890,1	5,2626	0,003132	2874,9	5,2374
590	0,003571	2975,3	5,3778	0,003458	2959,0	5,3507	0,003353	2943,3	5,3246	0,003256	2928,2	5,2995
600	0,003708	3026,7	5,4370	0,003590	3010,7	5,4103	0,003481	2995,2	5,3844	0,003379	2980,3	5,3595
610	0,003843	3076,7	5,4939	0,003721	3061,0	5,4675	0,003607	3045,7	5,4420	0,003501	3031,0	5,4173
620	0,003975	3125,0	5,5483	0,003849	3109,7	5,5225	0,003732	3094,9	5,4973	0,003622	3080,4	5,4729
630	0,004105	3171,8	5,6004	0,003976	3157,0	5,5751	0,003855	3142,5	5,5504	0,003741	3128,4	5,5264
640	0,004231	3217,0	5,6502	0,004099	3202,8	5,6255	0,003975	3188,8	5,6013	0,003859	3175,0	5,5777
650	0,004355	3260,9	5,6980	0,004220	3247,1	5,6738	0,004094	3233,6	5,6501	0,003974	3220,3	5,6270
660	0,004476	3303,4	5,7438	0,004339	3290,1	5,7201	0,004209	3277,0	5,6969	0,004088	3264,2	5,6743
670	0,004594	3344,7	5,7878	0,004455	3331,9	5,7647	0,004323	3319,3	5,7420	0,004199	3306,9	5,7198
680	0,004710	3385,0	5,8303	0,004568	3372,7	5,8076	0,004434	3360,5	5,7854	0,004308	3348,4	5,7636
690	0,004824	3424,3	5,8714	0,004679	3412,4	5,8491	0,004543	3400,6	5,8273	0,004414	3389,0	5,8060
700	0,004935	3462,8	5,9112	0,004788	3451,3	5,8893	0,004650	3439,9	5,8679	0,004519	3428,7	5,8470
710	0,005044	3500,6	5,9498	0,004895	3489,5	5,9283	0,004755	3478,5	5,9073	0,004622	3467,6	5,8867
720	0,005152	3537,7	5,9873	0,005001	3526,9	5,9663	0,004858	3516,3	5,9456	0,004723	3505,7	5,9253
730	0,005257	3574,3	6,0239	0,005104	3563,8	6,0032	0,004960	3553,5	5,9829	0,004823	3543,3	5,9629
740	0,005361	3610,3	6,0597	0,005206	3600,1	6,0392	0,005059	3590,1	6,0192	0,004921	3580,2	5,9996
750	0,005464	3645,8	6,0946	0,005306	3636,0	6,0744	0,005158	3626,3	6,0547	0,005017	3616,7	6,0354
760	0,005565	3681,0	6,1287	0,005405	3671,4	6,1089	0,005255	3662,0	6,0894	0,005113	3652,6	6,0704
770	0,005664	3715,7	6,1622	0,005503	3706,4	6,1426	0,005350	3697,3	6,1234	0,005206	3688,2	6,1046
780	0,005762	3750,1	6,1950	0,005599	3741,1	6,1757	0,005445	3732,2	6,1568	0,005299	3723,4	6,1382
790	0,005860	3784,2	6,2273	0,005694	3775,5	6,2082	0,005538	3766,8	6,1895	0,005390	3758,2	6,1711
800	0,005955	3818,0	6,2589	0,005788	3809,5	6,2401	0,005630	3801,1	6,2216	0,005481	3792,8	6,2034

Table 3. Water and Superheated Steam (Continuation) Wasser und überhitzter Dampf (Fortsetzung)

t	850 bar			900 bar			950 bar			1000 bar		
°C	v	h	s	v	h	s	v	h	s	v	h	s
0	0,0009621	82,1	−0,0044	0,0009602	86,7	−0,0052	0,0009583	91,3	−0,0059	0,0009565	95,9	−0,0067
10	0,0009642	121,2	0,1362	0,0009623	125,7	0,1349	0,0009605	130,1	0,1336	0,0009586	134,5	0,1323
20	0,0009670	160,9	0,2740	0,0009652	165,3	0,2724	0,0009634	169,7	0,2708	0,0009616	174,0	0,2692
30	0,0009704	201,0	0,4082	0,0009687	205,3	0,4064	0,0009669	209,5	0,4046	0,0009651	213,8	0,4027
40	0,0009743	241,1	0,5386	0,0009725	245,4	0,5366	0,0009708	249,6	0,5346	0,0009690	253,8	0,5325
50	0,0009785	281,4	0,6651	0,0009767	285,6	0,6629	0,0009750	289,7	0,6607	0,0009733	293,9	0,6585
60	0,0009831	321,7	0,7880	0,0009814	325,8	0,7856	0,0009796	329,9	0,7832	0,0009779	334,0	0,7808
70	0,0009882	362,1	0,9074	0,0009864	366,1	0,9048	0,0009846	370,1	0,9022	0,0009828	374,2	0,8996
80	0,0009936	402,5	1,0235	0,0009918	406,5	1,0207	0,0009900	410,4	1,0179	0,0009882	414,4	1,0152
90	0,0009994	443,0	1,1366	0,0009975	446,9	1,1336	0,0009957	450,8	1,1306	0,0009939	454,7	1,1277
100	0,0010057	483,6	1,2468	0,0010037	487,4	1,2436	0,0010018	491,2	1,2405	0,0009999	495,1	1,2373
110	0,0010123	524,2	1,3543	0,0010103	528,0	1,3509	0,0010083	531,7	1,3476	0,0010064	535,5	1,3442
120	0,0010193	564,9	1,4593	0,0010172	568,6	1,4557	0,0010152	572,3	1,4521	0,0010132	576,0	1,4486
130	0,0010267	605,8	1,5618	0,0010246	609,4	1,5580	0,0010225	613,0	1,5542	0,0010204	616,6	1,5505
140	0,0010346	646,7	1,6620	0,0010323	650,2	1,6580	0,0010301	653,7	1,6541	0,0010279	657,2	1,6502
150	0,0010428	687,7	1,7601	0,0010405	691,1	1,7559	0,0010382	694,6	1,7517	0,0010359	698,0	1,7476
160	0,0010515	728,8	1,8562	0,0010491	732,2	1,8518	0,0010467	735,5	1,8474	0,0010443	738,9	1,8431
170	0,0010607	770,1	1,9504	0,0010581	773,3	1,9457	0,0010556	776,6	1,9411	0,0010531	779,8	1,9366
180	0,0010703	811,5	2,0428	0,0010676	814,6	2,0379	0,0010649	817,8	2,0331	0,0010623	820,9	2,0283
190	0,0010804	853,1	2,1335	0,0010775	856,1	2,1284	0,0010747	859,1	2,1233	0,0010720	862,2	2,1183
200	0,0010910	894,8	2,2226	0,0010880	897,7	2,2172	0,0010850	900,6	2,2119	0,0010821	903,5	2,2067
210	0,0011022	936,7	2,3103	0,0010990	939,4	2,3046	0,0010958	942,2	2,2990	0,0010928	945,1	2,2935
220	0,0011139	978,8	2,3965	0,0011105	981,4	2,3905	0,0011072	984,1	2,3847	0,0011039	986,7	2,3789
230	0,0011263	1021,1	2,4815	0,0011226	1023,6	2,4752	0,0011191	1026,1	2,4690	0,0011156	1028,6	2,4630
240	0,0011393	1063,7	2,5652	0,0011354	1066,0	2,5586	0,0011316	1068,3	2,5522	0,0011279	1070,7	2,5458
250	0,0011529	1106,5	2,6479	0,0011488	1108,6	2,6409	0,0011447	1110,8	2,6342	0,0011408	1113,0	2,6275
260	0,0011674	1149,6	2,7295	0,0011629	1151,5	2,7222	0,0011585	1153,5	2,7151	0,0011543	1155,6	2,7081
270	0,0011826	1193,1	2,8102	0,0011777	1194,8	2,8025	0,0011730	1196,6	2,7950	0,0011684	1198,4	2,7877
280	0,0011986	1236,8	2,8901	0,0011933	1238,3	2,8820	0,0011882	1239,9	2,8741	0,0011833	1241,5	2,8663
290	0,0012156	1281,0	2,9692	0,0012098	1282,2	2,9606	0,0012043	1283,5	2,9522	0,0011990	1284,9	2,9441
300	0,0012335	1325,5	3,0475	0,0012273	1326,4	3,0384	0,0012213	1327,5	3,0296	0,0012155	1328,6	3,0210
310	0,0012526	1370,4	3,1253	0,0012457	1371,0	3,1156	0,0012392	1371,8	3,1062	0,0012329	1372,6	3,0971
320	0,0012729	1415,7	3,2023	0,0012654	1415,9	3,1920	0,0012582	1416,3	3,1820	0,0012514	1416,9	3,1724
330	0,0012947	1461,5	3,2788	0,0012864	1461,2	3,2677	0,0012785	1461,2	3,2570	0,0012710	1461,3	3,2467
340	0,0013182	1507,6	3,3546	0,0013090	1506,8	3,3426	0,0013003	1506,3	3,3311	0,0012921	1506,0	3,3200
350	0,0013438	1554,1	3,4299	0,0013335	1552,7	3,4169	0,0013239	1551,5	3,4043	0,0013148	1550,6	3,3922
360	0,001371	1607,6	3,5154	0,001360	1605,9	3,5020	0,001349	1604,5	3,4891	0,001339	1603,4	3,4767
370	0,001399	1652,7	3,5861	0,001386	1650,5	3,5719	0,001375	1648,7	3,5583	0,001363	1647,2	3,5454
380	0,001430	1703,5	3,6645	0,001415	1700,7	3,6492	0,001402	1698,3	3,6348	0,001390	1696,3	3,6211
390	0,001462	1756,0	3,7442	0,001446	1752,4	3,7279	0,001431	1749,4	3,7124	0,001417	1746,8	3,6979
400	0,001497	1809,0	3,8235	0,001479	1804,6	3,8059	0,001462	1800,8	3,7894	0,001446	1797,6	3,7738
410	0,001535	1862,4	3,9023	0,001514	1857,0	3,8832	0,001495	1852,3	3,8654	0,001478	1848,4	3,8487
420	0,001577	1916,3	3,9806	0,001553	1909,6	3,9597	0,001531	1903,9	3,9403	0,001511	1899,0	3,9223
430	0,001623	1970,5	4,0583	0,001595	1962,5	4,0354	0,001570	1955,6	4,0143	0,001547	1949,7	3,9948
440	0,001673	2025,3	4,1356	0,001641	2015,7	4,1105	0,001612	2007,4	4,0875	0,001587	2000,3	4,0664
450	0,001728	2080,7	4,2127	0,001691	2069,3	4,1852	0,001658	2059,6	4,1602	0,001629	2051,2	4,1373
460	0,001789	2136,9	4,2899	0,001745	2123,7	4,2599	0,001708	2112,4	4,2327	0,001675	2102,7	4,2079
470	0,001855	2194,0	4,3673	0,001805	2178,8	4,3346	0,001762	2165,9	4,3052	0,001724	2154,8	4,2785
480	0,001928	2252,2	4,4450	0,001869	2234,9	4,4095	0,001819	2220,3	4,3779	0,001777	2207,7	4,3492
490	0,002006	2311,3	4,5231	0,001938	2291,9	4,4848	0,001882	2275,5	4,4508	0,001833	2261,5	4,4202
500	0,002091	2371,5	4,6014	0,002013	2349,9	4,5602	0,001948	2331,7	4,5238	0,001893	2316,1	4,4913
510	0,002182	2432,3	4,6795	0,002093	2408,5	4,6355	0,002019	2388,5	4,5968	0,001957	2371,4	4,5624
520	0,002279	2493,4	4,7571	0,002177	2467,6	4,7105	0,002094	2445,7	4,6695	0,002024	2427,2	4,6331
530	0,002381	2554,3	4,8333	0,002267	2526,6	4,7845	0,002173	2503,1	4,7413	0,002094	2483,0	4,7031
540	0,002487	2614,4	4,9078	0,002361	2585,3	4,8570	0,002256	2560,2	4,8120	0,002168	2538,6	4,7719
550	0,002596	2673,4	4,9798	0,002458	2643,0	4,9276	0,002342	2616,6	4,8810	0,002246	2593,8	4,8393

Table 3. Water and Superheated Steam (Continuation) Wasser und überhitzter Dampf (Fortsetzung)

t	850 bar			900 bar			950 bar			1000 bar		
°C	v	h	s	v	h	s	v	h	s	v	h	s
550	0,002596	2673,4	4,9798	0,002458	2643,0	4,9276	0,002342	2616,6	4,8810	0,002246	2593,8	4,8393
560	0,002708	2730,8	5,0492	0,002558	2699,6	4,9960	0,002432	2672,2	4,9481	0,002326	2648,2	4,9050
570	0,002818	2785,7	5,1147	0,002659	2754,9	5,0619	0,002524	2726,7	5,0131	0,002409	2701,8	4,9689
580	0,002929	2839,6	5,1783	0,002760	2808,0	5,1246	0,002617	2780,1	5,0761	0,002493	2754,5	5,0311
590	0,003042	2892,9	5,2404	0,002863	2861,0	5,1863	0,002711	2832,0	5,1366	0,002580	2806,2	5,0914
600	0,003155	2945,3	5,3008	0,002967	2913,5	5,2468	0,002806	2884,3	5,1969	0,002668	2857,5	5,1505
610	0,003267	2996,5	5,3590	0,003070	2965,0	5,3055	0,002902	2936,2	5,2559	0,002758	2909,5	5,2097
620	0,003378	3046,3	5,4152	0,003172	3015,2	5,3620	0,002997	2986,8	5,3130	0,002847	2960,7	5,2674
630	0,003489	3094,9	5,4693	0,003274	3064,1	5,4165	0,003091	3036,1	5,3678	0,002934	3010,5	5,3228
640	0,003598	3142,2	5,5214	0,003375	3111,8	5,4689	0,003184	3083,9	5,4205	0,003020	3058,6	5,3758
650	0,003706	3188,2	5,5715	0,003476	3158,2	5,5195	0,003277	3130,5	5,4712	0,003106	3105,3	5,4267
660	0,003813	3233,0	5,6197	0,003576	3203,6	5,5684	0,003370	3176,1	5,5203	0,003192	3150,8	5,4757
670	0,003918	3276,6	5,6662	0,003675	3247,8	5,6156	0,003463	3220,6	5,5678	0,003277	3195,4	5,5232
680	0,004022	3319,1	5,7110	0,003773	3291,0	5,6611	0,003555	3264,3	5,6139	0,003364	3239,1	5,5693
690	0,004123	3360,6	5,7543	0,003869	3333,2	5,7052	0,003646	3307,0	5,6585	0,003450	3282,1	5,6142
700	0,004223	3401,2	5,7962	0,003964	3374,6	5,7479	0,003737	3348,9	5,7018	0,003536	3324,4	5,6579
710	0,004321	3440,9	5,8369	0,004058	3415,1	5,7893	0,003826	3390,1	5,7438	0,003621	3366,0	5,7004
720	0,004418	3479,9	5,8763	0,004150	3454,8	5,8295	0,003914	3430,5	5,7847	0,003705	3406,9	5,7418
730	0,004513	3518,2	5,9147	0,004241	3493,9	5,8686	0,004001	3470,2	5,8245	0,003789	3447,2	5,7822
740	0,004606	3555,9	5,9521	0,004331	3532,3	5,9067	0,004087	3509,3	5,8633	0,003871	3486,9	5,8216
750	0,004698	3593,1	5,9886	0,004419	3570,1	5,9439	0,004172	3547,8	5,9011	0,003952	3526,0	5,8600
760	0,004789	3629,7	6,0243	0,004505	3607,5	5,9802	0,004255	3585,8	5,9381	0,004032	3564,6	5,8976
770	0,004879	3665,9	6,0591	0,004591	3644,3	6,0157	0,004337	3623,3	5,9742	0,004111	3602,7	5,9342
780	0,004967	3701,8	6,0933	0,004675	3680,7	6,0505	0,004418	3660,3	6,0095	0,004189	3640,3	5,9701
790	0,005054	3737,2	6,1268	0,004759	3716,7	6,0845	0,004497	3696,9	6,0441	0,004266	3677,5	6,0053
800	0,005140	3772,3	6,1597	0,004841	3752,4	6,1179	0,004576	3733,1	6,0779	0,004341	3714,3	6,0397

Table 3a. Water and Superheated Steam with 1 K-Steps of Temperature in the Critical Range

Wasser und überhitzter Dampf mit 1 K-Stufung der Temperatur im kritischen Bereich

t	210 bar $t_s = 369{,}78\,°C$			215 bar $t_s = 371{,}76\,°C$			220 bar $t_s = 373{,}69\,°C$			225 bar		
	v	h	s	v	h	s	v	h	s	v	h	s
°C	m³/kg	kJ/kg	kJ/kg K	m³/kg	kJ/kg	kJ/kg K	m³/kg	kJ/kg	kJ/kg K	m³/kg	kJ/kg	kJ/kg K
350	0,0016503	1641,9	3,7197	0,0016430	1639,5	3,7145	0,0016361	1637,2	3,7095	0,0016294	1634,9	3,7046
351	0,0016615	1650,0	3,7330	0,0016537	1647,4	3,7276	0,0016463	1644,9	3,7223	0,0016393	1642,6	3,7172
352	0,0016729	1658,5	3,7467	0,0016646	1655,8	3,7410	0,0016569	1653,2	3,7355	0,0016495	1650,7	3,7302
353	0,0016848	1667,2	3,7605	0,0016761	1664,3	3,7545	0,0016678	1661,5	3,7488	0,0016600	1658,9	3,7433
354	0,0016974	1675,9	3,7744	0,0016881	1672,8	3,7682	0,0016793	1669,9	3,7622	0,0016710	1667,1	3,7564
355	0,0017106	1684,7	3,7885	0,0017006	1681,4	3,7818	0,0016913	1678,3	3,7755	0,0016825	1675,4	3,7695
356	0,0017247	1693,6	3,8027	0,0017139	1690,1	3,7957	0,0017039	1686,8	3,7890	0,0016945	1683,6	3,7827
357	0,0017396	1702,7	3,8171	0,0017279	1698,9	3,8097	0,0017172	1695,3	3,8026	0,0017071	1692,0	3,7960
358	0,0017555	1712,0	3,8318	0,0017428	1707,9	3,8239	0,0017312	1704,0	3,8165	0,0017204	1700,5	3,8094
359	0,0017725	1721,5	3,8469	0,0017587	1717,0	3,8384	0,0017460	1712,9	3,8305	0,0017344	1709,1	3,8231
360	0,0017908	1731,4	3,8625	0,0017756	1726,5	3,8533	0,0017619	1722,0	3,8449	0,0017492	1717,9	3,8370
361	0,0018107	1741,6	3,8786	0,0017939	1736,2	3,8687	0,0017787	1731,3	3,8596	0,0017650	1726,9	3,8512
362	0,0018324	1752,3	3,8954	0,0018136	1746,3	3,8846	0,0017968	1741,0	3,8748	0,0017818	1736,1	3,8658
363	0,0018563	1763,5	3,9131	0,0018350	1756,9	3,9012	0,0018164	1751,0	3,8905	0,0017998	1745,7	3,8808
364	0,0018830	1775,5	3,9319	0,0018586	1768,0	3,9186	0,0018376	1761,4	3,9069	0,0018191	1755,6	3,8964
365	0,0019132	1788,3	3,9521	0,0018847	1779,8	3,9371	0,0018608	1772,4	3,9242	0,0018401	1766,0	3,9126
366	0,0019482	1802,5	3,9742	0,0019142	1792,4	3,9570	0,0018865	1784,1	3,9424	0,0018630	1776,9	3,9297
367	0,0019900	1818,4	3,9991	0,0019481	1806,3	3,9786	0,0019153	1796,6	3,9619	0,0018883	1788,4	3,9477
368	0,0020424	1837,0	4,0281	0,0019880	1821,7	4,0027	0,0019481	1810,1	3,9831	0,0019164	1800,7	3,9670
369	0,0021142	1860,4	4,0647	0,0020370	1839,5	4,0304	0,0019864	1825,2	4,0066	0,0019483	1814,1	3,9878
370	0,0051418	2362,0	4,8447	0,0021013	1861,2	4,0642	0,0020326	1842,3	4,0333	0,0019852	1828,8	4,0107
371	0,0055422	2409,4	4,9184	0,0021980	1890,7	4,1100	0,0020914	1862,7	4,0649	0,0020291	1845,4	4,0365
372	0,0058332	2443,0	4,9704	0,0047120	2321,2	4,7776	0,0021736	1888,8	4,1054	0,0020836	1864,8	4,0665
373	0,0060725	2470,0	5,0123	0,0052029	2382,6	4,8727	0,0023169	1929,1	4,1679	0,0021561	1888,6	4,1034
374	0,0062802	2493,1	5,0480	0,0055243	2421,3	4,9325	0,0042851	2276,2	4,7044	0,0022662	1921,2	4,1538
375	0,0064660	2513,5	5,0795	0,0057795	2451,3	4,9788	0,0048860	2355,9	4,8276	0,0025085	1982,4	4,2483
376	0,0066355	2531,9	5,1079	0,0059970	2476,4	5,0175	0,0052377	2400,2	4,8958	0,0039426	2238,1	4,6426
377	0,0067923	2548,8	5,1339	0,0061893	2498,3	5,0512	0,0055079	2433,2	4,9466	0,0046027	2331,0	4,7856
378	0,0069389	2564,4	5,1579	0,0063634	2517,8	5,0813	0,0057342	2460,3	4,9883	0,0049771	2380,3	4,8612
379	0,0070769	2579,1	5,1804	0,0065235	2535,7	5,1086	0,0059324	2483,7	5,0241	0,0052590	2416,1	4,9162
380	0,0072076	2592,8	5,2015	0,0066725	2552,1	5,1338	0,0061105	2504,4	5,0559	0,0054924	2445,1	4,9606
381	0,0073322	2605,9	5,2214	0,0068123	2567,4	5,1573	0,0062735	2523,1	5,0846	0,0056952	2469,8	4,9985
382	0,0074512	2618,3	5,2404	0,0069444	2581,8	5,1792	0,0064245	2540,4	5,1109	0,0058765	2491,7	5,0318
383	0,0075673	2630,1	5,2585	0,0070700	2595,4	5,1999	0,0065659	2556,4	5,1353	0,0060416	2511,3	5,0618
384	0,0076741	2641,1	5,2753	0,0071898	2608,3	5,2196	0,0066991	2571,3	5,1580	0,0061942	2529,3	5,0891
385	0,0077778	2651,8	5,2915	0,0073045	2620,6	5,2382	0,0068254	2585,4	5,1795	0,0063366	2545,9	5,1144
386	0,0078787	2662,1	5,3072	0,0074158	2632,3	5,2561	0,0069457	2598,8	5,1997	0,0064706	2561,4	5,1380
387	0,0079769	2672,1	5,3223	0,0075191	2643,2	5,2727	0,0070608	2611,5	5,2190	0,0065974	2576,0	5,1601
388	0,0080726	2681,8	5,3371	0,0076196	2653,8	5,2887	0,0071713	2623,6	5,2374	0,0067180	2589,8	5,1809
389	0,0081660	2691,3	5,3514	0,0077173	2664,0	5,3042	0,0072773	2635,1	5,2548	0,0068333	2602,9	5,2007
390	0,0082572	2700,5	5,3652	0,0078126	2674,0	5,3192	0,0073771	2645,9	5,2711	0,0069438	2615,3	5,2196
391	0,0083463	2709,5	5,3788	0,0079054	2683,7	5,3338	0,0074742	2656,4	5,2869	0,0070501	2627,3	5,2376
392	0,0084335	2718,2	5,3919	0,0079961	2693,1	5,3480	0,0075688	2666,6	5,3022	0,0071504	2638,5	5,2545
393	0,0085188	2726,7	5,4047	0,0080846	2702,3	5,3617	0,0076611	2676,5	5,3171	0,0072467	2649,2	5,2706
394	0,0086023	2735,1	5,4172	0,0081712	2711,2	5,3751	0,0077511	2686,1	5,3315	0,0073405	2659,6	5,2861
395	0,0086842	2743,2	5,4294	0,0082560	2719,9	5,3882	0,0078390	2695,4	5,3455	0,0074320	2669,6	5,3012
396	0,0087645	2751,2	5,4413	0,0083390	2728,4	5,4009	0,0079249	2704,5	5,3591	0,0075212	2679,4	5,3158
397	0,0088433	2758,9	5,4529	0,0084203	2736,7	5,4133	0,0080090	2713,4	5,3724	0,0076084	2688,9	5,3301
398	0,0089207	2766,6	5,4643	0,0085000	2744,8	5,4254	0,0080913	2722,0	5,3853	0,0076936	2698,2	5,3439
399	0,0089967	2774,0	5,4754	0,0085782	2752,7	5,4372	0,0081720	2730,5	5,3979	0,0077769	2707,2	5,3573
400	0,0090714	2781,3	5,4863	0,0086549	2760,5	5,4487	0,0082510	2738,8	5,4102	0,0078585	2716,0	5,3704

Table 3a. Water and Superheated Steam with 1 K-Steps of Temperature in the Critical Range (Continuation)
Wasser und überhitzter Dampf mit 1 K-Stufung der Temperatur im kritischen Bereich (Fortsetzung)

t	230 bar			235 bar			240 bar			245 bar		
	v	h	s	v	h	s	v	h	s	v	h	s
°C	m³/kg	kJ/kg	kJ/kg K	m³/kg	kJ/kg	kJ/kg K	m³/kg	kJ/kg	kJ/kg K	m³/kg	kJ/kg	kJ/kg K
350	0,0016231	1632,8	3,6999	0,0016169	1630,8	3,6953	0,0016111	1628,8	3,6909	0,0016054	1627,0	3,6866
351	0,0016326	1640,4	3,7124	0,0016262	1638,2	3,7076	0,0016201	1636,2	3,7031	0,0016141	1634,2	3,6986
352	0,0016424	1648,4	3,7252	0,0016357	1646,1	3,7203	0,0016292	1644,0	3,7155	0,0016230	1641,9	3,7109
353	0,0016526	1656,4	3,7381	0,0016455	1654,1	3,7330	0,0016387	1651,8	3,7281	0,0016322	1649,7	3,7233
354	0,0016631	1664,5	3,7509	0,0016557	1662,0	3,7456	0,0016485	1659,7	3,7405	0,0016417	1657,4	3,7356
355	0,0016741	1672,6	3,7638	0,0016662	1670,0	3,7583	0,0016587	1667,5	3,7530	0,0016516	1665,1	3,7479
356	0,0016856	1680,7	3,7767	0,0016773	1677,9	3,7710	0,0016693	1675,3	3,7654	0,0016618	1672,8	3,7601
357	0,0016977	1688,9	3,7897	0,0016888	1685,9	3,7837	0,0016804	1683,1	3,7779	0,0016724	1680,5	3,7724
358	0,0017103	1697,1	3,8028	0,0017008	1694,0	3,7965	0,0016919	1691,1	3,7905	0,0016835	1688,3	3,7847
359	0,0017236	1705,5	3,8161	0,0017134	1702,2	3,8094	0,0017039	1699,1	3,8031	0,0016950	1696,1	3,7971
360	0,0017376	1714,1	3,8296	0,0017267	1710,5	3,8226	0,0017166	1707,2	3,8159	0,0017071	1704,0	3,8096
361	0,0017524	1722,8	3,8433	0,0017407	1719,0	3,8359	0,0017298	1715,4	3,8289	0,0017197	1712,1	3,8223
362	0,0017681	1731,7	3,8574	0,0017555	1727,6	3,8495	0,0017438	1723,8	3,8422	0,0017329	1720,2	3,8352
363	0,0017848	1740,9	3,8718	0,0017711	1736,5	3,8635	0,0017585	1732,4	3,8557	0,0017469	1728,6	3,8483
364	0,0018027	1750,4	3,8867	0,0017877	1745,6	3,8778	0,0017741	1741,2	3,8695	0,0017615	1737,1	3,8618
365	0,0018219	1760,2	3,9022	0,0018055	1755,0	3,8926	0,0017906	1750,3	3,8838	0,0017770	1745,9	3,8755
366	0,0018426	1770,5	3,9183	0,0018245	1764,8	3,9080	0,0018083	1759,7	3,8985	0,0017935	1755,0	3,8897
367	0,0018652	1781,3	3,9352	0,0018451	1775,1	3,9240	0,0018271	1769,4	3,9137	0,0018110	1764,3	3,9043
368	0,0018900	1792,7	3,9530	0,0018674	1785,8	3,9407	0,0018475	1779,6	3,9296	0,0018297	1774,1	3,9195
369	0,0019176	1805,0	3,9720	0,0018918	1797,1	3,9584	0,0018695	1790,3	3,9463	0,0018498	1784,2	3,9353
370	0,0019486	1818,1	3,9926	0,0019188	1809,2	3,9772	0,0018935	1801,6	3,9638	0,0018716	1794,9	3,9519
371	0,0019842	1832,6	4,0150	0,0019491	1822,3	3,9975	0,0019200	1813,6	3,9825	0,0018952	1806,1	3,9694
372	0,0020261	1848,7	4,0400	0,0019835	1836,5	4,0195	0,0019496	1826,5	4,0025	0,0019213	1818,1	3,9879
373	0,0020771	1867,2	4,0687	0,0020236	1852,2	4,0439	0,0019830	1840,5	4,0242	0,0019501	1830,9	4,0077
374	0,0021425	1889,3	4,1029	0,0020717	1870,0	4,0715	0,0020215	1855,9	4,0480	0,0019826	1844,7	4,0291
375	0,0022342	1917,6	4,1465	0,0021315	1890,8	4,1034	0,0020670	1873,1	4,0745	0,0020197	1859,7	4,0524
376	0,0023871	1959,3	4,2107	0,0022110	1916,0	4,1424	0,0021223	1892,6	4,1047	0,0020629	1876,3	4,0779
377	0,0027855	2049,3	4,3493	0,0023280	1949,6	4,1941	0,0021930	1915,8	4,1403	0,0021145	1894,9	4,1065
378	0,0037544	2219,8	4,6115	0,0025368	2002,1	4,2748	0,0022897	1944,7	4,1847	0,0021785	1916,3	4,1395
379	0,0043661	2310,2	4,7501	0,0029947	2098,4	4,4225	0,0024369	1984,2	4,2453	0,0022619	1942,1	4,1790
380	0,0047469	2362,5	4,8302	0,0036619	2215,1	4,6014	0,0026918	2044,0	4,3370	0,0023779	1974,8	4,2292
381	0,0050346	2400,5	4,8884	0,0041820	2294,9	4,7234	0,0031124	2128,9	4,4669	0,0025528	2019,0	4,2968
382	0,0052722	2431,1	4,9351	0,0045502	2347,6	4,8039	0,0036113	2216,6	4,6008	0,0028218	2079,2	4,3887
383	0,0054778	2457,0	4,9747	0,0048363	2386,8	4,8637	0,0040454	2285,3	4,7057	0,0031838	2150,8	4,4978
384	0,0056610	2479,8	5,0093	0,0050741	2418,5	4,9119	0,0043874	2336,0	4,7829	0,0035802	2221,0	4,6047
385	0,0058275	2500,2	5,0404	0,0052801	2445,4	4,9528	0,0046643	2375,3	4,8427	0,0039454	2280,4	4,6951
386	0,0059809	2518,9	5,0687	0,0054637	2468,9	4,9886	0,0048980	2407,6	4,8916	0,0042555	2327,9	4,7672
387	0,0061239	2536,1	5,0948	0,0056305	2490,0	5,0206	0,0051021	2435,1	4,9333	0,0045175	2366,3	4,8255
388	0,0062582	2552,1	5,1190	0,0057840	2509,3	5,0497	0,0052844	2459,2	4,9698	0,0047436	2398,5	4,8742
389	0,0063851	2567,1	5,1417	0,0059270	2527,0	5,0765	0,0054502	2480,8	5,0025	0,0049431	2426,2	4,9161
390	0,0065057	2581,3	5,1632	0,0060611	2543,4	5,1013	0,0056030	2500,5	5,0322	0,0051225	2450,7	4,9531
391	0,0066208	2594,8	5,1835	0,0061877	2558,9	5,1246	0,0057452	2518,7	5,0595	0,0052861	2472,7	4,9862
392	0,0067311	2607,6	5,2028	0,0063080	2573,4	5,1465	0,0058786	2535,5	5,0849	0,0054371	2492,7	5,0163
393	0,0068372	2619,9	5,2212	0,0064228	2587,2	5,1672	0,0060046	2551,3	5,1086	0,0055779	2511,2	5,0441
394	0,0069384	2631,6	5,2388	0,0065326	2600,4	5,1869	0,0061242	2566,2	5,1309	0,0057100	2528,3	5,0698
395	0,0070338	2642,4	5,2551	0,0066381	2612,9	5,2057	0,0062382	2580,3	5,1520	0,0058349	2544,4	5,0939
396	0,0071267	2653,0	5,2709	0,0067398	2625,0	5,2237	0,0063474	2593,7	5,1721	0,0059534	2559,6	5,1165
397	0,0072173	2663,2	5,2862	0,0068347	2636,2	5,2406	0,0064522	2606,5	5,1912	0,0060665	2573,9	5,1379
398	0,0073056	2673,2	5,3010	0,0069265	2646,9	5,2565	0,0065532	2618,7	5,2095	0,0061747	2587,5	5,1583
399	0,0073920	2682,8	5,3154	0,0070161	2657,3	5,2720	0,0066485	2630,4	5,2269	0,0062786	2600,6	5,1777
400	0,0074763	2692,3	5,3294	0,0071035	2667,3	5,2870	0,0067392	2641,2	5,2430	0,0063787	2613,0	5,1962

Table 4. International Skeleton Tables with Tolerances

Experimental results, derived from a number of authors, played a fundamental role in establishing the formulations of the properties of steam. The various results were plotted on a simple grid of pressure and temperature, and for each point on the grid a weighted average was calculated: each of these values has an associated tolerance, which specifies the range within which the actual point of state should lie. This critical evaluation of all known experimental data (international input) was carried out by an international working group set up by the IAPS. The Skeleton Tables are therefore a summarized documentation of experimental data on the properties of steam.

The Skeleton Tables 1985 for specific volume and specific enthalpy in single fluid phase and at saturation were accepted at the Tenth International Conference on the Properties of Steam 1984 [277]. The values of these Tables range from 0 °C to 800 °C and from 1 bar to 10000 bar. These Tables replace the Skeleton Tables 1963 [60].

The International Skeleton Tables for dynamic viscosity (Table 4d) and thermal conductivity (Table 4e) are published in the IAPS-Releases [278, 279] respectively: the values of Table 4d for dynamic viscosity correspond to the values of the earlier Release of 1975 [264].

Internationale Rahmentafeln mit Toleranzen

Bei der Aufstellung der Formulationen für die Eigenschaften des Wasserdampfes spielen die experimentellen Daten eine entscheidende Rolle. Um diese, von verschiedenen Autoren stammenden Daten, auswerten zu können, wurden sie auf ein einheitliches Raster von Druck und Temperatur umgerechnet und für jeden Rasterpunkt ein gewichteter Mittelwert gebildet. Zu jedem Mittelwert gehört eine Toleranz, die den Bereich kennzeichnet, innerhalb dessen der wahre Zustandspunkt liegen sollte. Diese kritische Auswertung aller bekannt gewordenen experimentellen Daten (international input) wurde durch eine internationale Arbeitsgruppe im Auftrag der IAPS durchgeführt. Die auf diese Weise entstandenen Rahmentafeln (Skeleton Tables) sind demnach Dokumentationen der experimentellen Daten über die Eigenschaften des Wasserdampfs.

Auf der Zehnten Internationalen Konferenz über die Eigenschaften des Wasserdampfs (September 1984) wurden die hier (in den Tafeln 4a, 4b und 4c) wiedergegebenen Rahmentafeln 1985 (Skeleton Tables 1985) für das spezifische Volumen und die spezifische Enthalpie im einphasigen und im Sättigungsbereich angenommen [277]. Die Werte dieser Tafeln reichen von 0 °C bis 800 °C und von 1 bar bis 10000 bar. Sie ersetzen die bisher gültigen Rahmentafeln 1963 (Skeleton Tables 1963) [60].

Die internationalen Rahmentafeln für die dynamische Viskosität (Tafel 4d) und die Wärmeleitfähigkeit (Tafel 4e) sind in den IAPS-Verlautbarungen [278] bzw. [279] publiziert, wobei die Werte der Tafel 4d für die dynamische Viskosität den Werten der früheren Verlautbarung [264] von 1975 entsprechen.

Table 4a. International Skeleton Table 1985 for Specific Volume *v* and Specific Enthalpy *h* at Saturation State with the Associated Tolerances

For each parameter the second column represents the tolerance plus/minus in the same unit as the parameter.

Internationale Rahmentafel 1985 für das spezifische Volumen *v* und die spezifische Enthalpie *h* im Sättigungszustand mit den zugehörigen Toleranzen

Die jeweils zweite Spalte enthält die Plus/Minus-Toleranz in der gleichen Einheit wie die Zustandsgröße.

t °C	p bar	±	v' dm³/kg	±	v'' dm³/kg	±	h' kJ/kg	±	h'' kJ/kg	±
0,01	0,00611659	0,00000010	1,000210	0,000010	206031	150	0,000611787	0,000000010	2500,3	1,6
5	0,0087246	0,0000005	1,000085	0,000010	147064	100	21,017	0,010	2509,8	1,6
10	0,0122792	0,0000009	1,000347	0,000010	106353	80	42,013	0,021	2519,2	1,6
15	0,0170528	0,0000017	1,000947	0,000010	77917	60	62,968	0,029	2528,4	1,6
20	0,0233849	0,0000029	1,001844	0,000010	57791	40	83,895	0,036	2537,6	1,6
25	0,031687	0,000005	1,003008	0,000015	43364	30	104,81	0,04	2546,7	1,6
30	0,042451	0,00006	1,004415	0,000015	32900	25	125,71	0,05	2555,7	1,6
35	0,056263	0,00008	1,006046	0,000015	25223	20	146,60	0,06	2564,7	1,6
40	0,073811	0,00010	1,007887	0,000015	19530	15	167,50	0,06	2573,7	1,6
45	0,095897	0,00012	1,009925	0,000015	15264	10	188,39	0,07	2582,6	1,6
50	0,123446	0,00015	1,012149	0,000015	12037	8	209,29	0,07	2591,4	1,6
55	0,157521	0,00017	1,014551	0,000015	9573	7	230,20	0,08	2600,2	1,6
60	0,199331	0,00021	1,017126	0,000015	7674	6	251,12	0,09	2609,0	1,6
65	0,250239	0,00024	1,019866	0,000015	6200	5	272,05	0,10	2617,7	1,6
70	0,311777	0,00028	1,022768	0,000015	5045	4	293,00	0,10	2626,3	1,6
75	0,385653	0,000033	1,025829	0,000020	4133,3	3,0	313,96	0,11	2634,8	1,6
80	0,473759	0,000038	1,029045	0,000020	3408,9	2,5	334,93	0,11	2643,2	1,6
85	0,57818	0,00004	1,032416	0,000020	2829,0	2,0	355,93	0,12	2651,6	1,6
90	0,70121	0,00005	1,035939	0,000020	2361,8	1,7	376,95	0,13	2659,8	1,6
95	0,84533	0,00005	1,039615	0,000020	1982,9	1,4	397,99	0,14	2667,9	1,6
100	1,01325		1,043442	0,000020	1673,7	1,2	419,07	0,14	2675,8	1,6
110	1,4324	0,0004	1,051558	0,000020	1210,7	0,9	461,30	0,18	2691,3	1,6
120	1,9848	0,0005	1,060296	0,000025	892,3	0,7	503,69	0,22	2706,2	1,6
125	2,3201	0,0006	1,064903	0,000025	770,9	0,6	524,94	0,24	2713,3	1,6
130	2,7002	0,0007	1,069674	0,000025	668,8	0,5	546,25	0,26	2720,3	1,6
140	3,6119	0,0009	1,079718	0,000025	509,0	0,4	589,01	0,30	2733,6	1,6
150	4,7571	0,0012	1,090460	0,000030	392,85	0,28	632,01	0,35	2746,0	1,6
160	6,1766	0,0015	1,10194	0,00005	307,08	0,22	675,3	0,4	2757,5	1,7
170	7,9147	0,0020	1,11422	0,00007	242,81	0,19	718,9	0,4	2767,9	1,8
175	8,9180	0,0022	1,12067	0,00009	216,77	0,17	740,8	0,4	2772,7	1,8
180	10,0193	0,0025	1,12734	0,00009	194,01	0,16	762,8	0,5	2777,2	1,9
190	12,5417	0,003	1,14139	0,00010	156,49	0,14	807,2	0,5	2785,2	1,9
200	15,537	0,004	1,15645	0,00015	127,31	0,12	852,0	0,5	2791,9	2,0
210	19,062	0,005	1,17262	0,00015	104,37	0,10	897,4	0,6	2797,2	2,1
220	23,178	0,006	1,19004	0,00020	86,15	0,09	943,3	0,6	2800,9	2,2
230	27,950	0,007	1,20886	0,00020	71,55	0,08	989,9	0,7	2802,8	2,4
240	33,446	0,008	1,22927	0,00025	59,75	0,07	1037,2	0,7	2802,9	2,5
250	39,735	0,010	1,25149	0,00030	50,12	0,06	1085,4	0,7	2800,9	2,6
260	46,892	0,012	1,27583	0,00035	42,20	0,06	1134,6	0,8	2796,5	2,8
270	54,996	0,014	1,3027	0,0004	35,64	0,05	1184,9	0,8	2789,6	3,0
280	64,127	0,016	1,3324	0,0005	30,17	0,04	1236,5	0,9	2779,7	3,1
290	74,375	0,018	1,3658	0,0005	25,57	0,04	1289,6	0,9	2766,5	3,2
300	85,831	0,022	1,4037	0,0006	21,671	0,033	1344,6	1,0	2749,4	3,4
310	98,597	0,025	1,4473	0,0007	18,344	0,033	1401,7	1,0	2727,7	3,6
320	112,784	0,028	1,4984	0,0007	15,479	0,033	1461,7	1,1	2700,3	3,8
330	128,515	0,032	1,5601	0,0008	12,987	0,032	1525,4	1,1	2666	4
340	145,93	0,04	1,6374	0,0009	10,790	0,029	1594,1	1,2	2622	4
350	165,21	0,04	1,7403	0,0010	8,812	0,026	1670,6	1,2	2564	5
360	186,57	0,05	1,894	0,008	6,957	0,035	1760,9	1,9	2482	6
370	210,33	0,05	2,215	0,020	4,96	0,04	1890,0	3,5	2334	7
371	212,86	0,05	2,280	0,020	4,71	0,04	1909,8	3,5	2309	7
372	215,42	0,05	2,365	0,020	4,43	0,04	1933,8	3,5	2277	7
373	218,02	0,05	2,496	0,025	4,07	0,04	1966	4	2233	10
373,99	220,64	0,05	3,106	0,030	3,106	0,030	2086	15	2086	15

Table 4b. Part 1. International Skeleton Table 1985 for Specific Volume

The specific volume values and their associated tolerances denoted in dm³/kg are represented as the upper figures and the lower figures, respectively.

Internationale Rahmentafel 1985 für das spezifische Volumen

Die Werte für das spezifische Volumen und ihre zugehörigen Toleranzen in dm³/kg sind durch die Ziffern der oberen bzw. der unteren Zeile wiedergegeben.

p bar	\multicolumn{12}{c}{temperature t in °C}
	0	25	50	75	100	125	150	175	200	250	300	350
1,01325	1,00016	1,00296	1,01211	1,02580	1,04344[1]	1792,9	1910,7	2027,7	2143,7	2374,4	2604,2	2833,2
	±0,00001[2]	0,00001	0,00001	0,00002	0,00002	1,2	1,2	1,2	1,2	1,4	1,4	1,4
5	0,99995	1,00278	1,01193	1,02560	1,04324	1,06474	1,09045	399,28	424,80	474,25	522,49	570,03
	0,00006	0,00006	0,00006	0,00006	0,00006	0,00007	0,00009	0,32	0,32	0,32	0,32	0,34
10	0,99969	1,00256	1,01170	1,02536	1,04299	1,06443	1,09010	1,12057	205,85	232,57	257,84	282,38
	0,00010	0,00010	0,00010	0,00010	0,00010	0,00011	0,00014	0,00017	0,21	0,20	0,20	0,20
25	0,99893	1,00188	1,01103	1,02466	1,04222	1,06358	1,08910	1,11934	1,15552	86,95	98,84	109,69
	0,00010	0,00010	0,00010	0,00010	0,00010	0,00013	0,00014	0,00018	0,00025	0,09	0,09	0,09
50	0,99767	1,00076	1,00992	1,02350	1,04096	1,06215	1,08744	1,11734	1,15296	1,24960	45,29	51,91
	0,00010	0,00010	0,00010	0,00010	0,00010	0,00015	0,00017	0,00020	0,00028	0,00031	0,07	0,08
75	0,99642	0,99965	1,00882	1,02236	1,03972	1,06075	1,08580	1,11537	1,15050	1,24520	26,71	32,41
	0,00010	0,00010	0,00010	0,00010	0,00010	0,00016	0,00021	0,00029	0,00032	0,00037	0,04	0,05
100	0,99518	0,99855	1,00774	1,02122	1,03848	1,05935	1,08417	1,11342	1,14810	1,2409	1,3975	22,42
	0,00010	0,00010	0,00010	0,00010	0,00010	0,00016	0,00021	0,00030	0,00034	0,0004	0,0006	0,04
125	0,99396	0,99745	1,00666	1,02010	1,03724	1,05798	1,08255	1,11152	1,1457	1,2367	1,3872	16,12
	0,00010	0,00010	0,00010	0,00010	0,00010	0,00016	0,00021	0,00031	0,0004	0,0004	0,0006	0,04
150	0,99274	0,99636	1,00559	1,01898	1,03603	1,05662	1,08097	1,10964	1,1434	1,2327	1,3776	11,470
	0,00010	0,00010	0,00010	0,00010	0,00010	0,00016	0,00021	0,00033	0,0004	0,0005	0,0006	0,034
175	0,99153	0,99527	1,00452	1,01787	1,03482	1,05526	1,07940	1,10780	1,1411	1,2288	1,3687	1,7144
	0,00010	0,00010	0,00010	0,00010	0,00010	0,00016	0,00021	0,00033	0,0004	0,0005	0,0006	0,0017
200	0,99032	0,99420	1,00346	1,01677	1,03362	1,05393	1,07786	1,10597	1,1389	1,2250	1,3604	1,6649
	0,00010	0,00010	0,00010	0,00010	0,00010	0,00016	0,00021	0,00033	0,0004	0,0005	0,0006	0,0016
225	0,98914	0,99313	1,00242	1,01569	1,03245	1,05261	1,07635	1,10413	1,1367	1,2214	1,3528	1,6286
	0,00010	0,00010	0,00010	0,00010	0,00010	0,00016	0,00021	0,00033	0,0004	0,0005	0,0006	0,0016
250	0,98796	0,99205	1,00139	1,01461	1,03128	1,05130	1,07485	1,10230	1,1345	1,2178	1,3453	1,5983
	0,00010	0,00010	0,00010	0,00010	0,00010	0,00016	0,00021	0,00033	0,0004	0,0005	0,0006	0,0014
275	0,98678	0,99100	1,00035	1,01353	1,03012	1,05000	1,07336	1,10055	1,1323	1,2143	1,3383	1,5733
	0,00010	0,00010	0,00010	0,00010	0,00012	0,00016	0,00024	0,00033	0,0004	0,0005	0,0007	0,0013
300	0,98562	0,98995	0,99932	1,01246	1,02897	1,04872	1,07189	1,09880	1,1302	1,2110	1,3316	1,5521
	0,00010	0,00010	0,00010	0,00010	0,00012	0,00016	0,00024	0,00033	0,0004	0,0005	0,0007	0,0012
350	0,98333	0,98789	0,99729	1,01036	1,02670	1,04620	1,06900	1,09540	1,1261	1,2045	1,3192	1,5168
	0,00010	0,00010	0,00010	0,00010	0,00012	0,00016	0,00024	0,00033	0,0004	0,0005	0,0007	0,0012
400	0,98108	0,98586	0,99528	1,00828	1,02446	1,04371	1,06616	1,09210	1,1221	1,1982	1,3078	1,4878
	0,00010	0,00010	0,00010	0,00010	0,00012	0,00017	0,00024	0,00033	0,0004	0,0005	0,0007	0,0012
450	0,97886	0,98385	0,99330	1,00623	1,02226	1,04128	1,06340	1,0888	1,1183	1,1923	1,2972	1,4634
	0,00010	0,00010	0,00010	0,00010	0,00012	0,00017	0,00024	0,0004	0,0004	0,0005	0,0007	0,0011

500	1,4420 0,0011	1,2874 0,0007	1,1866 0,0005	1,1145 0,0005	1,0857 0,0004	1,06069 0,00024	1,03889 0,00017	1,02009 0,00012	1,00421 0,00010	0,99136 0,00010	0,98186 0,00010	0,97660 0,00010
550	1,4232 0,0011	1,2781 0,0007	1,1812 0,0005	1,1109 0,0005	1,0826 0,0004	1,05804 0,00025	1,03654 0,00017	1,01796 0,00012	1,00223 0,00010	0,98944 0,00010	0,97992 0,00010	0,97451 0,00010
600	1,4063 0,0010	1,2695 0,0007	1,1760 0,0005	1,1074 0,0005	1,0796 0,0004	1,05545 0,00025	1,03424 0,00017	1,01586 0,00012	1,00027 0,00010	0,98755 0,00010	0,97799 0,00010	0,97240 0,00010
650	1,3910 0,0010	1,2614 0,0008	1,1710 0,0006	1,1039 0,0005	1,0767 0,0004	1,05290 0,00025	1,03197 0,00017	1,01380 0,00012	0,99834 0,00010	0,98568 0,00010	0,97609 0,00010	0,97031 0,00010
700	1,3770 0,0010	1,2537 0,0008	1,1661 0,0006	1,1006 0,0005	1,0739 0,0004	1,05040 0,00026	1,02975 0,00017	1,01178 0,00012	0,99644 0,00010	0,98383 0,00010	0,97422 0,00010	0,96826 0,00010
750	1,3642 0,0010	1,2464 0,0008	1,1615 0,0006	1,0974 0,0005	1,0711 0,0004	1,04795 0,00026	1,02756 0,00017	1,00978 0,00012	0,99457 0,00010	0,98202 0,00010	0,97237 0,00010	0,96624 0,00010
800	1,3523 0,0010	1,2394 0,0008	1,1570 0,0006	1,0942 0,0005	1,0683 0,0004	1,04554 0,00028	1,02541 0,00017	1,00781 0,00012	0,99272 0,00010	0,98022 0,00010	0,97056 0,00010	0,96426 0,00010
850	1,3411 0,0011	1,2328 0,0009	1,1526 0,0007	1,0911 0,0005	1,0656 0,0004	1,04318 0,00029	1,02329 0,00019	1,00587 0,00014	0,99090 0,00013	0,97845 0,00011	0,96875 0,00011	0,96230 0,00011
900	1,3307 0,0012	1,2264 0,0009	1,1484 0,0007	1,0881 0,0005	1,0630 0,0004	1,04086 0,00029	1,02121 0,00020	1,00396 0,00017	0,98910 0,00014	0,97670 0,00012	0,96698 0,00012	0,96037 0,00011
950	1,3209 0,0013	1,2203 0,0010	1,1443 0,0008	1,0851 0,0005	1,0604 0,0004	1,03858 0,00030	1,01916 0,00021	1,00207 0,00018	0,98732 0,00013	0,97497 0,00013	0,96522 0,00013	0,95848 0,00012
1000	1,3116 0,0013	1,2145 0,0010	1,1403 0,0008	1,0822 0,0005	1,0579 0,0004	1,03633 0,00031	1,01713 0,00024	1,00021 0,00021	0,98556 0,00016	0,97325 0,00013	0,96347 0,00015	0,95660 0,00015
1100	1,2944 0,0014	1,2033 0,0011	1,1326 0,0009	1,0766 0,0006	1,0529 0,0005	1,03190 0,00036	1,01317 0,00027	0,99653 0,00022	0,98208 0,00018	0,96985 0,00017	0,96004 0,00017	0,95290 0,00020
1200	1,2788 0,0015	1,1929 0,0012	1,1253 0,0009	1,0711 0,0006	1,0482 0,0005	1,0277 0,0004	1,00932 0,00031	0,99297 0,00025	0,97869 0,00020	0,96655 0,00020	0,95671 0,00020	0,94940 0,00030
1300	1,2646 0,0015	1,1833 0,0013	1,1184 0,0010	1,0660 0,0008	1,0436 0,0006	1,0235 0,0005	1,00558 0,00034	0,98950 0,00027	0,97540 0,00023	0,96335 0,00020	0,95347 0,00020	0,9460 0,0004
1400	1,2516 0,0016	1,1743 0,0013	1,1118 0,0011	1,0609 0,0007	1,0392 0,0006	1,0196 0,0005	1,00190 0,00035	0,98612 0,00029	0,97220 0,00025	0,96023 0,00023	0,95032 0,00022	0,9426 0,0005
1500	1,2395 0,0016	1,1657 0,0014	1,1055 0,0011	1,0561 0,0008	1,0349 0,0006	1,0157 0,0005	0,99830 0,00032	0,98282 0,00032	0,96907 0,00027	0,95720 0,00025	0,94725 0,00024	0,9394 0,0006
1600	1,2282 0,0017	1,1577 0,0014	1,0995 0,0011	1,0515 0,0009	1,0307 0,0007	1,0119 0,0005	0,99490 0,00039	0,97990 0,00034	0,96602 0,00029	0,95423 0,00030	0,94427 0,00029	0,9362 0,0007
1700	1,2177 0,0017	1,1500 0,0014	1,0938 0,0011	1,0470 0,0009	1,0267 0,0007	1,0083 0,0006	0,9915 0,0004	0,97645 0,00037	0,96302 0,00032	0,95131 0,00031	0,94134 0,00031	0,9331 0,0008
1800	1,2078 0,0017	1,1426 0,0014	1,0882 0,0011	1,0426 0,0009	1,0228 0,0007	1,0048 0,0006	0,9883 0,0005	0,9733 0,0004	0,96010 0,00033	0,94846 0,00033	0,93849 0,00033	0,9301 0,0010
1900	1,1984 0,0017	1,1356 0,0014	1,0828 0,0011	1,0384 0,0009	1,0190 0,0007	1,0013 0,0006	0,9851 0,0005	0,9704 0,0005	0,95725 0,00034	0,94568 0,00034	0,93571 0,00034	0,9272 0,0015
2000	1,1895 0,0017	1,1289 0,0014	1,0775 0,0011	1,0342 0,0009	1,0153 0,0007	0,9979 0,0006	0,9820 0,0006	0,9674 0,0005	0,95440 0,00037	0,94296 0,00036	0,93299 0,00035	0,9244 0,0020
2200	1,1730 0,0018	1,1162 0,0015	1,0675 0,0014	1,0261 0,0012	1,0080 0,0008	0,9913 0,0007	0,9759 0,0006	0,9618 0,0006	0,9490 0,0006	0,9376 0,0006	0,9277 0,0006	0,9189 0,0025
2400	1,1579 0,0020	1,1045 0,0016	1,0583 0,0014	1,0186 0,0012	1,0011 0,0010	0,9850 0,0009	0,9700 0,0009	0,9563 0,0008	0,9437 0,0008	0,9325 0,0008	0,9226 0,0008	0,9137 0,0030
2600	1,1443 0,0023	1,0937 0,0019	1,0497 0,0016	1,0116 0,0014	0,9947 0,0012	0,9790 0,0011	0,9645 0,0011	0,9510 0,0011	0,9387 0,0010	0,9276 0,0010	0,9177 0,0010	0,9088 0,0035

Table 4b. Part 1. International Skeleton Table 1985 for Specific Volume (Continuation)

Internationale Rahmentafel 1985 für das spezifische Volumen (Fortsetzung)

temperature t in °C

Each cell gives the specific volume value with its associated tolerance (value ± tolerance).

p bar	0	25	50	75	100	125	150	175	200	250	300	350
2800	0,904 ±0,004	0,9130 ±0,0012	0,9228 ±0,0012	0,9339 ±0,0012	0,9460 ±0,0013	0,9591 ±0,0013	0,9733 ±0,0013	0,9886 ±0,0014	1,0050 ±0,0015	1,0416 ±0,0018	1,0836 ±0,0021	1,1317 ±0,0026
3000	0,900 ±0,004	0,9085 ±0,0012	0,9183 ±0,0014	0,9292 ±0,0014	0,9411 ±0,0014	0,9540 ±0,0015	0,9678 ±0,0015	0,9828 ±0,0015	0,9988 ±0,0017	1,0339 ±0,0019	1,0740 ±0,0023	1,1197 ±0,0028
3200	0,895 ±0,005	0,9041 ±0,0014	0,9138 ±0,0015	0,9246 ±0,0015	0,9364 ±0,0016	0,9490 ±0,0016	0,9626 ±0,0016	0,9771 ±0,0017	0,9927 ±0,0019	1,0266 ±0,0021	1,0651 ±0,0025	1,1080 ±0,0030
3400	0,891 ±0,007	0,8999 ±0,0015	0,9095 ±0,0016	0,9202 ±0,0016	0,9319 ±0,0017	0,9443 ±0,0017	0,9575 ±0,0017	0,9717 ±0,0018	0,9869 ±0,0020	1,0198 ±0,0023	1,0568 ±0,0027	1,0980 ±0,0032
3600	0,887 ±0,008	0,8958 ±0,0016	0,9054 ±0,0017	0,9160 ±0,0017	0,9275 ±0,0017	0,9396 ±0,0017	0,9526 ±0,0018	0,9664 ±0,0019	0,9813 ±0,0020	1,0132 ±0,0024	1,0488 ±0,0029	1,0880 ±0,0034
3800	0,883 ±0,009	0,8918 ±0,0016	0,9013 ±0,0017	0,9119 ±0,0017	0,9233 ±0,0018	0,9352 ±0,0018	0,9478 ±0,0018	0,9613 ±0,0019	0,9759 ±0,0020	1,0069 ±0,0025	1,0413 ±0,0030	1,0790 ±0,0035
4000	0,880 ±0,010	0,8879 ±0,0017	0,8973 ±0,0017	0,9078 ±0,0017	0,9191 ±0,0018	0,9308 ±0,0018	0,9432 ±0,0019	0,9564 ±0,0019	0,9707 ±0,0020	1,0009 ±0,0025	1,0341 ±0,0031	1,0700 ±0,0036
4500	0,871 ±0,015	0,8788 ±0,0017	0,8881 ±0,0018	0,8984 ±0,0018	0,9093 ±0,0018	0,9205 ±0,0019	0,9333 ±0,0019	0,9449 ±0,0020	0,9583 ±0,0023	0,9867 ±0,0027	1,0170 ±0,0032	1,0500 ±0,0038
5000	0,863 ±0,015	0,8702 ±0,0017	0,8795 ±0,0018	0,8896 ±0,0018	0,9002 ±0,0019	0,9110 ±0,0019	0,9222 ±0,0020	0,9340 ±0,0022	0,9468 ±0,0025	0,9735 ±0,0030	1,0020 ±0,0036	1,033 ±0,004
5500	0,855 ±0,020	0,8620 ±0,0018	0,8715 ±0,0019	0,8814 ±0,0019	0,8915 ±0,0020	0,9020 ±0,0021	0,9128 ±0,0021	0,9241 ±0,0023	0,9361 ±0,0026	0,9614 ±0,0032	0,9880 ±0,0039	1,017 ±0,005
6000	0,848 ±0,020	0,8541 ±0,0018	0,8639 ±0,0019	0,8737 ±0,0020	0,8834 ±0,0021	0,8936 ±0,0022	0,9040 ±0,0022	0,9148 ±0,0024	0,9263 ±0,0027	0,9501 ±0,0033	0,975 ±0,004	1,003 ±0,005
6500	0,842 ±0,020	0,8465 ±0,0018	0,8567 ±0,0019	0,8664 ±0,0021	0,8759 ±0,0022	0,8857 ±0,0022	0,8958 ±0,0023	0,9062 ±0,0025	0,9171 ±0,0027	0,9398 ±0,0033	0,964 ±0,004	0,990 ±0,005
7000		0,8393 ±0,0018	0,8499 ±0,0019	0,8596 ±0,0021	0,8687 ±0,0022	0,8782 ±0,0023	0,8879 ±0,0024	0,8980 ±0,0025	0,9086 ±0,0027	0,9304 ±0,0033	0,953 ±0,004	0,978 ±0,005
7500		0,8326 ±0,0018	0,8435 ±0,0020	0,8532 ±0,0021	0,8620 ±0,0022	0,8712 ±0,0023	0,8806 ±0,0024	0,8903 ±0,0025	0,9007 ±0,0027	0,9215 ±0,0033	0,943 ±0,004	0,967 ±0,005
8000		0,8263 ±0,0019	0,8373 ±0,0021	0,8470 ±0,0022	0,8555 ±0,0023	0,8644 ±0,0024	0,8735 ±0,0025	0,8830 ±0,0027	0,8931 ±0,0029	0,9130 ±0,0035	0,934 ±0,004	0,956 ±0,005
8500		0,8200 ±0,0022	0,8310 ±0,0023	0,8410 ±0,0024	0,8490 ±0,0025	0,8580 ±0,0026	0,8660 ±0,0027	0,8760 ±0,0028	0,8860 ±0,0031	0,9050 ±0,0036	0,924 ±0,004	0,945 ±0,005
9000		0,8150 ±0,0024	0,8250 ±0,0025	0,8350 ±0,0026	0,8430 ±0,0027	0,8520 ±0,0028	0,8600 ±0,0028	0,8690 ±0,0030	0,8790 ±0,0032	0,8970 ±0,0037	0,916 ±0,004	0,936 ±0,005
9500			0,8200 ±0,0028	0,8300 ±0,0029	0,8380 ±0,0029	0,8460 ±0,0030	0,8540 ±0,0030	0,8630 ±0,0032	0,8720 ±0,0034	0,8900 ±0,0038	0,908 ±0,004	0,926 ±0,005
10000			0,815 ±0,004	0,825 ±0,004	0,833 ±0,004	0,841 ±0,004	0,849 ±0,004	0,857 ±0,004	0,866 ±0,004	0,883 ±0,004	0,900 ±0,005	0,918 ±0,006

[1] At this point, a set of the specific volume value and the associated tolerance is given for the saturated water. The value for the saturated steam is (1673.7 ± 1.2) dm³/kg.

[1] Hier sind die Werte für gesättigtes Wasser angegeben. Die Werte für gesättigten Dampf lauten (1673.7 ± 1,2) dm³/kg.

[2] Except this tolerance, the sign (±) of tolerance is omitted from the description.

[2] Außer an dieser Stelle ist das ± Zeichen der Toleranz weggelassen.

Table 4b. Part 2. International Skeleton Table 1985 for Specific Volume

The specific volume values and their associated tolerances denoted in dm³/kg are represented as the upper figures and the lower figures, respectively.

Internationale Rahmentafel 1985 für das spezifische Volumen

Die Werte für das spezifische Volumen und ihre zugehörigen Toleranzen in dm³/kg sind durch die Ziffern in der oberen bzw. der unteren Zeile wiedergegeben.

p bar	temperature t in °C											
	375	400	425	450	475	500	550	600	650	700	750	800
1,01325	2947,7	3062,3	3176,0	3290,5	3404,4	3518,8	3747,1	3975,2	4203,2	4431,2	4659,2	4887,1
	±1,5¹	1,5	1,6	1,6	1,7	1,8	1,9	2,0	2,0	2,0	2,0	2,0
5	593,68	617,23	640,72	664,2	687,6	710,9	757,5	804,0	850,5	896,9	943,3	989,6
	0,36	0,37	0,38	0,4	0,4	0,4	0,4	0,4	0,4	0,4	0,4	0,4
10	294,52	306,53	318,49	330,39	342,25	354,06	377,60	401,06	424,44	447,78	471,09	494,35
	0,20	0,20	0,20	0,20	0,20	0,20	0,20	0,20	0,20	0,20	0,20	0,20
25	114,93	120,05	125,10	130,09	135,03	139,93	149,64	159,24	168,78	178,28	187,74	197,16
	0,09	0,10	0,10	0,10	0,10	0,10	0,10	0,10	0,10	0,10	0,10	0,10
50	54,92	57,80	60,58	63,28	65,92	68,53	73,63	78,61	83,52	88,41	93,26	98,07
	0,08	0,09	0,09	0,09	0,09	0,10	0,10	0,10	0,10	0,10	0,10	0,10
75	34,76	36,93	38,99	40,95	42,85	44,70	48,28	51,74	55,12	58,48	61,79	65,05
	0,05	0,06	0,06	0,06	0,07	0,07	0,07	0,08	0,08	0,08	0,08	0,08
100	24,533	26,41	28,13	29,74	31,28	32,76	35,60	38,31	40,92	43,51	46,05	48,55
	0,037	0,04	0,04	0,04	0,06	0,06	0,07	0,07	0,07	0,07	0,07	0,07
125	18,248	20,007	21,564	22,980	24,320	25,59	27,99	30,25	32,41	34,54	36,61	38,65
	0,027	0,030	0,032	0,034	0,039	0,05	0,05	0,05	0,05	0,05	0,05	0,05
150	13,890	15,653	17,135	18,451	19,660	20,790	22,91	24,87	26,74	28,56	30,32	32,05
	0,025	0,023	0,026	0,028	0,031	0,037	0,04	0,04	0,04	0,04	0,05	0,05
175	10,556	12,452	13,921	15,181	16,313	17,361	19,280	21,03	22,69	24,29	25,84	27,35
	0,023	0,020	0,022	0,023	0,026	0,031	0,039	0,04	0,04	0,04	0,05	0,05
200	7,672	9,947	11,463	12,702	13,788	14,774	16,549	18,156	19,65	21,09	22,48	23,82
	0,018	0,018	0,021	0,021	0,022	0,027	0,030	0,035	0,04	0,04	0,05	0,05
225	2,44	7,866	9,503	10,750	11,809	12,753	14,422	15,914	17,295	18,60	19,86	21,08
	0,05	0,016	0,017	0,019	0,020	0,023	0,025	0,030	0,035	0,04	0,05	0,05
250	1,980	6,002	7,883	9,166	10,214	11,129	12,720	14,121	15,400	16,618	17,77	18,89
	0,005	0,013	0,015	0,017	0,018	0,020	0,020	0,020	0,030	0,035	0,04	0,05
275	1,8623	4,181	6,503	7,849	8,899	9,797	11,329	12,657	13,860	14,990	16,069	17,10
	0,0034	0,020	0,012	0,014	0,016	0,018	0,020	0,020	0,030	0,030	0,035	0,04
300	1,7917	2,794	5,301	6,735	7,795	8,682	10,168	11,438	12,581	13,640	14,651	15,61
	0,0026	0,014	0,010	0,012	0,014	0,016	0,018	0,020	0,020	0,030	0,030	0,04
350	1,7009	2,106	3,426	4,958	6,052	6,927	8,342	9,519	10,562	11,520	12,410	13,276
	0,0020	0,004	0,008	0,009	0,011	0,012	0,015	0,017	0,020	0,025	0,030	0,035
400	1,6406	1,9101	2,535	3,691	4,761	5,620	6,980	8,086	9,053	9,931	10,740	11,520
	0,0017	0,0031	0,005	0,006	0,008	0,010	0,014	0,016	0,018	0,020	0,025	0,030
450	1,5955	1,8029	2,1860	2,912	3,820	4,633	5,934	6,981	7,886	8,700	9,454	10,160
	0,0015	0,0022	0,0033	0,005	0,007	0,008	0,010	0,010	0,015	0,018	0,020	0,030
500	1,5592	1,7304	2,0081	2,4860	3,173	3,892	5,116	6,107	6,959	7,721	8,421	9,070
	0,0014	0,0018	0,0025	0,0038	0,005	0,007	0,010	0,010	0,014	0,018	0,018	0,025

Table 4b. Part 2. International Skeleton Table 1985 for Specific Volume
(Continuation)

Internationale Rahmentafel 1985 für das spezifische Volumen
(Fortsetzung)

temperature t in °C

p bar	375	400	425	450	475	500	550	600	650	700	750	800
550	1,5291	1,6762	1,8056	2,2410	2,749	3,347	4,469	5,404	6,208	6,925	7,581	8,196
	0,0014	0,0017	0,0024	0,0033	0,005	0,006	0,008	0,008	0,012	0,018	0,018	0,020
600	1,5032	1,6330	1,8158	2,0840	2,470	2,954	3,956	4,832	5,591	6,268	6,886	7,464
	0,0013	0,0016	0,0022	0,0030	0,004	0,005	0,007	0,008	0,010	0,016	0,018	0,018
650	1,4805	1,5970	1,7546	1,9747	2,2800	2,671	3,549	4,362	5,078	5,718	6,303	6,848
	0,0013	0,0015	0,0021	0,0028	0,0036	0,005	0,007	0,008	0,010	0,015	0,018	0,018
700	1,4604	1,5664	1,7054	1,8922	2,1430	2,463	3,226	3,975	4,648	5,253	5,807	6,323
	0,0012	0,0015	0,0020	0,0026	0,0034	0,004	0,006	0,007	0,009	0,015	0,015	0,018
750	1,4423	1,5398	1,6646	1,8271	2,0390	2,308	2,969	3,653	4,284	4,857	5,383	5,872
	0,0012	0,0015	0,0020	0,0025	0,0033	0,004	0,006	0,007	0,008	0,010	0,014	0,018
800	1,4259	1,5164	1,6298	1,7740	1,9579	2,1870	2,762	3,385	3,975	4,516	5,016	5,481
	0,0012	0,0015	0,0020	0,0024	0,0031	0,0039	0,006	0,006	0,008	0,010	0,010	0,016
850	1,4109	1,4953	1,5995	1,7294	1,8917	2,0910	2,595	3,160	3,710	4,222	4,696	5,139
	0,0013	0,0016	0,0020	0,0024	0,0030	0,0038	0,005	0,006	0,008	0,010	0,010	0,015
900	1,3970	1,4763	1,5728	1,6912	1,8365	2,0130	2,458	2,971	3,483	3,966	4,417	4,839
	0,0014	0,0017	0,0020	0,0024	0,0029	0,0036	0,005	0,006	0,007	0,008	0,010	0,015
950	1,3841	1,4590	1,5490	1,6578	1,7896	1,9477	2,344	2,810	3,286	3,742	4,170	4,573
	0,0015	0,0018	0,0021	0,0024	0,0029	0,0035	0,005	0,006	0,007	0,008	0,010	0,014
1000	1,3720	1,4430	1,5274	1,6282	1,7488	1,8919	2,248	2,672	3,115	3,545	3,952	4,336
	0,0016	0,0019	0,0021	0,0024	0,0028	0,0034	0,005	0,005	0,006	0,007	0,010	0,013
1100	1,3500	1,4143	1,4897	1,5779	1,6813	1,8018	2,096	2,452	2,835	3,217	3,585	3,935
	0,0016	0,0019	0,0022	0,0024	0,0027	0,0034	0,005	0,005	0,006	0,007	0,010	0,012
1200	1,3305	1,3895	1,4578	1,5365	1,6273	1,7315	1,983	2,285	2,619	2,959	3,291	3,612
	0,0017	0,0020	0,0022	0,0025	0,0027	0,0031	0,005	0,005	0,006	0,007	0,009	0,011
1300	1,3129	1,3677	1,4303	1,5015	1,5827	1,6748	1,893	2,155	2,447	2,751	3,053	3,347
	0,0018	0,0020	0,0023	0,0025	0,0027	0,0030	0,005	0,005	0,006	0,007	0,009	0,010
1400	1,2970	1,3482	1,4061	1,4713	1,5448	1,6273	1,821	2,051	2,310	2,582	2,857	3,126
	0,0018	0,0021	0,0023	0,0025	0,0027	0,0030	0,005	0,005	0,006	0,007	0,008	0,009
1500	1,2825	1,3306	1,3845	1,4447	1,5120	1,5869	1,760	1,966	2,196	2,442	2,692	2,941
	0,0019	0,0021	0,0023	0,0025	0,0027	0,0030	0,004	0,005	0,006	0,007	0,008	0,009
1600	1,2691	1,3145	1,3650	1,4210	1,4831	1,5519	1,710	1,894	2,102	2,324	2,553	2,782
	0,0019	0,0021	0,0023	0,0025	0,0027	0,0030	0,004	0,005	0,006	0,007	0,007	0,008
1700	1,2567	1,2996	1,3472	1,3996	1,4573	1,5210	1,665	1,834	2,022	2,225	2,435	2,647
	0,0019	0,0021	0,0023	0,0025	0,0027	0,0030	0,004	0,005	0,006	0,007	0,007	0,008
1800	1,2450	1,2858	1,3309	1,3801	1,4341	1,4934	1,627	1,781	1,953	2,139	2,332	2,529
	0,0019	0,0022	0,0023	0,0025	0,0027	0,0030	0,004	0,005	0,006	0,007	0,007	0,008
1900	1,2341	1,2730	1,3157	1,3622	1,4129	1,4685	1,593	1,735	1,893	2,064	2,243	2,426
	0,0019	0,0022	0,0023	0,0025	0,0027	0,0030	0,004	0,005	0,006	0,007	0,007	0,008

2000	2,336 / 0,009	2,165 / 0,008	1,998 / 0,007	1,840 / 0,006	1,694 / 0,005	1,562 / 0,004	1,4456 / 0,0030	1,3934 / 0,0027	1,3454 / 0,0025	1,3013 / 0,0023	1,2609 / 0,0022	1,2239 / 0,0019
2200	2,184 / 0,010	2,034 / 0,008	1,889 / 0,007	1,752 / 0,006	1,625 / 0,005	1,508 / 0,004	1,4050 / 0,0032	1,3580 / 0,0031	1,3152 / 0,0030	1,2752 / 0,0029	1,2387 / 0,0022	1,2049 / 0,0020
2400	2,061 / 0,010	1,928 / 0,008	1,800 / 0,007	1,680 / 0,006	1,568 / 0,005	1,463 / 0,004	1,370 / 0,004	1,3280 / 0,0035	1,2890 / 0,0034	1,2520 / 0,0029	1,2190 / 0,0025	1,1878 / 0,0022
2600	1,960 / 0,010	1,841 / 0,008	1,727 / 0,007	1,619 / 0,006	1,519 / 0,005	1,426 / 0,004	1,341 / 0,004	1,3030 / 0,0035	1,2660 / 0,0035	1,2320 / 0,0030	1,2010 / 0,0028	1,1724 / 0,0026
2800	1,877 / 0,010	1,770 / 0,008	1,666 / 0,007	1,569 / 0,006	1,478 / 0,005	1,393 / 0,005	1,316 / 0,004	1,2800 / 0,0038	1,2460 / 0,0036	1,2150 / 0,0034	1,1860 / 0,0032	1,1584 / 0,0029
3000	1,808 / 0,010	1,710 / 0,008	1,615 / 0,007	1,525 / 0,006	1,442 / 0,005	1,364 / 0,005	1,293 / 0,004	1,260 / 0,004	1,228 / 0,004	1,1980 / 0,0038	1,1710 / 0,0036	1,1451 / 0,0032
3200	1,748 / 0,010	1,658 / 0,008	1,571 / 0,007	1,488 / 0,006	1,410 / 0,006	1,338 / 0,005	1,272 / 0,005	1,241 / 0,005	1,211 / 0,004	1,183 / 0,004	1,1570 / 0,0038	1,1320 / 0,0035
3400	1,696 / 0,010	1,612 / 0,008	1,531 / 0,007	1,454 / 0,006	1,382 / 0,006	1,315 / 0,005	1,254 / 0,005	1,224 / 0,005	1,196 / 0,005	1,170 / 0,004	1,145 / 0,004	1,1210 / 0,0036
3600	1,650 / 0,011	1,572 / 0,008	1,496 / 0,007	1,424 / 0,006	1,357 / 0,006	1,294 / 0,005	1,236 / 0,005	1,209 / 0,005	1,182 / 0,005	1,157 / 0,005	1,133 / 0,004	1,1100 / 0,0038
3800	1,610 / 0,011	1,536 / 0,008	1,465 / 0,007	1,398 / 0,006	1,334 / 0,006	1,275 / 0,006	1,220 / 0,005	1,194 / 0,005	1,169 / 0,005	1,145 / 0,005	1,122 / 0,004	1,1000 / 0,0039
4000	1,572 / 0,011	1,504 / 0,008	1,437 / 0,007	1,374 / 0,006	1,314 / 0,006	1,257 / 0,006	1,205 / 0,005	1,181 / 0,005	1,157 / 0,005	1,134 / 0,005	1,111 / 0,004	1,090 / 0,004
4500	1,492 / 0,011	1,433 / 0,008	1,376 / 0,007	1,321 / 0,006	1,268 / 0,006	1,218 / 0,006	1,172 / 0,006	1,151 / 0,005	1,130 / 0,005	1,108 / 0,005	1,087 / 0,005	1,069 / 0,004
5000	1,429 / 0,011	1,377 / 0,008	1,327 / 0,008	1,278 / 0,007	1,231 / 0,007	1,186 / 0,007	1,145 / 0,006	1,125 / 0,006	1,106 / 0,006	1,086 / 0,005	1,067 / 0,005	1,049 / 0,005
5500	1,379 / 0,011	1,332 / 0,010	1,286 / 0,009	1,241 / 0,009	1,199 / 0,009	1,158 / 0,008	1,120 / 0,008	1,103 / 0,008	1,086 / 0,008	1,067 / 0,006	1,049 / 0,006	1,032 / 0,005
6000	1,335 / 0,012	1,293 / 0,010	1,251 / 0,010	1,210 / 0,010	1,171 / 0,009	1,134 / 0,009	1,099 / 0,009	1,083 / 0,009	1,067 / 0,008	1,050 / 0,007	1,033 / 0,006	1,017 / 0,006
6500	1,297 / 0,012	1,258 / 0,011	1,220 / 0,011	1,183 / 0,011	1,147 / 0,010	1,113 / 0,009	1,081 / 0,009	1,065 / 0,009	1,050 / 0,008	1,034 / 0,007	1,018 / 0,007	1,004 / 0,006
7000	1,263 / 0,013	1,228 / 0,013	1,193 / 0,012	1,159 / 0,012	1,125 / 0,011	1,093 / 0,010	1,063 / 0,009	1,048 / 0,009	1,034 / 0,008	1,020 / 0,007	1,005 / 0,007	0,992 / 0,006
7500	1,235 / 0,015	1,201 / 0,014	1,169 / 0,014	1,137 / 0,013	1,105 / 0,012	1,075 / 0,010	1,047 / 0,009	1,033 / 0,009	1,020 / 0,009	1,007 / 0,008	0,993 / 0,007	0,980 / 0,006
8000	1,209 / 0,019	1,178 / 0,017	1,147 / 0,016	1,116 / 0,015	1,086 / 0,012	1,057 / 0,010	1,031 / 0,009	1,018 / 0,009	1,006 / 0,009	0,993 / 0,008	0,980 / 0,007	0,968 / 0,006
8500	1,185 / 0,019	1,155 / 0,017	1,126 / 0,017	1,097 / 0,016	1,067 / 0,012	1,040 / 0,010	1,016 / 0,010	1,004 / 0,009	0,992 / 0,009	0,980 / 0,008	0,968 / 0,007	0,956 / 0,006
9000	1,161 / 0,019	1,134 / 0,017	1,106 / 0,016	1,079 / 0,016	1,051 / 0,012	1,026 / 0,011	1,002 / 0,010	0,991 / 0,009	0,979 / 0,009	0,968 / 0,008	0,956 / 0,007	0,946 / 0,006
9500	1,140 / 0,019	1,114 / 0,017	1,088 / 0,016	1,063 / 0,016	1,036 / 0,013	1,013 / 0,011	0,990 / 0,010	0,979 / 0,009	0,967 / 0,009	0,956 / 0,008	0,946 / 0,007	0,936 / 0,006
10000	1,120 / 0,022	1,096 / 0,022	1,071 / 0,019	1,047 / 0,016	1,022 / 0,015	1,000 / 0,012	0,979 / 0,010	0,968 / 0,010	0,957 / 0,010	0,946 / 0,009	0,936 / 0,009	0,927 / 0,007

¹ Except this tolerance, the sign (±) of tolerance is omitted from the description.
¹ Außer an dieser Stelle ist das ± Zeichen für die Toleranz weggelassen.

Table 4c. Part 1. International Skeleton Table 1985 for Specific Enthalpy

The specific enthalpy values and their associated tolerances denoted in kJ/kg are represented as the upper figures and the lower figures, respectively.

Internationale Rahmnentafel 1985 für die spezifische Enthalpie

Die Werte für die spezifische Enthalpie und ihre zugehörigen Toleranzen in kJ/kg sind durch die Ziffern der oberen bzw. der unteren Reihe wiedergegeben.

p bar	\multicolumn{12}{c}{temperature t in °C}
	0	25	50	75	100	125	150	175	200	250	300	350
1,01325	0,06	104,86	209,33	313,97	419,04[1]	2726,1	2775,7	2824,8	2874,3	2973,6	3073,7	3175,0
	±0,01[2]	0,07	0,10	0,11	0,15	2,0	2,0	2,0	2,0	2,0	3,0	3,0
5	0,47	105,23	209,67	314,29	419,34	525,11	632,00	2800,1	2854,4	2959,8	3063,0	3167
	0,01	0,10	0,16	0,16	0,20	0,33	0,5	2,0	2,0	3,0	3,6	4
10	0,98	105,69	210,10	314,69	419,71	525,48	632,3	740,9	2826,9	2941,5	3050,0	3157
	0,01	0,12	0,19	0,28	0,33	0,35	0,5	0,7	2,0	3,0	3,9	4
25	2,50	107,08	211,40	315,90	420,84	526,5	633,3	741,7	852,4	2879	3007	3125
	0,01	0,15	0,19	0,28	0,33	0,4	0,5	0,7	0,9	4	4	4
50	5,04	109,38	213,55	317,92	422,72	528,2	634,8	743,0	853,4	1085,4	2923	3067
	0,03	0,16	0,19	0,28	0,33	0,4	0,5	0,7	0,9	1,8	4	4
75	7,57	111,69	215,70	319,94	424,60	530,0	636,4	744,3	854,5	1085,4	2813	3000
	0,04	0,16	0,19	0,28	0,34	0,4	0,5	0,7	0,9	1,8	4	4
100	10,09	113,98	217,85	321,95	426,49	531,7	637,9	745,7	855,5	1085,5	1342,9	2922
	0,05	0,17	0,19	0,29	0,34	0,4	0,5	0,7	0,9	1,8	2,0	4
125	12,59	116,28	220,00	323,97	428,38	533,4	639,5	747,1	856,6	1085,6	1340,2	2825
	0,06	0,17	0,19	0,29	0,34	0,4	0,5	0,7	0,9	1,8	2,0	4
150	15,09	118,57	222,14	325,98	430,26	535,2	641,1	748,4	857,8	1085,8	1337,9	2691
	0,07	0,17	0,20	0,29	0,34	0,4	0,5	0,7	0,9	1,8	2,0	7
175	17,58	120,85	224,28	328,00	432,15	536,9	642,7	749,8	858,9	1086,1	1335,8	1662,2
	0,08	0,18	0,30	0,30	0,34	0,4	0,5	0,7	0,9	1,8	2,0	3,0
200	20,06	123,13	226,43	330,01	434,04	538,7	644,3	751,2	860,0	1086,4	1334,0	1645,6
	0,10	0,18	0,30	0,30	0,34	0,4	0,5	0,7	0,9	1,8	2,0	3,0
225	22,53	125,41	228,56	332,03	435,94	540,5	645,9	752,6	861,2	1086,7	1332,3	1633,3
	0,11	0,18	0,30	0,30	0,34	0,4	0,5	0,7	0,9	1,8	2,0	3,0
250	25,00	127,68	230,70	334,04	437,83	542,2	647,5	754,1	862,4	1087,1	1330,9	1623,4
	0,12	0,19	0,30	0,30	0,35	0,4	0,5	0,7	0,9	1,8	2,0	3,0
275	27,45	129,95	232,84	336,06	439,72	544,0	649,1	755,5	863,6	1087,6	1329,6	1615,3
	0,13	0,19	0,30	0,30	0,35	0,4	0,5	0,7	0,9	1,8	2,0	3,0
300	29,90	132,21	234,97	338,07	441,62	545,7	650,7	756,9	864,8	1088,1	1328,5	1608,4
	0,15	0,19	0,30	0,30	0,35	0,4	0,5	0,7	0,9	1,8	2,0	3,0
350	34,76	136,73	239,23	342,10	445,41	549,3	654,0	759,8	867,3	1089,2	1326,6	1597,2
	0,17	0,25	0,30	0,30	0,35	0,4	0,5	0,7	0,9	1,8	2,0	3,0
400	39,60	141,23	243,48	346,12	449,20	552,8	657,3	762,8	869,8	1090,4	1325,3	1588,5
	0,19	0,26	0,30	0,30	0,36	0,4	0,5	0,7	0,9	1,8	2,0	3,0
450	44,40	145,71	247,72	350,14	453,00	556,4	660,6	765,8	872,4	1091,8	1324,3	1581,4
	0,22	0,29	0,30	0,30	0,36	0,4	0,5	0,7	0,9	1,8	2,0	3,0

	(1)	(2)	(3)	(4)	(5)	(6)	(7)	(8)	(9)	(10)	(11)	(12)
500	49,17 / 0,24	150,17 / 0,31	251,95 / 0,30	354,16 / 0,30	456,80 / 0,36	560,0 / 0,4	663,9 / 0,5	768,8 / 0,7	875,0 / 0,9	1093,3 / 1,8	1323,6 / 2,0	1575,7 / 3,0
550	53,91 / 0,27	154,63 / 0,34	256,18 / 0,30	358,18 / 0,30	460,60 / 0,36	563,6 / 0,4	667,2 / 0,5	771,8 / 0,7	877,7 / 0,9	1094,9 / 1,8	1323,2 / 2,0	1571,0 / 2,9
600	58,62 / 0,29	159,06 / 0,36	260,40 / 0,35	362,20 / 0,35	464,40 / 0,36	567,2 / 0,4	670,6 / 0,5	774,9 / 0,7	880,4 / 0,9	1096,6 / 1,8	1323,1 / 2,0	1567,1 / 2,9
650	63,30 / 0,31	163,48 / 0,39	264,60 / 0,35	366,20 / 0,35	468,20 / 0,37	570,8 / 0,4	673,9 / 0,5	777,9 / 0,7	883,1 / 0,9	1098,4 / 1,8	1323,2 / 2,0	1563,9 / 2,9
700	67,96 / 0,34	167,9 / 0,4	268,8 / 0,5	370,2 / 0,5	472,1 / 0,5	574,4 / 0,5	677,3 / 0,5	781,0 / 0,7	885,9 / 0,9	1100,3 / 1,8	1323,6 / 2,0	1561,2 / 3,0
750	72,60 / 0,36	172,3 / 0,4	273,0 / 0,6	374,2 / 0,6	475,9 / 0,5	578,0 / 0,5	680,7 / 0,5	784,2 / 0,8	888,7 / 0,9	1102,3 / 1,8	1324,1 / 2,0	1559,1 / 3,0
800	77,20 / 0,38	176,7 / 0,4	277,2 / 0,6	378,2 / 0,6	479,7 / 0,7	581,6 / 0,7	684,1 / 0,7	787,3 / 0,8	891,6 / 0,9	1104,3 / 1,8	1324,7 / 2,0	1557,3 / 3,0
850	81,8 / 0,4	181,0 / 0,4	281,4 / 0,6	382,2 / 0,7	483,5 / 0,8	585,2 / 0,8	687,5 / 0,8	790,5 / 0,9	894,5 / 1,0	1106,4 / 1,8	1325,6 / 2,0	1555,9 / 3,0
900	86,3 / 0,4	185,4 / 0,4	285,5 / 0,7	386,2 / 0,7	487,3 / 0,9	588,9 / 0,9	690,9 / 0,9	793,7 / 1,0	897,4 / 1,0	1108,5 / 1,8	1326,5 / 2,0	1554,8 / 3,0
950	90,9 / 0,4	189,7 / 0,4	289,7 / 0,7	390,2 / 0,8	491,2 / 1,0	592,5 / 1,0	694,4 / 1,0	796,9 / 1,1	900,3 / 1,3	1110,8 / 1,9	1327,6 / 2,5	1554,1 / 3,2
1000	95,4 / 0,4	194,0 / 0,4	293,9 / 0,7	394,2 / 0,8	495,0 / 1,1	596,1 / 1,1	697,8 / 1,2	800,1 / 1,5	903,3 / 1,5	1113,0 / 2,0	1328,8 / 2,8	1553,5 / 3,3
1100	104,4 / 0,6	202,6 / 0,5	302,2 / 0,7	402,2 / 0,9	502,6 / 1,2	603,4 / 1,3	704,7 / 1,5	806,6 / 1,7	909,3 / 1,9	1117,7 / 2,4	1331,6 / 3,0	1553,2 / 3,4
1200	113,3 / 0,6	211,2 / 0,5	310,1 / 0,7	410,1 / 1,0	510,3 / 1,2	610,7 / 1,4	711,7 / 1,6	813,1 / 1,8	915,3 / 2,2	1122,6 / 2,6	1334,7 / 3,1	1553,1 / 3,5
1300	122,1 / 0,7	219,7 / 0,7	318,6 / 0,8	418,1 / 1,0	517,9 / 1,3	618,1 / 1,4	718,6 / 1,8	819,7 / 1,9	921,5 / 2,3	1127,7 / 2,7	1338,1 / 3,2	1554,1 / 3,5
1400	130,9 / 0,7	228,2 / 0,7	326,9 / 0,8	426,0 / 1,0	525,6 / 1,3	625,4 / 1,5	725,6 / 1,8	826,3 / 2,0	927,7 / 2,3	1132,9 / 2,7	1341,8 / 3,2	1555,7 / 3,6
1500	139,6 / 0,8	236,6 / 0,7	335,0 / 0,8	433,9 / 1,0	533,2 / 1,3	632,7 / 1,5	732,6 / 1,8	832,9 / 2,0	933,9 / 2,3	1138,2 / 2,8	1345,8 / 3,2	1557,7 / 3,6
1600	148,2 / 0,8	244,9 / 0,7	343,2 / 0,8	441,8 / 1,1	540,8 / 1,3	640,1 / 1,6	739,7 / 1,8	839,7 / 2,1	940,3 / 2,3	1143,6 / 2,8	1350,0 / 3,3	1560,1 / 3,6
1700	156,8 / 1,1	253,3 / 0,7	351,3 / 0,8	449,7 / 1,1	548,5 / 1,3	647,4 / 1,6	746,8 / 1,8	846,4 / 2,1	946,7 / 2,3	1149,1 / 2,8	1354,4 / 3,3	1562,9 / 3,6
1800	165,3 / 1,1	261,6 / 0,7	359,4 / 0,9	457,6 / 1,1	556,1 / 1,4	654,8 / 1,6	753,9 / 1,8	853,2 / 2,1	953,1 / 2,3	1154,8 / 2,8	1358,9 / 3,3	1565,9 / 3,6
1900	173,8 / 1,2	269,9 / 0,8	367,5 / 0,9	465,5 / 1,1	563,7 / 1,4	662,2 / 1,6	761,0 / 1,9	860,0 / 2,1	959,7 / 2,4	1160,5 / 2,8	1363,6 / 3,3	1569,3 / 3,6
2000	182,2 / 1,4	278,1 / 1,0	375,6 / 1,0	473,3 / 1,1	571,4 / 1,4	669,6 / 1,6	768,1 / 1,9	866,9 / 2,1	966,2 / 2,4	1166,3 / 2,8	1368,5 / 3,3	1572,8 / 3,6
2200	198,9 / 1,8	294,5 / 1,1	391,6 / 1,1	489,0 / 1,2	586,6 / 1,4	684,4 / 1,7	782,4 / 1,9	880,7 / 2,2	979,4 / 2,4	1178,2 / 2,9	1378,6 / 3,4	1580,7 / 3,7
2400	215,5 / 2,1	310,7 / 1,3	407,6 / 1,3	504,6 / 1,4	601,8 / 1,6	699,1 / 1,8	796,8 / 2,2	894,5 / 2,2	992,8 / 2,4	1190,3 / 2,9	1389,1 / 3,4	1589,2 / 3,7
2600	232,0 / 2,7	326,9 / 1,4	423,5 / 1,4	520,2 / 1,5	617,0 / 1,7	713,9 / 2,0	811,2 / 2,3	908,5 / 2,3	1006,2 / 2,5	1202,7 / 2,9	1400,1 / 3,4	1598,4 / 3,7

Table 4c. Part 1. International Skeleton Table 1985 for Specific Enthalpy
(Continuation)

Internationale Rahmentafel 1985 für die spezifische Enthalpie
(Fortsetzung)

p bar	temperature t in °C											
	0	25	50	75	100	125	150	175	200	250	300	350
2800	248,4	342,9	439,3	535,7	632,2	728,7	825,6	922,5	1019,8	1215,2	1411,3	1608,1
	3,7	1,8	1,6	1,6	1,8	2,0	2,3	2,5	2,7	3,0	3,4	3,7
3000	265	358,8	455,0	551,1	647,3	743,5	840,0	936,6	1033,4	1227,9	1422,9	1618,2
	5	2,0	1,8	1,8	1,8	2,1	2,4	2,7	2,8	3,1	3,5	3,8
3200	281	374,7	470,7	566,6	662,5	758,3	854,5	950,7	1047,2	1240,8	1434,8	1628,7
	7	2,1	1,8	1,8	1,9	2,1	2,4	2,7	2,8	3,1	3,5	3,8
3400	297	390,5	486,3	581,9	677,6	773,1	868,9	964,8	1061,0	1253,8	1446,8	1639,6
	9	2,1	1,9	1,9	1,9	2,2	2,5	2,8	2,9	3,1	3,5	3,8
3600	314	406,2	501,8	597,2	692,6	787,9	883,4	979,0	1074,8	1266,9	1459,0	1650,7
	11	2,2	2,0	2,0	2,0	2,2	2,5	2,8	2,9	3,1	3,6	3,8
3800	330	421,8	517,3	612,5	707,7	802,7	897,9	993,2	1088,7	1280,1	1471,4	1662,2
	13	2,5	2,0	2,0	2,0	2,3	2,6	2,8	3,0	3,2	3,6	3,9
4000	346	437,4	532,7	627,8	722,7	817,4	912,4	1007,5	1102,7	1293,4	1484,0	1673,8
	15	3,0	2,1	2,1	2,1	2,3	2,6	2,9	3,0	3,2	3,6	3,9
4500	387	476	570,9	665,7	760,1	854,2	948,6	1043,2	1137,7	1327,0	1516,0	1704
	19	4	2,2	2,2	2,2	2,4	2,7	3,0	3,1	3,3	3,7	4
5000	428	514	608,8	703,4	797,3	890,9	984,8	1078,9	1173,0	1361,1	1548,7	1735
	25	5	3,0	2,3	2,3	2,5	2,8	3,1	3,2	3,4	3,8	4
5500	469	552	646,3	740,8	834,4	927,6	1021,0	1114,7	1208,3	1395,4	1581,7	1767
	33	6	3,2	2,4	2,4	2,6	2,9	3,2	3,3	3,4	3,9	4
6000	510	590	683,4	778,0	871,3	964,1	1057,1	1150,6	1243,7	1430,0	1615,3	1799
	41	7	3,4	2,5	2,5	2,8	3,0	3,3	3,4	3,4	3,9	4
6500	550	627	720	814,9	908,1	1000,4	1093,1	1186,4	1279,2	1464,8	1649	1832
	50	8	4	2,7	2,7	3,0	3,2	3,4	3,6	3,7	4	4
7000		663	756	852	944,7	1036,7	1129,0	1222,2	1314,8	1499,6	1683	1866
		9	5	4	3,5	3,4	3,4	3,6	3,7	3,8	4	4
7500		700	792	888	981,2	1073,0	1165,0	1258,0	1350,2	1534,6	1718	1899
		10	6	4	3,9	3,8	3,8	3,9	3,9	3,9	4	4
8000		736	828	924	1017	1109	1201	1294	1386	1570	1752	1933
		12	7	6	5	5	5	4	4	4	4	4
8500		772	862	960	1054	1145	1236	1329	1421	1605	1787	1967
		14	8	6	5	5	5	4	4	4	4	4
9000		808	897	996	1089	1181	1272	1365	1457	1640	1821	2002
		16	8	7	7	7	5	4	4	4	4	4
9500			931	1031	1125	1216	1307	1401	1492	1675	1856	2036
			10	8	8	7	5	4	4	4	4	4
10000			965	1066	1161	1252	1343	1436	1527	1710	1891	2071
			15	10	9	8	6	6	5	5	5	5

[1] At this point, a set of the specific enthalpy value and the associated tolerance is given for the saturated water. The value for the saturated steam is (2675,8 ± 2,0) kJ/kg.

[1] Hier sind die Werte für gesättigtes Wasser angegeben. Die Werte für gesättigten Dampf lauten (2675,8 ± 2,0) kJ/kg.

[2] Except this tolerance, the sign (±) of toleranc is omitted from description.

Table 4c. Part 2. International Skeleton Table 1985 for Specific Enthalpy

The specific enthalpy values and their associated tolerances denoted in kJ/kg are represented as the upper figures and the lower figures, respectively.

Internationale Rahmentafel 1985 für die spezifische Enthalpie

Die Werte für die spezifische Enthalpie und ihre zugehörigen Toleranzen in kJ/kg sind durch die Ziffern der oberen bzw. der unteren Reihe wiedergegeben.

p bar	temperature t in °C											
	375	400	425	450	475	500	550	600	650	700	750	800
1,013 25	3226,2 / ±3,01	3277,7 / 3,0	3329,7 / 3,0	3382,0 / 3,0	3434,7 / 3,0	3487,9 / 3,0	3595,4 / 3,0	3704,7 / 3,0	3815,7 / 3,8	3928,5 / 3,9	4043 / 4	4159 / 4
5	3219 / 4	3271 / 4	3324 / 4	3377 / 4	3430 / 4	3484 / 4	3592 / 4	3702 / 4	3813 / 4	3926 / 4	4041 / 4	4157 / 4
10	3210 / 4	3263 / 4	3317 / 4	3370 / 4	3424 / 4	3478 / 4	3587 / 5	3698 / 5	3810 / 5	3923 / 5	4038 / 6	4155 / 6
25	3182 / 4	3239 / 4	3295 / 4	3350 / 4	3406 / 4	3462 / 4	3574 / 5	3686 / 5	3800 / 5	3915 / 5	4031 / 6	4149 / 6
50	3132 / 4	3195 / 4	3256 / 4	3316 / 4	3375 / 4	3434 / 4	3550 / 5	3666 / 5	3783 / 5	3900 / 5	4018 / 6	4137 / 6
75	3077 / 4	3147 / 4	3215 / 4	3279 / 4	3342 / 4	3404 / 4	3526 / 5	3646 / 5	3765 / 5	3885 / 5	4005 / 6	4126 / 6
100	3014 / 4	3095 / 4	3170 / 4	3241 / 4	3308 / 4	3374 / 4	3501 / 5	3625 / 5	3748 / 5	3870 / 6	3992 / 6	4114 / 8
125	2942 / 4	3038 / 4	3122 / 4	3200 / 4	3273 / 4	3342 / 5	3476 / 5	3604 / 5	3730 / 7	3855 / 7	3979 / 8	4103 / 10
150	2858 / 4	2974 / 4	3071 / 4	3156 / 4	3235 / 4	3309 / 5	3449 / 6	3583 / 6	3712 / 7	3839 / 8	3965 / 8	4091 / 10
175	2751 / 4	2901 / 4	3014 / 4	3110 / 4	3196 / 4	3275 / 5	3422 / 6	3561 / 7	3693 / 8	3824 / 9	3952 / 9	4080 / 11
200	2601 / 5	2816 / 5	2952 / 4	3060 / 4	3154 / 4	3240 / 5	3395 / 6	3538 / 8	3675 / 9	3808 / 9	3938 / 9	4068 / 11
225	1966 / 9	2713 / 5	2883 / 5	3007 / 5	3111 / 5	3203 / 5	3367 / 6	3515 / 8	3656 / 9	3792 / 10	3925 / 10	4056 / 12
250	1849 / 5	2578 / 5	2805 / 5	2950 / 5	3065 / 5	3164 / 5	3338 / 6	3492 / 8	3637 / 9	3776 / 10	3911 / 10	4044 / 13
275	1814 / 5	2380 / 5	2716 / 5	2888 / 5	3017 / 5	3124 / 5	3308 / 6	3469 / 8	3618 / 9	3760 / 10	3897 / 10	4032 / 13
300	1791 / 4	2152 / 5	2613 / 5	2821 / 5	2966 / 5	3083 / 5	3278 / 6	3445 / 8	3598 / 10	3744 / 10	3884 / 11	4021 / 13
350	1761,9 / 3,5	1988 / 4	2373 / 5	2672 / 5	2857 / 5	2997 / 5	3216 / 6	3397 / 8	3559 / 10	3711 / 11	3856 / 12	3997 / 13
400	1742,1 / 3,4	1931 / 4	2198 / 4	2512 / 5	2741 / 5	2906 / 5	3153 / 6	3348 / 8	3520 / 10	3678 / 11	3828 / 12	3973 / 13
450	1727,4 / 3,4	1897 / 4	2110 / 4	2377 / 5	2624 / 5	2813 / 5	3088 / 6	3299 / 8	3481 / 10	3646 / 11	3801 / 12	3949 / 13
500	1716,0 / 3,4	1874 / 4	2060 / 4	2284 / 5	2520 / 5	2723 / 5	3024 / 6	3250 / 8	3441 / 10	3613 / 11	3773 / 12	3926 / 13

Table 4c. Part 2. International Skeleton Table 1985 for Specific Enthalpy
(Continuation)

Internationale Rahmentafel 1985 für die spezifische Enthalpie
(Fortsetzung)

temperature t in °C

p bar	375	400	425	450	475	500	550	600	650	700	750	800
550	1706,8	1857	2026	2223	2438	2641	2961	3202	3403	3581	3746	3902
	3,4	4	4	4	5	5	6	8	10	11	12	13
600	1699,3	1843	2001	2180	2375	2571	2901	3155	3364	3549	3719	3879
	3,4	4	4	4	5	5	6	8	10	11	12	13
650	1693,0	1832	1982	2148	2328	2513	2845	3109	3327	3518	3693	3856
	3,4	4	4	4	5	5	6	8	10	11	12	13
700	1687,8	1822	1967	2123	2291	2466	2795	3066	3291	3488	3667	3834
	3,4	4	4	4	5	5	6	8	10	11	12	13
750	1683,4	1815	1954	2104	2262	2428	2749	3025	3256	3458	3641	3812
	3,4	4	4	4	5	5	6	8	10	11	12	13
800	1679,7	1808,3	1944	2087	2239	2397	2710	2987	3223	3430	3617	3791
	3,5	3,9	4	4	5	5	6	8	10	11	12	13
850	1676,5	1802,8	1935	2074	2220	2371	2675	2952	3192	3403	3593	3770
	3,5	3,9	4	4	4	5	6	8	10	11	12	13
900	1673,9	1798,0	1927	2063	2204	2349	2645	2920	3163	3377	3570	3750
	3,5	3,9	4	4	4	5	6	8	10	11	12	13
950	1671,7	1794,0	1921	2053	2190	2331	2618	2891	3135	3352	3548	3730
	3,5	3,9	4	4	4	5	6	8	10	11	12	13
1000	1669,9	1790,5	1915	2045	2178	2316	2595	2864	3109	3327	3525	3710
	3,6	3,9	4	4	4	5	6	8	10	11	12	13
1100	1667,3	1785,1	1906	2031	2159	2290	2557	2819	3063	3284	3486	3674
	3,7	3,9	4	4	4	5	6	8	10	11	12	13
1200	1665,8	1781,2	1899	2021	2145	2271	2527	2781	3023	3245	3450	3640
	3,7	3,9	4	4	4	5	6	8	10	11	12	13
1300	1665,2	1778,6	1894	2013	2133	2256	2504	2751	2989	3211	3418	3610
	3,7	3,9	4	4	4	5	6	8	10	11	12	13
1400	1665,3	1777,0	1891	2007	2124	2244	2485	2726	2960	3181	3389	3582
	3,8	3,9	4	4	4	5	6	8	10	11	12	13
1500	1666,0	1776,2	1888	2002	2118	2234	2469	2705	2935	3155	3363	3556
	3,8	3,9	4	4	4	5	6	8	10	11	12	13
1600	1667,3	1776,2	1887	1999	2112	2226	2457	2688	2915	3133	3341	3534
	3,8	3,9	4	4	4	5	6	8	10	11	12	13
1700	1669,1	1776,7	1886	1996	2108	2220	2447	2674	2898	3114	3321	3515
	3,8	3,9	4	4	4	5	6	8	10	11	13	13
1800	1671,2	1777,8	1886	1995	2105	2216	2438	2662	2883	3097	3303	3497
	3,8	3,9	4	4	4	5	6	8	10	11	12	13
1900	1673,7	1779,3	1886	1994	2103	2212	2432	2652	2870	3083	3288	3481
	3,8	3,9	4	4	4	5	6	8	10	11	12	13

2000	1676,5 / 3,8	1781,2 / 3,9	1887 / 4	1994 / 4	2101 / 4	2209 / 5	2426 / 6	2644 / 8	2860 / 10	3071 / 11	3275 / 12	3467 / 13
2200	1682,9 / 3,8	1786,1 / 3,9	1890 / 4	1995 / 4	2101 / 4	2206 / 5	2418 / 6	2631 / 8	2843 / 10	3051 / 11	3252 / 12	3443 / 13
2400	1690,3 / 3,8	1792,2 / 3,9	1895 / 4	1998 / 4	2102 / 4	2206 / 5	2414 / 6	2623 / 8	2831 / 10	3036 / 11	3235 / 12	3424 / 13
2600	1698,4 / 3,8	1799,1 / 3,9	1900 / 4	2002 / 4	2105 / 4	2207 / 5	2412 / 6	2618 / 8	2823 / 10	3025 / 11	3223 / 12	3410 / 13
2800	1707,2 / 3,9	1806,9 / 3,9	1907 / 4	2008 / 4	2109 / 4	2210 / 5	2412 / 6	2615 / 8	2817 / 10	3017 / 11	3213 / 12	3399 / 13
3000	1716,5 / 3,9	1815 / 4	1915 / 4	2014 / 4	2114 / 5	2214 / 5	2414 / 6	2614 / 8	2815 / 10	3012 / 11	3207 / 12	3391 / 13
3200	1726,3 / 3,9	1824 / 4	1923 / 4	2022 / 5	2121 / 5	2220 / 5	2417 / 6	2616 / 8	2814 / 10	3010 / 11	3202 / 12	3386 / 13
3400	1736,5 / 3,9	1834 / 4	1931 / 5	2029 / 5	2128 / 5	2226 / 6	2422 / 7	2618 / 9	2814 / 10	3009 / 11	3200 / 12	3382 / 13
3600	1747 / 4	1844 / 4	1941 / 5	2038 / 5	2135 / 5	2233 / 6	2427 / 7	2622 / 9	2816 / 10	3009 / 11	3199 / 12	3380 / 17
3800	1758 / 4	1854 / 4	1950 / 5	2047 / 5	2144 / 5	2240 / 6	2433 / 7	2626 / 9	2819 / 10	3011 / 11	3200 / 12	3380 / 20
4000	1769 / 4	1865 / 4	1960 / 5	2056 / 5	2152 / 5	2248 / 6	2440 / 7	2632 / 9	2824 / 10	3014 / 11	3202 / 16	3381 / 20
4500	1798 / 4	1892 / 4	1987 / 5	2081 / 5	2176 / 5	2271 / 6	2459 / 7	2648 / 9	2837 / 10	3025 / 12	3211 / 16	3388 / 23
5000	1828 / 4	1921 / 4	2015 / 5	2108 / 5	2202 / 6	2295 / 6	2482 / 7	2668 / 9	2855 / 10	3041 / 15	3224 / 19	3400 / 23
5500	1859 / 4	1952 / 4	2044 / 5	2137 / 5	2230 / 6	2322 / 6	2506 / 8	2691 / 9	2876 / 10	3060 / 15	3241 / 19	3415 / 24
6000	1891 / 4	1983 / 4	2074 / 5	2167 / 5	2258 / 6	2350 / 7	2533 / 8	2716 / 9	2899 / 10	3081 / 15	3261 / 19	3433 / 24
6500	1923 / 4	2014 / 4	2106 / 5	2197 / 6	2288 / 6	2379 / 7	2560 / 8	2742 / 9	2923 / 11	3104 / 15	3282 / 19	3454 / 27
7000	1956 / 4	2047 / 5	2137 / 5	2228 / 6	2319 / 6	2409 / 7	2589 / 8	2769 / 11	2949 / 14	3128 / 15	3306 / 26	3476 / 34
7500	1989 / 4	2079 / 5	2170 / 5	2260 / 6	2350 / 6	2440 / 7	2619 / 9	2797 / 11	2976 / 14	3154 / 21	3330 / 33	3500 / 40
8000	2023 / 4	2113 / 5	2203 / 5	2292 / 6	2382 / 6	2471 / 7	2649 / 9	2827 / 14	3004 / 18	3181 / 25	3360 / 40	3520 / 50
8500	2057 / 4	2146 / 5	2236 / 5	2325 / 6	2414 / 7	2503 / 7	2680 / 10	2857 / 17	3033 / 24	3208 / 31	3380 / 40	3550 / 60
9000	2091 / 4	2180 / 5	2269 / 6	2358 / 6	2447 / 9	2536 / 10	2711 / 13	2887 / 20	3062 / 30	3240 / 38	3410 / 50	3570 / 60
9500	2125 / 4	2214 / 5	2303 / 7	2392 / 7	2480 / 12	2568 / 12	2743 / 16	2918 / 23	3090 / 40	3260 / 50	3430 / 50	3600 / 70
10000	2159 / 5	2248 / 5	2337 / 8	2425 / 9	2514 / 13	2601 / 15	2776 / 22	2949 / 29	3120 / 50	3290 / 60	3460 / 70	3620 / 80

¹ Except this tolerance, the sign (±) of tolerance is omitted from the description.
² Außer an dieser Stelle ist das ± Zeichen der Toleranz weggelassen.

Table 4d. International Skeleton Table for Dynamic Viscosity of Water and Steam in 10^{-6} kg/s m

Of each pair of figures the upper represents the adopted value of dynamic viscosity and the lower the tolerance in 10^{-6} kg/s m

Internationale Rahmentafel der dynamischen Viskosität von Wasser und Dampf in 10^{-6} kg/s m

Von je zwei übereinanderstehenden Zahlen gibt die obere den vereinbarten Wert der dynamischen Viskosität und die untere die Toleranz in 10^{-6} kg/s m an

p bar	temperature t in °C										
	0	25	50	75	100	150	200	250	300	350	375
1	1791	890,9	547,1	377,3	12,42	14,29	16,26	18,30	20,36	22,43	23,45
	18	8,9	5,5	3,8	0,25	0,29	0,33	0,37	0,41	0,45	0,47
5	1790	891,2	546,7	378,0	281,7	182,3	16,05	18,16	20,25	22,32	23,43
	18	8,9	5,5	3,8	2,8	1,8	0,32	0,36	0,41	0,45	0,47
10	1789	891,1	546,8	378,2	281,9	182,4	15,92	18,09	20,21	22,29	23,40
	18	8,9	5,5	3,8	2,8	1,8	0,32	0,36	0,40	0,45	0,47
25	1786	890,8	547,1	378,5	283,3	182,8	134,6	17,85	20,07	22,22	23,37
	18	8,9	5,5	3,8	2,8	1,8	1,4	0,36	0,40	0,44	0,47
50	1780	890,3	547,7	379,2	283,1	183,4	135,2	106,5	19,88	22,15	23,33
	18	8,9	5,5	3,8	2,8	1,8	1,4	1,1	0,40	0,44	0,47
75	1774	889,8	548,3	379,8	283,8	184,1	135,9	107,2	19,75	22,12	23,34
	18	8,9	5,5	3,8	2,8	1,8	1,4	1,1	0,40	0,44	0,47
100	1768	889,4	548,7	380,4	284,7	184,7	136,4	107,8	87,1	22,16	23,39
	18	8,9	5,5	3,8	2,9	1,9	1,4	1,1	1,7	0,44	0,47
125	1762	889,1	549,1	381,0	285,3	185,3	137,0	108,5	88,0	22,35	23,57
	18	8,9	5,5	3,8	2,9	1,9	1,4	1,1	1,8	0,45	0,47
150	1756	888,7	549,5	381,6	286,0	186,0	137,6	109,1	89,0	22,84	23,88
	18	8,9	5,5	3,8	2,9	1,9	1,4	1,1	1,8	0,46	0,48
175	1750	888,5	550,0	382,3	286,7	186,9	138,2	109,8	89,9	67,3	24,49
	18	8,9	5,5	3,8	2,9	1,9	1,4	1,1	1,8	2,0	0,49
200	1744	888,2	550,4	382,9	287,4	187,3	138,8	110,4	90,8	69,5	25,85
	17	8,9	5,5	3,8	2,9	1,9	1,4	1,1	1,8	2,1	0,52
225	1738	887,9	550,9	383,5	288,0	187,9	138,4	111,1	91,6	71,4	48,2
	17	8,9	5,5	3,8	2,9	1,9	1,4	1,1	1,8	2,1	3,9
250	1733	887,6	551,3	384,2	288,7	188,5	140,0	111,7	92,4	73,0	58,8
	17	8,9	5,5	3,8	2,9	1,9	1,4	1,1	1,9	2,2	1,2
275	1728	887,4	551,8	384,8	289,4	189,1	140,6	112,3	93,1	74,4	62,4
	17	8,9	5,5	3,9	2,9	1,9	1,4	1,1	1,9	2,2	1,2
300	1723	887,2	552,3	385,5	290,0	189,8	141,2	112,9	93,9	75,7	64,9
	17	8,9	5,5	3,9	2,9	1,9	1,4	1,1	1,9	2,3	1,3
350	1713	886,8	553,3	386,7	291,4	191,0	142,3	114,1	95,3	78,0	68,6
	17	8,9	5,5	3,9	2,9	1,9	1,4	1,1	1,9	2,3	1,4
400	1705	886,6	554,3	388,0	292,7	192,2	143,5	115,3	96,5	79,9	71,3
	17	8,9	5,5	3,9	2,9	1,9	1,4	1,2	1,9	2,4	1,4
450	1697	886,5	555,3	389,3	294,0	193,4	144,6	116,4	97,8	81,7	73,7
	17	8,9	5,6	3,9	2,9	1,9	1,5	1,2	2,0	2,5	1,5
500	1690	886,4	556,3	390,6	295,4	194,6	145,8	117,6	99,0	83,4	75,9
	17	8,9	5,6	3,9	3,0	2,0	1,5	1,2	2,0	2,5	2,3
550	1684	886,5	557,4	392,0	296,7	195,8	146,9	118,7	100,2	84,9	77,8
	17	8,9	5,6	3,9	3,0	2,0	1,5	1,2	2,0	2,6	2,3
600	1679	886,7	558,5	393,3	298,0	197,0	148,0	119,7	101,3	86,3	79,5
	17	8,9	5,6	3,9	3,0	2,0	1,5	1,2	2,0	2,6	2,4
650	1674	886,9	559,7	394,6	299,4	198,2	149,0	120,8	102,5	87,7	81,0
	17	8,9	5,6	4,0	3,0	2,0	1,5	1,2	2,1	2,6	2,4
700	1670	887,3	560,9	395,9	300,7	199,4	150,1	121,9	103,6	89,0	82,5
	17	8,9	5,6	4,0	3,0	2,0	1,5	1,2	2,1	2,7	2,5
750	1666	887,7	562,0	397,3	302,0	200,6	151,2	122,9	104,6	90,3	83,9
	17	8,9	5,6	4,0	3,0	2,0	1,5	1,2	2,1	2,7	2,5
800	1662	888,3	563,3	398,6	303,4	201,8	152,3	123,9	105,6	91,4	85,2
	17	8,9	5,6	4,0	3,0	2,0	1,5	1,2	2,1	2,7	2,6
850	1659	888,8	564,5	400,0	304,6	203,0	153,3	124,9	106,6	92,6	86,4
	17	8,9	5,7	4,0	3,1	2,0	1,5	1,3	2,1	2,8	2,6
900	1656	889,5	565,8	401,4	305,9	204,2	154,3	125,9	107,6	93,7	87,5
	17	8,9	5,7	4,0	3,1	2,0	1,5	1,3	2,2	2,8	2,6
950	1653	890,3	567,1	402,8	307,3	205,4	155,4	126,9	108,6	94,7	88,7
	17	8,9	5,7	4,0	3,1	2,1	1,6	1,3	2,2	2,8	2,7
1000	1651	891,1	568,4	404,2	308,6	206,5	156,4	127,9	109,6	95,8	89,8
	17	8,9	5,7	4,0	3,1	2,1	1,6	1,3	2,2	2,9	2,7

Table 4d. International Skeleton Table for Dynamic Viscosity of Water and Steam in 10^{-6} kg/s m
(Continuation)

Internationale Rahmentafel der dynamischen Viskosität von Wasser und Dampf in 10^{-6} kg/s m
(Fortsetzung)

p bar	temperature t in °C										
	400	425	450	475	500	550	600	650	700	750	800
1	24,47	25,49	26,50	27,51	28,52	30,53	32,55	34,6	36,6	38,6	40,5
	0,49	0,51	0,53	0,55	0,86	0,92	0,98	1,0	1,1	1,2	1,2
5	24,44	25,49	26,53	27,57	28,64	30,67	32,77	34,7	36,7	38,5	40,3
	0,49	0,51	0,53	0,55	0,86	0,92	0,98	1,0	1,1	1,2	1,2
10	24,43	25,49	26,53	27,58	28,65	30,68	32,79	34,8	36,8	38,5	40,4
	0,49	0,51	0,53	0,55	0,86	0,92	0,98	1,0	1,1	1,2	1,2
25	24,41	25,49	26,54	27,59	28,66	30,72	32,84	34,8	36,8	38,6	40,4
	0,49	0,51	0,53	0,55	0,86	0,92	0,99	1,0	1,1	1,2	1,2
50	24,42	25,52	26,60	27,66	28,73	30,82	32,77	34,9	36,9	38,7	40,6
	0,49	0,51	0,53	0,55	0,86	0,92	0,98	1,1	1,1	1,2	1,2
75	24,46	25,58	26,68	27,76	28,81	30,94	32,87	34,9	37,0	38,8	40,7
	0,49	0,51	0,53	0,56	0,86	0,93	0,99	1,1	1,1	1,2	1,2
100	24,52	25,65	26,75	27,82	28,95	31,08	33,02	35,1	37,2	39,0	40,9
	0,49	0,51	0,53	0,56	0,87	0,93	0,99	1,1	1,1	1,2	1,2
125	24,69	25,81	26,91	27,98	29,09	31,19	35,2	35,2	37,4	39,2	41,1
	0,49	0,52	0,54	0,56	0,87	0,94	1,0	1,1	1,1	1,2	1,2
150	24,98	26,06	27,13	28,18	29,30	31,44	33,4	35,5	37,6	39,4	41,2
	0,50	0,52	0,54	0,56	0,88	0,94	1,0	1,1	1,1	1,2	1,2
175	25,37	26,38	27,42	28,42	29,49	31,70	33,7	35,7	37,8	39,6	41,4
	0,51	0,53	0,55	0,57	0,88	0,95	1,0	1,1	1,1	1,2	1,2
200	26,03	26,83	27,80	28,76	29,81	31,98	33,9	35,9	38,0	39,8	41,6
	0,52	0,54	0,56	0,58	0,89	0,96	1,0	1,1	1,1	1,2	1,3
225	27,11	27,50	28,31	29,17	30,17	32,38	34,2	36,2	38,2	39,8	41,9
	0,54	0,55	0,57	0,58	0,91	0,97	1,0	1,1	1,2	1,2	1,3
250	29,10	28,43	28,99	29,70	30,56	32,73	34,6	36,5	38,5	40,2	41,9
	0,58	0,57	0,58	0,59	0,92	0,98	1,0	1,1	1,2	1,2	1,3
275	33,88	29,81	29,84	30,33	31,08	33,11	34,9	36,8	38,7	40,4	42,2
	0,68	0,60	0,60	0,61	0,93	0,99	1,1	1,1	1,2	1,2	1,3
300	43,97	31,84	30,97	31,06	31,68	33,6	35,3	37,2	39,0	40,7	42,5
	0,89	0,64	0,62	0,62	0,95	1,0	1,1	1,1	1,2	1,2	1,3
350	56,4	39,47	34,19	33,17	33,10	34,6	36,1	37,9	39,8	41,3	43,0
	1,1	0,79	0,68	0,66	0,99	1,0	1,1	1,1	1,2	1,2	1,3
400	62,1	49,26	39,16	36,06	35,2	35,7	35,5	38,8	40,4	42,0	43,7
	1,2	0,99	0,78	0,72	1,1	1,1	1,1	1,2	1,2	1,3	1,3
450	65,8	55,6	44,87	39,90	37,6	37,4	38,6	40,0	41,2	43,1	44,4
	1,3	1,1	0,90	0,80	1,1	1,1	1,2	1,2	1,2	1,3	1,3
500	68,2	60,1	50,5	44,0	40,5	39,1	40,0	40,6	42,2	43,7	45,3
	2,0	1,8	1,5	1,3	1,2	1,2	1,2	1,2	1,3	1,3	1,4
550	70,9	63,6	55,3	48,4	43,9	41,0	41,4	41,8	42,5	44,6	45,9
	2,1	1,9	1,7	1,5	1,3	1,2	1,2	1,3	1,3	1,3	1,4
600	73,1	66,1	59,2	52,3	47,6	43,1	41,7	42,9	43,2	44,8	46,6
	2,2	2,0	1,8	1,6	1,4	1,3	1,3	1,3	1,3	1,3	1,4
650	75,2	68,1	62,3	55,5	50,8	45,1	43,2	43,9	44,2	45,4	46,8
	2,3	2,0	1,9	1,7	1,5	1,4	1,3	1,3	1,3	1,4	1,4
700	76,9	70,5	64,9	58,8	53,7	47,5	44,8	44,3	44,4	46,2	47,4
	2,3	2,1	2,0	1,8	1,6	1,4	1,3	1,3	1,3	1,4	1,4
750	78,5	72,2	66,9	61,3	56,2	49,7	45,7	45,5	45,6	46,8	48,1
	2,4	2,2	2,0	1,8	1,7	1,5	1,4	1,4	1,4	1,4	1,4
800	79,9	74,0	68,3	63,6	58,7	52,1	47,4	47,0	46,6	47,3	48,6
	2,4	2,2	2,1	1,9	1,8	1,6	1,4	1,4	1,4	1,4	1,4
850	81,4	75,8	70,2	65,5	60,8	54,0	49,9	47,6	47,6	48,1	49,0
	2,4	2,3	2,1	2,0	1,8	1,6	1,5	1,4	1,4	1,4	1,5
900	82,7	77,2	72,3	67,3	62,8	55,8	51,4	48,9	49,1	48,9	49,7
	2,5	2,3	2,2	2,0	1,9	1,7	1,5	1,5	1,5	1,5	1,5
950	83,6	78,6	73,8	69,1	64,6	57,7	53,6	50,9	49,5	49,8	50,3
	2,5	2,4	2,2	2,1	1,9	1,7	1,6	1,5	1,5	1,5	1,5
1000	85,0	79,8	74,6	69,8	66,1	59,3	55,1	52,1	50,5	51,1	51,0
	2,6	2,4	2,2	2,1	2,0	1,8	1,7	1,6	1,5	1,5	1,5

Table 4e. International Skeleton Table for Thermal Conductivity of Water and Steam in mW/K m

Of each pair of figures the upper represents the adopted value of thermal conductivity and the lower the tolerance in mW/Km

Internationale Rahmentafel für die Wärmeleitfähigkeit von Wasser und Dampf in mW/K m

Von je zwei übereinanderstehenden Zahlen gibt die obere den vereinbarten Wert der Wärmeleitfähigkeit und die untere die Toleranz in mW/Km an

p bar	temperature t in °C										
	0	25	50	75	100	150	200	250	300	350	375
1	563	610	643	664	25,0	28,9	33,3	38,1	43,3	49,0	52,0
	11	9	9	10	0,5	0,6	0,7	0,8	0,9	1,0	1,0
5	563	610	643	664	680	688	34,1	38,7	43,7	49 1	52,6
	11	9	9	10	10	10	1,0	1,2	1,3	1,5	1,6
10	564	611	643	666	681	689	35,9	39,5	44,3	49,5	53,0
	11	9	9	10	10	10	1,4	1,2	1,3	1,5	1,6
25	566	611	644	666	682	690	668	43,8	46,5	50,9	54,7
	11	9	9	10	10	10	10	1,4	1,4	1,5·	1,6
50	567	613	645	668	683	691	671	625	52,7	54,1	56,5
	11	12	12	13	13	13	13	12	1,6	1,9	1,7
75	570	614	647	669	685	694	673	628	63,6	59,6	60,5
	11	12	12	13	13	13	13	12	1,9	1,8	1,8
100	571	615	648	669	686	695	675	631	557	68,2	65,3
	11	12	13	13	13	13	13	12	11	2,0	2,1
125	571	616	649	672	687	697	678	634	562	81,2	73,6
	11	12	13	13	13	13	13	12	11	2,4	2,2
150	573	617	650	673	689	700	680	638	566	107,5	84,8
	11	12	13	13	13	14	13	12	11	6,7	2,5
175	573	618	651	674	691	701	682	639	571	452	104,2
	11	12	13	13	13	14	13	12	11	13	3,1
200	574	619	653	676	691	703	684	641	576	465	144,0
	11	12	13	13	13	14	13	12	11	14	4,7
225	574	620	654	678	692	705	686	646	581	476	478
	11	12	13	13	13	14	13	12	11	14	39
250	577	621	655	679	694	707	689	648	588	482	400
	11	12	13	13	13	14	13	13	11	14	14
275	578	622	656	680	696	708	690	651	589	490	413
	11	12	13	13	13	14	13	13	11	14	14
300	578	623	658	681	697	710	692	653	593	498	426
	11	12	13	13	13	14	13	13	11	15	13
350	580	625	660	684	700	714	696	660	601	511	453
	11	12	13	13	14	14	13	13	12	15	13
400	583	626	662	686	702	717	700	664	608	526	471
	11	12	13	13	14	14	14	13	12	15	14
450	584	629	664	690	705	721	704	670	615	537	486
	11	12	13	13	14	14	14	13	12	16	14
500	586	630	666	692	708	724	708	673	621	547	498
	11	12	13	13	14	14	14	13	12	44	40
550	589	633	667	694	710	726	712	678	629	558	510
	11	12	13	13	14	14	14	13	12	45	41
600	590	635	670	697	713	729	715	682	634	566	525
	11	12	13	13	14	14	14	13	12	45	42
650	592	638	673	699	715	733	718	688	639	574	535
	11	12	13	14	14	14	14	13	12	46	43
700	597	639	674	702	718	735	721	691	645	582	546
	11	12	13	14	14	14	14	13	12	47	44
750	599	641	675	705	720	738	725	696	648	589	554
	12	12	13	14	14	14	14	13	13	47	44
800	599	645	677	707	723	739	729	699	653	598	564
	12	12	13	14	14	14	14	14	13	48	45
850	601	646	680	706	726	742	732	702	659	604	571
	12	12	13	14	14	14	14	14	13	48	46
900	604	648	681	710	728	745	735	707	665	611	578
	12	13	13	14	14	14	14	14	13	49	46
950	608	650	685	713	731	748	739	711	669	615	586
	12	13	13	14	14	15	14	14	13	49	47
1000	609	650	686	716	735	749	742	715	672	624	594
	12	13	13	14	14	15	14	14	13	50	47

Table 4e. International Skeleton Table for Thermal Conductivity of Water and Steam in mW/K m
(Continuation)

Internationale Rahmentafel für die Wärmeleitfähigkeit von Wasser und Dampf in mW/K m
(Fortsetzung)

p bar	\multicolumn{11}{c}{temperature t in °C}										
	400	**425**	**450**	**475**	**500**	**550**	**600**	**650**	**700**	**750**	**800**
1	54,9	57,9	60,6	63,8	67,1	73,1	79,9	86,4	93,4	100,5	107,5
	1,1	1,2	1,2	1,3	1,3	1,5	2,4	2,6	2,8	3,0	3,2
5	55,5	58,5	61,4	64,5	67,7	74,0	80,5	87,2	93,8	100,9	108,0
	1,7	1,8	1,8	1,9	2,0	2,2	3,2	3,5	3,8	4,0	4,3
10	56,0	58,6	61,7	64,7	68,0	74,3	81,0	87,7	94,3	101,4	108,6
	1,7	1,8	1,9	1,9	2,0	2,2	3,2	3,5	3,8	4,1	4,3
25	56,9	59,6	62,6	65,6	68,7	75,1	81,5	88,8	95,3	102,4	109,5
	1,7	1,8	1,9	2,0	2,1	2,3	3,3	3,6	3,8	4,1	4,4
50	58,6	60,9	64,0	66,4	69,3	75,4	81,5	91,4	95,7	103,6	109,6
	1,8	1,8	1,9	2,0	2,1	2,3	3,3	3,7	3,8	4,1	4,4
75	62,7	64,0	66,7	69,5	73,3	80,0	87,3	96,4	101,0	108,1	112,4
	1,9	1,9	2,0	2,1	2,2	2,4	3,5	5,3	4,0	4,3	4,5
100	66,9	67,4	69,4	72,1	75,6	82,5	89,4	97,5	102,9	111,2	118,1
	2,0	2,0	2,1	2,2	2,3	2,5	3,6	4,6	4,1	5,1	5,2
125	72,4	72,0	74,1	76,1	79,4	85,0	90,7	97,9	102,9	109,9	116,3
	2,2	2,2	2,2	2,3	2,4	2,6	3,6	3,9	4,1	4,4	4,7
150	79,9	77,8	78,4	79,3	82,4	87,5	93,4	100,3	105,6	112,7	118,0
	2,4	2,3	2,4	2,4	2,5	2,6	3,7	4,0	4,2	4,5	4,7
175	90,0	84,8	84,0	84,2	85,7	90,2	96,2	102,5	106,0	114,4	119,7
	2,7	2,5	2,5	2,5	2,6	2,7	3,8	4,1	4,2	4,6	4,8
200	104,9	93,7	90,8	90,1	91,6	94,9	98,6	105,5	109,3	116,8	122,7
	3,1	2,8	2,7	2,7	2,7	3,0	3,9	4,2	4,4	4,7	4,9
225	124,1	105,9	98,6	95,9	96,0	98,1	102,6	107,6	112,1	119,2	123,7
	4,6	3,2	3,0	2,9	2,9	2,9	4,1	4,3	4,5	4,8	4,9
250	166,4	120,6	108,3	102,8	101,5	102,3	105,7	110,7	114,5	121,5	126,2
	6,7	3,6	3,2	3,1	3,0	3,1	4,2	4,4	4,6	4,9	5,0
275	240,8	139,2	120,3	111,1	107,3	106,1	108,7	113,0	118,0	123,4	127,8
	8,4	6,3	3,6	3,3	3,2	3,2	4,3	4,5	4,7	4,9	5,1
300	337	175,0	133,8	119,4	114,1	110,6	112,3	116,2	119,9	125,7	130,2
	12	8,1	4,0	3,6	3,4	3,3	4,5	4,6	4,8	5,0	5,2
350	384	260,5	176,3	144,3	129,7	121,1	119,8	122,7	125,1	130,0	134,6
	12	7,8	5,5	4,3	3,9	3,6	4,8	4,9	5,0	5,2	5,4
400	399	331	233,2	178,9	152,9	133,9	129,2	129,5	131,8	135,8	139,3
	16	11	7,2	5,5	4,6	4,0	5,2	5,2	5,3	5,4	5,6
450	425	365	287	219,0	180,1	148,2	138,5	136,4	137,7	141,1	144,5
	12	11	12	7,9	5,4	4,4	5,5	5,5	5,5	5,6	5,8
500	444	381	325	263	211	164	150	145	145	146	149
	36	30	26	21	17	13	12	12	12	12	12
550	461	401	354	297	244	184	162	154	152	153	155
	37	32	28	24	20	15	13	12	12	12	12
600	476	423	366	322	277	207	176	164	159	159	161
	38	34	29	26	22	16	14	13	13	13	13
650	489	438	387	332	299	228	191	175	168	166	167
	39	35	31	26	24	18	15	14	13	13	13
700	499	453	406	355	322	253	205	186	178	173	173
	40	36	32	28	26	21	16	15	14	14	14
750	511	467	421	376	327	269	218	198	186	180	178
	41	37	34	30	26	22	17	16	15	14	15
800	521	480	435	393	346	298	235	209	(196)	(190)	(185)
	42	38	35	31	28	34	19	17	16	15	15
850	532	488	448	410	366	312	246	222	(206)	(196)	(194)
	43	39	36	33	29	33	20	18	17	16	15
900	544	500	460	424	385	308	259	233	(215)	(205)	(201)
	44	40	37	34	31	25	21	19	17	16	16
950	553	510	473	434	396	322	273	243	(226)	(214)	(207)
	44	41	38	35	32	26	22	19	18	17	17
1000	561	519	484	445	412	338	288	255	(236)	(221)	(215)
	45	42	39	36	33	27	23	20	19	18	17

Extrapolated values distinguished by parantheses
Extrapolierte Werte in Klammern

Table 5. Specific Heat Capacity at Constant Pressure c_p in kJ/kg K
Spezifische Wärmekapazität bei konstantem Druck c_p in kJ/kg K

p bar	t in °C → 0	50	100	120	140	150	160	180	200	220	240	250	260	280	300
0,1	4,217	1,893	1,903	1,909	1,916	1,920	1,924	1,933	1,943	1,953	1,964	1,969	1,975	1,987	1,998
1	4,217	4,181	2,026	2,005	1,991	1,986	1,983	1,979	1,979	1,982	1,986	1,989	1,993	2,001	2,010
5	4,215	4,180	4,215	4,244	4,285	4,310	2,291	2,216	2,161	2,123	2,097	2,088	2,080	2,071	2,066
10	4,212	4,179	4,214	4,243	4,283	4,308	4,337	2,593	2,446	2,340	2,263	2,233	2,208	2,170	2,145
20	4,207	4,177	4,211	4,240	4,280	4,305	4,334	4,403	4,494	2,918	2,694	2,608	2,534	2,419	2,337
25	4,204	4,175	4,210	4,239	4,279	4,304	4,332	4,401	4,491	4,612	2,966	2,840	2,734	2,569	2,451
30	4,201	4,174	4,209	4,238	4,277	4,302	4,330	4,399	4,488	4,608	3,282	3,108	2,963	2,738	2,578
40	4,196	4,172	4,207	4,235	4,275	4,299	4,327	4,394	4,483	4,600	4,761	4,866	3,528	3,139	2,873
50	4,191	4,170	4,205	4,233	4,272	4,296	4,323	4,390	4,477	4,592	4,750	4,853	4,977	3,659	3,234
60	4,186	4,167	4,203	4,230	4,269	4,293	4,320	4,386	4,471	4,585	4,739	4,839	4,961	4,375	3,691
70	4,181	4,165	4,200	4,228	4,266	4,290	4,317	4,382	4,466	4,577	4,729	4,826	4,944	5,274	4,298
75	4,178	4,164	4,199	4,227	4,265	4,288	4,315	4,380	4,463	4,574	4,724	4,820	4,936	5,260	4,686
80	4,175	4,163	4,198	4,226	4,263	4,287	4,313	4,378	4,461	4,570	4,718	4,814	4,928	5,247	5,155
90	4,170	4,161	4,196	4,223	4,261	4,284	4,310	4,373	4,455	4,563	4,708	4,801	4,913	5,221	5,741
100	4,165	4,158	4,194	4,221	4,258	4,281	4,307	4,369	4,450	4,556	4,698	4,789	4,898	5,196	5,692
110	4,160	4,156	4,192	4,218	4,255	4,278	4,303	4,365	4,445	4,549	4,689	4,777	4,883	5,172	5,645
120	4,155	4,154	4,190	4,216	4,253	4,275	4,300	4,361	4,440	4,542	4,679	4,766	4,869	5,149	5,601
125	4,153	4,153	4,189	4,215	4,251	4,273	4,299	4,359	4,437	4,539	4,674	4,760	4,862	5,137	5,580
130	4,151	4,152	4,188	4,214	4,250	4,272	4,297	4,357	4,435	4,535	4,670	4,755	4,855	5,126	5,560
140	4,146	4,150	4,185	4,212	4,247	4,269	4,294	4,354	4,430	4,529	4,661	4,744	4,842	5,105	5,521
150	4,141	4,148	4,183	4,209	4,245	4,266	4,291	4,350	4,425	4,522	4,652	4,733	4,829	5,084	5,483
160	4,136	4,145	4,181	4,207	4,242	4,263	4,288	4,346	4,420	4,516	4,643	4,722	4,816	5,064	5,448
170	4,131	4,143	4,179	4,205	4,239	4,261	4,285	4,342	4,415	4,510	4,634	4,712	4,804	5,044	5,414
175	4,129	4,142	4,178	4,204	4,238	4,259	4,283	4,340	4,413	4,507	4,630	4,707	4,797	5,035	5,397
180	4,127	4,141	4,177	4,203	4,237	4,258	4,282	4,338	4,411	4,504	4,626	4,702	4,791	5,026	5,381
190	4,122	4,139	4,175	4,200	4,234	4,255	4,279	4,335	4,406	4,497	4,618	4,692	4,780	5,007	5,350
200	4,117	4,137	4,173	4,198	4,232	4,252	4,276	4,331	4,401	4,491	4,609	4,683	4,768	4,990	5,321
210	4,113	4,135	4,171	4,196	4,229	4,250	4,273	4,327	4,397	4,485	4,601	4,673	4,757	4,973	5,292
220	4,108	4,133	4,169	4,194	4,227	4,247	4,270	4,324	4,392	4,480	4,594	4,664	4,746	4,956	5,265
225	4,106	4,132	4,168	4,193	4,226	4,246	4,268	4,322	4,390	4,477	4,590	4,659	4,740	4,948	5,252
230	4,104	4,131	4,167	4,192	4,224	4,244	4,267	4,320	4,388	4,474	4,586	4,655	4,735	4,940	5,239
240	4,099	4,129	4,165	4,189	4,222	4,242	4,264	4,317	4,383	4,468	4,578	4,646	4,724	4,924	5,213
250	4,095	4,127	4,163	4,187	4,219	4,239	4,261	4,313	4,379	4,463	4,571	4,637	4,714	4,909	5,189
260	4,090	4,125	4,161	4,185	4,217	4,236	4,258	4,310	4,375	4,457	4,563	4,629	4,704	4,894	5,166
270	4,086	4,123	4,159	4,183	4,215	4,234	4,255	4,306	4,370	4,452	4,556	4,620	4,694	4,879	5,143
275	4,084	4,122	4,158	4,182	4,213	4,232	4,254	4,305	4,368	4,449	4,553	4,616	4,689	4,872	5,132
280	4,082	4,121	4,157	4,181	4,212	4,231	4,253	4,303	4,366	4,446	4,549	4,612	4,684	4,865	5,121
290	4,077	4,119	4,155	4,179	4,210	4,229	4,250	4,300	4,362	4,441	4,542	4,604	4,674	4,852	5,100
300	4,073	4,117	4,153	4,177	4,207	4,226	4,247	4,296	4,358	4,436	4,535	4,596	4,665	4,838	5,080
310	4,069	4,115	4,151	4,175	4,205	4,224	4,244	4,293	4,354	4,430	4,529	4,588	4,656	4,825	5,060
320	4,065	4,113	4,150	4,172	4,203	4,221	4,241	4,290	4,350	4,425	4,522	4,580	4,647	4,812	5,041
330	4,060	4,111	4,148	4,170	4,200	4,219	4,239	4,286	4,346	4,420	4,515	4,573	4,638	4,800	5,022
340	4,056	4,109	4,146	4,168	4,198	4,216	4,236	4,283	4,342	4,415	4,509	4,565	4,629	4,788	5,004
350	4,052	4,107	4,144	4,166	4,196	4,214	4,233	4,280	4,338	4,410	4,502	4,558	4,621	4,776	4,987
360	4,048	4,106	4,142	4,164	4,194	4,211	4,231	4,277	4,334	4,406	4,496	4,551	4,612	4,764	4,970
370	4,044	4,104	4,140	4,162	4,191	4,209	4,228	4,274	4,330	4,401	4,490	4,544	4,604	4,753	4,953
380	4,040	4,102	4,138	4,160	4,189	4,206	4,226	4,271	4,326	4,396	4,484	4,537	4,596	4,742	4,937
390	4,036	4,100	4,137	4,158	4,187	4,204	4,223	4,268	4,323	4,391	4,478	4,530	4,588	4,731	4,921
400	4,032	4,098	4,135	4,156	4,185	4,202	4,220	4,265	4,319	4,387	4,472	4,523	4,581	4,720	4,906
420	4,024	4,095	4,131	4,152	4,180	4,197	4,215	4,258	4,312	4,378	4,461	4,510	4,565	4,700	4,877
440	4,016	4,091	4,127	4,149	4,176	4,192	4,210	4,253	4,304	4,369	4,449	4,497	4,551	4,680	4,849
450	4,013	4,089	4,126	4,147	4,174	4,190	4,208	4,250	4,301	4,364	4,444	4,491	4,544	4,671	4,836
460	4,009	4,087	4,124	4,145	4,172	4,188	4,205	4,247	4,297	4,360	4,438	4,485	4,537	4,661	4,823
480	4,001	4,084	4,120	4,141	4,167	4,183	4,200	4,241	4,290	4,352	4,428	4,473	4,523	4,643	4,797
500	3,994	4,081	4,117	4,137	4,163	4,179	4,196	4,235	4,284	4,343	4,417	4,461	4,510	4,626	4,773

Table 5. Specific Heat Capacity at Constant Pressure c_p in kJ/kg K (Continuation)
Spezifische Wärmekapazität bei konstantem Druck c_p in kJ/kg K (Fortsetzung)

p bar	t in °C → 300	320	340	350	360	380	400	420	425	440	450	460	480	500
0,1	1,998	2,011	2,023	2,029	2,036	2,049	2,062	2,075	2,078	2,089	2,095	2,102	2,116	2,129
1	2,010	2,020	2,031	2,037	2,043	2,055	2,067	2,080	2,083	2,093	2,099	2,106	2,119	2,132
5	2,066	2,066	2,069	2,071	2,074	2,081	2,090	2,100	2,102	2,110	2,116	2,122	2,133	2,146
10	2,145	2,129	2,120	2,118	2,116	2,117	2,120	2,126	2,128	2,133	2,138	2,142	2,152	2,162
20	2,337	2,279	2,239	2,224	2,213	2,196	2,186	2,182	2,182	2,182	2,183	2,185	2,190	2,197
25	2,451	2,367	2,308	2,286	2,267	2,240	2,222	2,212	2,211	2,208	2,207	2,207	2,210	2,214
30	2,578	2,464	2,383	2,352	2,326	2,287	2,261	2,244	2,241	2,235	2,232	2,231	2,230	2,233
40	2,873	2,686	2,553	2,502	2,459	2,392	2,346	2,314	2,308	2,293	2,285	2,280	2,273	2,270
50	3,234	2,949	2,751	2,675	2,611	2,512	2,441	2,392	2,381	2,356	2,344	2,333	2,319	2,310
60	3,691	3,264	2,980	2,874	2,784	2,646	2,547	2,476	2,462	2,426	2,407	2,391	2,367	2,352
70	4,298	3,649	3,247	3,101	2,980	2,795	2,663	2,569	2,550	2,501	2,475	2,453	2,420	2,397
75	4,686	3,879	3,398	3,228	3,088	2,876	2,726	2,619	2,597	2,542	2,512	2,486	2,447	2,420
80	5,155	4,141	3,565	3,365	3,203	2,961	2,792	2,671	2,646	2,583	2,549	2,520	2,475	2,444
90	5,741	4,793	3,953	3,677	3,461	3,146	2,932	2,781	2,750	2,671	2,629	2,592	2,535	2,494
100	5,692	5,693	4,441	4,058	3,765	3,354	3,086	2,899	2,862	2,766	2,714	2,669	2,598	2,547
110	5,645	7,083	5,071	4,532	4,132	3,592	3,255	3,028	2,982	2,867	2,804	2,750	2,665	2,602
120	5,601	6,486	5,903	5,136	4,585	3,869	3,443	3,166	3,112	2,975	2,901	2,837	2,736	2,661
125	5,580	6,433	6,427	5,501	4,853	4,025	3,545	3,240	3,181	3,032	2,951	2,882	2,772	2,691
130	5,560	6,383	7,061	5,920	5,155	4,195	3,653	3,317	3,253	3,090	3,003	2,929	2,810	2,722
140	5,521	6,291	9,082	6,973	5,885	4,588	3,893	3,482	3,405	3,214	3,112	3,025	2,888	2,787
150	5,483	6,206	8,088	8,543	6,841	5,066	4,168	3,663	3,571	3,346	3,228	3,128	2,971	2,854
160	5,448	6,128	7,775	11,896	8,156	5,658	4,488	3,864	3,754	3,488	3,351	3,236	3,057	2,925
170	5,414	6,056	7,519	9,677	10,212	6,399	4,864	4,090	3,956	3,642	3,483	3,351	3,148	2,998
175	5,397	6,022	7,407	9,301	11,905	6,840	5,077	4,212	4,066	3,724	3,552	3,412	3,195	3,036
180	5,381	5,989	7,304	8,988	15,646	7,340	5,309	4,343	4,182	3,809	3,624	3,474	3,243	3,075
190	5,350	5,927	7,120	8,494	13,406	8,559	5,840	4,631	4,435	3,991	3,776	3,603	3,342	3,155
200	5,321	5,869	6,961	8,117	11,233	10,199	6,476	4,958	4,721	4,192	3,941	3,742	3,447	3,238
210	5,292	5,814	6,820	7,816	10,012	13,390	7,241	5,333	5,044	4,412	4,119	3,891	3,557	3,324
220	5,265	5,763	6,695	7,569	9,203	19,659	8,167	5,761	5,410	4,655	4,313	4,050	3,673	3,414
225	5,252	5,738	6,638	7,461	8,889	26,581	8,702	5,998	5,612	4,787	4,416	4,134	3,733	3,461
230	5,239	5,715	6,583	7,361	8,617	43,400	9,292	6,252	5,826	4,925	4,524	4,221	3,795	3,508
240	5,213	5,669	6,482	7,183	8,166	73,716	10,673	6,813	6,296	5,222	4,755	4,406	3,924	3,605
250	5,189	5,626	6,389	7,027	7,814	25,665	13,504	7,453	6,828	5,552	5,007	4,605	4,060	3,707
260	5,166	5,585	6,304	6,889	7,570	17,235	17,018	8,180	7,427	5,915	5,281	4,820	4,204	3,813
270	5,143	5,546	6,226	6,766	7,395	13,875	22,092	9,004	8,098	6,315	5,580	5,051	4,356	3,923
275	5,132	5,527	6,189	6,710	7,318	12,841	24,858	9,455	8,462	6,529	5,739	5,173	4,435	3,980
280	5,121	5,509	6,153	6,656	7,250	12,038	26,990	9,932	8,847	6,753	5,904	5,300	4,516	4,038
290	5,100	5,473	6,086	6,555	7,123	10,863	27,424	11,775	9,676	7,229	6,254	5,566	5,686	4,158
300	5,080	5,439	6,022	6,463	7,002	10,040	24,484	13,300	11,281	7,744	6,630	5,851	4,861	4,283
310	5,060	5,407	5,963	6,378	6,890	9,425	20,834	14,867	12,529	8,298	7,032	6,154	5,053	4,412
320	5,041	5,376	5,907	6,299	6,788	8,947	17,637	16,159	13,760	8,887	7,459	6,474	5,249	4,547
330	5,022	5,346	5,855	6,226	6,682	8,562	15,215	16,842	14,824	9,511	7,909	6,810	5,455	4,687
340	5,004	5,317	5,805	6,158	6,590	8,244	13,454	16,844	15,355	10,164	8,379	7,161	5,668	4,831
350	4,987	5,290	5,758	6,094	6,507	7,976	12,161	16,361	15,464	11,263	8,868	7,526	5,889	4,979
360	4,970	5,263	5,713	6,034	6,424	7,747	11,187	15,622	15,199	11,806	9,371	7,902	6,116	5,132
370	4,953	5,237	5,670	5,977	6,351	7,548	10,432	14,774	14,706	12,180	9,884	8,286	6,347	5,287
380	4,937	5,213	5,630	5,924	6,280	7,374	9,833	13,899	14,097	12,364	10,452	8,678	6,583	5,445
390	4,921	5,189	5,591	5,873	6,212	7,219	9,346	13,053	13,437	12,377	10,737	9,073	6,820	5,604
400	4,906	5,166	5,553	5,824	6,150	7,081	8,942	12,267	12,769	12,257	10,909	9,470	7,059	5,764
420	4,877	5,121	5,483	5,734	6,030	6,845	8,310	10,931	11,510	11,773	10,953	9,820	7,536	6,082
440	4,849	5,080	5,418	5,651	5,917	6,640	7,838	9,901	10,442	11,137	10,719	9,892	8,006	6,390
450	4,836	5,060	5,387	5,612	5,862	6,538	7,643	9,482	9,986	10,800	10,542	9,865	8,180	6,539
460	4,823	5,041	5,357	5,574	5,814	6,449	7,470	9,115	9,580	10,462	10,341	9,799	8,332	6,684
480	4,797	5,004	5,301	5,498	5,719	6,282	7,174	8,506	8,896	9,808	9,899	9,584	8,440	6,959
500	4,773	4,969	5,248	5,422	5,628	6,132	6,930	8,027	8,350	9,210	9,437	9,302	8,381	7,218

Table 5. Specific Heat Capacity at Constant Pressure c_p in kJ/kg K (Continuation)
Spezifische Wärmekapazität bei konstantem Druck c_p in kJ/kg K (Fortsetzung)

p bar	t in °C → 500	520	540	550	560	580	600	620	640	650	660	680	700	750	800
0,1	2,129	2,143	2,157	2,164	2,171	2,185	2,199	2,212	2,226	2,233	2,240	2,253	2,266	2,299	2,331
1	2,132	2,146	2,160	2,166	2,173	2,187	2,201	2,214	2,228	2,235	2,241	2,255	2,268	2,301	2,332
5	2,146	2,158	2,171	2,177	2,184	2,197	2,210	2,223	2,236	2,243	2,249	2,262	2,275	2,307	2,337
10	2,162	2,174	2,185	2,191	2,197	2,209	2,221	2,234	2,246	2,252	2,259	2,271	2,284	2,314	2,344
20	2,197	2,205	2,214	2,219	2,224	2,234	2,245	2,256	2,267	2,272	2,278	2,289	2,301	2,329	2,357
25	2,214	2,221	2,229	2,233	2,237	2,246	2,256	2,267	2,277	2,282	2,288	2,299	2,310	2,337	2,364
30	2,233	2,237	2,243	2,247	2,251	2,259	2,268	2,278	2,287	2,292	2,298	2,308	2,318	2,344	2,370
40	2,270	2,271	2,274	2,276	2,279	2,285	2,292	2,300	2,308	2,313	2,317	2,326	2,336	2,360	2,383
50	2,310	2,306	2,305	2,306	2,307	2,311	2,316	2,322	2,329	2,333	2,337	2,345	2,353	2,375	2,397
60	2,352	2,343	2,338	2,337	2,337	2,338	2,341	2,345	2,351	2,354	2,357	2,364	2,371	2,390	2,410
70	2,397	2,382	2,372	2,369	2,367	2,366	2,366	2,369	2,372	2,375	2,377	2,383	2,389	2,405	2,423
75	2,420	2,402	2,390	2,386	2,383	2,380	2,379	2,380	2,383	2,385	2,387	2,392	2,398	2,413	2,430
80	2,444	2,422	2,408	2,403	2,399	2,394	2,392	2,392	2,394	2,396	2,397	2,402	2,407	2,421	2,437
90	2,494	2,465	2,445	2,437	2,432	2,423	2,418	2,416	2,417	2,417	2,418	2,421	2,425	2,436	2,450
100	2,547	2,510	2,484	2,474	2,465	2,453	2,445	2,441	2,439	2,439	2,439	2,440	2,443	2,452	2,464
110	2,602	2,557	2,524	2,511	2,501	2,484	2,473	2,466	2,462	2,461	2,460	2,460	2,461	2,468	2,477
120	2,661	2,606	2,566	2,550	2,537	2,516	2,502	2,492	2,486	2,483	2,482	2,480	2,480	2,483	2,491
125	2,691	2,632	2,588	2,570	2,556	2,533	2,516	2,505	2,497	2,495	2,493	2,490	2,489	2,491	2,498
130	2,722	2,658	2,610	2,591	2,575	2,549	2,531	2,518	2,509	2,506	2,504	2,500	2,498	2,499	2,504
140	2,787	2,711	2,655	2,633	2,614	2,583	2,561	2,545	2,534	2,529	2,526	2,521	2,517	2,515	2,518
150	2,854	2,767	2,703	2,677	2,654	2,618	2,592	2,572	2,558	2,553	2,548	2,541	2,536	2,531	2,532
160	2,925	2,826	2,752	2,722	2,696	2,654	2,623	2,600	2,583	2,577	2,571	2,562	2,556	2,547	2,546
170	2,998	2,887	2,802	2,769	2,739	2,691	2,653	2,629	2,609	2,601	2,594	2,583	2,575	2,564	2,559
175	3,036	2,918	2,829	2,792	2,761	2,710	2,672	2,644	2,622	2,613	2,606	2,594	2,585	2,572	2,566
180	3,075	2,950	2,855	2,817	2,784	2,730	2,689	2,658	2,635	2,626	2,618	2,605	2,595	2,580	2,573
190	3,155	3,015	2,909	2,867	2,830	2,769	2,723	2,688	2,662	2,651	2,642	2,626	2,615	2,596	2,587
200	3,238	3,083	2,966	2,918	2,877	2,809	2,758	2,719	2,689	2,677	2,666	2,648	2,635	2,613	2,601
210	3,324	3,153	3,024	2,971	2,925	2,850	2,793	2,750	2,716	2,702	2,690	2,671	2,655	2,630	2,615
220	3,414	3,225	3,083	3,026	2,975	2,893	2,830	2,781	2,744	2,729	2,715	2,693	2,676	2,646	2,630
225	3,461	3,263	3,114	3,054	3,001	2,914	2,848	2,797	2,758	2,742	2,728	2,704	2,686	2,655	2,637
230	3,508	3,300	3,145	3,082	3,027	2,936	2,867	2,813	2,772	2,756	2,741	2,716	2,696	2,663	2,644
240	3,605	3,378	3,208	3,139	3,079	2,981	2,905	2,846	2,801	2,783	2,766	2,739	2,717	2,680	2,658
250	3,707	3,458	3,273	3,199	3,133	3,026	2,943	2,880	2,830	2,810	2,792	2,762	2,738	2,697	2,672
260	3,813	3,541	3,340	3,259	3,189	3,072	2,983	2,914	2,860	2,838	2,818	2,786	2,760	2,714	2,687
270	3,923	3,626	3,409	3,322	3,245	3,120	3,023	2,948	2,890	2,866	2,845	2,809	2,781	2,732	2,701
275	3,980	3,670	3,444	3,353	3,274	3,144	3,043	2,966	2,905	2,880	2,858	2,821	2,792	2,740	2,708
280	4,038	3,714	3,480	3,385	3,303	3,168	3,064	2,983	2,921	2,895	2,872	2,833	2,803	2,749	2,716
290	4,158	3,806	3,552	3,451	3,362	3,217	3,105	3,019	2,951	2,924	2,899	2,857	2,824	2,767	2,730
300	4,283	3,899	3,626	3,517	3,422	3,267	3,147	3,055	2,983	2,953	2,926	2,882	2,846	2,784	2,745
310	4,412	3,996	3,702	3,586	3,484	3,318	3,190	3,091	3,014	2,982	2,954	2,906	2,868	2,802	2,759
320	4,547	4,096	3,781	3,656	3,547	3,370	3,233	3,128	3,046	3,012	2,982	2,931	2,891	2,819	2,774
330	4,687	4,199	3,860	3,727	3,611	3,423	3,278	3,165	3,078	3,042	3,010	2,956	2,913	2,837	2,789
340	4,831	4,305	3,942	3,800	3,677	3,476	3,322	3,203	3,111	3,072	3,038	2,981	2,935	2,855	2,803
350	4,979	4,413	4,026	3,874	3,743	3,531	3,367	3,241	3,143	3,103	3,067	3,006	2,958	2,873	2,818
360	5,132	4,524	4,111	3,950	3,811	3,586	3,413	3,280	3,176	3,133	3,095	3,032	2,981	2,891	2,833
370	5,287	4,638	4,198	4,027	3,880	3,642	3,459	3,319	3,210	3,164	3,124	3,057	3,003	2,909	2,848
380	5,445	4,753	4,287	4,106	3,950	3,699	3,506	3,358	3,243	3,196	3,153	3,083	3,026	2,927	2,863
390	5,604	4,870	4,377	4,186	4,022	3,756	3,554	3,398	3,277	3,227	3,183	3,108	3,049	2,945	2,877
400	5,764	4,988	4,468	4,267	4,194	3,814	3,601	3,438	3,311	3,258	3,212	3,134	3,072	2,963	2,892
420	6,082	5,225	4,653	4,431	4,241	3,933	3,698	3,518	3,379	3,322	3,271	3,185	3,118	2,999	2,922
440	6,390	5,458	4,839	4,598	4,391	4,054	3,797	3,600	3,448	3,385	3,330	3,237	3,164	3,035	2,952
450	6,539	5,571	4,932	4,682	4,466	4,115	3,847	3,641	3,483	3,417	3,360	3,263	3,187	3,053	2,967
460	6,684	5,681	5,023	4,761	4,542	4,177	3,897	3,683	3,517	3,449	3,389	3,289	3,209	3,071	2,981
480	6,959	5,889	5,200	4,929	4,692	4,300	3,999	3,766	3,587	3,514	3,449	3,341	3,255	3,106	3,011
500	7,218	6,077	5,367	5,087	4,838	4,424	4,101	3,850	3,657	3,578	3,508	3,392	3,301	3,142	3,040

Table 5. Specific Heat Capacity at Constant Pressure c_p in kJ/kg K (Continuation)
Spezifische Wärmekapazität bei konstantem Druck c_p in kJ/kg K (Fortsetzung)

p bar ↓	t in °C → 0	50	100	120	140	150	160	180	200	220	240	250	260	280	300
500	3,994	4,081	4,117	4,137	4,163	4,179	4,196	4,235	4,284	4,343	4,417	4,461	4,510	4,626	4,773
520	3,986	4,077	4,113	4,133	4,159	4,174	4,191	4,230	4,277	4,335	4,407	4,450	4,497	4,609	4,750
540	3,979	4,074	4,110	4,130	4,155	4,170	4,186	4,224	4,270	4,327	4,397	4,438	4,484	4,592	4,728
550	3,975	4,072	4,108	4,128	4,153	4,168	4,184	4,221	4,267	4,323	4,392	4,433	4,478	4,584	4,717
560	3,971	4,071	4,107	4,126	4,151	4,165	4,182	4,219	4,264	4,319	4,388	4,428	4,472	4,577	4,707
580	3,964	4,067	4,103	4,122	4,147	4,161	4,177	4,213	4,258	4,312	4,378	4,417	4,460	4,561	4,686
600	3,957	4,064	4,100	4,119	4,143	4,157	4,172	4,208	4,251	4,304	4,369	4,407	4,449	4,547	4,667
620	3,949	4,061	4,097	4,115	4,139	4,153	4,168	4,203	4,245	4,297	4,360	4,397	4,438	4,532	4,648
640	3,942	4,058	4,093	4,112	4,135	4,149	4,164	4,198	4,239	4,289	4,351	4,387	4,427	4,519	4,630
650	3,938	4,056	4,092	4,110	4,133	4,147	4,161	4,195	4,236	4,286	4,347	4,382	4,421	4,512	4,621
700	3,920	4,049	4,084	4,102	4,124	4,136	4,151	4,183	4,222	4,268	4,326	4,359	4,396	4,480	4,579
750	3,902	4,042	4,076	4,093	4,114	4,127	4,140	4,171	4,207	4,252	4,306	4,337	4,372	4,450	4,540
800	3,883	4,035	4,068	4,085	4,105	4,117	4,130	4,159	4,194	4,236	4,287	4,317	4,349	4,422	4,505
850	3,864	4,028	4,061	4,077	4,097	4,108	4,120	4,148	4,181	4,221	4,269	4,297	4,328	4,396	4,472
900	3,844	4,022	4,054	4,069	4,088	4,099	4,110	4,137	4,168	4,206	4,252	4,278	4,307	4,371	4,441
950	3,823	4,016	4,046	4,061	4,079	4,090	4,101	4,126	4,156	4,192	4,236	4,261	4,288	4,348	4,412
1000	3,801	4.010	4,039	4,054	4,071	4,081	4,092	4,116	4,144	4,178	4,220	4,244	4,269	4,326	4,385

p bar ↓	t in °C → 300	320	340	350	360	380	400	420	425	440	450	460	480	500
500	4,773	4,969	5,248	5,422	5,628	6,132	6,930	8,027	8,350	9,210	9,437	9,302	8,381	7,218
520	4,750	4,936	5,198	5,360	5,546	6,006	6,715	7,640	7,910	8,683	8,982	8,986	8,333	7,359
540	4,728	4,905	5,151	5,297	5,469	5,892	6,521	7,323	7,548	8,228	8,554	8,656	8,232	7,420
550	4,717	4,890	5,128	5,269	5,431	5,841	6,427	7,184	7,391	8,025	8,353	8,490	8,168	7,422
560	4,707	4,875	5,106	5,240	5,397	5,789	6,335	7,057	7,248	7,838	8,162	8,327	8,096	7,420
580	4,686	4,846	5,064	5,197	5,330	5,697	6,178	6,820	6,994	7,504	7,810	8,009	7,936	7,400
600	4,667	4,819	5,023	5,137	5,270	5,603	6,042	6,616	6,777	7,217	7,499	7,711	7,761	7,358
620	4,648	4,792	4,985	5,090	5,209	5,527	5,917	6,332	6,570	6,977	7,242	7,457	7,576	7,288
640	4,630	4,767	4,948	5,047	5,157	5,453	5,808	6,268	6,386	6,774	7,017	7,220	7,388	7,210
650	4,621	4,755	4,930	5,024	5,133	5,414	5,758	6,192	6,300	6,686	6,919	7,110	7,321	7,169
700	4,579	4,697	4,846	4,939	5,032	5,271	5,558	5,901	5,995	6,291	6,482	6,650	6,920	6,927
750	4,540	4,645	4,770	4,855	4,937	5,146	5,393	5,674	5,748	5,980	6,138	6,290	6,534	6,632
800	4,505	4,597	4,723	4,790	4,864	5,050	5,261	5,493	5,557	5,748	5,878	6,009	6,202	6,350
850	4,472	4,566	4,678	4,737	4,803	4,966	5,157	5,353	5,415	5,581	5,686	5,800	5,977	6,110
900	4,441	4,529	4,633	4,687	4,748	4,893	5,063	5,242	5,290	5,434	5,525	5,623	5,784	5,907
950	4,412	4,497	4,596	4,648	4,702	4,829	4,978	5,144	5,181	5,308	5,382	5,467	5,618	5,722
1000	4,385	4,461	4,559	4,610	4,659	4,773	4,904	5,062	5,090	5,205	5,268	5,341	5,462	5,568

p bar ↓	t in °C → 500	520	540	550	560	580	600	620	640	650	660	680	700	750	800
500	7,218	6,077	5,367	5,087	4,839	4,424	4,101	3,850	3,657	3,578	3,508	3,392	3,301	3,142	3,040
520	7,359	6,244	5,518	5,234	4,980	4,547	4,203	3,935	3,728	3,643	3,568	3,443	3,346	3,177	3,070
540	7,420	6,389	5,651	5,367	5,112	4,667	4,305	4,020	3,798	3,707	3,627	3,494	3,390	3,212	3,099
550	7,422	6,454	5,710	5,428	5,173	4,726	4,356	4,062	3,833	3,739	3,657	3,520	3,413	3,229	3,113
560	7,420	6,515	5,764	5,484	5,232	4,783	4,406	4,104	3,869	3,772	3,686	3,545	3,435	3,246	3,127
580	7,400	6,627	5,857	5,582	5,337	4,891	4,505	4,188	3,939	3,836	3,745	3,596	3,479	3,280	3,156
600	7,358	6,711	5,946	5,683	5,432	4,991	4,600	4,271	4,008	3,900	3,804	3,646	3,522	3,314	3,184
620	7,288	6,757	6,022	5,772	5,522	5,080	4,691	4,352	4,077	3,963	3,862	3,695	3,565	3,347	3,211
640	7,210	6,752	6,089	5,844	5,598	5,157	4,775	4,430	4,145	4,025	3,920	3,744	3,608	3,379	3,238
650	7,169	6,740	6,110	5,871	5,630	5,180	4,815	4,469	4,178	4,056	3,948	3,769	3,629	3,395	3,252
700	6,927	6,631	6,197	5,969	5,730	5,300	4,953	4,624	4,338	4,205	4,087	3,888	3,732	3,473	3,317
750	6,632	6,493	6,196	5,979	5,767	5,365	5,046	4,723	4,478	4,341	4,215	4,001	3,832	3,548	3,379
800	6,350	6,320	6,108	5,947	5,762	5,401	5,106	4,839	4,590	4,455	4,328	4,106	3,927	3,620	3,438
850	6,110	6,132	6,007	5,876	5,726	5,437	5,154	4,893	4,651	4,540	4,418	4,197	4,014	3,690	3,495
900	5,907	5,945	5,874	5,783	5,670	5,422	5,171	4,928	4,696	4,590	4,478	4,270	4,090	3,758	3,549
950	5,722	5,773	5,739	5,672	5,595	5,402	5,178	4,952	4,721	4,607	4,505	4,318	4,151	3,824	3,603
1000	5,568	5,620	5,591	5,554	5,502	5,366	5,172	4,947	4,717	4,604	4,501	4,336	4,192	3,887	3,656

Table 5a. Specific Heat Capacity c_{p0} and Enthalpy h_0 at Zero Pressure
Spezifische Wärmekapazität c_{p0} und Enthalpie h_0 beim Drucke Null

t	c_{p0}	h_0	t	c_{p0}	h_0	t	c_{p0}	h_0	t	c_{p0}	h_0
°C	kJ/kg K	kJ/kg	°C	kJ/kg K	kJ/kg	°C	kJ/kg K	kJ/kg	°C	kJ/kg K	kJ/kg
0	1,8516	2501,78	200	1,9391	2880,11	400	2,0613	3279,74	600	2,1983	3705,61
10	1,8549	2520,31	210	1,9446	2899,53	410	2,0679	3300,38	610	2,2052	3727,63
20	1,8583	2538,88	220	1,9501	2919,00	420	2,0746	3321,09	620	2,2121	3749,71
30	1,8618	2557,48	230	1,9557	2938,53	430	2,0813	3341,87	630	2,2190	3771,87
40	1,8654	2576,11	240	1,9614	2958,12	440	2,0881	3362,72	640	2,2258	3794,09
50	1,8692	2594,79	250	1,9672	2977,76	450	2,0949	3383,64	650	2,2326	3816,39
60	1,8731	2613,50	260	1,9730	2997,46	460	2,1017	3404,62	660	2,2394	3838,75
70	1,8771	2632,25	270	1,9789	3017,22	470	2,1085	3425,67	670	2,2462	3861,17
80	1,8812	2651,04	280	1,9849	3037,04	480	2,1154	3446,79	680	2,2529	3883,67
90	1,8855	2669,87	290	1,9910	3056,92	490	2,1223	3467,98	690	2,2596	3906,23
100	1,8898	2688,75	300	1,9971	3076,86	500	2,1292	3489,24	700	2,2663	3928,86
110	1,8943	2707,67	310	2,0033	3096,86	510	2,1361	3510,56	710	2,2729	3951,56
120	1,8989	2726,63	320	2,0095	3116,93	520	2,1430	3531,96	720	2,2795	3974,32
130	1,9035	2745,65	330	2,0158	3137,05	530	2,1499	3553,42	730	2,2861	3997,15
140	1,9083	2764,70	340	2,0222	3157,24	540	2,1568	3574,95	740	2,2926	4020,04
150	1,9132	2783,81	350	2,0286	3177,50	550	2,1637	3596,56	750	2,2991	4043,00
160	1,9182	2802,97	360	2,0350	3197,81	560	2,1707	3618,23	760	2,3055	4066,02
170	1,9233	2822,18	370	2,0415	3218,20	570	2,1776	3639,97	770	2,3119	4089,11
180	1,9285	2841,44	380	2,0480	3238,64	580	2,1845	3661,78	780	2,3182	4112,26
190	1,9338	2860,75	390	2,0546	3259,16	590	2,1914	3683,66	790	2,3244	4135,47
200	1,9391	2880,11	400	2,0613	3279,74	600	2,1983	3705,61	800	2,3306	4158,75

Table 5b. Specific Heat Capacity c_p' and c_p'' in State of Saturation
Spezifische Wärmekapazität c_p' und c_p'' im Sättigungszustand

t	p	c_p'	c_p''	t	p	c_p'	c_p''
°C	bar	kJ/kg K		°C	bar	kJ/kg K	
0	0,006108	4,217	1,854	200	15,549	4,497	2,843
10	0,012270	4,193	1,860	210	19,077	4,551	2,988
20	0,02337	4,182	1,866	220	23,198	4,613	3,150
30	0,04241	4,179	1,875	230	27,976	4,685	3,331
40	0,07375	4,179	1,885	240	33,478	4,769	3,536
50	0,12335	4,181	1,899	250	39,776	4,867	3,772
60	0,19920	4,185	1,916	260	46,943	4,983	4,047
70	0,3116	4,190	1,936	270	55,058	5,122	4,373
80	0,4736	4,197	1,962	280	64,202	5,290	4,767
90	0,7011	4,205	1,992	290	74,461	5,499	5,253
100	1,0133	4,216	2,028	300	85,927	5,762	5,863
110	1,4327	4,229	2,070	310	98,700	6,104	6,650
120	1,9854	4,245	2,120	320	112,89	6,565	7,722
130	2,7013	4,263	2,176	330	128,63	7,219	9,361
140	3,614	4,285	2,241	340	146,05	8,233	12,21
150	4,760	4,310	2,314	350	165,35	10,11	17,15
160	6,181	4,339	2,398	360	186,75	14,58	25,12
170	7,920	4,371	2,491	370	210,54	43,17	76,92
180	10,027	4,408	2,596	373	218,20	166,0	297,8
190	12,551	4,449	2,713				

Table 5c. Specific Heat Capacity, Diagram
Spezifische Wärmekapazität, Diagramm

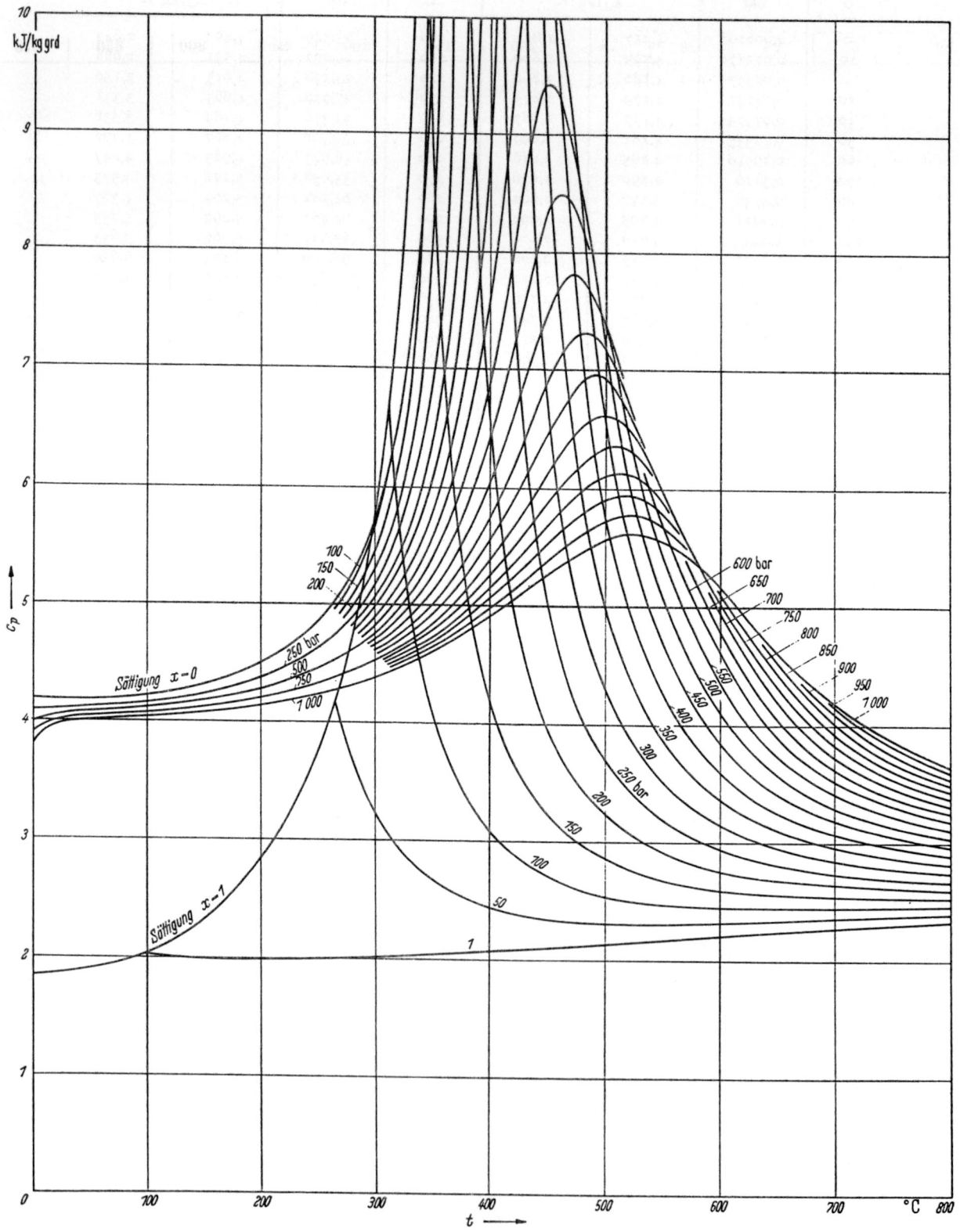

Table 6a. Prandtl Number as Function of Pressure p and Temperature t

The Prandtl number is calculated using Eqs. (C II,1), (C II,3) and the 1967 IFC Formulation for Industrial Use (Section C I).

Prandtl-Zahl als Funktion von Druck p und Temperatur t

Die Prandtl-Zahl ist mit Hilfe der Gln. (C II,1), (C II,3) und der 1967 IFC Formulation for Industrial Use (Abschn. C I) berechnet worden.

p bar	temperature t in °C										
	0	25	50	75	100	150	200	250	300	350	375
1	13,45	6,125	3,571	2,388	1,003	0,978	0,959	0,947	0,938	0,930	0,927
5	13,43	6,120	3,569	2,387	1,753	1,150	1,013	0,976	0,953	0,938	0,933
10	13,41	6,115	3,568	2,387	1,752	1,150	1,078	1,015	0,973	0,949	0,941
25	13,33	6,098	3,562	2,384	1,751	1,149	0,9099	1,149	1,05	0,987	0,968
50	13,21	6,071	3,553	2,381	1,749	1,148	0,9083	0,8341	1,21	1,07	1,03
75	13,10	6,044	3,544	2,377	1,748	1,148	0,9067	0,8296	1,44	1,18	1,11
100	12,98	6,018	3,535	2,374	1,746	1,147	0,9053	0,8255	0,898	1,31	1,21
125	12,87	5,992	3,527	2,370	1,745	1,146	0,9039	0,8216	0,881	1,52	1,34
150	12,76	5,967	3,518	2,367	1,743	1,145	0,9026	0,8180	0,867	1,88	1,52
175	12,65	5,943	3,510	2,364	1,742	1,145	0,9013	0,8146	0,854	1,41	1,80
200	12,55	5,919	3,502	2,360	1,740	1,144	0,9001	0,8114	0,843	1,24	2,30
225	12,45	5,896	3,494	2,357	1,739	1,143	0,8990	0,8084	0,833	1,14	9,72
250	12,35	5,873	3,486	2,354	1,737	1,143	0,8979	0,8056	0,824	1,08	2,12
275	12,25	5,851	3,479	2,351	1,736	1,142	0,8969	0,8030	0,816	1,03	1,59
300	12,16	5,829	3,472	2,348	1,735	1,142	0,8959	0,8005	0,808	0,994	1,36
350	11,98	5,787	3,458	2,343	1,732	1,141	0,8941	0,7960	0,795	0,939	1,16
400	11,80	5,748	3,444	2,337	1,730	1,140	0,8925	0,7919	0,784	0,898	1,05
450	11,64	5,710	3,431	2,332	1,728	1,139	0,8911	0,7882	0,774	0,866	0,985
500	11,48	5,674	3,419	2,328	1,726	1,138	0,8897	0,7849	0,766	0,839	0,940
550	11,33	5,639	3,408	2,323	1,724	1,138	0,8885	0,7819	0,758	0,817	0,905
600	11,19	5,607	3,396	2,319	1,723	1,137	0,8875	0,7792	0,752	0,798	0,879
650	11,05	5,576	3,386	2,315	1,721	1,137	0,8865	0,7767	0,746	0,781	0,857
700	10,92	5,546	3,376	2,311	1,720	1,136	0,8856	0,7745	0,740	0,765	0,840
750	10,79	5,519	3,367	2,307	1,718	1,136	0,8848	0,7724	0,735	0,751	0,825
800	10,66	5,492	3,358	2,304	1,717	1,136	0,8841	0,7705	0,731	0,738	0,812
850	10,54	5,467	3,350	2,301	1,716	1,136	0,8834	0,7688	0,727	0,727	0,801
900	10,42	5,443	3,342	2,298	1,715	1,135	0,8828	0,7672	0,723	0,716	0,791
950	10,31	5,421	3,334	2,295	1,714	1,135	0,8823	0,7657	0,720	0,705	0,782
1000	10,19	5,399	3,327	2,293	1,713	1,135	0,8819	0,7643	0,717	0,696	0,775

p bar	temperature t in °C										
	400	425	450	475	500	550	600	650	700	750	800
1	0,924	0,920	0,917	0,914	0,911	0,905	0,898	0,892	0,886	0,880	0,874
5	0,928	0,924	0,920	0,917	0,913	0,906	0,900	0,893	0,887	0,881	0,875
10	0,934	0,929	0,924	0,920	0,916	0,908	0,902	0,895	0,889	0,882	0,876
25	0,954	0,944	0,936	0,929	0,924	0,915	0,907	0,900	0,892	0,886	0,879
50	1,00	0,976	0,960	0,948	0,939	0,925	0,915	0,907	0,899	0,891	0,883
75	1,06	1,02	0,991	0,971	0,956	0,937	0,924	0,914	0,905	0,896	0,888
100	1,12	1,07	1,03	0,999	0,977	0,949	0,933	0,921	0,910	0,901	0,892
125	1,20	1,12	1,07	1,03	1,00	0,963	0,942	0,927	0,916	0,905	0,896
150	1,30	1,19	1,11	1,06	1,03	0,979	0,951	0,934	0,921	0,910	0,899
175	1,44	1,26	1,16	1,10	1,05	0,995	0,962	0,941	0,926	0,914	0,903
200	1,63	1,36	1,22	1,14	1,08	1,01	0,972	0,948	0,932	0,918	0,906
225	1,90	1,48	1,29	1,18	1,11	1,03	0,983	0,955	0,937	0,922	0,910
250	2,46	1,63	1,36	1,23	1,15	1,05	0,994	0,962	0,941	0,926	0,913
275	3,60	1,82	1,45	1,28	1,18	1,07	1,00	0,968	0,946	0,930	0,916
300	3,26	2,02	1,55	1,33	1,21	1,09	1,02	0,975	0,950	0,933	0,918
350	1,81	2,40	1,77	1,45	1,29	1,12	1,04	0,987	0,958	0,939	0,923
400	1,37	1,93	1,89	1,56	1,35	1,16	1,05	0,997	0,965	0,944	0,927
450	1,18	1,54	1,71	1,65	1,40	1,19	1,07	1,01	0,969	0,947	0,930
500	1,07	1,30	1,51	1,51	1,42	1,21	1,09	1,01	0,972	0,949	0,932
550	1,00	1,16	1,34	1,40	1,37	1,21	1,10	1,02	0,973	0,949	0,932
600	0,951	1,06	1,20	1,29	1,31	1,20	1,11	1,02	0,972	0,947	0,932
650	0,913	0,994	1,10	1,20	1,24	1,17	1,11	1,02	0,969	0,945	0,930
700	0,882	0,943	1,02	1,11	1,18	1,13	1,10	1,02	0,966	0,940	0,926
750	0,858	0,903	0,963	1,04	1,11	1,10	1,08	1,02	0,962	0,935	0,922
800	0,837	0,870	0,917	0,984	1,05	1,08	1,06	1,01	0,957	0,930	0,917
850	0,820	0,843	0,880	0,938	1,00	1,05	1,03	0,997	0,950	0,924	0,911
900	0,806	0,820	0,850	0,901	0,962	1,01	1,02	0,980	0,942	0,918	0,905
950	0,793	0,801	0,824	0,870	0,927	0,981	0,997	0,959	0,932	0,912	0,899
1000	0,782	0,783	0,801	0,844	0,897	0,950	0,973	0,937	0,919	0,906	0,894

Table 6b. Dynamic Viscosity, Thermal Conductivity and Prandtl Number in State of Saturation

Dynamische Viskosität, Wärmeleitfähigkeit und Prandtlzahl im Sättigungszustand

t	p	η'	η''	λ'	λ''	Pr'	Pr''
°C	bar	10^{-6} kg/s m		mW/K m		—	
0,00	0,006108	1793	9,216	561,9	16,49	13,46	1,036
0,01	0,006112	1792	9,216	561,9	16,49	13,45	1,036
10	0,01227	1306	9,461	582,0	17,21	9,412	1,022
20	0,02337	1002	9,727	599,6	17,95	6,989	1,011
30	0,04241	797,7	10,01	615,0	18,70	5,420	1,003
40	0,07375	653,2	10,31	628,6	19,48	4,343	0,9975
50	0,12335	547,0	10,62	640,5	20,28	3,571	0,9940
60	0,19920	466,5	10,93	650,7	21,10	3,000	0,9925
70	0,3116	404,0	11,26	659,4	21,96	2,567	0,9928
80	0,4736	354,4	11,59	666,8	22,86	2,230	0,9948
90	0,7011	314,4	11,93	672,7	23,80	1,965	0,9984
100	1,01325	281,7	12,27	677,5	24,79	1,753	1,004
110	1,433	254,6	12,61	680,9	25,84	1,581	1,010
120	1,985	232,0	12,96	683,3	26,96	1,441	1,019
130	2,701	212,8	13,30	684,4	28,15	1,325	1,028
140	3,614	196,4	13,65	684,5	29,42	1,230	1,039
150	4,760	182,4	13,99	683,6	30,77	1,150	1,052
160	6,181	170,2	14,34	681,5	32,22	1,083	1,067
170	7,920	159,5	14,68	678,5	33,77	1,028	1,083
180	10,03	150,1	15,02	674,5	35,42	0,9810	1,101
190	12,55	141,8	15,37	669,4	37,20	0,9423	1,121
200	15,55	134,3	15,71	663,4	39,10	0,9106	1,143
210	19,08	127,6	16,06	656,3	41,14	0,8850	1,166
220	23,20	121,6	16,41	648,3	43,35	0,8650	1,192
230	27,98	116,0	16,76	639,3	45,74	0,8503	1,221
240	33,48	110,9	17,12	629,2	48,34	0,8406	1,253
250	39,78	106,2	17,49	618,1	51,18	0,8360	1,289
260	46,94	101,8	17,87	605,9	54,33	0,8368	1,332
270	55,06	97,56	18,27	592,6	57,84	0,8432	1,382
280	64,20	93,56	18,70	578,0	61,82	0,8563	1,442
290	74,46	89,69	19,15	562,2	66,40	0,8773	1,515
300	85,93	85,91	19,65	545,0	71,78	0,9082	1,605
310	98,70	82,16	20,21	526,4	78,26	0,9527	1,717
320	112,9	78,37	20,84	506,3	86,34	1,016	1,864
330	128,6	74,48	21,60	484,5	96,93	1,110	2,086
340	146,1	70,37	22,55	461,1	111,8	1,256	2,464
350	165,4	65,83	23,82	436,3	134,5	1,525	3,036
360	186,8	60,31	25,73	411,8	176,8	2,169	3,656
370	210,5	52,10	29,63	416,4	306,4	5,440	7,438
371	213,1	50,79	30,40	429,0	342,4	6,793	9,003
372	215,6	49,16	31,41	454,7	396,1	9,217	11,95
373	218,2	46,90	32,92	517,5	490,2	15,04	20,00

6 c. Diagrams of Transport Properties
Diagramme der Transportgrößen

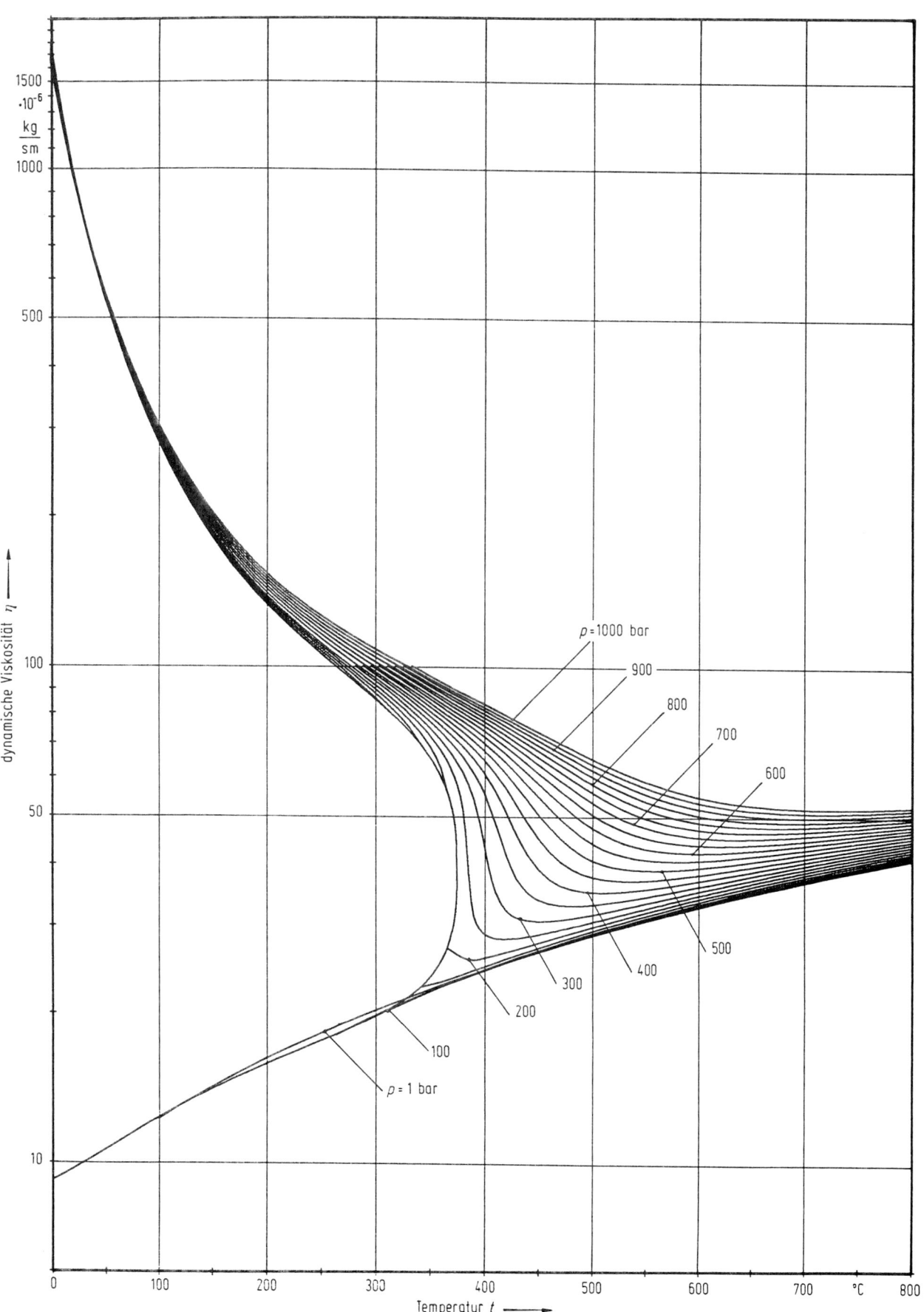

Fig. 1. η, **t-Diagram**

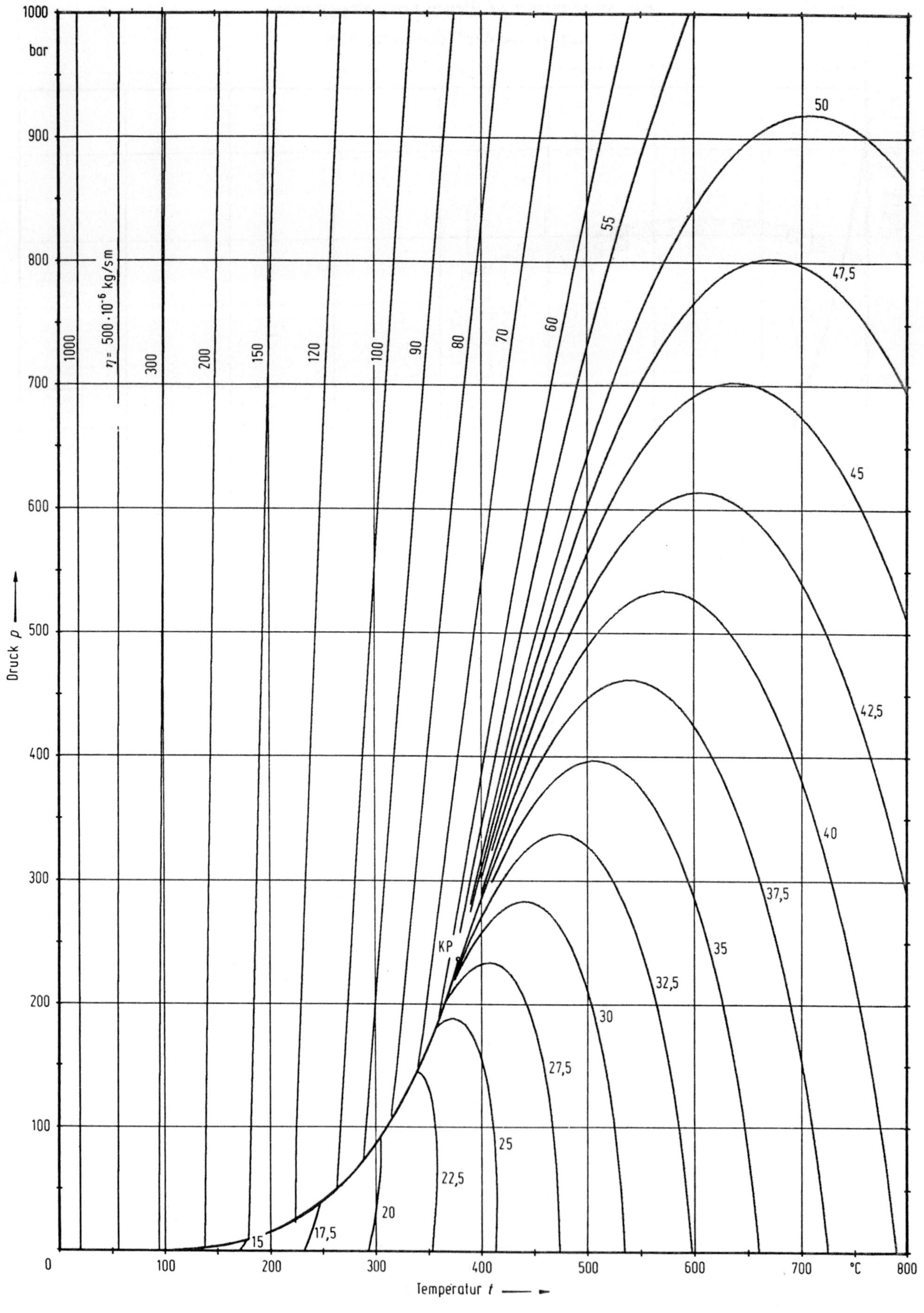
1000
bar
900
800
700
600
500
400
300
200
100
0
Druck p
100
200
300
400
500
600
700
°C
800
Temperatur t
η = 500·10⁻⁶ kg/sm
1000
300
200
150
120
100
90
80
70
60
55
50
47,5
45
42,5
40
37,5
35
32,5
30
27,5
25
22,5
20
17,5
15
KP

Fig. 2. η in the p, t-Diagram

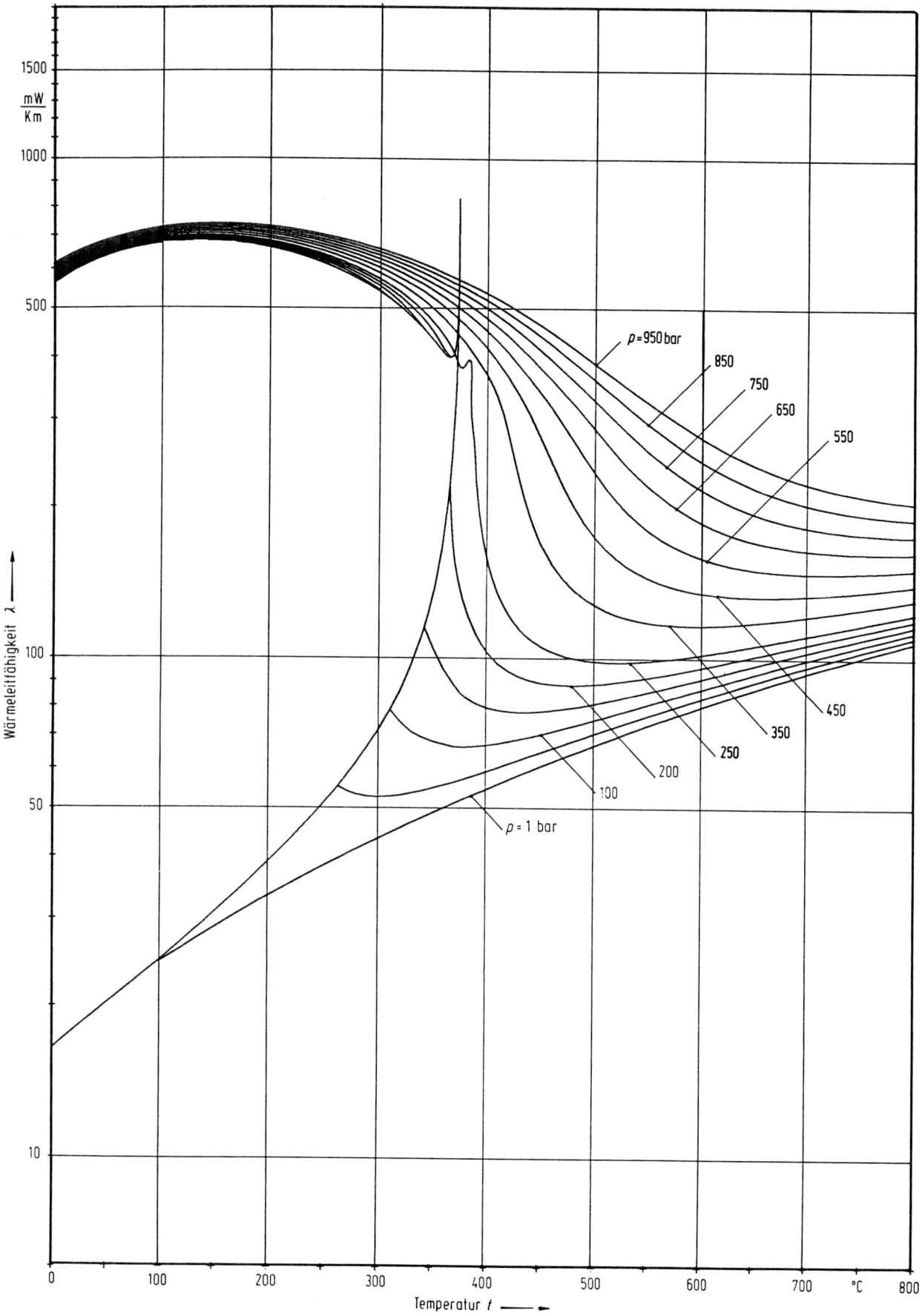

Fig. 3. λ, t-Diagram

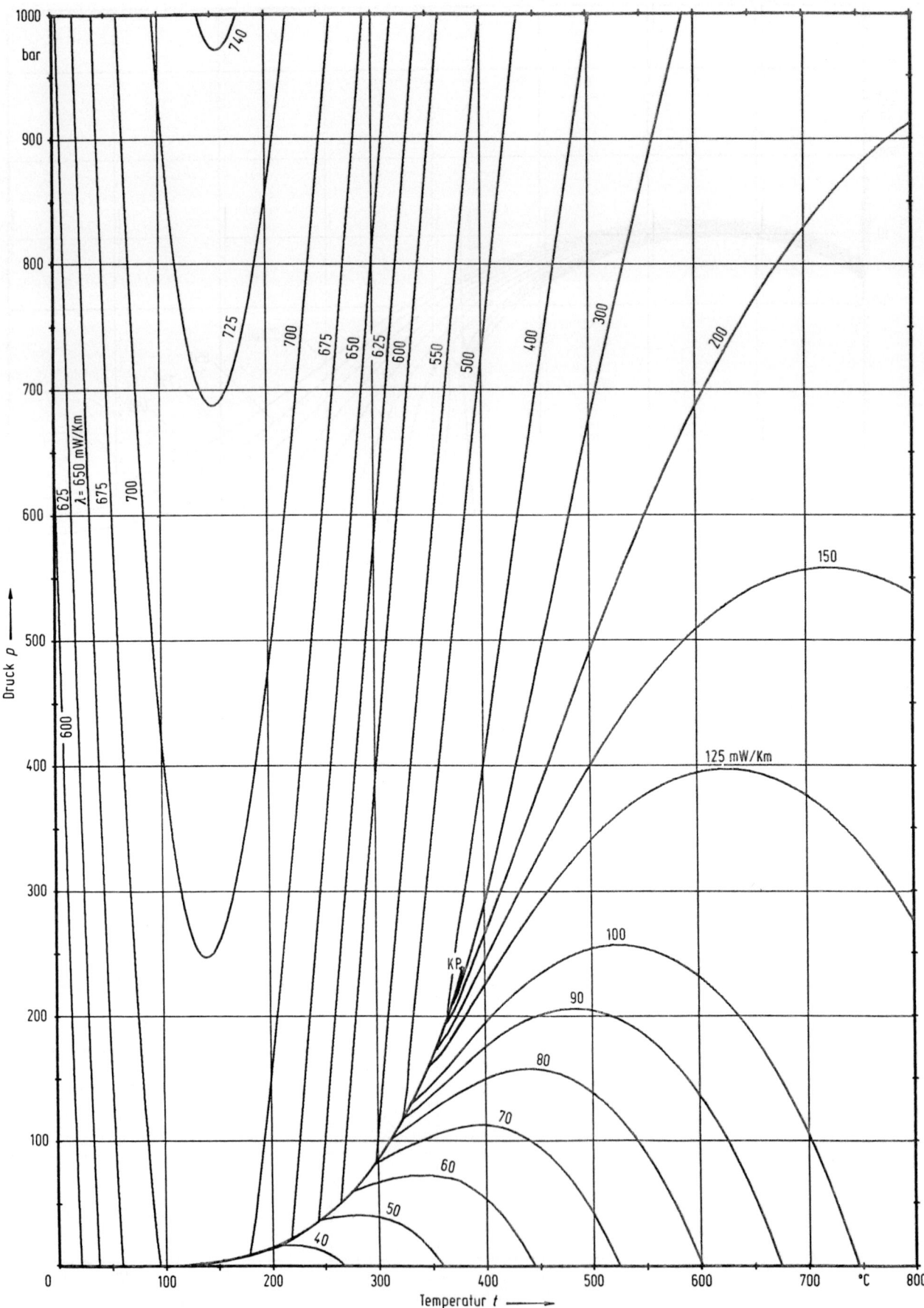

Fig. 4. λ in the p, t-Diagram

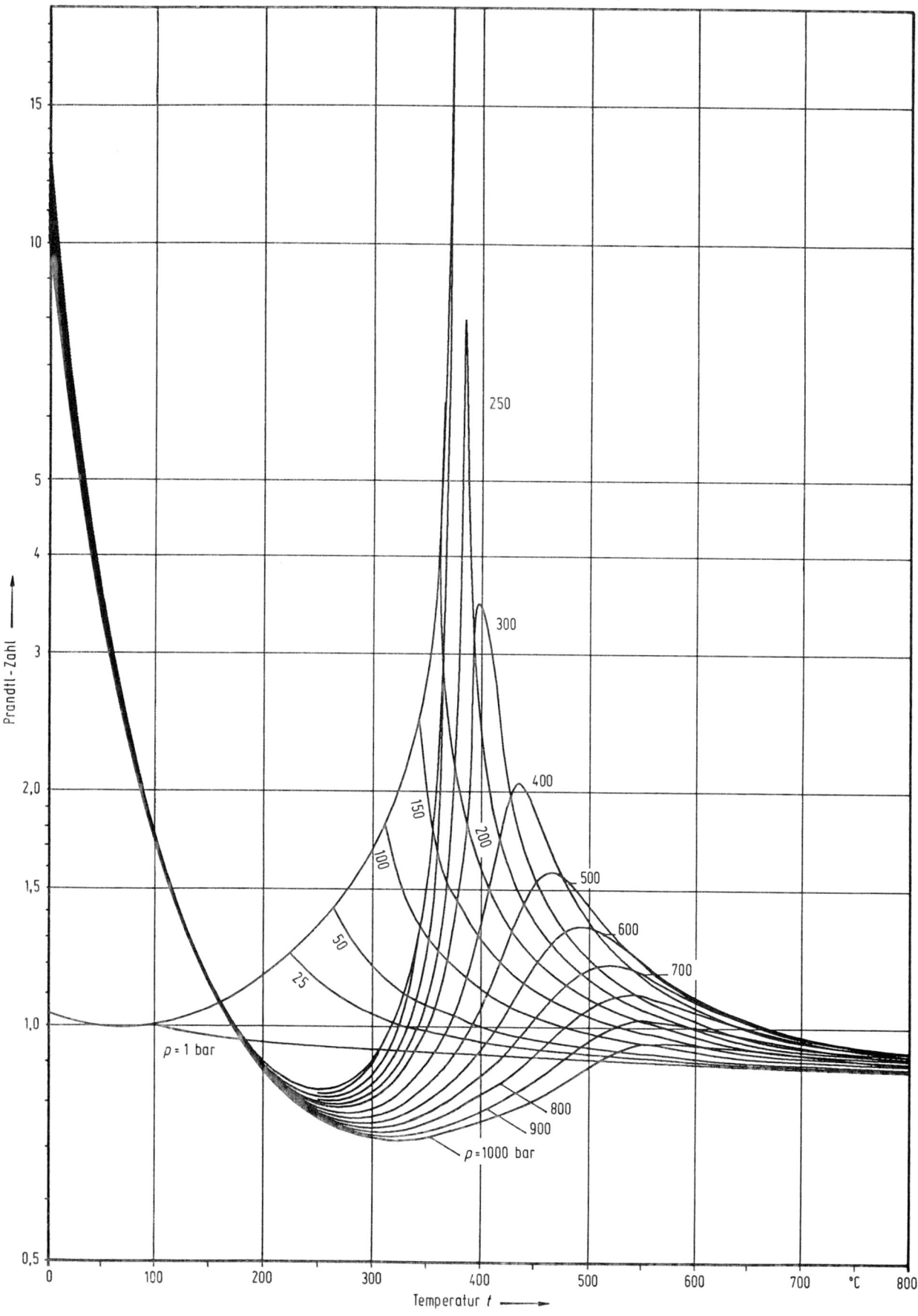

Fig. 5. *Pr, t*-Diagram

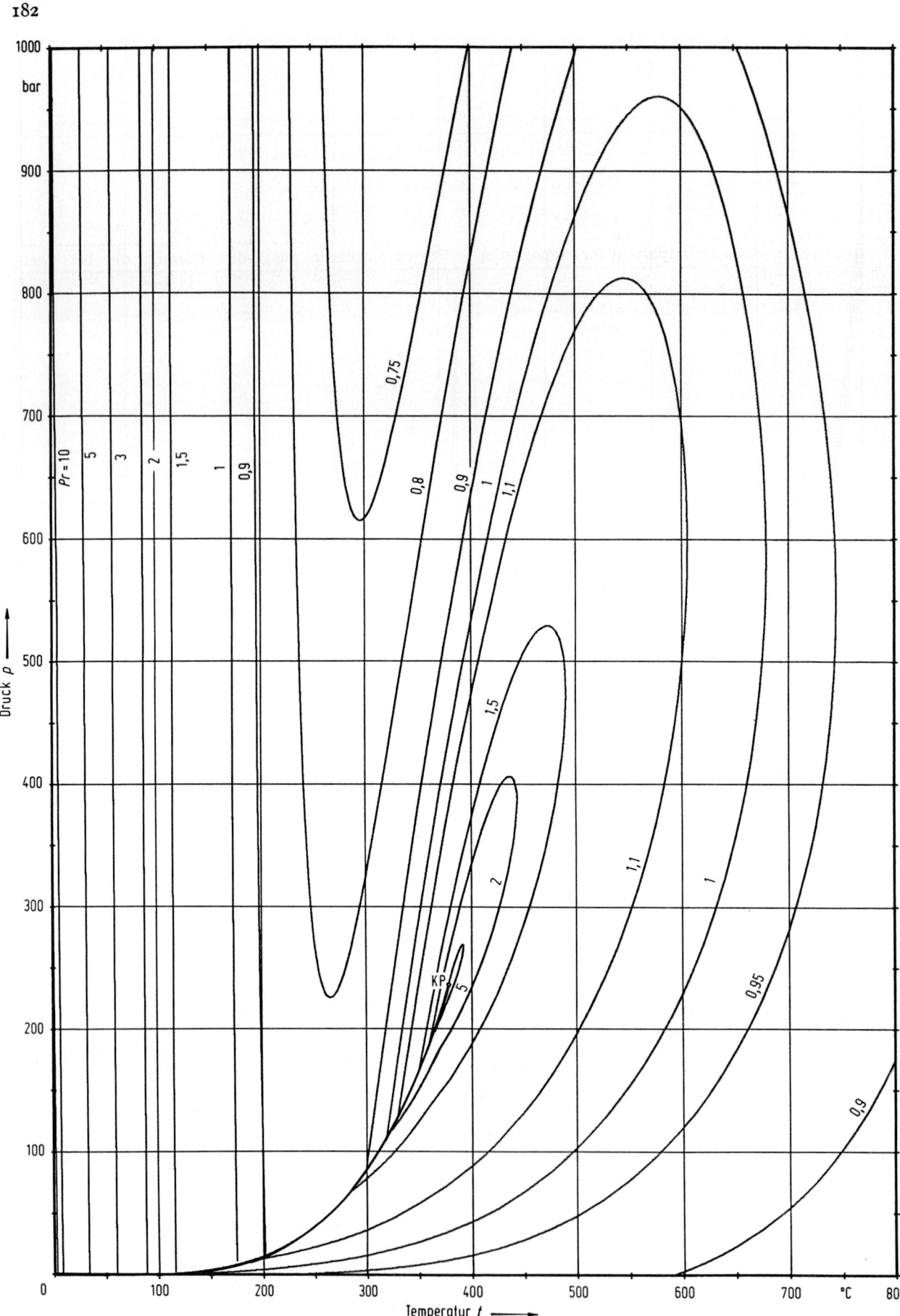

Fig. 6. *Pr* in the *p, t*-Diagram

Table 7. Surface Tension and Laplace Coefficient
Oberflächenspannung und Laplace-Koeffizient

The surface tension σ and the Laplace Coefficient a (Laplace Constant) are quantities of the interface of liquid and vapour phase which are important for evapouration processes for example.

The Skeleton Table of Surface Tension was revised by the Working Group III of the International Association for the Properties of Steam and internationally standardized in 1975 in the form of a release [263]. The equation

Die Oberflächenspannung σ und der Laplace-Koeffizient a (Laplace-Konstante) sind Zustandsgrößen der Phasengrenzfläche flüssig-dampfförmig, die bei zweiphasigen Systemen z. B. Siedevorgängen von Bedeutung sind.

Die Rahmentafel der Oberflächenspannung wurde von der Arbeitsgruppe III der ,,International Association for the Properties of Steam" überarbeitet und 1975 in Form einer Empfehlung zum internationalen Standard erhoben [263]. Die Gleichung

$$\sigma = B \cdot \left(\frac{T_c - T}{T_c}\right)^{\mu} \cdot \left(1 + b \cdot \left(\frac{T_c - T}{T_c}\right)\right) \tag{1}$$

with the constants

mit den Konstanten

$$B = 235{,}8 \cdot 10^{-3}\ \text{N/m}, \quad b = -0{,}625, \quad \mu = 1{,}256$$

is valid from the triple point to the critical point and gives the surface tension σ in N/m within the tolerances of the Skeleton Table. For the critical temperature the value $T_c = 647{,}15$ K has to be used. The tolerances of the table values are 0,5% from 0,01 to 180 °C and increase steadily up to 3% at 360 °C.

The Laplace Coefficient is defined by the equation

gibt die Oberflächenspannung σ in N/m vom Tripelpunkt bis zum kritischen Punkt innerhalb der Toleranzen der Rahmentafel wieder. Die kritische Temperatur ist dabei zu $T_c = 647{,}15$ K einzusetzen. Die Toleranzen der Tafelwerte betragen 0,5% von 0,01 bis 180 °C und steigen bis 360 °C stetig auf 3% an.

Der Laplace-Koeffizient ist durch die Gleichung

$$a = \sqrt{\frac{\sigma}{g \cdot (\varrho' - \varrho'')}} \tag{2}$$

with g as local acceleration of gravity (Standard value $g = 9{,}80665$ m/s²) and ϱ', ϱ'' as densities of saturated liquid and steam respectively. The approximation

mit g als örtlicher Schwerebeschleunigung (Normwert $g = 9{,}80665$ m/s²) und ϱ' und ϱ'' als Sättigungsdichten von Flüssigkeit und Dampf definiert. Die Näherungsgleichung

$$a = \left\{C \cdot \left(\frac{T_c - T}{T_c}\right)^{\zeta} \cdot \left(1 + c \cdot \left(\frac{T_c - T}{T_c}\right)\right)\right\}^{0,5} \tag{3}$$

with the constants

mit den Konstanten

$$C = 1{,}66 \cdot 10^{-5}\ \text{m}^2, \quad c = -0{,}4083, \quad \zeta = 0{,}91,$$
$$T_c = 647{,}16\ \text{K}$$

gives the defined Laplace Coefficient of eq. (2) with σ and $(\varrho' - \varrho'')$ from the Skeleton Tables with an accuracy of higher than 0,5%.

gibt den durch die Gl. (2) definierten Laplace-Koeffizienten mit σ und $(\varrho' - \varrho'')$ aus den Rahmentafeln mit einer Genauigkeit höher als 0,5% wieder.

Skeleton Table of Surface Tension and Laplace Coefficient
Rahmentafel der Oberflächenspannung und des Laplace-Koeffizienten

t °C	σ 10^{-3} N/m	a 10^{-3} m	t °C	σ 10^{-3} N/m	a 10^{-3} m
0,01	75,64	2,778	210	35,41	2,068
10	74,23	2,752	220	33,10	2,017
20	72,75	2,726	230	30,77	1,963
30	71,20	2,700	240	28,42	1,906
40	69,60	2,675	250	26,06	1,846
50	67,94	2,648	260	23,67	1,782
60	66,24	2,621	270	21,30	1,715
70	64,47	2,594	280	18,94	1,643
80	62,67	2,565	290	16,61	1,566
90	60,82	2,535	300	14,30	1,482
100	58,91	2,505	310	12,04	1,392
110	56,96	2,473	320	9,81	1,292
120	54,96	2,439	330	7,66	1,181
130	52,93	2,405	340	5,59	1,053
140	50,85	2,369	350	3,65	0,900
150	48,74	2,332	360	1,90	0,705
160	46,58	2,292	370	0,45	0,399
170	44,40	2,252	374	0,00	0,000
180	42,19	2,209			
190	39,95	2,164			
200	37,69	2,118			

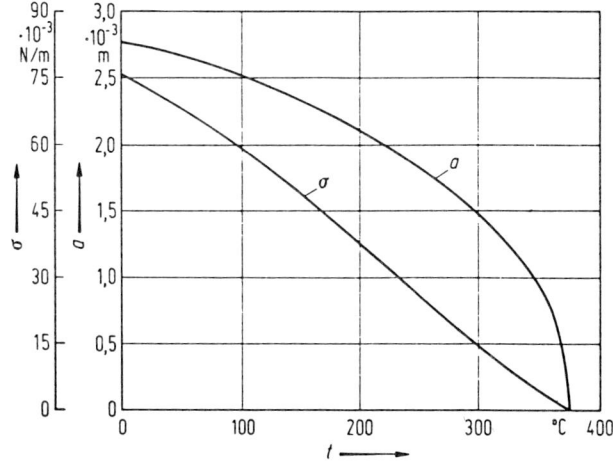

184

Table 8. Isentropic Exponent

Isentropenexponent

The isentropic exponent *k* is defined by the equation

Der Isentropenexponent *k* ist definiert durch die Gleichung

$$k = -\frac{v}{p}\left(\frac{\partial p}{\partial v}\right)_s$$

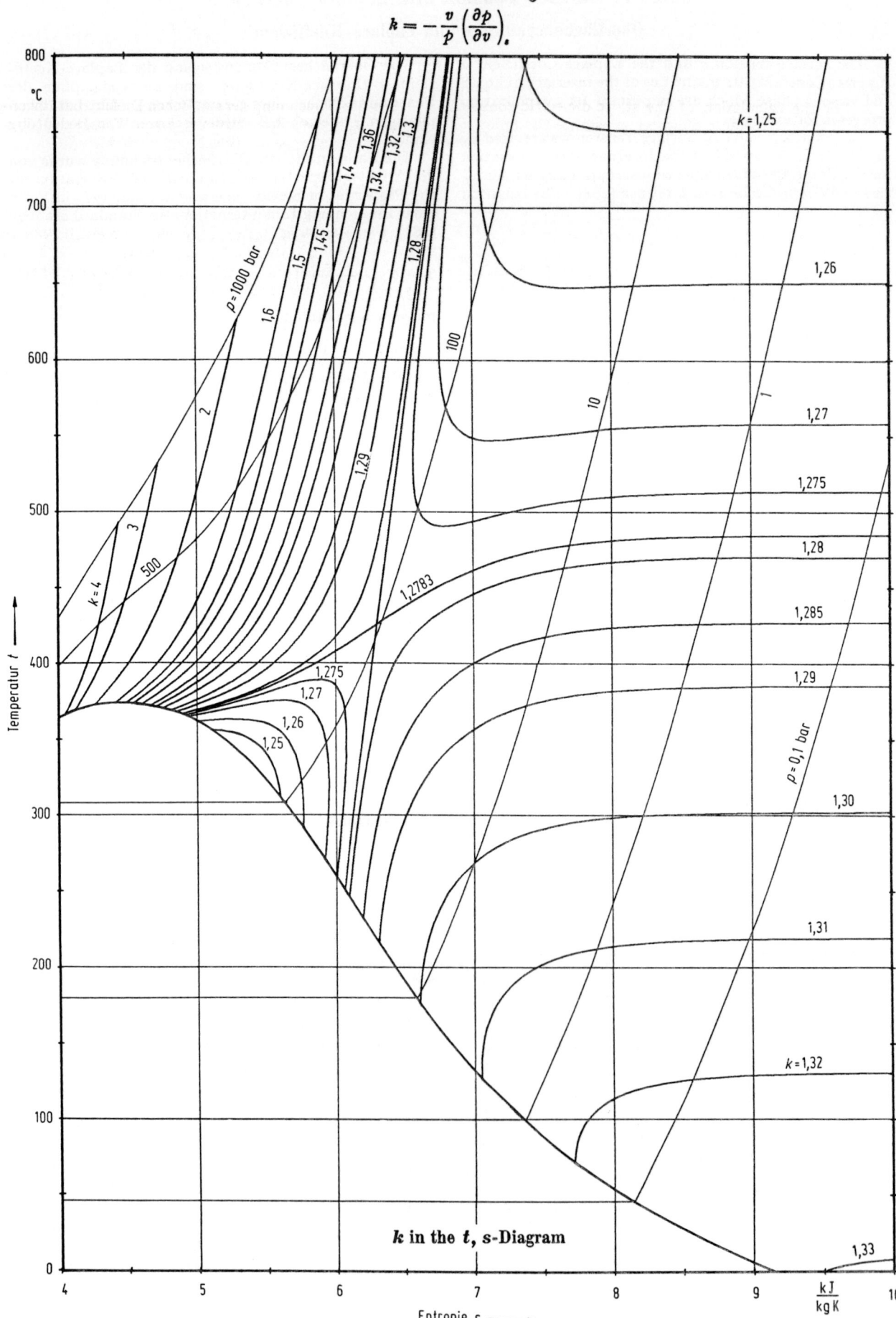

k in the t, s-Diagram

Table 9. Static Dielectric Constant (Static Permittivity)

Statische Dielektrizitätskonstante (statische Permittivität)

For the calculation of the static dielectric constant (static permittivity) ε of water substance the following equation is recommended [266]:

Für die Berechnung der statischen Dielektrizitätskonstante (statischen Permittivität) ε von Wasser und Wasserdampf wird folgende Gleichung empfohlen [266]:

$$\varepsilon_r = 1 + (A/T^*)\varrho^* + (B/T^* + C + DT^*)\varrho^{*2} + (E/T^* + FT^* + GT^{*2})\varrho^{*3} + (H/T^{*2} + I/T^* + K)\varrho^{*4}, \quad (1)$$

where
$\varepsilon_r = \varepsilon/\varepsilon_0$ relative static dielectric constant (relative static permittivity),
ε_0 electric constant (permittivity of vacuum) with the present best value [267]

wobei bedeuten
$\varepsilon_r = \varepsilon/\varepsilon_0$ die relative statische Dielektrizitätskonstante (relative statische Permittivität),
ε_0 die elektrische Feldkonstante (Permittivität des Vakuum) mit dem heutigen Bestwert [267]

$$\varepsilon_0 = (8{,}854\ 187\ 82 \pm 0{,}000\ 000\ 07) \cdot 10^{-12}\ \text{As/Vm},$$

T temperature,
$T^* = T/T_0$ with $T_0 = 298{,}15$ K,
ϱ density,
$\varrho^* = \varrho/\varrho_0$ with $\varrho_0 = 1000$ kg/m³.
The constants A to K in eq. (1) have the following values:

T Temperatur,
$T^* = T/T_0$ mit $T_0 = 298{,}15$ K,
ϱ Dichte,
$\varrho^* = \varrho/\varrho_0$ mit $\varrho_0 = 1000$ kg/m³.
Die Konstanten A bis K in Gl. (1) haben folgende Werte:

A =	7,625 71		F =	41,790 9
B =	244,003		G =	- 10,209 9
C =	- 140,569		H =	- 45,205 9
D =	27,784 1		I =	84,639 5
E =	- 96,280 5		K =	- 35,864 4

The density ϱ should be calculated for pressures up to 1000 bar with the "1968 IFC Formulation for Scientific and General Use" [262], for higher pressures with an equation of state proposed by JUZA [268].

Die Dichte ϱ soll für Drücke bis 1000 bar mit der „1968 IFC Formulation for Scientific and General Use" [262], für höhere Drücke mit einer von JUZA [268] angegebenen Zustandsgleichung berechnet werden.

The relative uncertainties of ε_r calculated by eq. (1) are shown in Fig. 1 [269].

Die relativen Unsicherheiten von ε_r nach Gl. (1) sind in Fig. 1 dargestellt [269].

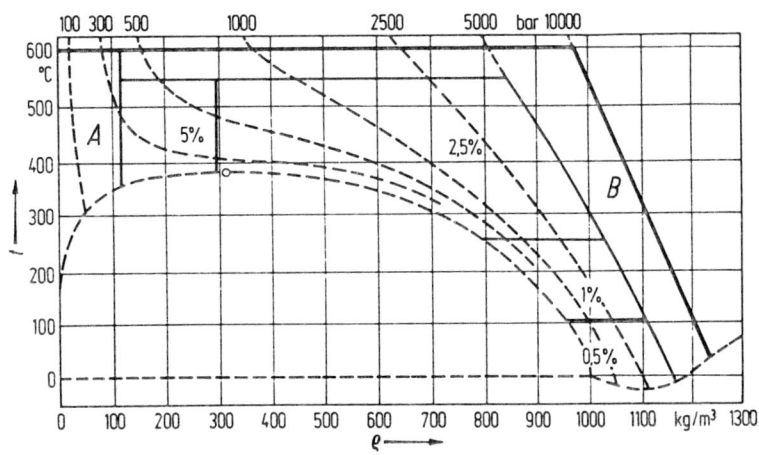

Fig. 1

Relative uncertainty of ε_r from eq. (1) in a temperature-density-plot.
A: more than 5%;
B: can be extrapolated with reduced reliability

Relative Unsicherheit von ε_r nach Gl. (1) in einem Temperatur-Dichte-Diagramm.
A: über 5%;
B: Extrapolation nur mit verringerter Zuverlässigkeit möglich.

Relative static dielectric constant (relative static permittivity) from eq. (1)
Relative statische Dielektrizitätskonstante (relative statische Permittivität) nach Gl. (1)

p bar	t in °C ⟶								
	0	25	50	75	100	125	150	175	200
100	88,28	78,85	70,27	62,59	55,76	49,70	44,30	39,47	35,11
200	88,75	79,24	70,63	62,94	56,11	50,05	44,66	39,85	35,52
300	89,20	79,63	70,98	63,28	56,44	50,39	45,01	40,22	35,91
400	89,64	80,00	71,32	63,61	56,77	50,72	45,34	40,56	36,28
500	90,07	80,36	71,66	63,93	57,08	51,03	45,67	40,89	36,63
600	90,49	80,72	71,98	64,24	57,39	51,34	45,98	41,21	36,96
700	90,90	81,07	72,30	64,54	57,69	51,64	46,28	41,52	37,28
800	91,29	81,42	72,62	64,84	57,98	51,93	46,57	41,82	37,59
900	91,67	81,75	72,92	65,13	58,27	52,21	46,86	42,11	37,89
1000	92,04	82,08	73,22	65,42	58,55	52,49	47,14	42,39	38,17
1250	92,89	82,84	73,93	66,09	59,19	53,12	47,78	43,05	38,86
1500	93,71	83,57	74,62	66,74	59,82	53,75	48,40	43,68	39,50
1750	94,48	84,28	75,27	67,36	60,42	54,34	48,98	44,27	40,10
2000	95,20	84,94	75,89	67,95	61,00	54,90	49,54	44,83	40,66
2250	95,87	85,58	76,50	68,53	61,55	55,44	50,08	45,36	41,20
2500	96,51	86,20	77,08	69,08	62,08	55,96	50,59	45,87	41,70
3000	97,69	87,34	78,17	70,14	63,10	56,94	51,55	46,82	42,65
3500	98,75	88,40	79,19	71,12	64,05	57,86	52,45	47,70	43,52
4000	99,72	89,39	80,13	72,03	64,94	58,74	53,30	48,53	44,33
4500	100,60	90,30	81,02	72,89	65,78	59,56	54,10	49,31	45,10
5000	101,42	91,16	81,84	73,69	66,57	60,33	54,85	50,05	45,82

p bar	t in °C ⟶								
	225	250	275	300	350	400	450	500	550
100	31,13	27,43	23,90	20,39	1,23	1,17	1,14	1,11	1,10
200	31,58	27,95	24,54	21,24	14,07	1,64	1,42	1,32	1,26
300	32,01	28,43	25,11	21,95	15,66	5,91	2,07	1,68	1,51
400	32,40	28,87	25,61	22,56	16,72	10,46	3,84	2,34	1,90
500	32,78	29,28	26,08	23,10	17,55	12,16	6,57	3,45	2,48
600	33,13	29,67	26,50	23,58	18,24	13,28	8,53	4,90	3,26
700	33,47	30,03	26,90	24,02	18,84	14,16	9,87	6,31	4,20
800	33,79	30,37	27,27	24,43	19,37	14,88	10,88	7,50	5,16
900	34,10	30,70	27,62	24,81	19,85	15,50	11,70	8,47	6,06
1000	34,40	31,01	27,95	25,17	20,29	16,05	12,39	9,29	6,88
1250	35,13	31,78	28,76	26,03	21,26	17,21	13,77	10,88	8,53
1500	35,78	32,46	29,47	26,77	22,09	18,16	14,85	12,07	9,80
1750	36,39	33,09	30,12	27,45	22,83	18,98	15,74	13,04	10,81
2000	36,97	33,67	30,72	28,07	23,49	19,69	16,51	13,86	11,65
2250	37,51	34,22	31,28	28,64	24,09	20,33	17,19	14,56	12,38
2500	38,02	34,74	31,81	29,17	24,65	20,91	17,80	15,19	13,01
3000	38,97	35,69	32,77	30,15	25,65	21,94	18,85	16,25	14,07
3500	39,83	36,56	33,64	31,02	26,53	22,83	19,74	17,14	14,93
4000	40,64	37,36	34,43	31,81	27,32	23,62	20,52	17,89	15,66
4500	41,38	38,09	35,16	32,54	28,04	24,32	21,20	18,55	16,28
5000	42,09	38,78	35,84	33,21	28,70	24,96	21,82	19,14	16,83

Table 10. Ion Product

Ionenprodukt

The Ion Product of water substance K_w is usually defined as the product of the activity of hydrogen ions $a(H^+)$ and the activity of hydroxide ions $a(OH^-)$. Since in pure water the concentrations of the ions are small, the activities may be replaced by the molalities:

Das Ionenprodukt von Wasser und Wasserdampf K_w ist üblicherweise als Produkt der Aktivität der Wasserstoff-Ionen $a(H^+)$ und der Aktivität der Hydroxid-Ionen $a(OH^-)$ definiert. Da in reinem Wasser die Ionenkonzentrationen klein sind, kann man die Aktivitäten durch die Molalitäten ersetzen:

$$K_w = m(H^+) \cdot m(OH^-),$$

where
$m(H^+)$ molality of hydrogen ions,
$m(OH^-)$ molality of hydroxide ions.

For the calculation of the ion product K_w the following equation is recommended [270]:

worin bedeuten
$m(H^+)$ die Molalität der Wasserstoff-Ionen,
$m(OH^-)$ die Molalität der Hydroxid-Ionen.

Zur Berechnung des Ionenprodukts K_w wird folgende Gleichung empfohlen [270]:

$$\log_{10}K_w^* = A + B/T^* + C/T^{*2} + D/T^{*3} + (E + F/T^* + G/T^{*2})\log_{10}\varrho^*, \tag{1}$$

where
$K_w^* = K_w/K_{w_0}$ with $K_{w_0} = 1$ $(mol/kg)^2$,
T Temperature in the 1968 International Practical Temperature Scale,
$T^* = T/T_0$ with $T_0 = 298{,}15$ K,
ϱ density,
$\varrho^* = \varrho/\varrho_0$ with $\varrho_0 = 1000$ kg/m³.
The constants A to G in eq. (1) have the following values:

wobei
$K_w^* = K_w/K_{w_0}$ mit $K_{w_0} = 1$ $(mol/kg)^2$,
T Temperatur in der Internationalen Praktischen Temperaturskala von 1968,
$T^* = T/T_0$ mit $T_0 = 298{,}15$ K,
ϱ Dichte,
$\varrho^* = \varrho/\varrho_0$ mit $\varrho_0 = 1000$ kg/m³.
Die Konstanten A bis G in Gl. (1) haben folgende Werte:

$$A = -4{,}098$$
$$B = -10{,}884$$
$$C = 2{,}5156$$
$$D = -1{,}5032$$
$$E = 13{,}957$$
$$F = -4{,}2338$$
$$G = 9{,}6341$$

The density ϱ should be calculated for pressures up to 1000 bar with the "1967 IFC Formulation for Industrial Use" [261], for higher pressures see [270].

The uncertainties Δ of $\log_{10}K_w^*$ are [271]:

Die Dichte ϱ soll für Drücke bis 1000 bar mit der „1967 IFC Formulation for Industrial Use" [261] berechnet werden, für höhere Drücke siehe [270].

Die Unsicherheiten Δ von $\log_{10}K_w^*$ betragen [271]:

Δ	p in bar	t in °C
± 0,01	for saturation vapor pressures bei Sättigungsdrücken	
± 0,03	< 10000	< 250
± 0,03 to ± 0,05	< 10000	$250 \leqq (t/°C) < 1000$
± 0,05 to ± 0,3	≈ 10000	≈ 1000

Values of − $\log_{10}K_w{}^*$ calculated by eq. (1) as a function of pressure p and temperature t

Werte von − $\log_{10}K_w{}^*$ nach Gl. (1) als Funktion des Druckes p und der Temperatur t

p bar ↓	t in °C →								
	0	25	50	75	100	150	200	250	300
Saturated vapor	14,938	13,995	13,275	12,712	12,265	11,638	11,289	11,191	11,406
250	14,83	13,90	13,19	12,63	12,18	11,54	11,16	11,01	11,14
500	14,72	13,82	13,11	12,55	12,10	11,45	11,05	10,85	10,86
750	14,62	13,73	13,04	12,48	12,03	11,36	10,95	10,72	10,66
1000	14,53	13,66	12,96	12,41	11,96	11,29	10,86	10,60	10,50
1500	14,34	13,53	12,85	12,29	11,84	11,16	10,71	10,43	10,26
2000	14,21	13,40	12,73	12,18	11,72	11,04	10,57	10,27	10,08
2500	14,08	13,28	12,62	12,07	11,61	10,92	10,45	10,12	9,91
3000	13,97	13,18	12,53	11,98	11,53	10,83	10,34	9,99	9,76
3500	13,87	13,09	12,44	11,90	11,44	10,74	10,24	9,88	9,63
4000	13,77	13,00	12,35	11,82	11,37	10,66	10,16	9,79	9,52
5000	13,60	12,83	12,19	11,66	11,22	10,52	10,00	9,62	9,34
6000	13,44	12,68	12,05	11,53	11,09	10,39	9,87	9,48	9,18
7000	13,31	12,55	11,93	11,41	10,97	10,27	9,75	9,35	9,04
8000	13,18	12,43	11,82	11,30	10,86	10,17	9,64	9,24	8,93
9000	13,04	12,31	11,71	11,20	10,77	10,07	9,54	9,13	8,82
10000	12,91	12,21	11,62	11,11	10,68	9,98	9,45	9,04	8,71

p bar ↓	t in °C →								
	350	400	450	500	600	700	800	900	1000
Saturated vapor	12,30	—	—	—	—	—	—	—	—
250	11,77	19,43	21,59	22,40	23,27	23,81	24,23	24,59	24,93
500	11,14	11,88	13,74	16,13	18,30	19,29	19,92	20,39	20,80
750	10,79	11,17	11,89	13,01	15,25	16,55	17,35	17,93	18,39
1000	10,54	10,77	11,19	11,81	13,40	14,70	15,58	16,22	16,72
1500	10,22	10,29	10,48	10,77	11,59	12,50	13,30	13,97	14,50
2000	9,98	9,98	10,07	10,23	10,73	11,36	11,98	12,54	12,97
2500	9,79	9,74	9,77	9,86	10,18	10,63	11,11	11,59	12,02
3000	9,61	9,54	9,53	9,57	9,78	10,11	10,49	10,89	11,24
3500	9,47	9,37	9,33	9,34	9,48	9,71	10,02	10,35	10,62
4000	9,34	9,22	9,16	9,15	9,23	9,41	9,65	9,93	10,13
5000	9,13	8,99	8,90	8,85	8,85	8,95	9,11	9,30	9,42
6000	8,96	8,80	8,69	8,62	8,57	8,61	8,72	8,86	8,97
7000	8,81	8,64	8,51	8,42	8,34	8,34	8,40	8,51	8,64
8000	8,68	8,50	8,36	8,25	8,13	8,10	8,13	8,21	8,38
9000	8,57	8,37	8,22	8,10	7,95	7,89	7,89	7,95	8,12
10000	8,46	8,25	8,09	7,96	7,78	7,70	7,68	7,70	7,85

C. Formulations and Equations

I. The 1967 IFC Formulation for Industrial Use

A Formulation of the thermodynamic properties of ordinary water substance prepared by the International Formulation Committee (IFC) of the Sixth International Conference on the Properties of Steam.

This formulation allowed for unrestricted publication is given here almost verbally with only some slight reductions not of interest for the user.

C. Formulationen und Gleichungen

I. Die 1967 IFC Formulation für industriellen Gebrauch

Eine Formulation der thermodynamischen Eigenschaften von Wasser und Wasserdampf, aufgestellt vom Internationalen Formulation Committee (IFC) der Sechsten Internationalen Dampftafel-Konferenz.

Diese zur unbeschränkten Veröffentlichung freigegebene Formulation ist hier fast wörtlich abgedruckt mit nur geringen, den Benutzer nicht interessierenden Kürzungen.

Contents

Inhalt

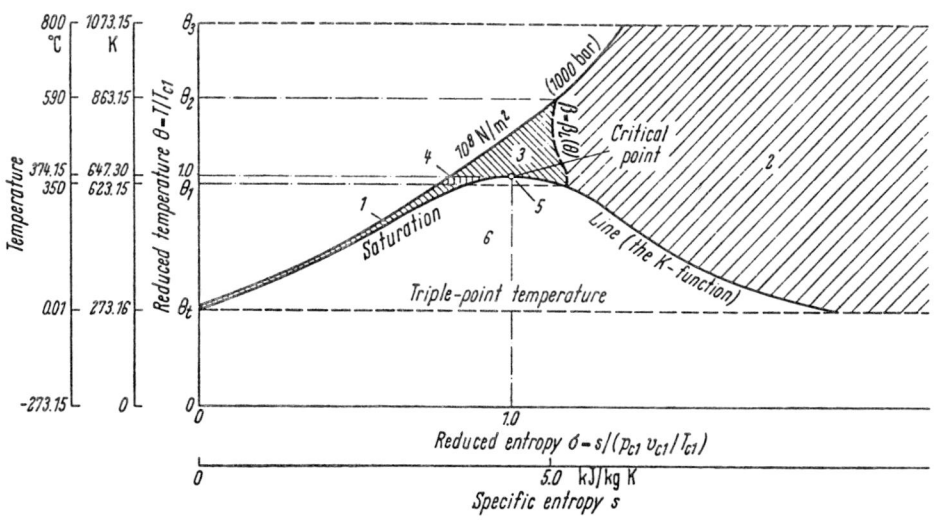

Fig. 1

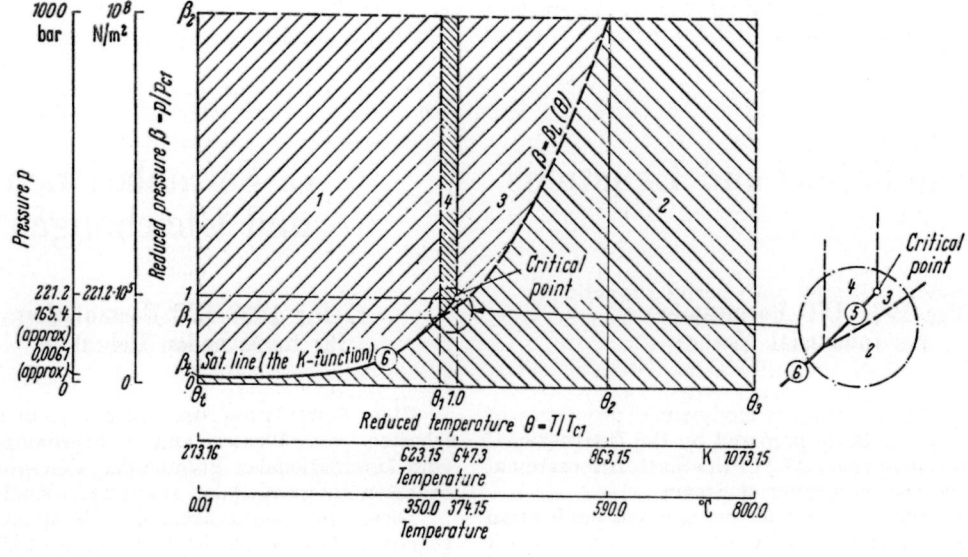

Fig. 2

Introduction

With the increasing use of digital computers, particularly in complicated calculations relating to plant design and cycle optimisation, it has become necessary to have a *formulation* of the thermodynamic properties of water substance convenient for industrial use.

The various thermodynamic properties are not independent of each other. For example, when the pressure p and temperature T are chosen as the independent variables of the formulation, then expressions (here called *derived functions*) for the specific volume, entropy, enthalpy and all other thermodynamic properties may be derived directly by partial differentiation of the so-called *canonical (or characteristic) function* $g = g(p, T)$, where g is the specific free enthalpy (Gibbs function). Similarly, when the specific volume v and temperature T are chosen as the independent variables, then expressions for the pressure, specific entropy, enthalpy and all other thermodynamic properties may be derived directly by partial differentiation of the *canonical function* $f = f(v, T)$, where f is the specific free energy (Helmholtz function). The formulation is presented in terms of these canonical functions, thereby maintaining thermodynamic consistency.

The canonical functions provide the definitive expression of the formulation. The *derived* functions are for practical use and are secondary to the canonical functions.

The formulation presented herein describes the thermodynamic properties of ordinary water substance throughout the whole of the region that extends in pressure from the ideal gas limit (at zero pressure) to a pressure of 10^8 N/m² (1000 bar), and that extends in temperature from 273,16 K (0,01 °C) to 1073,15 K (800 °C).

This whole region is divided into six *sub-regions*, numbered 1 to 6 and shown on the temperature-entropy plane in Fig. 1 (p. 175) and on the pressure-temperature plane in Fig. 2 (p. 176).

Section 1 of this Statement lists the *physical quantities*, defines the *quantity symbols* and *units* used in the formulation, and defines certain *constant quantities* with the aid of which the expressions are presented in terms of reduced dimensionless variables.

Section 2 presents the *reduced dimensionless quantities* and also the required *thermodynamic relations* by means of which expressions for the derived functions can be obtained from the given canonical functions.

Einführung

Mit dem zunehmenden Gebrauch von Rechenanlagen für komplizierte Rechnungen, besonders beim Entwurf von Kraftanlagen und der Optimierung ihrer Kreisprozesse, wurde es nötig, eine für den Gebrauch der Industrie geeignete *Formulation* der thermodynamischen Eigenschaften des Wassers und Wasserdampfes zu haben.

Wenn im folgenden der Einfachheit halber von Wasserdampf gesprochen wird, so ist darin der flüssige Zustand einbezogen.

Die verschiedenen thermodynamischen Eigenschaften sind nicht unabhängig voneinander. Wenn Druck p und Temperatur T als unabhängige Veränderliche einer Formulation gewählt werden, können Ausdrücke (hier *abgeleitete* Funktionen genannt) für spezifisches Volumen v, spezifische Entropie s, spezifische Enthalpie h und damit alle anderen thermodynamischen Eigenschaften durch partielle Differentiation der sogenannten *kanonischen (oder charakteristischen) Funktion* $g = g(p, T)$ abgeleitet werden, die man spezifische freie Enthalpie (Gibbs-Funktion) nennt. In ähnlicher Weise erhält man für v und T als unabhängige Veränderliche Ausdrücke für p, s, h usw. durch partielle Differentiation der *kanonischen Funktion* $f = f(v, T)$, die man spezifische freie Energie (Helmholtz-Funktion) nennt. Die folgende Formulation ist in kanonischen Funktionen angegeben und damit thermodynamisch konsistent.

Die kanonischen Funktionen sind die vollständige Formulation. Die daraus *abgeleiteten* Funktionen sind für den praktischen Gebrauch bestimmt und von sekundärer Art.

Die folgende Formulation beschreibt die thermodynamischen Eigenschaften des gewöhnlichen Wasserdampfes (d. h. natürlicher Isotopenzusammensetzung) vom Zustand des idealen Gases (bei $p = 0$) bis zum Drucke 10^8 N/m² (1000 bar) und für den Temperaturbereich von 273,16 K (0,01 °C) bis 1073,15 K (800 °C).

Das ganze Gebiet ist in sechs *Bereiche* 1 bis 6 aufgeteilt, wie das Fig. 1 (S. 175) im T, s- und Fig. 2 (S. 176) im p, t-Diagramm darstellt.

Abschnitt 1 dieser Darstellung enthält ein Verzeichnis der *physikalischen Größen*, ihrer *Bezeichnungen* und *Einheiten* und definiert gewisse *konstante Größen*, durch welche die Ausdrücke mit reduzierten dimensionslosen Veränderlichen dargestellt werden.

Abschnitt 2 bringt die *reduzierten dimensionslosen Größen* und die *thermodynamischen Beziehungen*, mit denen man die reduzierten Funktionen aus den angegebenen kanonischen Funktionen erhält.

Section 3 specifies the sub-regions, which are identified by numbers, and gives information relating to equations which define the boundaries between sub-regions. These equations are identified by the letters K and L, the K-function being the equation for the saturation line and the L-function being the equation for a boundary between two sub-regions in the single-phase region.

Section 4 gives the specification for the *sub-formulation* to be used in each sub-region. Each such sub-formulation comprises the canonical function relevant to the sub-region, together with derived functions.

Section 5 presents the function giving the saturation line, which also serves as a boundary between sub-regions. This function is identified by the letter K.

Section 6 presents the canonical functions, which are identified by the letters A, B, C and D. The canonical parts of the sub-formulations set out in Section 4 each comprise one or more of these principal canonical functions.

Section 7 gives the values of the constants of the formulation. Most of these values are given numerically; a few, which are derived from other constants, are given symbolically.

The material in Sections 1 to 7 is sufficient and necessary to specify the formulation.

Section 8 gives the numerical values of the derived constants and a derived form of the L-function convenient for computer use.

Section 9 presents those derived functions which are of practical importance.

Section 10 gives information on the magnitudes of small discontinuities in property values which occur at some of the boundaries between sub-regions and draws attention to the need for caution when making certain calculations.

Abschnitt 3 beschreibt die durch Zahlen gekennzeichneten Bereiche und gibt Gleichungen für Grenzen zwischen ihnen. Diese Gleichungen sind durch die Buchstaben K und L bezeichnet, wobei die K-Funktion die Gleichung der Sättigungslinie, die L-Funktion die Gleichung der Grenze zweier Bereiche im Einphasengebiet ist.

Abschnitt 4 gibt die Unterformulationen für jeden Bereich. Jede Unterformulation enthält die kanonische Funktion des Bereiches und die abgeleiteten Funktionen.

Abschnitt 5 zeigt die Funktion der Sättigungslinie, die zugleich als Bereichsgrenze dient, gekennzeichnet durch den Buchstaben K.

Abschnitt 6 enthält die kanonischen Funktionen, gekennzeichnet durch die Buchstaben A, B, C und D. Die kanonischen Teile der Unterformulationen von Abschnitt 4 umfassen eine oder mehrere der grundlegenden kanonischen Funktionen.

Abschnitt 7 gibt die Werte der Konstanten der Formulation meist als Zahlen; einige wenige, von anderen Konstanten abgeleitete auch als Gleichungen.

Das Material der Abschnitte 1 bis 7 ist hinreichend und notwendig für die Formulation.

Abschnitt 8 gibt die Zahlenwerte der abgeleiteten Konstanten und eine abgeleitete L-Funktion für die Rechenmaschine.

Abschnitt 9 enthält die praktisch wichtigen abgeleiteten Funktionen.

Abschnitt 10 zeigt die Größe von kleinen Unstetigkeiten der Zustandsgrößen an einigen der Bereichsgrenzen und weist auf die bei manchen Berechnungen nötige Vorsicht hin.

1. Physical quantities, quantity symbols, units and defined constant quantities

1. Physikalische Größen, ihre Bezeichnungen und ihre Einheiten, definierte konstante Größen

1.1 Physical quantities (properties) — Physikalische Größen (Eigenschaften)

The following physical quantities are given the symbols listed:
specific free energy (Helmholtz function)

specific free enthalpy (Gibbs function)
specific enthalpy
specific entropy
specific volume
pressure
temperature (thermodynamic temperature)
specific isochoric heat capacity
specific isobaric heat capacity
quantities at the critical point
quantities for the saturated liquid
quantities for the saturated vapour
increments in quantities for evaporation from liquid to vapour

quantities at the triple point
specific idealgas constant
saturation pressure
saturation temperature

Die folgenden physikalischen Größen erhalten die Bezeichnungen:

spezifische freie Energie (Helmholtz-Funktion)	f
spezifische freie Enthalpie (Gibbs-Funktion)	g
spezifische Enthalpie	h
spezifische Entropie	s
spezifisches Volumen	v
Druck	p
Temperatur (thermodynamische)	T
spezifische isochore Wärmekapazität	c_v
spezifische isobare Wärmekapazität	c_p
Zustandsgrößen am kritischen Punkt	v_c, p_c, T_c
Zustandsgrößen für gesättigtes Wasser	f_f, h_f, s_f, v_f
Zustandsgrößen für gesättigten Dampf	h_g, s_g, v_g
Indices für die Verdampfung	h_{fg}, s_{fg}, v_{fg}
Zustandsgrößen am Tripelpunkt	f_{tt}, s_{tt}, p_t, T_t
Gaskonstante des Wasserdampfes	R
Sättigungsdruck	p_s
Sättigungstemperatur	T_s

1.2 Units

The units of the Système International d'Unités (SI units) are used and have the definitions assigned to them by the Conférence Générale des Poids et Mesures (CGMP).

These SI units are

1.2 Einheiten

Es werden die Einheiten des „Système International d'Unités" (SI-Einheiten) benutzt, wie sie die Conférence Générale des Poids et Mesures" (CGPM) definiert hat.

Diese SI-Einheiten sind:

Quantities	Units	Unit Symbol	Größen	Einheiten	Einheitszeichen
f, g, h	joule per kilogramme	J/kg	f, g, h	Joule per Kilogramm	J/kg
v	metre cubed per kilogramme	m³/kg	v	Kubikmeter per Kilogramm	m³/kg
$p^1)$	newton per metre squared	N/m²	$p^1)$	Newton per Quadratmeter	N/m²
	joule per metre cubed	J/m³		Joule per Kubikmeter	J/m³
	pascal	Pa		Pascal	Pa
T	kelvin	K	T	Kelvin	K
s, R, c_v, c_p	joule per kilogramme kelvin	J/kg K	s, R, c_v, c_p	Joule per Kilogramm Kelvin	J/kg K

The International Organisation for Standardisation (ISO/R 31) has provided equations for other units in terms of SI units.

The definitions given by the CGPM and the ISO imply that

(exactly)
$$T_t = 273{,}16 \text{ K} \qquad \text{(genau)}$$

and that the (thermodynamic) Celsius temperature is exactly $T - T_0$, where

(exactly)
$$T_0 = 273{,}15 \text{ K.} \qquad \text{(genau)}$$

The symbol T in this Statement refers throughout to thermodynamic (absolute) temperature. Temperatures on the International Practical Scale of Temperature (1948) provide a closely approximate realisation of the numerical values on the thermodynamic Celsius scale. The constants listed in Section 7 are appropriate for use when the International Practical Scale and the thermodynamic Celsius scale are treated as being identical.

¹) The names and unit symbols given here are synonyms for the same unit of pressure.

Die Internationale Standard Organisation (ISO/R 31) hat Gleichungen für andere Einheiten in SI-Einheiten angegeben.

Die Definitionen der CGPM enthalten auch

und daß die (thermodynamische) Celsiustemperatur genau $T - T_0$ ist, mit

Das Symbol T ist immer die thermodynamische (absolute) Temperatur. Temperaturen der Internationalen Praktischen Temperaturskala (1948) sind eine recht genaue Näherung der thermodynamischen Celsiusskala. Die Konstanten in Abschnitt 7 sind brauchbar, wenn die Internationale Praktische Skala und die thermodynamische Celsiusskala als identisch angesehen werden.

¹) Die hier aufgeführten Namen und Einheitensymbole sind Synonyme für dieselbe Druckeinheit.

1.3 Defined constant quantities

In accordance with the decisions of the ICPS (5th International Conference, London, 1956):

1.3 Definierte konstante Größen

Übereinstimmend mit den Entscheidungen der ICPS (5. Internationale Konferenz, London, 1956) ist am Tripelpunkt

$$s_{tt} = 0, \qquad f_{tt} = 0.$$

The IFC, at its First Meeting in Prague, 1965, defined certain symbols for certain constant quantities. Among these are:

Das IFC definierte auf seiner ersten Sitzung in Prag 1965 unter anderen folgende konstante Größen:

$$p_{t1} = 611{,}2 \text{ N/m}^2 = 611{,}2 \text{ J/m}^3$$

$$T_{c1} = 647{,}3 \text{ K}$$

$$p_{c1} = 22\,120\,000 \text{ N/m}^2 = 22\,120\,000 \text{ J/m}^3$$

$$v_{c1} = 0{,}003\,17 \text{ m}^3/\text{kg}$$

$$R_1 = 461{,}51 \text{ J/kg K.}$$

Note: Subscripts t and c, appearing alone, would refer to the actual values at the actual triple and critical points respectivly; these values are not known exactly. The further subscript 1 refers to the above constant quantities, which coincide with the values adopted by the 6th International Conference, New York, 1963, as the nearest estimates, at that time, of the true values. It is stressed that the constant listed in Section 7 are those appropriate for use when the defined constant quantities are as given above, and that no alterations to these defined constant quantities can be made without reviewing the values of the constants listed.

Bemerkung: Allein auftretende Indizes t und c beziehen sich auf die wirklichen nicht genau bekannten Werte vom Tripelpunkt und kritischem Punkt. Der weitere Index 1 bezieht sich auf die obigen konstanten Größen, die von der 6. Internationalen Konferenz New York 1963 angenommen wurden als die zu dieser Zeit besten Schätzungen der wahren Werte. Es sei betont, daß die Konstanten des Abschnittes 7 gültig sind für die oben definierten Konstanten und daß deren Änderung auch eine Nachprüfung der ersteren nötig macht.

2. Reduced dimensionless quantities, and thermodynamic relations

2. Reduzierte dimensionslose Größen, thermodynamische Beziehnungen

2.1 Reduced dimensionless quantities

2.1 Reduzierte dimensionslose Größen

a) In accord with IFC:

a) Nach IFC ist:

the reduced pressure	$p/p_{o1} = \beta$	der reduzierte Druck
the reduced temperature	$T/T_{o1} = \Theta$	die reduzierte Temperatur
the reduced volume	$v/v_{o1} = \chi$	das reduzierte Volumen
the reduced enthalpy	$h/(p_{o1}v_{o1}) = \varepsilon$	die reduzierte Enthalpie
the reduced entropy	$s/(p_{o1}v_{o1}/T_{o1}) = \sigma$	die reduzierte Entropie

b) It has been found expedient to add

b) Es erwies sich als zweckmäßig hinzuzufügen:

the reduced free enthalpy (Gibbs function) $g/(p_{o1}v_{o1}) = \varepsilon - \Theta\,\sigma = \zeta$ die reduzierte freie Enthalpie (Gibbs-Funktion)

the reduced free energy (Helmholtz function) $f/(p_{o1}v_{o1}) = \zeta - \beta\,\chi = \psi$ die reduzierte freie Energie (Helmholtz-Funktion)

the reduced ideal-gas constant $R_1 T_{o1}/(p_{1o}v_{o1}) = I_1$ die reduzierte Gaskonstante

Use is also made of

Gebraucht werden weiter:

the reduced saturation pressure, where $p_{\bullet} = p_{\bullet}(T)$ $p_{\bullet}/p_{o1} = \beta_k(\Theta)$ der reduzierte Sättigungsdruck, wobei $p_{\bullet} = p_{\bullet}(T)$

the reduced saturation temperature, where $T_{\bullet} = T_{\bullet}(p)$ $T_{\bullet}/T_{o1} = \Theta_K(\beta)$ die reduzierte Sättigungstemperatur, wobei $T_{\bullet} = T_{\bullet}(p)$

the reduced triple-point temperature $T_t/T_{o1} = \Theta_t$ die reduzierte Tripelpunkts-Temperatur

the reduced triple-point pressure $p_t/p_{o1} = \beta_t = \beta_K(\Theta_t)$ der reduzierte Tripelpunkts-Druck

Numerical values for Θ_t, β_t, $p_{o1}v_{o1}$, $p_{o1}v_{o1}/T_{o1}$ and I_1 are given in Section 8.

Zahlenwerte für Θ_t, β_t, $p_{o1}v_{o1}$, $p_{o1}v_{o1}/T_{o1}$ und I_1 enthält Abschnitt 8

2.2 Thermodynamic relations

2.2 Thermodynamische Beziehungen

The known thermodynamic relations,

Die bekannten thermodynamischen Beziehungen

$$s = -(\partial g/\partial T)_p = -(\partial f/\partial T)_v, \qquad p = -(\partial f/\partial v)_T,$$
$$v = +(\partial g/\partial p)_T, \qquad h = g + Ts = f + pv + Ts$$

when written in terms of the reduced dimensionless quantities become:

als reduzierte dimensionslose Größen geschrieben lauten:

$$\sigma = -(\partial\zeta/\partial\Theta)_\beta = -(\partial\psi/\partial\Theta)_\chi, \qquad \beta = -(\partial\psi/\partial\chi)_\Theta,$$
$$\chi = +(\partial\zeta/\partial\beta)_\Theta, \qquad \varepsilon = \zeta + \Theta\,\sigma = \psi + \beta\,\chi + \Theta\,\sigma.$$

The reduced specific heat-capacities are given by:

Die reduzierten spezifischen Wärmekapazitäten sind:

$$\frac{c_p T_{c1}}{p_{c1}v_{c1}} = -\Theta\left(\frac{\partial^2\zeta}{\partial\Theta^2}\right)_\beta = -\Theta\left(\frac{\partial^2\psi}{\partial\Theta^2}\right)_\chi + \Theta\left(\frac{\partial^2\psi}{\partial\chi\,\partial\Theta}\right)^2\Big/\left(\frac{\partial^2\psi}{\partial\chi^2}\right)_\Theta,$$

$$\frac{c_v T_{c1}}{p_{c1}v_{c1}} = -\Theta\left(\frac{\partial^2\psi}{\partial\Theta^2}\right)_\chi = -\Theta\left(\frac{\partial^2\zeta}{\partial\Theta^2}\right)_\beta + \Theta\left(\frac{\partial^2\zeta}{\partial\Theta\,\partial\beta}\right)^2\Big/\left(\frac{\partial^2\zeta}{\partial\beta^2}\right)_\Theta.$$

3. Specification of the sub-regions

8.1 Sub-regions. The sub-regions are specified in the following table and illustrated in Figs. 1 and 2

3. Spezifizierung der Unterbereiche

8.1 Bereiche. Die Bereiche sind in der folgenden Tabelle und in Fig. 1 und 2 angegeben

Temperature range Temperaturbereich	Pressure range Druckbereich	Sub-region Unterbereich
	$0 \leqq \beta < \beta_K(\Theta)$	2
$\Theta_t \leqq \Theta \leqq \Theta_1$	$\beta = \beta_K(\Theta)$	6
	$\beta_K(\Theta) < \beta \leqq \beta_2$	1
	$0 \leqq \beta \leqq \beta_L(\Theta)$	2
$\Theta_1 < \Theta < 1$	$\beta_L(\Theta) < \beta < \beta_K(\Theta)$	3
	$\beta = \beta_K(\Theta)$	5
	$\beta_K(\Theta) < \beta \leqq \beta_2$	4
$1 \leqq \Theta < \Theta_2$	$0 \leqq \beta \leqq \beta_L(\Theta)$	2
	$\beta_L(\Theta) < \beta \leqq \beta_2$	3
$\Theta_2 \leqq \Theta \leqq \Theta_3$	$0 \leqq \beta \leqq \beta_2$	2

The functions $\beta_K(\Theta)$ and $\beta_L(\Theta)$ are equations for boundaries between sub-regions, the K-function being the equation for the saturation line and the L-function the equation for the boundary between sub-region 2 and 3. These functions, and the constants required to complete the specification of the sub-regions in the table, are specified in Sections 3.2 and 3.3 respectively.

Die Funktionen $\beta_K(\Theta)$ und $\beta_L(\Theta)$ sind Gleichungen der Grenzen zwischen Bereichen, dabei ist die K-Funktion die Sättigungslinie und die L-Funktion die Grenze zwischen den Unterbereichen 2 und 3. Diese Funktionen und die Konstanten zur Festlegung der Unterbereiche der vorstehenden Tabelle sind in Abschnitt 3.2 und 3.3 enthalten.

8.2 Equations for boundaries between sub-regions

3.2.1 The K-function
Reduced saturation pressure
This function is given in Section 5.
3.2.2 The L-function
Reduced pressure along the boundary between sub-regions 2 and 3.

8.2 Gleichungen der Grenzen zwischen Unterbereichen

3.2.1 Die K-Funktion
Der reduzierte Sättigungsdruck
Diese Funktion enthält Abschnitt 5.
3.2.2 Die L-Funktion
Reduzierter Druck längs der Grenze zwischen Unterbereich 2 und 3.

$$\beta_L = \beta_L(\Theta) = \frac{(\Theta_2 - \Theta)\beta_1 + (\Theta - \Theta_1)\beta_2 - L(\Theta_2 - \Theta)(\Theta - \Theta_1)}{\Theta_2 - \Theta_1},$$

whence wobei
$$\frac{d\beta_L}{d\Theta} = \beta'_L = \beta'_L(\Theta) = \frac{\beta_2 - \beta_1 - L(\Theta_2 - 2\Theta + \Theta_1)}{\Theta_2 - \Theta_1}.$$

Derived forms for β_L and β'_L, convenient for computer use, are given in Section 8.2.

Abgeleitete Formen für β_L und β'_L für Rechenmaschinen enthält Abschnitt 8.2.

8.3 Constants relating to boundaries between sub-regions

3.3.1 Primary constants
The constant L and the constants relating to the K-function are given in Section 7.1.

3.3.2 Expressions for values of derived constants
Expressions for the values of derived constants are given in Section 7.2.

3.3.3 Numerical values of derived constants
The numerical values of derived constants are given in Section 8.1 and the numerical values of the constants Selating to the derived forms for β_L and β'_L are given in rection 8.2.

8.3 Konstanten der Grenzen zwischen den Unterbereichen

3.3.1 Primäre Konstanten
Die Konstante L und die Konstanten der K-Funktion enthält Abschnitt 7.1.

3.3.2 Ausdrücke für die Werte der abgeleiteten Konstanten
Ausdrücke für die Werte der abgeleiteten Konstanten enthält Abschnitt 7.2.

3.3.3 Numerische Werte der abgeleiteten Konstanten
Die numerischen Werte der abgeleiteten Konstanten enthält Abschnitt 8.1 und die numerischen Werte der abgeleiteten Formen für β_L und β'_L sind angegeben in Abschnitt 8.2.

4. Sub-formulations

For each sub-region there are set out below
(1) the canonical function, and
(2) the derived functions, and the relations between the canonical and derived functions.

The functions of each sub-formulation are identified by the same number as that identifying the sub-region. The functions $\zeta_A(\Theta, \beta)$, $\zeta_B(\Theta, \beta)$, $\psi_C(\Theta, \chi)$ and $\psi_D(\Theta, \chi)$ are given in Section 6.

The purpose of introducing the terms α_0 and $\alpha_1\Theta$ is explained in Section 7.2.

4. Unterformulationen

Für jeden Bereich sind im folgenden angegeben
(1) die kanonische Funktion und
(2) die abgeleiteten Funktionen und die Beziehungen zwischen kanonischen und abgeleiteten Funktionen.

Die Funktionen jeder Unterformulation sind gekennzeichnet durch die gleiche Zahl wie der Unterbereich. Die Funktionen $\zeta_A(\Theta, \beta)$, $\zeta_B(\Theta, \beta)$, $\psi_C(\Theta, \chi)$ und $\psi_D(\Theta, \chi)$ sind angegeben in Abschnitt 6.

Der Sinn der Einführung der Terme α_0 und $\alpha_1\Theta$ ist in Abschnitt 7.2 erläutert.

4.1 Sub-region 1 — Unterbereich 1

$$\zeta = \zeta_1(\Theta, \beta) = \zeta_A(\Theta, \beta) + \alpha_0 + \alpha_1\,\Theta,$$

$$\chi = \chi_1(\Theta, \beta) = (\partial\zeta_1/\partial\beta)_\Theta, \qquad \sigma = \sigma_1(\Theta, \beta) = -(\partial\zeta_1/\partial\Theta)_\beta, \qquad \varepsilon = \varepsilon_1(\Theta, \beta) = \zeta_1 + \sigma_1\,\Theta.$$

4.2 Sub-region 2 — Unterbereich 2

$$\zeta = \zeta_2(\Theta, \beta) = \zeta_B(\Theta, \beta) + \alpha_0 + \alpha_1\,\Theta,$$

$$\chi = \chi_2(\Theta, \beta) = (\partial\zeta_2/\partial\beta)_\Theta, \qquad \sigma = \sigma_2(\Theta, \beta) = -(\partial\zeta_2/\partial\Theta)_\beta, \qquad \varepsilon = \varepsilon_2(\Theta, \beta) = \zeta_2 + \sigma_2\,\Theta.$$

4.3 Sub-region 3 — Unterbereich 3

$$\psi = \psi_3(\Theta, \chi) = \psi_C(\Theta, \chi) + \alpha_0 + \alpha_1\,\Theta,$$

$$\beta = \beta_3(\Theta, \chi) = -(\partial\psi_3/\partial\chi)_\Theta, \qquad \varepsilon = \varepsilon_3(\Theta, \chi) = \psi_3 + \sigma_3\,\Theta + \beta_3\,\chi,$$

$$\sigma = \sigma_3(\Theta, \chi) = -(\partial\psi_3/\partial\Theta)_\chi, \qquad \zeta = \zeta_3(\Theta, \chi) = \psi_3 + \beta_3\,\chi.$$

Expressions having Θ and β as the independent variables are needed later.

The equation $\beta = \beta_3(\Theta, \chi)$, when solved for χ, gives $\chi = \chi_3(\Theta, \beta)$.

Then

Ausdrücke mit Θ und β als unabhängigen Variablen werden später gebraucht.

Die Gleichung $\beta = \beta_3(\Theta, \chi)$, aufgelöst nach χ, ergibt $\chi = \chi_3(\Theta, \beta)$.

Dann ist

$$\sigma = \sigma_3[\Theta, \chi_3(\Theta, \beta)] = \sigma_3(\Theta, \beta), \qquad \varepsilon = \varepsilon_3[\Theta, \chi_3(\Theta, \beta)] = \varepsilon_3(\Theta, \beta), \qquad \zeta = \zeta_3[\Theta, \chi_3(\Theta, \beta)] = \zeta_3(\Theta, \beta).$$

4.4 Sub-region 4 — Unterbereich 4

$$\psi = \psi_4(\Theta, \chi) = \psi_C(\Theta, \chi) + \alpha_0 + \alpha_1\,\Theta + \psi_D(\Theta, \chi),$$

$$\beta = \beta_4(\Theta, \chi) = -(\partial\psi_4/\partial\chi)_\Theta, \qquad \varepsilon = \varepsilon_4(\Theta, \chi) = \psi_4 + \sigma_4\,\Theta + \beta_4\,\chi,$$

$$\sigma = \sigma_4(\Theta, \chi) = -(\partial\psi_4/\partial\Theta)_\chi, \qquad \zeta = \zeta_4(\Theta, \chi) = \psi_4 + \beta_4\,\chi.$$

Expressions having Θ and β as the independent variables are needed later.

The equation $\beta = \beta_4(\Theta, \chi)$ when solved for χ, gives $\chi = \chi_4(\Theta, \beta)$.

Then

Ausdrücke mit Θ und β als unabhängigen Veränderlichen werden später gebraucht.

Die Gleichung $\beta = \beta_4(\Theta, \chi)$ nach χ aufgelöst, ergibt $\chi = \chi_4(\Theta, \beta)$.

Damit wird

$$\sigma = \sigma_4[\Theta, \chi_4(\Theta, \beta)] = \sigma_4(\Theta, \beta), \qquad \varepsilon = \varepsilon_4[\Theta, \chi_4(\Theta, \beta)] = \varepsilon_4(\Theta, \beta), \qquad \zeta = \zeta_4[\Theta, \chi_4(\Theta, \beta)] = \zeta_4(\Theta, \beta).$$

4.5, 6 Sub-region 5 and 6 — Unterbereich 5 und 6

$$\beta = \beta_K(\Theta)$$

Dryness fraction $x = \dfrac{\chi - \chi_f}{\chi_g - \chi_f} = \dfrac{\sigma - \sigma_f}{\sigma_g - \sigma_f} = \dfrac{\varepsilon - \varepsilon_f}{\varepsilon_g - \varepsilon_f}$, where the subscripts f and g refer respectively to the liquid and gaseous phases and the quantities bearing these subscripts are given below.

Feuchtegrad $x = \dfrac{\chi - \chi_f}{\chi_g - \chi_f} = \dfrac{\sigma - \sigma_f}{\sigma_g - \sigma_f} = \dfrac{\varepsilon - \varepsilon_f}{\varepsilon_g - \varepsilon_f}$, wobei die Indices f und g sich auf die flüssige bzw. gasförmige Phase beziehen, und die entsprechenden Größen sind die folgenden.

4.5 Sub-region 5 — Unterbereich 5

$$\chi_f = \chi_4[\Theta, \beta_K(\Theta)], \qquad \chi_g = \chi_3[\Theta, \beta_K(\Theta)],$$

$$\sigma_f = \sigma_4[\Theta, \beta_K(\Theta)], \qquad \sigma_g = \sigma_3[\Theta, \beta_K(\Theta)],$$

$$\varepsilon_f = \varepsilon_4[\Theta, \beta_K(\Theta)], \qquad \varepsilon_g = \varepsilon_3[\Theta, \beta_K(\Theta)].$$

4.6 Sub-region 6 — Unterbereich 6

$$\chi_t = \chi_1[\Theta, \beta_K(\Theta)], \qquad \chi_g = \chi_2[\Theta, \beta_K(\Theta)],$$
$$\sigma_t = \sigma_1[\Theta, \beta_K(\Theta)], \qquad \sigma_g = \sigma_2[\Theta, \beta_K(\Theta)],$$
$$\varepsilon_t = \varepsilon_1[\Theta, \beta_K(\Theta)], \qquad \varepsilon_g = \varepsilon_2[\Theta, \beta_K(\Theta)].$$

The definitions given in Sections 4.1 to 4.4 of the derived functions $\chi_i, \sigma_i, \varepsilon_i$ (i = 1 to 4) are hereby extended to include $\beta = \beta_K(\Theta)$.

The function $\beta_K(\Theta)$ is given in Section 5.

Die Definitionen in Abschnitt 4.1 bis 4.4 der ab geleiteten Funktionen $\chi_i, \sigma_i, \varepsilon_i$ (i = 1 bis 4) sind hier durch erweitert, so daß sie auch $\beta = \beta_K(\Theta)$ einschließen

Die Funktion $\beta_K(\Theta)$ ist in Abschnitt 5 angegeben

5. The K-function (Saturation line) Reduced saturation pressure

This function gives the saturation line, which is also a boundary between sub-regions.

The equation for the reduced saturation pressure, β_K, as a function of the reduced temperature, Θ, is

5. Die K-Funktion (Sättigungslinie) Reduzierter Sättigungsdruck

Diese Funktion gibt die Sättigungslinie, die zugleich eine Grenze zwischen Unterbereichen ist.

Die Gleichung für den reduzierten Sättigungsdruck β_K als Funktion der reduzierten Temperatur Θ ist

$$\beta_K(\Theta) = \exp\left[\frac{1}{\Theta} \frac{\sum\limits_{\nu=1}^{5} k_\nu (1-\Theta)^\nu}{1 + k_6(1-\Theta) + k_7(1-\Theta)^2} - \frac{(1-\Theta)}{k_8(1-\Theta)^2 + k_9}\right].$$

The constants of the K-function are given in Section 7.1.

Die Konstanten der K-Funktion sind angegeben in Abschnitt 7.1.

6. Canonical functions

6.1 The A-function Reduced free enthalpy (Gibbs function)

6. Kanonische Funktionen

6.1 Die A-Funktion Reduzierte freie Enthalpie (Gibbs-Funktion)

$$\zeta_A(\Theta, \beta) = A_0 \Theta(1 - \ln\Theta) + \sum_{\nu=1}^{10} A_\nu \Theta^{\nu-1} + A_{11}\left(\frac{17}{29}Z - \frac{17}{12}Y\right)Z^{12/17} + \{A_{12} + A_{13}\Theta + A_{14}\Theta^2 + A_{15}(a_6 - \Theta)^{10} +$$
$$+ A_{16}(a_7 + \Theta^{19})^{-1}\}\beta - (a_8 + \Theta^{11})^{-1}(A_{17}\beta + A_{18}\beta^2 + A_{19}\beta^3) - A_{20}\Theta^{18}(a_9 + \Theta^2)\{(a_{10} + \beta)^{-3} + a_{11}\beta\} +$$
$$+ A_{21}(a_{12} - \Theta)\beta^3 + A_{22}\Theta^{-20}\beta^4,$$

where

wobei

$$Z = Y + (a_3 Y^2 - 2a_4\Theta + 2a_5\beta)^{\frac{1}{2}} \quad \text{and und} \quad Y = 1 - a_1\Theta^2 - a_2\Theta^{-6}.$$

6.2 The B-function Reduced free enthalpy (Gibbs function)

6.2 Die B-Funktion Reduzierte freie Enthalpie (Gibbs-Funktion)

$$\zeta_B(\Theta, \beta) = I_1\Theta\ln\beta + B_0\Theta(1 - \ln\Theta) + \sum_{\nu=1}^{5} B_{0\nu}\Theta^{\nu-1} - (B_{11}X^{13} + B_{12}X^3)\beta - (B_{21}X^{18} + B_{22}X^2 + B_{23}X)\beta^2 -$$
$$- (B_{31}X^{18} + B_{32}X^{10})\beta^3 - (B_{41}X^{25} + B_{42}X^{14})\beta^4 - (B_{51}X^{32} + B_{52}X^{28} + B_{53}X^{24})\beta^5 -$$
$$- \frac{(B_{61}X^{12} + B_{62}X^{11})\beta^4}{1 + b_{61}X^{14}\beta^4} - \frac{(B_{71}X^{24} + B_{72}X^{18})\beta^5}{1 + b_{71}X^{19}\beta^5} - \frac{(B_{81}X^{24} + B_{82}X^{14})\beta^6}{1 + (b_{81}X^{54} + b_{82}X^{27})\beta^6} + \beta\left(\frac{\beta}{\beta_L}\right)^{10}\sum_{\nu=0}^{6} B_{9\nu}X^\nu,$$

where $X = \exp[b(1-\Theta)]$ and $\beta_L = \beta_L(\Theta)$, the expression for which is given in Section 3.2.2.

The B-function may also be expressed more compactly as follows:

wobei $X = \exp[b(1-\Theta)]$ und $\beta_L = \beta_L(\Theta)$, wofür der Ausdruck in Abschnitt 3.2.2 angegeben ist.

Die B-Funktion kann kürzer auch wie folgt geschrieben werden:

$$\zeta_B(\Theta, \beta) = I_1\Theta\ln\beta + B_0\Theta(1 - \ln\Theta) + \sum_{\nu=1}^{5} B_{0\nu}\Theta^{\nu-1} - \sum_{\mu=1}^{5}\left\{\beta^\mu \sum_{\nu=1}^{n(\mu)} B_{\mu\nu}X^{z(\mu,\nu)}\right\} -$$
$$- \sum_{\mu=6}^{8} \frac{\sum\limits_{\nu=1}^{n(\mu)} B_{\mu\nu}X^{z(\mu,\nu)}}{\beta^{2-\mu} + \sum\limits_{\lambda=1}^{l(\mu)} b_{\mu\lambda}X^{x(\mu,\lambda)}} + \beta\left(\frac{\beta}{\beta_L}\right)^{10}\sum_{\nu=0}^{6} B_{9\nu}X^\nu.$$

The numbers of terms $n(\mu)$ and $l(\mu)$, and the exponents $z(\mu, \nu)$ and $x(\mu, \lambda)$ are as follows:

Die Anzahl der Glieder $n(\mu)$ und $l(\mu)$ sowie der Exponenten $z(\mu, \nu)$ und $x(\mu, \lambda)$ zeigt die folgende Tabelle:

μ	$n(\mu)$	$z(\mu, \nu)$			$l(\mu)$	$x(\mu, \lambda)$		μ
		$\nu = 1$	$\nu = 2$	$\nu = 3$		$\lambda = 1$	$\lambda = 2$	
1	2	13	3	—	—	—	—	1
2	3	18	2	1	—	—	—	2
3	2	18	10	—	—	—	—	3
4	2	25	14	—	—	—	—	4
5	3	32	28	24	—	—	—	5
6	2	12	11	—	1	14	—	6
7	2	24	18	—	1	19	—	7
8	2	24	14	—	2	54	27	8

6.3 The C-function
Reduced free energy (Helmholtz function)

6.3 Die C-Funktion
Reduzierte freie Energie (Helmholtz-Funktion)

$$\psi_C(\Theta, \chi) = C_{00} + C_{01}\chi + \sum_{\nu=2}^{11} C_{0\nu}\chi^{1-\nu} + C_{012}\ln\chi + \left[C_{11}\chi + \sum_{\nu=2}^{6} C_{1\nu}\chi^{1-\nu} + C_{17}\ln\chi\right](\Theta - 1) +$$

$$+ \left[C_{21}\chi + \sum_{\nu=2}^{7} C_{2\nu}\chi^{1-\nu} + C_{28}\ln\chi\right](\Theta - 1)^2 + \left[C_{31}\chi + \sum_{\nu=2}^{9} C_{3\nu}\chi^{1-\nu} + C_{310}\ln\chi\right](\Theta - 1)^3 +$$

$$+ (C_{40} + C_{41}\chi^{-5})\Theta^{-23}(\Theta - 1) + C_{50}\Theta\ln\Theta + \chi^6 \sum_{\nu=0}^{4} C_{6\nu}\Theta^{-2-\nu} + \sum_{\nu=0}^{8} C_{7\nu}(\Theta - 1)^{\nu+1}.$$

6.4 The D-function
Reduced free energy (Helmholtz function)

6.4 Die D-Funktion
Reduzierte freie Energie (Helmholtz-Funktion)

$$\psi_D(\Theta, \chi) = \sum_{\mu=3}^{4} \sum_{\nu=0}^{4} D_{\mu\nu} y^\mu \chi^{-\nu} + y^{32} \sum_{\nu=0}^{2} D_{5\nu}\chi^\nu, \quad \text{wobei where} \quad y = (1 - \Theta)/(1 - \Theta_1).$$

6.5 Constants relating to canonical functions

The values of the constants introduced in Section 6.1, 6.2, 6.3 and 6.4 are given in Section 7.1.

6.5 Konstanten der kanonischen Funktionen

Die Werte der in Abschnitt 6.1, 6.2, 6.3 und 6.4 eingeführten Konstanten bringt Abschnitt 7.1.

7. Values of the constants

7.1 Numerical values of the primary constants

7. Werte der Konstanten

7.1 Zahlenwerte der primären Konstanten

7.1.1 Sub-region 1 — Unterbereich 1

$A_0 = 6,824\,687\,741 \cdot 10^3$
$A_1 = -5,422\,063\,673 \cdot 10^2$
$A_2 = -2,096\,666\,205 \cdot 10^4$
$A_3 = 3,941\,286\,787 \cdot 10^4$
$A_4 = -6,733\,277\,739 \cdot 10^4$
$A_5 = 9,902\,381\,028 \cdot 10^4$
$A_6 = -1,093\,911\,774 \cdot 10^5$
$A_7 = 8,590\,841\,667 \cdot 10^4$
$A_8 = -4,511\,168\,742 \cdot 10^4$
$A_9 = 1,418\,138\,926 \cdot 10^4$
$A_{10} = -2,017\,271\,113 \cdot 10^3$

$A_{11} = 7,982\,692\,717 \cdot 10^0$
$A_{12} = -2,616\,571\,843 \cdot 10^{-2}$
$A_{13} = 1,522\,411\,790 \cdot 10^{-3}$
$A_{14} = 2,284\,279\,054 \cdot 10^{-2}$
$A_{15} = 2,421\,647\,003 \cdot 10^2$
$A_{16} = 1,269\,716\,088 \cdot 10^{-10}$
$A_{17} = 2,074\,838\,328 \cdot 10^{-7}$
$A_{18} = 2,174\,020\,350 \cdot 10^{-8}$
$A_{19} = 1,105\,710\,498 \cdot 10^{-9}$
$A_{20} = 1,293\,441\,934 \cdot 10^1$
$A_{21} = 1,308\,119\,072 \cdot 10^{-5}$
$A_{22} = 6,047\,626\,338 \cdot 10^{-14}$

$a_1 = 8,438\,375\,405 \cdot 10^{-1}$
$a_2 = 5,362\,162\,162 \cdot 10^{-4}$
$a_3 = 1,720\,000\,000 \cdot 10^0$
$a_4 = 7,342\,278\,489 \cdot 10^{-2}$
$a_5 = 4,975\,858\,870 \cdot 10^{-2}$
$a_6 = 6,537\,154\,300 \cdot 10^{-1}$
$a_7 = 1,150\,000\,000 \cdot 10^{-6}$
$a_8 = 1,510\,800\,000 \cdot 10^{-5}$
$a_9 = 1,418\,800\,000 \cdot 10^{-1}$
$a_{10} = 7,002\,753\,165 \cdot 10^0$
$a_{11} = 2,995\,284\,926 \cdot 10^{-4}$
$a_{12} = 2,040\,000\,000 \cdot 10^{-1}$

7.1.2 Sub-region 2 — Unterbereich 2

$B_0 = 1,683\,599\,274 \cdot 10^1$
$B_{01} = 2,856\,067\,796 \cdot 10^1$
$B_{02} = -5,438\,923\,329 \cdot 10^1$
$B_{03} = 4,330\,662\,834 \cdot 10^{-1}$
$B_{04} = -6,547\,711\,697 \cdot 10^{-1}$
$B_{05} = 8,565\,182\,058 \cdot 10^{-2}$
$B_{11} = 6,670\,375\,918 \cdot 10^{-2}$
$B_{12} = 1,388\,983\,801 \cdot 10^0$
$B_{21} = 8,390\,104\,328 \cdot 10^{-2}$
$B_{22} = 2,614\,670\,893 \cdot 10^{-2}$
$B_{23} = -3,373\,439\,453 \cdot 10^{-2}$
$B_{31} = 4,520\,918\,904 \cdot 10^{-1}$

$B_{32} = 1,069\,036\,614 \cdot 10^{-1}$
$B_{41} = -5,975\,336\,707 \cdot 10^{-1}$
$B_{42} = -8,847\,535\,804 \cdot 10^{-2}$
$B_{51} = 5,958\,051\,609 \cdot 10^{-1}$
$B_{52} = -5,159\,303\,373 \cdot 10^{-1}$
$B_{53} = 2,075\,021\,122 \cdot 10^{-1}$
$B_{61} = 1,190\,610\,271 \cdot 10^{-1}$
$B_{62} = -9,867\,174\,132 \cdot 10^{-2}$
$B_{71} = 1,683\,998\,803 \cdot 10^{-1}$
$B_{72} = -5,809\,438\,001 \cdot 10^{-2}$
$B_{81} = 6,552\,390\,126 \cdot 10^{-3}$
$B_{82} = 5,710\,218\,649 \cdot 10^{-4}$

$B_{90} = 1,936\,587\,558 \cdot 10^2$
$B_{91} = -1,388\,522\,425 \cdot 10^3$
$B_{92} = 4,126\,607\,219 \cdot 10^3$
$B_{93} = -6,508\,211\,677 \cdot 10^3$
$B_{94} = 5,745\,984\,054 \cdot 10^3$
$B_{95} = -2,693\,088\,365 \cdot 10^3$
$B_{96} = 5,235\,718\,623 \cdot 10^2$
$b = 7,633\,333\,333 \cdot 10^{-1}$
$b_{61} = 4,006\,073\,948 \cdot 10^{-1}$
$b_{71} = 8,636\,081\,627 \cdot 10^{-2}$
$b_{81} = -8,532\,322\,921 \cdot 10^{-1}$
$b_{82} = 3,460\,208\,861 \cdot 10^{-1}$

7.1.3 Sub-region 3 — Unterbereich 3

$$C_{00} = -6,839\,90000 \cdot 10^0$$
$$C_{01} = -1,722\,604\,20 \cdot 10^{-2}$$
$$C_{02} = -7,771\,750\,39 \cdot 10^0$$
$$C_{03} = 4,204\,607\,52 \cdot 10^0$$
$$C_{04} = -2,768\,070\,38 \cdot 10^0$$
$$C_{05} = 2,104\,197\,07 \cdot 10^0$$
$$C_{06} = -1,146\,495\,88 \cdot 10^0$$
$$C_{07} = 2,231\,380\,85 \cdot 10^{-1}$$
$$C_{08} = 1,162\,503\,63 \cdot 10^{-1}$$
$$C_{09} = -8,209\,005\,44 \cdot 10^{-2}$$
$$C_{010} = 1,941\,292\,39 \cdot 10^{-2}$$
$$C_{011} = -1,694\,705\,76 \cdot 10^{-3}$$
$$C_{012} = -4,311\,577\,033 \cdot 10^0$$
$$C_{11} = 7,086\,360\,85 \cdot 10^{-1}$$
$$C_{12} = 1,236\,794\,55 \cdot 10^1$$
$$C_{13} = -1,203\,890\,04 \cdot 10^1$$
$$C_{14} = 5,404\,374\,22 \cdot 10^0$$
$$C_{15} = -9,938\,650\,43 \cdot 10^{-1}$$
$$C_{16} = 6,275\,231\,82 \cdot 10^{-2}$$

$$C_{17} = -7,747\,430\,16 \cdot 10^0$$
$$C_{21} = -4,298\,850\,92 \cdot 10^0$$
$$C_{22} = 4,314\,305\,38 \cdot 10^1$$
$$C_{23} = -1,416\,193\,13 \cdot 10^1$$
$$C_{24} = 4,041\,724\,59 \cdot 10^0$$
$$C_{25} = 1,555\,463\,26 \cdot 10^0$$
$$C_{26} = -1,665\,689\,35 \cdot 10^0$$
$$C_{27} = 3,248\,811\,58 \cdot 10^{-1}$$
$$C_{28} = 2,936\,553\,25 \cdot 10^1$$
$$C_{31} = 7,948\,418\,42 \cdot 10^{-6}$$
$$C_{32} = 8,088\,597\,47 \cdot 10^1$$
$$C_{33} = -8,361\,533\,80 \cdot 10^1$$
$$C_{34} = 3,586\,365\,17 \cdot 10^1$$
$$C_{35} = 7,518\,959\,54 \cdot 10^0$$
$$C_{36} = -1,261\,606\,40 \cdot 10^1$$
$$C_{37} = 1,097\,174\,62 \cdot 10^0$$
$$C_{38} = 2,121\,454\,92 \cdot 10^0$$
$$C_{39} = -5,465\,295\,66 \cdot 10^{-1}$$

$$C_{310} = 8,328\,754\,13 \cdot 10^0$$
$$C_{40} = 2,759\,717\,76 \cdot 10^{-6}$$
$$C_{41} = -5,090\,739\,85 \cdot 10^{-4}$$
$$C_{50} = 2,106\,363\,32 \cdot 10^2$$
$$C_{60} = 5,528\,935\,335 \cdot 10^{-2}$$
$$C_{61} = -2,336\,365\,955 \cdot 10^{-1}$$
$$C_{62} = 3,697\,071\,420 \cdot 10^{-1}$$
$$C_{63} = -2,596\,415\,470 \cdot 10^{-1}$$
$$C_{64} = 6,828\,087\,013 \cdot 10^{-2}$$
$$C_{70} = -2,571\,600\,553 \cdot 10^2$$
$$C_{71} = -1,518\,783\,715 \cdot 10^2$$
$$C_{72} = 2,220\,723\,208 \cdot 10^1$$
$$C_{73} = -1,802\,039\,570 \cdot 10^2$$
$$C_{74} = 2,357\,096\,220 \cdot 10^3$$
$$C_{75} = -1,462\,335\,698 \cdot 10^4$$
$$C_{76} = 4,542\,916\,630 \cdot 10^4$$
$$C_{77} = -7,053\,556\,432 \cdot 10^4$$
$$C_{78} = 4,381\,571\,428 \cdot 10^4$$

7.1.4 Sub-region 4 — Unterbereich 4

$$D_{30} = -1,717\,616\,747 \cdot 10^0$$
$$D_{31} = 3,526\,389\,875 \cdot 10^0$$
$$D_{32} = -2,690\,899\,373 \cdot 10^0$$
$$D_{33} = 9,070\,982\,605 \cdot 10^{-1}$$
$$D_{34} = -1,138\,791\,156 \cdot 10^{-1}$$

$$D_{40} = 1,301\,023\,613 \cdot 10^0$$
$$D_{41} = -2,642\,777\,743 \cdot 10^0$$
$$D_{42} = 1,996\,765\,362 \cdot 10^0$$
$$D_{43} = -6,661\,557\,013 \cdot 10^{-1}$$

$$D_{44} = 8,270\,860\,589 \cdot 10^{-2}$$
$$D_{50} = 3,426\,663\,535 \cdot 10^{-4}$$
$$D_{51} = -1,236\,521\,258 \cdot 10^{-3}$$
$$D_{52} = 1,155\,018\,309 \cdot 10^{-3}$$

7.1.5 Saturation line — Sättigungslinie

$$k_1 = -7,691\,234\,564 \cdot 10^0$$
$$k_2 = -2,608\,023\,696 \cdot 10^1$$
$$k_3 = -1,681\,706\,546 \cdot 10^2$$
$$k_4 = 6,423\,285\,504 \cdot 10^1$$

$$k_5 = -1,189\,646\,225 \cdot 10^2$$
$$k_6 = 4,167\,117\,320 \cdot 10^0$$
$$k_7 = 2,097\,506\,760 \cdot 10^1$$

$$k_8 = 10^9$$
$$k_9 = 6$$

7.1.6 Boundary between sub-regions 2 and 3 — Grenze zwischen Unterbereich 2 und 3

$$L = 7,160\,997\,524 \cdot 10^0.$$

A derived form of the L function and the values of the resulting derived constants are given in Section 8.2.

Eine abgeleitete Form der L-Funktion und die Werte der zugehörigen abgeleiteten Konstanten enthält Abschnitt 8.2.

7.2 Expressions for values of derived constants

$$\Theta_t = 27316/64730, \qquad \Theta_2 = 86315/64730,$$
$$\Theta_1 = 62315/64730, \qquad \Theta_3 = 107315/64730,$$

The numerical values of the above 8 constants are given to 10 digits in Section 8.1.

The constants α_0 and α_1 may be taken each to be zero. If it be desired that the calculated values of the internal energy and entropy at the reference state (the liquid phase at the triple point) each approximate to zero with the highest precision, then these constants should be evaluated, to suit the computer in use, by means of the following expressions:

7.2 Ausdrücke für die Werte der abgeleiteten Konstanten

$$\beta_2 = 10000/2212, \qquad \beta_t = \beta_K'(\Theta_t),$$
$$\beta_1 = \beta_K(\Theta_1), \qquad I_1 = R_1 T_{c1}/(p_{c1} v_{c1}).$$

Die Zahlenwerte der obigen 8 Konstanten sind mit 10 Stellen in Abschnitt 8.1 angegeben.

Die Konstanten α_0 und α_1 können beide zu Null angenommen werden. Wenn die berechneten Werte von innerer Energie und Entropie am Bezugszustand (flüssige Phase im Tripelpunkt) mit höchster Genauigkeit Null werden sollen, sind sie gemäß dem benutzten Computer mit Hilfe der folgenden Ausdrücke zu berechnen:

$$\alpha_0 = [-\zeta_A + \beta(\partial\zeta_A/\partial\beta)_\Theta + \Theta(\partial\zeta_A/\partial\Theta)_\beta]_{\Theta=\Theta_t,\ \beta=\beta_t}, \qquad \alpha_1 = [-(\partial\zeta_A/\partial\Theta)_\beta]_{\Theta=\Theta_t,\ \beta=\beta_t}.$$

8. Derived constants

8.1 Numerical values of derived constants

8. Abgeleitete Konstanten

8.1 Zahlenwerte der abgeleiteten Konstanten

$$\alpha_0 = 0, \qquad \beta_2 = 4,520\,795\,660 \cdot 10^0, \qquad \Theta_3 = 1,657\,886\,606 \cdot 10^0,$$
$$\alpha_1 = 0, \qquad \Theta_t = 4,219\,990\,731 \cdot 10^{-1}, \qquad I_1 = 4,260\,321\,148 \cdot 10^0,$$
$$\Theta_1 = 9,626\,911\,787 \cdot 10^{-1}, \qquad \beta_1 = 7,475\,191\,707 \cdot 10^{-1},$$
$$\Theta_2 = 1,333\,462\,073 \cdot 10^0, \qquad \beta_t = 2,763\,311\,032 \cdot 10^{-5}.$$

For convenience the adopted constant quantities are repeated here:

Zur Bequemlichkeit sind die vereinbarten Konstanten hier wiederholt:

$$T_{c1} = 647,3\ \text{K (genau) (exactly)}, \qquad p_{c1} = 22\,120\,000\ \text{N/m}^2 \text{ (genau) (exactly)}, \qquad v_{c1} = 0,003\,17\ \text{m}^3/\text{kg (exactly) (genau)},$$

whence the constant quantities given below are derived: woraus die folgenden konstanten Größen abgeleitet sind:

$$p_{c1} v_{c1} = 70\,120,4\ \text{J/kg (exactly) (genau)}, \qquad p_{c1} v_{c1}/T_{c1} = 108,327\,5143\ \text{J/kg K}.$$

8.2 Derived form of the L-function and values of the constants relating thereto

8.2 Abgeleitete Form der L-Funktion und Werte der zugehörigen Konstanten

When the L-function is rearranged to give

Wenn die L-Funktion umgeformt wird in

$$\beta_L = \beta_L(\Theta) = L_0 + L_1\,\Theta + L_2\,\Theta^2, \qquad \frac{d\beta_L}{d\Theta} = \beta_L' = \beta_L'(\Theta) = L_1 + 2L_2\,\Theta,$$

then the derived constants L_0, L_1 and L_2 have the numerical values

haben die zugehörigen Konstanten L_0, L_1 und L_2 die Zahlenwerte

$$L_0 = 1{,}574\,373\,327 \cdot 10^1, \qquad L_1 = -3{,}417\,061\,978 \cdot 10^1, \qquad L_2 = 1{,}931\,380\,707 \cdot 10^1.$$

9. Derived functions

9. Abgeleitete Funktionen

9.1 Sub-region 1 — Unterbereich 1

Reduced volume — Reduziertes Volumen

$$v/v_{c1} = \chi_1 = (\partial\zeta_A/\partial\beta)_\Theta,$$

$$\chi_1 = A_{11}\,a_5\,Z^{-5/17} + \{A_{12} + A_{13}\,\Theta + A_{14}\,\Theta^2 + A_{15}(a_6 - \Theta)^{10} + A_{16}(a_7 + \Theta^{19})^{-1}\} -$$
$$- (a_8 + \Theta^{11})^{-1}(A_{17} + 2A_{18}\,\beta + 3A_{19}\,\beta^2) - A_{20}\,\Theta^{18}(a_9 + \Theta^2)\{-3(a_{10} + \beta)^{-4} + a_{11}\} +$$
$$+ 3A_{21}(a_{12} - \Theta)\,\beta^2 + 4A_{22}\,\Theta^{-20}\,\beta^3,$$

where

wobei

$$Z = Y + (a_3\,Y^2 - 2a_4\,\Theta + 2a_5\,\beta)^{\frac12}, \qquad Y = 1 - a_1\,\Theta^2 - a_2\,\Theta^{-6}.$$

Reduzierte Entropie — Reduced entropy

$$s/(p_{c1}\,v_{c1}/T_{c1}) = \sigma_1 = -(\partial\zeta_A/\partial\Theta)_\beta - \alpha_1,$$

$$\sigma_1 = -\alpha_1 + A_0\ln\Theta - \sum_{\nu=2}^{10}(\nu-1)\,A_\nu\,\Theta^{\nu-2} + A_{11}\left[\left\{\frac{5}{12}Z - (a_3-1)\,Y\right\}Y' + a_4\right]Z^{-5/17} +$$
$$+ \{-A_{13} - 2A_{14}\,\Theta + 10A_{15}(a_6 - \Theta)^9 + 19A_{16}(a_7 + \Theta^{19})^{-2}\,\Theta^{18}\}\,\beta - 11(a_8 + \Theta^{11})^{-2}\,\Theta^{10}(A_{17}\,\beta + A_{18}\,\beta^2 + A_{19}\,\beta^3) +$$
$$+ A_{20}\,\Theta^{17}(18a_9 + 20\,\Theta^2)\{(a_{10} + \beta)^{-3} + a_{11}\,\beta\} + A_{21}\,\beta^3 + 20A_{22}\,\Theta^{-21}\,\beta^4,$$

where

wobei

$$Y' = -2a_1\,\Theta + 6a_2\,\Theta^{-7}.$$

Reduced enthalpy — Reduzierte Enthalpie

$$h/(p_{c1}\,v_{c1}) = \varepsilon_1 = \zeta_A + \alpha_0 + \alpha_1\,\Theta + \Theta\,\sigma_1,$$

$$\varepsilon_1 = \alpha_0 + A_0\,\Theta - \sum_{\nu=1}^{10}(\nu-2)\,A_\nu\,\Theta^{\nu-1} + A_{11}\left[Z\left\{17\left(\frac{Z}{29} - \frac{Y}{12}\right) + 5\,\Theta\frac{Y'}{12}\right\} + a_4\,\Theta - (a_3-1)\,\Theta\,Y\,Y'\right]Z^{-5/17} +$$
$$+ \{A_{12} - A_{14}\,\Theta^2 + A_{15}(9\,\Theta + a_6)(a_6 - \Theta)^9 + A_{16}(20\,\Theta^{19} + a_7)(a_7 + \Theta^{19})^{-2}\}\,\beta -$$
$$- (12\,\Theta^{11} + a_8)(a_8 + \Theta^{11})^{-2}(A_{17}\,\beta + A_{18}\,\beta^2 + A_{19}\,\beta^3) + A_{20}\,\Theta^{18}(17a_9 + 19\,\Theta^2)\{(a_{10} + \beta)^{-3} + a_{11}\,\beta\} +$$
$$+ A_{21}\,a_{12}\,\beta^3 + 21A_{22}\,\Theta^{-20}\,\beta^4.$$

9.2 Sub-region 2 — Unterbereich 2

Reduced volume — Reduziertes Volumen

$$v/v_{c1} = \chi_2 = (\partial\zeta_B/\partial\beta)_\Theta,$$

$$\chi_2 = I_1\,\Theta/\beta - \sum_{\mu=1}^{5}\mu\,\beta^{\mu-1}\sum_{\nu=1}^{n(\mu)}B_{\mu\nu}\,X^{z(\mu,\nu)} - \sum_{\mu=6}^{8}\frac{(\mu-2)\,\beta^{1-\mu}\sum_{\nu=1}^{n(\mu)}B_{\mu\nu}\,X^{z(\mu,\nu)}}{\left\{\beta^{2-\mu} + \sum_{\lambda=1}^{l(\mu)}b_{\mu\lambda}\,X^{x(\mu,\lambda)}\right\}^2} + 11\left(\frac{\beta}{\beta_L}\right)^{10}\sum_{\nu=0}^{6}B_{9\nu}\,X^\nu,$$

where

wobei

$$X = \exp[b(1 - \Theta)] \quad \text{und and} \quad \beta_L = \beta_L(\Theta),$$

the expression for which is given in Section 3.2.2, and the numbers of terms $n(\mu)$ and $l(\mu)$, and the exponents $z(\mu,\nu)$ and $x(\mu,\lambda)$, are listed in Section 6.2.

wofür der Ausdruck in Abschnitt 3.2.2 gegeben ist, und die Glieder $n(\mu)$ und $l(\mu)$ sowie die Exponenten $z(\mu,\nu)$ und $x(\mu,\lambda)$ in Abschnitt 6.2 enthalten sind.

Reduced entropy — Reduzierte Entropie

$$s/(p_{c1}\,v_{c1}/T_{c1}) = \sigma_2 = -(\partial\zeta_B/\partial\Theta)_\beta - \alpha_1,$$

$$\sigma_2 = -\alpha_1 - I_1\ln\beta + B_0\ln\Theta - \sum_{\nu=1}^{5}(\nu-1)\,B_{0\nu}\,\Theta^{\nu-2} - b\sum_{\mu=1}^{5}\beta^\mu\sum_{\nu=1}^{n(\mu)}z(\mu,\nu)\,B_{\mu\nu}\,X^{z(\mu,\nu)} -$$
$$- b\sum_{\mu=6}^{8}\frac{\sum_{\nu=1}^{n(\mu)}B_{\mu\nu}\,X^{z(\mu,\nu)}\left[z(\mu,\nu) - \dfrac{\sum_{\lambda=1}^{l(\mu)}x(\mu,\lambda)\,b_{\mu\lambda}\,X^{x(\mu,\lambda)}}{\beta^{2-\mu} + \sum_{\lambda=1}^{l(\mu)}b_{\mu\lambda}\,X^{x(\mu,\lambda)}}\right]}{\beta^{2-\mu} + \sum_{\lambda=1}^{l(\mu)}b_{\mu\lambda}\,X^{x(\mu,\lambda)}} + \beta\left(\frac{\beta}{\beta_L}\right)^{10}\sum_{\nu=0}^{6}\left[\left\{\frac{10\beta_L'}{\beta_L} + \nu\,b\right\}B_{9\nu}\,X^\nu\right].$$

$$Reduced\ enthalpy\ —\ Reduzierte\ Enthalpie$$

$$h/(p_{o1}\, v_{o1}) = \varepsilon_2 = \zeta_B + \alpha_0 + \alpha_1\, \Theta + \Theta\, \sigma_2,$$

$$\varepsilon_2 = \alpha_0 + B_0\, \Theta - \sum_{\nu=1}^{5} B_{0\nu}(\nu - 2)\, \Theta^{\nu-1} - \sum_{\mu=1}^{5} \beta^\mu \sum_{\nu=1}^{n(\mu)} B_{\mu\nu}\{1 + z(\mu,\nu)\, b\, \Theta\}\, X^{z(\mu,\nu)} -$$

$$- \sum_{\mu=6}^{8} \frac{\sum\limits_{\nu=1}^{n(\mu)} B_{\mu\nu}\, X^{z(\mu,\nu)}\left[\{1 + z(\mu,\nu)\, b\, \Theta\} - \dfrac{b\, \Theta \sum\limits_{\lambda=1}^{l(\mu)} x(\mu,\lambda)\, b_{\mu\lambda}\, X^{z(\mu,\lambda)}}{\beta^{2-\mu} + \sum\limits_{\lambda=1}^{l(\mu)} b_{\mu\lambda}\, X^{z(\mu,\lambda)}}\right]}{\beta^{2-\mu} + \sum\limits_{\lambda=1}^{l(\mu)} b_{\mu\lambda}\, X^{z(\mu,\lambda)}} + \beta\left(\frac{\beta}{\beta_L}\right)^{10} \sum_{\nu=0}^{6}\left[\left\{1 + \Theta\left(\frac{10\beta_L'}{\beta_L} + \nu\, b\right)\right\} B_{9\nu}\, X^\nu\right].$$

9.3 Sub-region 3 — Unterbereich 3

$$Reduced\ pressure\ —\ Reduzierter\ Druck$$

$$p/p_{o1} = \beta_3 = -(\partial\, \psi_0/\partial\, \chi)_\Theta,$$

$$\beta_3 = -\left\{C_{01} + \sum_{\nu=2}^{11}(1-\nu)\, C_{0\nu}\, \chi^{-\nu} + C_{012}\, \chi^{-1}\right\} - \left\{C_{11} + \sum_{\nu=2}^{6}(1-\nu)\, C_{1\nu}\, \chi^{-\nu} + C_{17}\, \chi^{-1}\right\}(\Theta - 1) -$$

$$- \left\{C_{21} + \sum_{\nu=2}^{7}(1-\nu)\, C_{2\nu}\, \chi^{-\nu} + C_{28}\, \chi^{-1}\right\}(\Theta - 1)^2 - \left\{C_{31} + \sum_{\nu=2}^{9}(1-\nu)\, C_{3\nu}\, \chi^{-\nu} + C_{310}\, \chi^{-1}\right\}(\Theta - 1)^3 +$$

$$+ 5C_{41}\, \chi^{-6}\, \Theta^{-23}(\Theta - 1) - 6\chi^5 \sum_{\nu=0}^{4} C_{6\nu}\, \Theta^{-2-\nu}.$$

$$Reduced\ entropy\ —\ Reduzierte\ Entropie$$

$$s/(p_{o1}\, v_{o1}/T_{o1}) = \sigma_3 = -(\partial\, \psi_0/\partial\, \Theta)_\chi - \alpha_1,$$

$$\sigma_3 = -\alpha_1 - \left\{C_{11}\, \chi + \sum_{\nu=2}^{6} C_{1\nu}\, \chi^{1-\nu} + C_{17}\, \ln\chi + C_{50}\right\} - 2\left\{C_{21}\, \chi + \sum_{\nu=2}^{7} C_{2\nu}\, \chi^{1-\nu} + C_{28}\, \ln\chi\right\}(\Theta - 1) -$$

$$- 3\left\{C_{31}\, \chi + \sum_{\nu=2}^{9} C_{3\nu}\, \chi^{1-\nu} + C_{310}\, \ln\chi\right\}(\Theta - 1)^2 + (C_{40} + C_{41}\, \chi^{-5})(22\, \Theta^{-23} - 23\, \Theta^{-24}) - C_{50}\, \ln\Theta +$$

$$+ \chi^6 \sum_{\nu=0}^{4}\{(\nu + 2)\, C_{6\nu}\, \Theta^{-3-\nu}\} - \sum_{\nu=0}^{8}\{(\nu + 1)\, C_{7\nu}(\Theta - 1)^\nu\}.$$

$$Reduced\ enthalpy\ —\ Reduzierte\ Enthalpie$$

$$h/(p_{o1}\, v_{o1}) = \varepsilon_3 = \psi_0 + \alpha_0 + \alpha_1\, \Theta + \Theta\, \sigma_3 + \chi\, \beta_3,$$

$$\varepsilon_3 = \alpha_0 + \left\{(C_{00} - C_{012} - C_{50}) - C_{11}\, \chi + \sum_{\nu=2}^{11} \nu\, C_{0\nu}\, \chi^{1-\nu} - \sum_{\nu=2}^{6} C_{1\nu}\, \chi^{1-\nu} + (C_{012} - C_{17})\, \ln\chi\right\} +$$

$$+ \left\{(-C_{17} - C_{50}) - (C_{11} + 2C_{21})\, \chi + \sum_{\nu=2}^{6}(\nu - 1)\, C_{1\nu}\, \chi^{1-\nu} - 2\sum_{\nu=2}^{7} C_{2\nu}\, \chi^{1-\nu} - 2C_{28}\, \ln\chi\right\}(\Theta - 1) +$$

$$+ \left\{-C_{28} - (2C_{21} + 3C_{31})\, \chi + \sum_{\nu=2}^{7}(\nu - 2)\, C_{2\nu}\, \chi^{1-\nu} - 3\sum_{\nu=2}^{9} C_{3\nu}\, \chi^{1-\nu} - (C_{28} + 3C_{310})\, \ln\chi\right\}(\Theta - 1)^2 +$$

$$+ \left\{-C_{310} - 3C_{31}\, \chi + \sum_{\nu=2}^{9}(\nu - 3)\, C_{3\nu}\, \chi^{1-\nu} - 2C_{310}\, \ln\chi\right\}(\Theta - 1)^3 +$$

$$+ (23C_{40} + 28C_{41}\, \chi^{-5})\, \Theta^{-22} - (24C_{40} + 29C_{41}\, \chi^{-5})\, \Theta^{-23} + \chi^6 \sum_{\nu=0}^{4}\{(\nu - 3)\, C_{6\nu}\, \Theta^{-2-\nu}\} - \sum_{\nu=0}^{8}\{C_{7\nu}(1 + \nu\, \Theta)(\Theta - 1)^\nu\}.$$

9.4 Sub-region 4 — Unterbereich 4

$$Reduced\ pressure\ —\ Reduzierter\ Druck$$

$$p/p_{o1} = \beta_4 = \beta_3 - (\partial\, \psi_D/\partial\, \chi)_\Theta,$$

$$\beta_4 = \beta_3 + \sum_{\mu=3}^{4} \sum_{\nu=0}^{4} \nu\, D_{\mu\nu}\, y^\mu\, \chi^{-\nu-1} - y^{32} \sum_{\nu=0}^{2} \nu\, D_{5\nu}\, \chi^{\nu-1}.$$

$$Reduced\ entropy\ —\ Reduzierte\ Entropie$$

$$s/(p_{o1}\, v_{o1}/T_{o1}) = \sigma_4 = \sigma_3 - (\partial\, \psi_D/\partial\, \Theta)_\chi,$$

$$\sigma_4 = \sigma_3 + \frac{\sum\limits_{\mu=3}^{4} \sum\limits_{\nu=0}^{4} \mu\, D_{\mu\nu}\, y^{\mu-1}\, \chi^{-\nu} + 32y^{31} \sum\limits_{\nu=0}^{2} D_{5\nu}\, \chi^\nu}{1 - \Theta_1}.$$

$$Reduced\ enthalpy\ —\ Reduzierte\ Enthalpie$$

$$h/(p_{o1}\, v_{o1}) = \varepsilon_4 = \varepsilon_3 + \psi_D + (\sigma_4 - \sigma_3)\, \Theta + (\beta_4 - \beta_3)\, \chi,$$

$$\varepsilon_4 = \varepsilon_3 + \sum_{\mu=3}^{4} \sum_{\nu=0}^{4} D_{\mu\nu}\{(1 - \mu + \nu)\, y + \mu/(1 - \Theta_1)\}\, y^{\mu-1}\, \chi^{-\nu} - y^{31} \sum_{\nu=0}^{2} D_{5\nu}\{(31 + \nu)\, y - 32/(1 - \Theta_1)\}\, \chi^\nu.$$

Note: In the three derived expressions for sub-region 4:   *Beachte:* In den drei abgeleiteten Ausdrücken für Unterbereich 4 ist:

$$y = (1 - \Theta)/(1 - \Theta_1).$$

10. Cautionary notes

10.1 Discontinuities at boundaries between sub-regions

It should be noted that there are discontinuities in property values at the boundary between sub-regions 2 and 3 and at the boundary between subregions 1 and 4. Nevertheless, for the specific volume, specific enthalpy and specific free enthalpy, the discontinuities are less than the maximum values which the First Meeting of the International Formulation Committee (Prague, 1965) recommended should not be exceeded. However, when values of the specific entropy are listed at 5-kelvin intervals, there are four points on the boundary between sub-regions 2 and 3 at which the discontinuity has a value of 0,3 J/kg K, and one point having a value of 0,4 J/kg K, compared with the recommended maximum acceptable value of 0,2 J/kg K.

For the specific heat-capacity, the discontinuities at the boundary between sub-regions 1 and 4 do not exceed the recommended maximum value of 1%, but at the boundary between sub-regions 2 and 3 they are in general in excess of 1%, the greatest discontinuity being 6,5%. However, in this area near the critical point measurements are rather uncertain.

The magnitudes of these discontinuities are not significant industrially.

10.2 Precautions to be observed in numerical computation

(1) It will be clearly evident that, in order to attain an adequately precise result from numerical inversion in sub-regions 3 and 4, it is necessary to control the pressure error or the volume error, or both, within appropriate bounds. When pressure alone is controlled, very severe conditions must be imposed (possibly as fine as 1 part in 10^6 or, near the critical point, much finer).

(2) Note should be taken of the fact that errors may arise if finite-difference techniques be used with too small an interval.

10. Zu beachtende Bemerkungen

10.1 Diskontinuitäten an den Grenzen zwischen Subregionen

Es ist zu beachten, daß Diskontinuitäten der Eigenschaftswerte (Zustandsgrößen) an der Grenze zwischen Subregion 2 und 3 sowie an der Grenze zwischen Subregion 1 und 4 auftreten. Immerhin sind diese Diskontinuitäten für spezifisches Volumen, spezifische Enthalpie und spezifische freie Enthalpie kleiner als die auf der ersten Sitzung des IFC (Prag, 1965) empfohlenen, nicht zu überschreitenden Maximalwerte. Aber wenn Werte der spezifischen Entropie in 5 K-Schritten ausgerechnet werden, so beträgt an vier Punkten der Grenze zwischen Subregion 2 und 3 die Diskontinuität 0,3 J/kg K und an einem Punkt 0,4 J/kg K im Vergleich zu dem empfohlenen Maximalwert von 0,2 J/kg K.

Für die spezifische Wärmekapazität an der Grenze zwischen Subregion 1 und 4 überschreiten die Diskontinuitäten nicht den empfohlenen Maximalwert von 1%, aber an der Grenze zwischen Subregion 2 und 3 sind sie im allgemeinen größer als 1%, wobei ihr größter Wert 6,5% ist. Aber in diesem Gebiet in der Nähe des kritischen Punktes sind auch die Messungen recht unsicher.

Die Größe dieser Diskontinuitäten ist aber ohne Bedeutung für industrielle Anwendungen.

10.2 Bei der numerischen Rechnung zu beachten

(1) Es ist evident, daß man, um genügend genaue Werte bei der numerischen Inversion in Subregion 3 und 4 zu erhalten, die Unsicherheit von Druck und Volumen oder beide in angemessener Weise beachten muß. Wenn der Fehler des Druckes allein berücksichtigt wird, muß man sehr scharfe Bedingungen einhalten (möglicherweise bis zu 1 auf 10^6 und in der Nähe des kritischen Punktes noch mehr).

(2) Es ist auch zu beachten, daß Fehler beim Rechnen mit endlichen Differenzen auftreten können, wenn die Intervalle zu klein gewählt werden.

II. Equations for Dynamic Viscosity and Thermal Conductivity

II. Gleichungen für die dynamische Viskosität und Wärmeleitfähigkeit

1. Dynamic Viscosity

The values in Table 4d may be reproduced within the stated tolerances by the following equation for industrial use. This equation does not take into account the critical anomaly of the viscosity [278].

1. Dynamische Viskosität

Die Werte der Tafel 4d können innerhalb der angebenen Toleranzen durch folgende Gleichung für industrielle Berechnungen wiedergegeben werden. Diese Gleichung berücksichtigt nicht die kritische Anomalie der Viskosität [278].

$$\bar{\eta} = \bar{\eta}_0(\bar{T}) \cdot \exp\left[\bar{\varrho} \sum_{i=0}^{5} \sum_{j=0}^{6} H_{ij} \left(\frac{1}{\bar{T}} - 1\right)^i (\bar{\varrho} - 1)^j\right], \tag{1}$$

$$\eta_0(\bar{T}) = \sqrt{\bar{T}} \left(\sum_{i=0}^{3} \frac{H_i}{\bar{T}^i}\right)^{-1} \tag{2}$$

with the dimensionless quantities

mit den dimensionslosen Größen

$$\bar{T} = T/T^*, \quad \bar{\varrho} = \varrho/\varrho^*, \quad \bar{p} = p/p^*, \quad \bar{\eta} = \eta/\eta^*$$

and the reference constants

und den Bezugsgrößen

$$T^* = 647{,}27 \text{ K},$$
$$\varrho^* = 317{,}763 \text{ kg/m}^3,$$
$$p^* = 221{,}15 \text{ bar},$$
$$\eta^* = 55{,}071 \text{ } \mu\text{Pa s}.$$

The three reference constants T^*, ϱ^*, and p^* are close to but not identical with the critical constants.

The range of validity of Eq. (1) for the viscosity is given by

Die drei Bezugsgrößen T^*, ϱ^* und p^* sind nahezu gleich aber nicht identisch mit den kritischen Größen.

Gl. (1) für die Viskosität gilt in folgendem Bereich:

$$p \leq 5000 \text{ bar} \quad \text{for} \quad 0\,°C \leq t \leq 150\,°C,$$
$$p \leq 3500 \text{ bar} \quad \text{for} \quad 150\,°C < t \leq 600\,°C,$$
$$p \leq 3000 \text{ bar} \quad \text{for} \quad 600\,°C < t \leq 900\,°C.$$

The coefficients H_{ij} from Eq. (1) and H_i from Eq. (2) are given in Table II,1 and II,2 respectively.

Die Koeffizienten H_{ij} aus Gl. (1) und H_i aus Gl. (2) sind in Tabelle II,1 bzw. Tabelle II,2 wiedergegeben.

Table II,1. Coefficients H_{ij} from Eq. (1)

i	j	H_{ij}
0	0	$H_{00} = 0{,}513\,2047$
1	0	$H_{10} = 0{,}320\,5656$
4	0	$H_{40} = -0{,}778\,2567$
5	0	$H_{50} = 0{,}188\,5447$
0	1	$H_{01} = 0{,}215\,1778$
1	1	$H_{11} = 0{,}731\,7883$
2	1	$H_{21} = 1{,}241\,044$
3	1	$H_{31} = 1{,}476\,783$
0	2	$H_{02} = -0{,}281\,8107$
1	2	$H_{12} = -1{,}070\,786$
2	2	$H_{22} = -1{,}263\,184$
0	3	$H_{03} = 0{,}177\,8064$
1	3	$H_{13} = 0{,}460\,5040$
2	3	$H_{23} = 0{,}234\,0379$
3	3	$H_{33} = -0{,}492\,4179$
0	4	$H_{04} = -0{,}041\,766\,10$
3	4	$H_{34} = 0{,}160\,0435$
1	5	$H_{15} = -0{,}015\,783\,86$
3	6	$H_{36} = -0{,}003\,629\,481$

Coefficients H_{ij} omitted from the table are all equal to zero identically.

In der obigen Tabelle nicht aufgeführte Koeffizienten sind identisch Null.

Table II,2. Coefficients H_i from Eq. (2)

$$H_0 = 1{,}000\,000$$
$$H_1 = 0{,}978\,197$$
$$H_2 = 0{,}579\,829$$
$$H_3 = -0{,}202\,354$$

Values of the viscosity at saturation state in Table 6 b are calculated with the aid of Eq. (1), saturation pressures and saturation densities from the 1967 IFC Formulation for Industrial Use [261].

For the purpose of program verification the tabular entries contain more digits than justified by the tolerances listed in the Skeleton Table.

Werte der Viskosität im Sättigungszustand in Tafel 6 b sind mit Hilfe von Gl. (1) berechnet, Sättigungsdrücke und Sättigungsdichten aus der 1967 IFC Formulation for Industrial Use [261].

Um die Prüfung des Computer-Programms zu erleichtern, sind in Tafel 6 b mehr Dezimalstellen angegeben, als durch die Toleranzen der Rahmentafel gerechtfertigt wären.

2. Thermal Conductivity

The values in Table 4e may be reproduced within the stated tolerances by the following equation for industrial use. This equation does not take into account the critical anomaly of the termal conductivity and yields a finite value at the critical point [279].

2. Wärmeleitfähigkeit

Die Werte der Tafel 4e können innerhalb der angegebenen Toleranzen durch folgende Gleichung für industrielle Berechnungen wiedergegeben werden. Diese Gleichung berücksichtigt nicht die kritische Anomalie der Wärmeleitfähigkeit und ergibt einen endlichen Wert im kritischen Punkt [279].

$$\bar{\lambda} = \lambda_0(\bar{T}) + \lambda_1(\bar{\varrho}) + \lambda_2(\bar{T}, \bar{\varrho}), \tag{3}$$

$$\lambda_0(\bar{T}) = \sqrt{\bar{T}} \sum_{k=0}^{3} a_k \bar{T}^k, \tag{4}$$

$$\lambda_1(\bar{\varrho}) = b_0 + b_1 \bar{\varrho} + b_2 \exp\left[B_1(\bar{\varrho} + B_2)^2\right], \tag{5}$$

$$\lambda_2(\bar{T}, \bar{\varrho}) = \left[\frac{d_1}{\bar{T}^{10}} + d_2\right] \bar{\varrho}^{9/5} \exp\left[C_1(1 - \bar{\varrho}^{14/5})\right] + d_3 S \bar{\varrho}^Q \exp\left[\frac{Q}{1+Q}(1 - \bar{\varrho}^{1+Q})\right] + d_4 \exp\left[C_2 \bar{T}^{3/2} + \frac{C_3}{\bar{\varrho}^5}\right] \tag{6}$$

with the dimensionless quantities

mit den dimensionslosen Größen

$$\bar{T} = T/T^*, \quad \bar{\varrho} = \varrho/\varrho^*, \quad \bar{\lambda} = \lambda/\lambda^*$$

and thereference constants

und den Bezugsgrößen

$$T^* = 647,3 \text{ K},$$
$$\varrho^* = 317,7 \text{ kg/m}^3,$$
$$\lambda^* = 1 \text{ W/K m}.$$

The two reference constants T^* and ϱ^* are close to but not identical with the critical constants.

The range of validity of Eq. (3) for the thermal conductivity is given by

Die beiden Bezugsgrößen T^* und ϱ^* sind nahezu gleich aber nicht identisch mit den kritischen Größen.

Gl. (3) für die Wärmeleitfähigkeit gilt in folgendem Bereich:

$p \leq 1000$ bar	for	$0\,°C \leq t \leq 500\,°C$,
$p \leq 700$ bar	for	$500\,°C < t \leq 650\,°C$,
$p \leq 400$ bar	for	$650\,°C < t \leq 800\,°C$.

In Eq. (6), Q and S are functions of

In Gl. (6) sind Q und S Funktionen von

$$\Delta\bar{T} = |\bar{T} - 1| + C_4 \tag{7}$$

where

mit

$$Q = 2 + \frac{C_5}{\Delta\bar{T}^{3/5}}, \tag{8}$$

$$S = \begin{cases} \dfrac{1}{\Delta\bar{T}} & \text{for} \quad \bar{T} \geq 1 \\[2mm] \dfrac{C_6}{\Delta\bar{T}^{3/5}} & \text{for} \quad \bar{T} < 1. \end{cases} \tag{9}$$

The coefficients a_k from Eq. (4), b_i and B_i from Eq. (5) and d_i and C_i from Eqs. (6) to (9) are given in Table II,3, II,4, II,5 respectively.

Die Koeffizienten a_k aus Gl. (4), b_i und B_i aus Gl. (5) und d_i und C_i aus Gln. (6) bis (9) sind in den Tabellen II,3, II,4 und II,5 wiedergegeben.

Table II,3. Coefficients a_k from Eq. (4)

$a_0 =$	0,0102811
$a_1 =$	0,0299621
$a_2 =$	0,0156146
$a_3 =$	−0,00422464

Table II,4. Coefficients b_i and B_i from Eq. (5)

$b_0 = -0,397070$	$B_1 = -0,171587$
$b_1 = 0,400302$	$B_2 = 2,392190$
$b_2 = 1,060000$	

Table II,5. Coefficients d_i and C_i from Eqs. (6) to (9)

$d_1 = 0,0701309$	$C_1 = 0,642857$
$d_2 = 0,0118520$	$C_2 = -4,11717$
$d_3 = 0,00169937$	$C_3 = -6,17937$
$d_4 = -1,0200$	$C_4 = 0,00308976$
	$C_5 = 0,0822994$
	$C_6 = 10,0932$

Values of the thermal conductivity at saturation state in Table 6b are calculated with the aid of Eq. (3), saturation pressures and saturation densities from the 1967 IFC Formulation for Industrial Use [261].

Users should be aware of the fact that the above equation is subject to exponential underflow which most computers set to zero; this causes no errors in the final result.

To assist in programming the tabular entries contain more significant digits than is justified by the tolerances listed in the Skeleton Table.

Werte der Wärmeleitfähigkeit im Sättigungszustand in Tafel 6b sind mit Hilfe von Gl. (3) berechnet, Sättigungsdrücke und Sättigungsdichten aus der 1967 IFC Formulation for Industrial Use [261].

Bei der Berechnung nach obigen Gleichungen kann exponentieller Unterlauf eintreten, den die meisten Rechner als Null behandeln; Fehler im Endergebnis treten dadurch nicht auf.

Um das Programmieren zu erleichtern, enthalten die Tafelwerte mehr Dezimalstellen, als durch die Toleranzen der Rahmentafel gerechtfertigt wären.

III. Equations for the Thermodynamic Properties of Ordinary Water Substance at Saturation

III. Gleichungen für die thermodynamischen Zustandsgrößen von Wasser und Dampf bei Sättigung

A supplementary release of IAPS [280] contains a set of simple equations which yield for ordinary water substance the vapor pressure as well as the density, specific enthalpy and specific entropy of the saturated vapor and liquid. The values calculated from these equations for the vapor pressure and for the density and specific enthalpy of the vapor and liquid at saturation are identical to the values tabulated for these properties in the Skeleton Tables 1985.

More details about the following equations can be found in an article published in 1987 [281].

Eine ergänzende Verlautbarung der IAPS [280] enthält einen Satz von einfachen Gleichungen, mit denen man für gewöhnliches Wasser sowohl den Dampfdruck als auch die Dichten, die spezifischen Enthalpien und die spezifischen Entropien von Wasser und Dampf bei Sättigung berechnen kann. Die mit diesen Gleichungen berechneten Werte für den Dampfdruck und die Dichten und spezifischen Enthalpien von Wasser und Dampf bei Sättigung sind identisch mit jenen Werten, die in den IAPS-Rahmentafeln 1985 für diese Größen tabelliert sind.

Weitere Einzelheiten über die im folgenden mitgeteilten Gleichungen können einer im Jahre 1987 publizierten Arbeit entnommen werden [281].

1. Nomenclature

In addition to the quantities mentioned in Section A.2 the following quantities are used:

specific internal energy
auxiliary quantity for specific enthalpy
auxiliary quantity for specific entropy

1. Bezeichnungen

Neben den in Abschnitt A.2 genannten Größen werden noch folgende benötigt:

spezifische innere Energie	u
Hilfsgröße für die spezifische Enthalpie	α
Hilfsgröße für die spezifische Entropie	Φ

$$\theta = T/T_c$$

$$\tau = 1 - \theta$$

value at the critical point
value at the triple point

Wert am kritischen Punkt	Index c
Wert am Tripelpunkt	Index t

Reference constants

$T_c = 647,14$ K
$p_c = 220,64$ bar
$\varrho_c = 322$ kg/m³

Bezugsgrößen

$\alpha_0 = 1000$ J/kg
$\Phi_0 = \alpha_0/T_c$

for T_c, p_c, ϱ_c see Section A.4

The following equations are valid in the entire range of vapor-liquid equlibrium which corresponds to

für T_c, p_c und ϱ_c siehe Abschnitt A.4

Die folgenden Gleichungen gelten für den ganzen Bereich des Dampf-Flüssigkeits-Gleichgewichts, das heißt für

$$273{,}16 \text{ K} \leq T \leq 647{,}14 \text{ K}.$$

2. Vapor pressure
2. Dampfdruck

$$\ln(p/p_c) = (T_c/T)(a_1\tau + a_2\tau^{1,5} + a_3\tau^3 + a_4\tau^{3,5} + a_5\tau^4 + a_6\tau^{7,5}) \tag{1}$$

with

$a_1 = -7{,}85823$
$a_2 = +1{,}83991$
$a_3 = -11{,}7811$

mit

$a_4 = +22{,}6705$
$a_5 = -15{,}9393$
$a_6 = +1{,}77516$

3. Densities
3. Dichten

3.1 Saturated liquid
3.1 Gesättigte Flüssigkeit

$$\varrho'/\varrho_c = 1 + b_1\tau^{1/3} + b_2\tau^{2/3} + b_3\tau^{5/3} + b_4\tau^{16/3} + b_5\tau^{43/3} + b_6\tau^{110/3} \tag{2}$$

with

$b_1 = +1{,}99206$
$b_2 = +1{,}10123$
$b_3 = -0{,}512506$

mit

$b_4 = -1{,}75263$
$b_5 = -45{,}4485$
$b_6 = -6{,}75615 \cdot 10^5$

3.2 Saturated Vapor
3.2 Gesättigter Dampf

$$\ln(\varrho''/\varrho_c) = c_1\tau^{2/6} + c_2\tau^{4/6} + c_3\tau^{8/6} + c_4\tau^{18/6} + c_5\tau^{37/6} + c_6\tau^{71/6} \tag{3}$$

with

$c_1 = -2{,}02957$
$c_2 = -2{,}68781$
$c_3 = -5{,}38107$

mit

$c_4 = -17{,}3151$
$c_5 = -44{,}6384$
$c_6 = -64{,}3486$

4. Specific enthalpy and entropy
4. Spezifische Enthalpie und Entropie

4.1 Auxiliary equations
4.1 Hilfsgleichungen

$$\varkappa/\varkappa_0 = d_\varkappa + d_1\theta^{-19} + d_2\theta + d_3\theta^{4,5} + d_4\theta^5 + d_5\theta^{54,5}, \tag{4}$$

$$\Phi/\Phi_0 = d_\Phi + \frac{19}{20}d_1\theta^{-20} + d_2\ln\theta + \frac{9}{7}d_3\theta^{3,5} + \frac{5}{4}d_4\theta^4 + \frac{109}{107}d_5\theta^{53,5} \tag{5}$$

with

$d_1 = -5{,}71756 \cdot 10^{-8}$
$d_2 = +2689{,}81$
$d_3 = +129{,}889$
$d_4 = -137{,}181$

mit

$d_5 = +0{,}968874$
$d_\varkappa = -1135{,}481615639$
$d_\Phi = +2318{,}9142$

4.2 Specific enthalpy of the saturated liquid
4.2 Spezifische Enthalpie der gesättigten Flüssigkeit

$$h' = \varkappa + \frac{T}{\varrho'}\frac{dp}{dT}. \tag{6}$$

Eq. (6) yields the specific enthalpy of the saturated liquid when used with Eqs. (1), (2) and (4).

Gl. (6) ergibt die spezifische Enthalpie der gesättigten Flüssigkeit in Verbindung mit den Gln. (1), (2) und (4).

Note

Because u_t' and s_t' are zero by definition (see Section A.4) the specific enthalpy of the liquid at the triple point assumes the value

Bemerkung

Da u_t' und s_t' nach Definition Null sind (vgl. Abschnitt A4), nimmt die spezifische Enthalpie der Flüssigkeit am Tripelpunkt den Wert

$$h_t' = u_t' + p_t/\varrho_t' = 0{,}611787 \text{ J/kg}$$

In order to reproduce this numerical value for h_t', one needs to retain 13 figures for the constant d_α as quoted above. A decrease of the number of decimals in d_α affects the enthalpy of the satured liquid near the triple point, but does not affect the values of h'' significantly.

an. Um diesen Zahlenwert von h_t' zu erzeugen, muß man für d_α die oben angegebenen 13 Stellen beibehalten. Verringert man die Stellenzahl für d_α, so wird dadurch die Enthalpie der gesättigten Flüssigkeit in der Nähe des Tripelpunkts beeinflußt, aber nicht wesentlich die Werte von h''.

4.3 Specific enthalpy of the saturated vapor

4.3 Spezifische Enthalpie des gesättigten Dampfes

$$h'' = \alpha + \frac{T}{\varrho''}\frac{\mathrm{d}p}{\mathrm{d}T}. \tag{7}$$

Eq. (7) yields the specific enthalpy of the saturated vapor when used in conjunction with Eqs. (1), (3) and (4).

Gl. (7) ergibt die spezifische Enthalpie des gesättigten Dampfes in Verbindung mit den Gln. (1), (3) und (4).

4.4 Specific entropy of the saturated liquid

4.4 Spezifische Entropie der gesättigten Flüssigkeit

$$s' = \Phi + \frac{1}{\varrho'}\frac{\mathrm{d}p}{\mathrm{d}T}. \tag{8}$$

Eq. (8) yields the specific entropy of the saturated liquid when used in conjunction with Eqs. (1), (2) and (5).

Gl. (8) ergibt die spezifische Entropie der gesättigten Flüssigkeit in Verbindung mit den Gln. (1), (2) und (5).

4.5 Specific entropy of the saturated vapor

4.5 Spezifische Entropie des gesättigten Dampfes

$$s'' = \Phi + \frac{1}{\varrho''}\frac{\mathrm{d}p}{\mathrm{d}T}. \tag{9}$$

Eq. (9) yields the specific entropy of the saturated vapor when used in conjunction with Eqs. (1), (3) and (5).

Gl. (9) ergibt die spezifische Entropie des gesättigten Dampfes in Verbindung mit den Gln. (1), (3) und (5).

5. Computer-program verification

5. Prüfung des Computer-Programms

To assist the user in computer-program verification, Table 1 lists values for p, $\mathrm{d}p/\mathrm{d}T$, ϱ', ϱ'', α, h', h'', Φ, s' and s'' calculated at three temperatures. The results quoted in Table 1 were obtained with the aid of a computer having 14 significant figures and with the values of d_α and d_Φ given in Section 4.1. If the calculations are performed with the aid of a computer with seven significant figures, the results will be within the estimated uncertainty except for h' very close to the triple point.

Um dem Benutzer die Prüfung des Computer-Programms zu erleichtern, sind in Tabelle 1 Werte für p, $\mathrm{d}p/\mathrm{d}T$, ϱ', ϱ'', α, h', h'', Φ, s' und s'' für drei Temperaturen zusammengestellt. Die Ergebnisse wurden mit einem Computer mit 14 signifikanten Ziffern und mit den Werten d_α und d_Φ nach Abschnitt 4.1 berechnet. Rechnet man mit einem Computer mit sieben signifikanten Ziffern, bleiben die Ergebnisse · innerhalb der geschätzten Toleranzen außer bei den Werten von h' in unmittelbarer Nähe des Tripelpunktes.

Table 1. Thermodynamic Property Values at Saturation State Calculated at Three Selected Temperatures

Thermodynamische Zustandsgrößen bei Sättigung, berechnet bei drei ausgewählten Temperaturen

Symbol	$T = 273{,}16$ K	$T = 373{,}15$ K	$T = 647{,}14$ K	Unit/Einheit
p	$611{,}659$	$0{,}101\,325 \cdot 10^6$	$22{,}064 \cdot 10^6$	Pa
$\mathrm{d}p/\mathrm{d}T$	$44{,}426\,617$	$3{,}616 \cdot 10^3$	$267{,}9 \cdot 10^3$	Pa/K
ϱ'	$999{,}790$	$958{,}366$	322	kg/m^3
ϱ''	$0{,}004\,853\,64$	$0{,}597\,462$	322	kg/m^3
α	$-11{,}526\,335$	$417{,}66 \cdot 10^3$	$1\,548 \cdot 10^3$	J/kg
h'	$0{,}611\,787$	$419{,}07 \cdot 10^3$	$2\,086 \cdot 10^3$	J/kg
h''	$2\,500{,}3 \cdot 10^3$	$2\,675{,}8 \cdot 10^3$	$2\,086 \cdot 10^3$	J/kg
Φ	$-0{,}04$	$1{,}303 \cdot 10^3$	$3{,}578 \cdot 10^3$	J/kg K
s'	0	$1{,}307 \cdot 10^3$	$4{,}410 \cdot 10^3$	J/kg K
s''	$9{,}153 \cdot 10^3$	$7{,}355 \cdot 10^3$	$4{,}410 \cdot 10^3$	J/kg K